★ 双语版 ★

THE END OF SCIENCE
科学的终结

［美］约翰·霍根（John Horgan）著
孙雍君　张武军　译

清华大学出版社
北　京

北京市版权局著作权合同登记号　图字：01-2016-6620

John Horgan
The End of Science
ISBN:978-0-465-06592-9(ppk.)
Copyright© 1996 by John Horgan
"Preface to the 2015 Edition" copyright ©2015 by John Horgan
This edition published by arrangement with Basic Books, an imprint of Perseus Books, LL, a
subsidiary of Hachette Book Group, Inc., New York, USA. All rights reserved.

图书在版编目（CIP）数据

　　科学的终结：双语版：英、汉 /（美）约翰·霍根（John Horgan）著；孙雍君，张武军
译. —北京：清华大学出版社，2020.5
　　书名原文：The End of Science
　　ISBN 978-7-302-51954-6

　　Ⅰ.①科…　Ⅱ.①约…　②孙…　③张…　Ⅲ.①自然科学史—研究—西方国家—英、
汉　Ⅳ.①N095

　　中国版本图书馆 CIP 数据核字（2018）第 299570 号

责任编辑：张立红
封面设计：梁　洁
版式设计：方加青
责任校对：郭熙凤
责任印制：杨　艳

出版发行：清华大学出版社
　　　　　网　　　址：http://www.tup.com.cn，http://www.wqbook.com
　　　　　地　　　址：北京清华大学学研大厦 A 座　　　　　邮　　　编：100084
　　　　　社 总 机：010-62770175　　　　　　　　　　　　邮　　　购：010-62786544
　　　　　投稿与读者服务：010-62776969，c-service@tup.tsinghua.edu.cn
　　　　　质 量 反 馈：010-62772015，zhiliang@tup.tsinghua.edu.cn
印 装 者：大厂回族自治县彩虹印刷有限公司
经　　销：全国新华书店
开　　本：170mm×240mm　　印　　张：45　　字　　数：890 千字
版　　次：2020 年 7 月第 1 版　　印　　次：2020 年 7 月第 1 次印刷
定　　价：128.00 元

产品编号：073616-01

内容介绍

　　《科学美国人》是美国历史上最长的、一直连续出版的杂志，也是著名的《科学》（Science）的姊妹刊，是高水平的大众科学杂志，有151位诺贝尔奖得主撰稿。作为《科学美国人》杂志的资深撰稿人，约翰·霍根对于当代科学有着卓越的领悟，因为他具有常人无可企及的优越条件，能借工作的便利经常性地接触科学界的名家，诸如林恩·马古利斯、罗杰·彭罗斯、弗朗西斯·克里克、理查德·道金斯、弗里曼·戴森、默里·盖尔曼、斯蒂芬·杰伊·古尔德、斯蒂芬·霍金、托马斯·库恩、克里斯托弗·兰顿、卡尔·波普尔、史蒂文·温伯格以及爱德华·威尔逊等，并能得心应手地刺探他们内心深处的思想。

　　在《科学的终结》一书中，霍根以才华横溢的笔触描写了这些大名鼎鼎的人物平凡的一面和活跃的思想。科学家"在面对认识的限度时……更像常人一样，易受到自己的恐惧和欲望的左右"。

　　这种隐秘的恐惧，正是霍根在本书中所着力探讨的：是否所有重大的问题都已经被解决了？所有值得追求的知识都已被掌握了吗？是否存在着某种标志着科学之终结的"万物至理"？重大发现的时代一去不复返了吗？今天的科学是否已衰退到只能解答细枝末节的问题、只能修补现有理论的地步？

　　对于诸如此类的敏感问题，霍根在走访弗雷德·霍伊尔、诺姆·乔姆斯基、约翰·惠勒、克利福德·格尔茨及其他数十位杰出学者之后，在与这些名人就"上帝""星际旅行""超弦""夸克""混杂学""意识""神经达尔文主义""马克思的进步观""库恩的革命观""元胞自动机""机器人"以及"欧米加点"等话题进行了深入讨论的基础上，抽象出了率真的答案。由此引发的评述，混合着霍根对"终结论"所做的机智而又精辟的辩护，以及他对于整个科学事业的睿智、独到而又深刻的领悟，读来既让人兴奋激动，又给人带来衷心的愉悦。

科学家通常自以为与其他学者不同，因为他们坚信自己不是在建构真理，而是在发现真理；其工作是要揭示经验世界存在的规律，而不仅仅是去解释它。但科学的力量却存在着自身固有的限度：狭义相对论把物理运动（甚至信息转换）的速度限制在光速之下，量子力学显示出不确定性，而混沌理论则进一步证明完全的预见是不可能的。同时，科学合理性本身正受到新勒德派分子、保护动物权利活动分子、宗教极端主义者以及新时代信徒的攻击。

霍根强调，对科学的最大威胁可能来自科学规范的丧失，越来越多的理论工作者致力于玩弄被霍根称为"反讽科学"的理论，科学规范已被逐渐削弱成某种近似文学批评的东西。基于对当代思想大师们的采访与思索，霍根在提出对科学质疑的同时，也表达了他对科学的崇敬。

书中霍根描述了科学技术的飞速发展，读者可以迅速且全面地了解20世纪几乎所有领域中最先进、最震撼人心的各种科学技术进步和发明创造，同时如果想深入了解，霍根非常严肃地列出了相关依据和参考文献。与此同时，在这些喜悦的语言中，霍根非常犀利地指出这些震撼人心的科学技术不过是在前人的革命性科学成果的基础上的延伸和小创造，远远达不到日心说、DNA和宇宙大爆炸等革命性伟大发现的级别。所以，虽然科学技术在随时随地给我们的生活带来根本性的改变，我们却更加需要理性地思考：我们还能不能再有DNA那样伟大的科学发现了？

Praise for John Horgan's
The End of Science

In this **wonderful, provocative** book ... Horgan's approach is to take us along while he buttonholes several dozen of earth's crankiest, most opinionated, most exasperating scientists to get their views on where science is and where it's going.... They all come to life in Horgan's narrative.

—*Washington Post Book World*, front page review

Thanks to Mr. Horgan's smooth prose style, puckish sense of humor, and wicked eye for detail, these encounters make for zesty reading. Frequently they are hilarious.... **A thumping good book**.

—*Wall Street Journal*

A deft wordsmith and **keen observer**, Horgan offers lucid expositions of everything from superstring theory and Thomas Kuhn's analysis of scientific revolutions to the origin of life and sociobiology.

—*Business Week*

[The book's] greatest pleasures flow from Horgan's encounters with the characters who have made science their lives.

—*San Francisco Chronicle*

Rich in provocative ideas and **insightful** anecdotes.

—*Library Journal*

The End of Science, a lively, witty book by science writer John Horgan, exemplifies the genre's virtues and faults: **It is fun to read. It will make people think**.

—*Reason*

It is a sweeping argument, admirably brought off.

—*Washington Times*

John Horgan has everybody talking. Probably no science book of this year has generated as much comment.

—*Rocky Mountain News*

The End of Science is a revealing glimpse into the minds of some of our leading scientists and philosophers. Read it. **Enjoy it, learn from it**.

—*Hartford Courant*

在这本雄辩的佳作中……霍根引领着我们，旁观他对数十位科学家的采访。他的采访策略是揪住这些科学家不放，一定要他们坦露自己对于科学发展到了何等地步、最终将发展向何处等问题的看法；而这些科学家们，则是当今地球上最特立独行、最有思想，当然也是最难缠的人物……他们都在霍根娓娓道来的行文中，栩栩如生地跃入我们眼帘。

——《华盛顿邮报书评世界》

由于霍根那流畅的文笔、带有恶作剧色彩的幽默感和善于捕捉细节的犀利眼光，他笔下所描述的一系列采访令人读来饶有趣味，有时甚至会令你忍俊不禁……这绝对是一部好书。

——《华尔街日报》

作为一位技巧娴熟的语言大师和目光犀利的观察家，霍根为我们呈现了一个清晰而又流畅的世界，从超弦理论和托马斯·库恩对科学革命的分析，到生命起源和社会生物学。

——《商业周刊》

该书最大的阅读乐趣，源自霍根与那些以科学为终生目标的人们之间的遭遇战。

——《旧金山纪事报》

引人入胜的观点和极富洞察力的描述比比皆是。

——《图书馆学报》

科学写手约翰·霍根的《科学的终结》是一本生动有趣的书，堪称该文体的代表之作，也有着该文体所固有的优点和缺点。读来引人入胜，且能引人思考。

——《美国理性杂志》

这是一场盛大的雄辩盛宴，以令人敬佩的方式展现出来。

——《华盛顿时报》

约翰·霍根刺激得每个人都不吐不快。也许近年来描写科学的著作中，没有任何一本书能引发如此之多的评论。

——《落基山新闻》

《科学的终结》一书深入揭示了当代诸多顶级科学家和哲学家的内心世界。阅读它，享受它，并且感悟它吧！

——《哈特福德新闻报》

Science and Its Skeptics

Gary Marcus

Science has been taking a lot of punches lately. A recent cover story for *The Economist* argued, with cause, that "modern scientists have done too much trusting, and not enough verifying." A few days ago, the science writer-provocateur John Horgan wrote a dark reflection, in *Scientific American*, on a litany of failures in science that he has seen over his thirty-year career. Reporting on an "archaeological dig into the strata" of his career, which he says justifies why he's "so critical of science," Horgan finds himself struck by all the "breakthroughs" and "revolutions" that have failed to live up to their hype: string theory and other supposed "theories of everything," self-organized criticality and other theories of complexity, anti-angiogenesis drugs and other potential "cures" for cancer, drugs that can make depressed patients "better than well," "genes for" alcoholism, homosexuality, high IQ and schizophrenia.

All of this is true. String theory hasn't yet lived up to its promise (and may never). Complexity theory hasn't, either. People still get cancer and the antidepressants known as S.S.R.I.s help far fewer people than the early hype suggested. The locution a "gene for X" is, in most cases, a verbal sloppiness that leads only to false expectations. Scientists, and those who would report on them, have sometimes promised more than they can deliver.

Yet some depressed patients really do respond to S.S.R.I.s. And some forms of cancer, especially when discovered early, can be cured, or even prevented altogether with vaccination. Over the course of Horgan's career, H.I.V. has gone from being universally fatal to routinely treatable (in nations that can afford adequate drugs), while molecular biologists working in the nineteen eighties, when Horgan began writing, would be astounded both by the tools that have recently been developed, like whole-genome-sequencing, and the detail with which many molecular mechanisms are now understood: reading a biology textbook from 1983 is like reading a modern history text written before the Second World War. Then there is the tentative confirmation of the Higgs boson; the sequencing of Neanderthal DNA; the discovery of FOXP2, which is the first gene decisively tied to human language; the invention of optogenetics; and definitive proof that exoplanets exist. All of these are certifiable breakthroughs.

The problem with some of these critical observations is not that they are wrong, but that they are one-sided. When Horgan writes that "the biggest meta-story in science over the last few years—and one that caught me by surprise—is that much of the peer-reviewed scientific literature is rotten," it's not just that he is arguably overstating things, it's that he's missing half the *story*.

There *is* a crisis in replicability, as both Horgan and *The Economist* suggest (and as I noted last December). But there is also a huge, rapidly growing movement to address it. When I revisited the topic a few months later, I reported at least five new efforts focussed on increasing replicability. Since then, the list has continued to grow. There were at least three new announcements in the last few weeks: a new initiative to validate the fifty most important studies in cancer biology; a new pilot program that allows critical comments to be published at pubmed.gov (one of the leading portals for scientific research); and a new badge system for papers that share their materials and data, designed to promote replicability, at a leading journal. As the Cambridge (U.K.) scientist Rogier Kievit put it to me in an e-mail: "the avenues for (constructive) criticism of science are so much better now than they were even five years ago…. The half-life of nonsensical findings has decreased enormously, sometimes even to before the paper has officially been published." The wholesale shift in the culture of how scientists think about their craft is at least as significant a meta-story as the replicability crisis itself. But the prophets of doom never let their readers in on this happy secret.

It is absolutely correct for onlookers to call for increased skepticism and clearer thinking in science writing. I've sometimes heard it said, with a certain amount of condescension, that this or that field of science "needs its popularizers." But what science really needs is greater enthusiasm for those people who are willing to invest the time to try to sort the truth from hype and bring that to the public. Academic science does far too little to encourage such voices.

At the same time, it is facile to dismiss science itself. The most careful scientists, and the best science journalists, realize that all science is provisional. There will always be things that we haven't figured out yet, and even some that we get wrong. But science is not just about conclusions, which are occasionally incorrect. It's about a methodology for investigation, which includes, at its core, a relentless drive towards questioning that which came before. You can both love science and question it. As my father, who passed away earlier this year, taught me, there is no contradiction between the two.

November 6, 2013

(Gary Marcus)

现代科学越来越夸夸其谈?

盖瑞·马库斯

最近,"科学"受到了不少冲击。《经济学人》最近一期的封面文章称:"新潮的科学家们设想了太多东西,却没有证明多少。"几天前,科学界的记者兼卧底约翰·霍根在《科学美国人》杂志上发表了一篇灰色调的反思文章,历数其30年职业生涯中所目睹的种种科学失败。他自称对其个人职业生涯的这番"考古性的发掘",已足以证明自己为什么"对科学如此挑剔"。因为霍根吃惊地发现,所有的"突破"与"革命"都不过是浮云:弦理论和其他所谓的"万有理论",自组织临界现象和其他复杂性理论,抗血管生成药物和其他可能"治愈"癌症的疗法,可以使抑郁症患者"心情愉悦"的药物,导致酗酒、同性恋、高智商和精神分裂症的"基因",诸如此类。

这一切都是真的。弦理论还没有(也许永远不会)像人们期待的那样,解决基本粒子的问题。复杂性理论也一样没有解决。人们还是会得癌症,选择性血清素再吸收抑制剂(SSRI)这样的抗抑郁药远远没有像想象中的那样对很多人管用。"X基因"这样的惯用语,沦为了一种空洞的口头禅。科学家和那些写科学报道的记者许诺了太多东西,却无法一一兑现。

当然,SSRI对有些抑郁症患者的确有用。有些癌症,特别是发现得早的话,是可以治愈的,甚至可以通过接种疫苗来预防。在霍根的职业生涯中,艾滋病也从普遍致命发展到了可以进行常规治疗(只要所在国家能提供足够的药物)。在霍根开始写作的20世纪80年代,分子生物学会被后来的发展大大震惊,比如,全基因组测序,以及分子机制的细节。阅读一本1983年的生物学课本,就好像阅读二战前编写的现代史教科书。还有,希格斯玻色子(Higgs boson)的初步证实,尼安德特人基因序列的测定,首个控制语言能力发展的基因FOXP2的发现,光遗传学技术的发明,关于系外行星存在的确凿证据。所有这些都是确定无疑的重大突破。

霍根对这些重大进展的管窥之见，若说有问题，并不在于它们是错误的，而在于它们是片面的。当霍根感慨："令我震惊的是，过去几年里，科学叙事下最重大的真相在于，很多经过同行评议发表的科学文献都已过时。"他并未夸大其词，而是只说了一半的事实。

霍根和《经济学人》都认为科学的可重复性存在危机（我在2012年12月也提出过这一点）。但是，针对这个问题的解决方法也在大量涌现并快速增长。数月后，当我再次关注这个话题时，发现至少有五种新的方法来增加可重复性。此后，这个名单一直在增加。过去几个星期起码有三项新的突破：一项新的倡议，提倡使50项最重要的癌症生物学研究合法化；一个试点项目，允许在pubmed.gov（一个主要的科研门户网站）发表批评观点；一项对在一流杂志上分享材料、数据的论文推行奖励的新制度，该制度旨在促进可重复性。剑桥大学科学家罗吉尔·基维特在电子邮件里告诉我："科学批评的途径比五年前好多了……无聊的发现的半衰期比以前短得多，有些甚至在论文发表前就已灰飞烟灭。"在文化层面，科学家们对于应如何看待自己手稿的认识，已经整体上产生了转变，这至少与可重复性危机一样，同属科学叙事下的重大真相。可惜的是，末日预言家永远也不会引导其读者关注这等会令人感到开心的隐秘真相。

旁观者们呼吁科学写作应保持怀疑和清醒思考，这无疑是正确的。我有时听到人们轻描淡写地说，科学的这个或那个领域"需要大众化"。但是，科学真正需要的是投入大量热情和大量时间，从各种夸夸其谈中发现真理，并将其呈献给公众。学术科学在促发类似声音方面的作为，终究还是太少了。

与此同时，对科学本身进行否定是极其轻率的。最细心的科学家和最周到的科学记者都发现，所有的科学都是暂时的，永远都会有我们尚未弄清的东西，有时甚至会出现错误。但是，科学并不仅仅是给出结论，因为结论总是会出现这样那样的错误。科学是一套研究方法，在其内核中就包含着对前人的结论不遗余力地提出质疑。你可以同时热爱科学，并质疑它，两者之间并无矛盾，这是我那已于年初辞世的父亲曾给予我的教诲。（摘自《纽约客》（*The New Yorker*），原文标题Science and Its Skeptics）

2013年11月6日

（盖瑞·马库斯，纽约大学心理系教授）

哲学、科学、"终结"与科学家眼中的科学

刘兵

　　大约20年前，一本名为《科学的终结》的翻译著作，或许是因其书名的刺激和内容的有趣，曾畅销一时，引人关注，并引发了不少相关的讨论。然而，时过境迁，这本书在市面上已经很难找到了，20年前的许多读者在此书中译本出版后这些年进入了学术领域或关心类似问题，现在读过这本书的人也不多了。如今，清华大学出版社将此书重新出版，是很有意义的，这可以让许多更年轻的读者有机会阅读此书，并思考那些因阅读而引申出来的重要问题。

　　也恰恰因为此书书名的刺激、内容的有趣，以及作者观念的与众不同，此书无论是原版还是中译本在出版后，都招来了众多的争论，并且以批评反对的声音居多，甚至连中译本的译者在1997年写的一篇评介中，也是以这样的说法来结尾的："虽然作为一本较严肃的科普读物，《科学的终结》仍值得一读，它也确实具有较强的可读性，但正如李政道和郝柏林两位先生所云：此书在本质上却是'一本坏书'。因此，在阅读这本书时，'别忘了带上你的鞭子。'"

　　但是，经过了20年的沉淀，在这本书重新出版中译本时，还是可以对这本书以及作者在书中的观点重新进行一些简要的梳理和反思的。这种梳理和反思，或许能为读者提供一种理解和思考这本书的观点的视角（当然不是唯一的视角），或思路。

　　这样做的另一个理由，是该书作者其实并未把很多话讲得很直白，或者说，讲得并不那么明确，这也会在一定程度上影响阅读。在此，可以略加分析。

　　讲"科学的终结",关键在于"终结"的概念,即"终结"是指什么。在此书中,作者似乎也未明确地定义终结,但大致是指如下几方面的内容:是否所有重大的问题都已经被解决了,或者说,关于"宇宙以及我们在其中的位置"的终极的、根本的、纯粹的真理,已经为科学所揭示,因而"科学发现的伟大时代已经一去不复返了";今天的科学家们实际上是在追求那些琐屑、浅显而又枯燥的科学,或者转向失去了科学规范的所谓"反讽的科学",而这种"反讽的科学"却是思辨的、后经验的,像文学批评一样,不能向真理收敛,不能提供可检验的新奇见解;科学过去曾揭示了重要的真理,但科学的力量却存在着自身固有的限度;科学正在受到多方的敌对力量的大力攻击,如此等等。

　　在这样理解的"终结"之下,作者采访了数十位科学界及科学哲学界著名的学者,通过紧逼盯人般的问答方式,用采访材料来支持其"终结说"。至于这些科学家们所说的内容是否真的有力支持了这种终结说,读者自己完全可以作出自己的判断。

　　从以上的简要介绍可以看出,其实此书作者所说的"终结",并非常规语义中的终结,而是另有特别的含义,但是,这是否就等同于一般人心中所理解的终结呢?显然不是。而且,即使在作者所说的那几种意义上,也是完全可以有不同的观点存在的。例如,也正像许多在此书出版后出现的评论所指出的,在19世纪末,在物理学领域,人们也曾有类似的看法,但19—20世纪之交的物理学革命,恰恰带来了全新的量子力学和相对论。说"科学发现的伟大时代已经一去不复返了",这只能是一种推测,而且是无人能可靠地保证其一定会如此的推测,尽管也许有人在研究中会有这样的感觉。

　　与此同时,作者提到当下科学发展成所谓的"反讽的科学",即那种思辨的、无法以经验验证的、像文学批评一样的科学,这里也隐含了其将标准的和理想的西方近现代科学作为唯一的科学的某种意识,是一种典型的一元论,而且是以西方科学为中心的科学观。甚至于就在西方科学的范围内,除少数像近现代物理学那样的标准科学之外,许多其他科学的分支也并不严格地符合那种非常理想化的科学模式。更何况,多元的科学观,现在已经不再像过去那样被看作可怕的洪水猛兽。因此可以说,作为记者的本书作者的科学观,毕竟没有达到像专业的科学研究领域中的学者那样的水准,还只是比较朴素和相对陈旧的那种科学观。而作者所持的科学在终结的观点,偏偏也不是科学家的专利,而更接近于一个哲学问题。

但是，此书在出版后又确实引起了很大的反响，这也应该有其道理。作为《科学美国人》杂志的专职撰稿人，作者确实具备了优秀的记者和作家的素质，并且，能够联系到和采访上如此众多一流的大科学家，能够按照自己的意愿大胆地追问，并获得了在通常的文献中所见不多的这些科学家们评论自己的学说和当下科学研究问题的信息。这确实是很有意思的采访材料。而且，部分地也与作者预设的观点相关，这些科学家们在访谈中，确实也谈及了当下科学和科学研究中存在的问题，这些问题从这些大科学家口中说出，确实又有着特殊的分量。

前些年，甚至从更早开始，关于"终结"的书并不少见，如《历史的终结》等，都曾是畅销一时的著作。如果我们放平心态，不过分纠缠于"科学""终结"与否，而是更多地关注这本奇书中透露的在其他著作中并不常见的那许多科学家在采访中表达出来的鲜活的观点，则或许会更多一些收获。

年轻的读者再读此书，其学术环境、需求和心态都会与20年前的读者大有不同，也不一定能体会到那时读者的阅读感受，但一本有新意、有趣的书仍然是值得人们继续读下去的，这远比那些无趣的陈词滥调更有营养。你不必非得同意作者的观点，哪怕你觉得作者预设的观点不能在书中自圆其说也无妨，在新奇的观点下所引出的那些更有新意的关于科学的信息，仍然有其不可替代的价值和生命力。

2016年10月6日

（刘兵，清华大学社会科学学院科学技术与社会研究所教授，
博士生导师，中国科协—清华大学科技传播与普及研究中心主任）

尽现科学众生相
——闲话《科学的终结》

吴国盛

　　该书作者霍根是一位美国科学记者、《科学美国人》的资深撰稿人，具有西方记者通常具有的那种尖酸刻薄和咄咄逼人，因为西方记者是无冕之王嘛。这样的记者风格我们领教得不多，特别是把这种风格带入科学人物报道方面。在我们中国，科学技术是第一生产力，科学享有崇高的威望，连气功大师们都愿意往科学方面靠。科学人物报道往往是表扬性的，说科学家如何辛苦、如何积极工作为国争光、如何不计名利，等等。这本书令人大开眼界，充分暴露了科学家丰富的人性——既是人，就有优点和缺点。给人深刻印象的是，在霍根的笔下，那些大腕科学家们许多都是自以为是、妄自尊大、装腔作势、心胸狭隘的，而这些性格缺陷又同科学大师们强劲的思想魅力交织在一起，构成了一幅多姿多彩的科学众生相。

　　不知道霍根受的是怎样的教育，他大学是学文的还是学理的。介绍上说他毕业于哥伦比亚大学新闻学院，按我们中国人的说法那他就是学文出身了。他对人物外貌和性格的出色的描画本领，显示了他在语言艺术上的造诣不浅。可他对科学前沿的熟悉又格外令人吃惊，他的这本书几乎是近20年来重大科学进展的一个全面而生动的介绍。联想到前几年美国记者格莱克写的《混沌：开创新科学》，真使我们窥见了美国通才教育的巨大优越性。我国也出过不少写科学家的报告文学，从20多年前的《哥德巴赫猜想》到不久前的《中国863》，大部分都只写到了科学家的外部社会生活，没有深入他们的科学思想中去。不知道是我国的科学家都没有什么自己的思想，还是我们的作家不懂、不会写。读这本书的时候我倒是经常想，什么时候我们中国的记者也能以我们中国的科学家为对象，写出这么一本科学思想型的报告文学就好了。

　　书的内容读者得亲自去读，大致是通过对数十个著名科学家的访谈，得出结论说，以发现真理为目标的纯科学已经终结，因为大发现的时代已经过去，剩下的或者只是技术活，或者就是一些既不能证实也不能证伪的玄玄乎乎的幻想。他说的这些现象我都承认，但我不同意他的结论。因为，在科学发展的任何一个历史时期，科学界的状况大概都是如此——有的死干一些技术活，有的瞎想——可科学从来就是在"技术活"和"玄想"之间或缓或急地进步着。（原载《中国图书商报》1998年9月25日，本文为节选）

1998年9月25日

（吴国盛，清华大学人文学院教授）

译者前言

那还是去年暑热渐消的时候，清华大学出版社的张立红主任来访，说借霍根《科学的终结》新版在美国再度发行之机，想在国内重新出版该书的中译本。说实话，我起初对此事热情并不高，因为在今天的喧闹氛围中，还有多少人有闲暇看书呢？何况又是如此缺乏实用价值的一本，何况又是"炒冷饭"！但之所以最终还是答应下来，是考虑到如下三条理由：

其一，国内毕竟还是有不少爱读书，尤其是爱读此书的朋友在。在2009年前后，国内曾出过两份颇具影响力的书单，一个是中国出版集团主办的"改革开放30年最具影响力的300本书"，另一个是由中国图书商报和中国出版科研所联合主办的"新中国60年中国最具影响力的600本书"；而本书1997年的旧译本，有幸被两份书单收录在内。这的确说明国内不乏抬爱此书的读者；何况随着时间推移，还会催生出一茬茬喜欢此书的新读者！为了这些读者朋友，也应该支持清华大学出版社的善举。

其二，为此书对国内学界了解20世纪后半叶科技史所具有的潜在价值。历史的真貌，只有在拉开距离后才看得清楚。所以，国内学界对科技发展历史的印象，大抵是截止到20世纪中叶，甚至都延伸不到冷战结束的时候。霍根此书，以科技人物传略的方式，对20世纪后半叶的科技发展情况进行了整体勾勒——这当然还不能算是严肃的科技史，但若视之为这一时段科技发展的口述史，则与事实相去不会太远。所以，此书对国内学界认识20世纪后半叶科技发展的历史，有着不容忽视的价值——当然，别忘了戴上批判的眼镜。

其三，此书对国内公众理解科学具有实际价值。科学对于现代社会和民族国家来说，无疑是一种重要的强大力量，对它的利用必须建立在理解的基础之上；而不能理解的强大力量所带来的，更多的是恐惧。比如说数字技术乃至大数据，公众对之又了解多少呢？所以发达国家对于互联网经济大多持审慎态度。目前愈演愈烈的网络经济犯罪现象，仅仅是我们为公众理解严重落后于信息技术发展所付出的最初代价！科学技术的公众理解之重要，可见一斑。在美国等科技发达国家，公众对科技事业的关注、理解，早已形成一种文化传统；其在科技领域原始创新成果的大量涌现，以及20世纪转折之际蔓延整个西方的"科学大战"（science war）的爆发，都离不开公众对科学的理解这一文化基础的支撑。中国要实现建设创新型国家的目标，科技创新的水平、速度及其对社会发展的贡献率都必须优先得到提升，这一切同样离不开公众对科学的理解和支持。而霍根此书，堪称很好的公众理解科学的读物之一。

新的译本是在1997年旧译本的基础上，重新翻译加工而成的。很多朋友都曾为旧译本的完成付出过心血，包括（按姓名拼音顺序）：白奚、陈花胜、段彦新、胡泳涛、李中华、刘勇、陆丁、马宝建、潘涛和邹锐；我和他们中的绝大多数那时都还在北京大学读研，正是意气风发的年纪。为了翻译此书，大家精诚合作，嬉笑怒骂，挥斥方遒。可惜就在北大百年校庆前夕，大家各奔前程，许多人出了国并断了音讯；仍留在国内发展的，也联系渐稀。值此新译本付梓之际，向当年的老朋友们致以诚挚的谢意和问候。

由于科学自身的发展以及语言的流变，20年后重新出版此书的中译本，本着对读者负责的态度，采取了在旧译本基础上重新翻译的做法。新版的前言是旧译本所没有的，"跋 未尽的终结"也有不少地方与旧译本不同，现在的跋保留了1997年平装本再版的原貌，而旧译本的跋则是霍根先生为当时的中文译本所特意修订的产物。旧译本中因时间紧迫和年轻人的毛糙所留下的种种瑕疵，也尽力予以修正，细心的读者们自然不难发现，就不一一指出了。新版的翻译校订由我和张武军先生共同完成，他主要负责第二、三、八、九、十章的译校工作，我负责完成其余部分，并对全书译稿做了校订。

尽管又花了近一年的时间对译文进行反复校订，疏漏处仍在所难免，衷心期盼读者们不吝指正。

孙雍君
2016年8月15日

Preface to the 2015 Edition
Rebooting *The End of Science*

H ere's what a fanatic I am: When I have a captive audience of innocent youths, I expose them to my evil meme.

Since 2005, I've taught history of science to undergraduates at Stevens Institute of Technology, an engineering school perched on the bank of the Hudson River. After we get through ancient Greek "science," I make my students ponder this question: Will our theories of the cosmos seem as *wrong* to our descendants as Aristotle's theories seem to us?

I assure them there is no correct answer, then tell them the answer is "No," because Aristotle's theories were wrong and our theories are right. The earth orbits the sun, not vice versa, and our world is made not of earth, water, fire and air but of hydrogen, carbon and other elements that are in turn made of quarks and electrons.

Our descendants will learn much more about nature, and they will invent gadgets even cooler than smartphones. But their scientific version of reality will resemble ours, for two reasons: First, ours— as sketched out, for example, by Neil deGrasse Tyson in his terrific re-launch of *Cosmos*— is in many respects true; most new knowledge will merely extend and fill in our current maps of reality rather than forcing radical revisions. Second, some major remaining mysteries— Where did the universe come from? How did life begin? How, exactly, does a chunk of meat make a mind?—might be unsolvable.

That's my end-of-science argument in a nutshell, and I believe it as much today as I did when I was finishing my book twenty years ago. That's why I keep writing about my thesis, and why I make my students ponder it—even though I hope I'm wrong, and I'm oddly relieved when my students reject my pessimistic outlook.

My meme has spread far and wide, in spite of its unpopularity, and often intrudes even in upbeat discussions about science's future. A few brave souls have suggested that maybe there's something to this end-of-science idea.[1] But most pundits reject it in ritualistic fashion, as if performing an exorcism.[2]

Allusions to my book often take this form: "In the mid-1990s, this guy John Horgan said science is ending. But look at all that we've learned lately, and all that we still don't know! Horgan is as wrong as those fools at the end of the nineteenth century who said physics was finished!"

This preface to a new edition of *The End of Science* updates my argument, going beyond "Loose Ends," the Afterword for the 1997 paperback, which fended off the first volley of attacks. Science has accomplished a lot over the past eignteen years, from the completion of the Human Genome Project to the discovery of the Higgs boson, so it's not surprising that folks only sketchily familiar with my argument find it preposterous.

But so far my prediction that there would be no great "revelations or revolutions"—no insights into nature as cataclysmic as heliocentrism, evolution, quantum mechanics, relativity, the big bang—has held up just fine.

In some ways, science is in even *worse* shape today than I would have guessed back in the 1990s. In *The End of Science*, I predicted that scientists, as they struggle to overcome their limitations, would become increasingly desperate and prone to hyperbole. This trend has become more severe and widespread than I anticipated. In my thirty-plus years of covering science, the gap between the ideal of science and its messy, all-too-human reality has never been greater than it is today.

Over the past decade, statistician John Ioannidis has analyzed the peer-reviewed scientific literature and concluded that "most current published research findings are false."[3] Ioannidis blames the problem on increasingly fierce competition of scientists for attention and grants. "Much research is conducted for reasons other than the pursuit of truth," he wrote in 2011. "Conflicts of interest abound, and they influence outcomes."[4]

Since 1975 the number of biomedical and life-science articles retracted because of fraud has surged tenfold, according to a 2012 study.[5] Many more papers probably should be retracted but aren't because their flaws never come to light. "There are errors in a lot more of the scientific papers being published, written about and acted on than anyone would normally suppose, or like to think," *The Economist* announced in a 2013 cover story, "Trouble at the Lab."[6]

Science always has adversaries, from religious fundamentalists to global-warming deniers, but lately it has become its own worst enemy. Scientists are beginning to resemble lawyers and politicians, who too often abandon their noble ideals for the crass pursuit of wealth, power, and prestige.

Even setting aside the enormous problems of misconduct and unreliability of results, some prominent scientists have become arrogant, and dismissive of criticism, in ways that ill-serve their profession. Some seem intent on transforming science from a truth-seeking method into an ideology—scientism—which harshly denigrates nonscientific ways of engaging with the world, in a manner all too reminiscent of religious fundamentalism.

In his 2011 bestseller *Thinking, Fast and Slow*, psychologist Daniel Kahneman reveals all the ways in which our minds can fool us, confirming our preconceptions. Scientists like to cite Kahneman's findings when they bash creationists and climate-change skeptics. But as Kahneman himself emphasizes, scientists, too, can succumb to confirmation bias.[7]

I remain hopeful that science can regain its ethical bearings and help us solve our social problems: poverty, climate change, overpopulation, militarism. I'm often called a pessimist, but I'm an optimist about what matters most. I'll return to what matters most at the end of this essay—after explaining how my thesis accommodates advances in physics, biology, neuroscience, and chaoplexity (my snarky portmanteau term for chaos and complexity).

Physics Stumbles as the Cosmos Speeds Up

Particle physics, the grand goals of which entranced me when I started covering science more than three decades ago, is in such disarray that I almost hesitate to knock it. In 2012 the Large Hadron Collider finally yielded solid evidence of the Higgs boson, which confers mass to other particles and was first postulated in the early 1960s.

The Higgs has been dubbed, with hype that serves as its own parody, the "God particle." Its detection was almost comically anti-climactic, measured against physics' lofty ambitions. The Higgs serves as the capstone of the standard model of particle physics, a quantum description of electromagnetism and the nuclear forces. But the Higgs doesn't take us any closer to physics' ultimate goal than climbing a tree takes us to the moon.

For more than half a century, physicists have dreamed of finding a unified theory—sometimes called a "theory of everything"—that wraps up all of nature's forces, including gravity. They hoped this theory might solve the most profound of all mysteries: How did the universe come to be, and why does it take the form we observe and not some other form? As Einstein put it, did God have any choice in making the universe?

In *The End of Science*, I bash string theory, the leading candidate for a theory of everything, because it postulates particles too small to be observed by any conceivable experiment. In 2002, I bet Michio Kaku $1,000 that by 2020 no one would win a Nobel Prize for string theory or any other unified theory.[8] Does anyone else want a piece of this action?

Research on string theory—sometimes called M (for membrane) theory—has if anything regressed over the last decade. Physicists realized that the theory comes in virtually infinite versions (10^{500}, according to one estimate), each of which "predicts" a different universe.

Leonard Susskind and other string advocates have brazenly redefined this flaw as a feature, claiming that all of the universes allowed by string theory actually exist; our observable cosmos is just a minuscule bubble in an oceanic "multiverse."

Asked why we happen to live in this particular cosmos, multiverse enthusiasts cite the anthropic principle, which decrees that our cosmos must have the form we observe, because otherwise we wouldn't be here to observe it.[9] The anthropic principle is tautology masquerading as explanation.

Philosopher Karl Popper, whom I profile in *The End of Science*, denounced psychoanalysis and Marxism for being so vague and flexible that they could not be proven wrong, or "falsified," by any conceivable finding. Falsification has been widely embraced as our best criterion for distinguishing science from pseudo-science.[10]

Confronted with complaints that string and multiverse theories cannot be falsified, true believers like Sean Carroll now assert that falsification is overrated. Other high-profile physicists—notably Stephen Hawking and Lawrence Krauss—denigrate philosophy as worthless.[11]

Physicists' hubris is epitomized by Krauss's 2012 book *The Universe from Nothing*, which asserts that physicists have *already* discovered why there is something rather than nothing. Krauss's "answer" is just stale, hand-wavy speculation that leans heavily on quantum uncertainty, which clever theorists can employ to explain away anything.

Moreover, Krauss begs the question of where his version of "nothing"— the primordial quantum vacuum that supposedly spawned our world—came from.[12] In a blurb he must regret, biologist Richard Dawkins compares Krauss to Darwin. "If *On the Origin of Species* was biology's deadliest blow to supernaturalism," Dawkins gushes, "we may come to see *A Universe from Nothing* as the equivalent from cosmology."

I understand the desire of atheists like Dawkins, Krauss, and Hawking to supplant faith-based explanations of creation with empirical ones—and, more broadly, to counter superstition with reason. But when they defend pseudo-scientific nonsense like multiverses and the anthropic principle, they hurt their cause more than they help it.

Physics has produced genuine thrills since *The End of Science* was published, including the most dramatic scientific discovery of the last twenty years. In the late 1990s, astronomers peering at supernovas in distant galaxies—which provide a measure of the universe's size and expansion—discovered that the rate of expansion is growing.

I didn't expect this astonishing finding to hold up, but it did; the discoverers of the acceleration of the universe won the Nobel Prize in 2011. Early on, some physicists predicted that cosmic acceleration might be analogous to radioactivity and other anomalies that triggered revolutionary shifts in physics at the dawn of the twentieth century. But so far, the cosmic acceleration remains just a fascinating twist of big bang cosmology.

We Are Still Alone

Another major discovery to emerge from astronomy is that many stars have planets orbiting them. When I started my career, astronomers had not confirmed the existence of a single exoplanet. Now, with the help of exquisitely sensitive telescopes and ingenious signal-processing methods, they have discerned thousands. Extrapolating from these data, astronomers estimate that the Milky Way may harbor as many planets as stars.

But we still can't answer our most urgent question: Are we alone? As I said in *The End of Science*, I fervently hope extraterrestrial life is detected in my lifetime, because that could blow science wide open. But given the enormous distances to even the closest exoplanets—which would take millennia for our fastest spacecraft to reach—chances of such a discovery remain vanishingly slight.[13]

Scientists still can't figure out how life arose on Earth more than three billion years ago. A 2011 report in the *New York Times* on origin-of-life research highlights the same theories I examined in *The End of Science*.[14] Directed panspermia, which holds that aliens planted the seeds of life on earth—and which comes awfully close to intelligent design—is apparently making a comeback.

Where Are All the Designer Babies?

On some fronts, of course, biology is barreling along at a prodigious clip. In 2003, fifty years after Francis Crick and James Watson deciphered the double helix, the Human Genome Project finished (more or less) decoding all our DNA. Francis Collins, the project's director, called it "an amazing adventure into ourselves," as a result of which medical advances would be "significantly accelerated."[15]

The federal project was completed ahead of time and under budget, because of the surging power and plummeting costs of gene-sequencing machines and other biotechnologies. These advances have inspired triumphal rhetoric about how we are on the verge of a profound new understanding of and control over ourselves—as well as the usual hand-wringing about designer babies.

But so far the medical advances promised by Francis Collins have not materialized. Over the past twenty-five years, researchers have carried out some 2,000 clinical trials of gene therapies, which treat diseases by tweaking patients' DNA.[16] At this writing, no gene therapies have been approved for commercial sale in the United States (although in 2012 one was approved in Europe). Tellingly, cancer mortality rates—in spite of countless alleged "breakthroughs"—also remain extremely high.[17]

We still lack basic understanding of how genes make us who we are. Behavioral genetics seeks to identify variations in genes that underpin variations between people—in, for example, temperament and susceptibility to mental illnesses. The field has announced countless dramatic "discoveries"— genes for homosexuality, high IQ, impulsivity, violent aggression, schizophrenia—but so far it has not produced a single robust finding.[18]

Other lines of investigations, while intriguing, have complicated rather than clarified our understanding of how genes work. Researchers have shown that evolution and embryonic development can interact in surprising ways (evo devo); that subtle environmental factors can affect genes' expression

23

(epigenetics); and that supposedly inert sections of our genome (junk DNA) might actually serve some purpose after all.

Biology's plight recalls that of physics in the 1950s, when accelerators churned out weird new particles hard to reconcile with conventional quantum physics. Confusion yielded to clarity after Murray Gell-Mann and others conjectured that neutrons and protons are composed of triplets of strange particles called "quarks."

Could some novel idea—as bold as the quark—help clarify genetics, perhaps in ways that help the field fulfill its medical potential? I hope so. Could biology be on the verge of a profound paradigm shift, akin to the quantum revolution in physics?

I doubt it. Biology's framework of neo-Darwinian theory plus DNA-based genetics has proven to be remarkably resilient, and my guess is that it will easily absorb any surprises, just as quantum field theory absorbed quarks.

Ironically, as research undermines simple biological models, some prominent biologists have advocated extremely deterministic, reductionist models of human behavior. Jerry Coyne asserts that free will is an illusion,[19] and Richard Wrangham and Edward Wilson contend that humans are innately warlike.[20]

Racist theories of intelligence have even resurfaced. In 2007 James Watson—the grand old man of modern genetics—said that the problems of blacks stem from their innate inferiority.[21] When it comes to arrogance and overreaching, physicists are hardly alone.

The Neural Code, Panpsychism, and DARPA

Of all the initial objections to *The End of Science*, the one I took most to heart was that I didn't delve deeply enough into research on the human mind, which has more potential to produce "revelations or revolutions" than any other scientific endeavor. I agreed, so much so that I devoted my next two books to mind-related topics.

The Undiscovered Mind (1999) critiques psychiatry, neuroscience, evolutionary psychology, artificial intelligence and other fields that seek to explain the mind and heal its disorders. *Rational Mysticism* (2003) explores mystical states of consciousness, with which I've been obsessed since my acid-addled youth.

I've continued tracking neuroscience over the past decade. I'm especially intrigued by efforts to crack the "neural code."[22] This is the algorithm, or set of rules, that supposedly transforms electrochemical impulses in our brains into perceptions, memories, emotions, decisions.

A solution to the neural code could have profound consequences, intellectual and practical. It could help solve ancient philosophical conundrums such as the mind-body problem, with which Plato wrestled, and the riddle of free will. It could lead to better theories of and treatments for brain-based disorders, from schizophrenia to Alzheimer's.

The neural code is arguably the most important problem in science—and the hardest. Like the origin of life, the neural code has spawned a surfeit of theories with contradictory claims and assumptions. Whereas all organisms share a remarkably similar genetic code, the neural code of humans may differ from the codes of other animals. In fact, each individual brain may employ many neural codes, which keep evolving in response to new stimuli.

A solution to the neural code could provide mind-science with the unifying principle it desperately needs. But for now mind-science remains in what philosopher Thomas Kuhn called a "pre-scientific" state, lurching from one fad to another. Even battered old paradigms like psychoanalysis and behaviorism still have adherents.

It gets worse. Christof Koch, who for decades has led the effort to explain consciousness in neural terms—and who used to be my go-to source on brain research—recently started espousing panpsychism.[23] This idea, which dates back to Buddha and Plato, holds that consciousness pervades even nonliving matter. Like rocks.

In *The End of Science*, I coined the term "ironic science" to describe propositions so speculative and fuzzy—string and multiverse theories come to mind—that they should be viewed as fictions. I predicted that as conventional science yields diminishing returns, ironic science will proliferate. Panpsychism represents a splendid fulfillment of my prediction.

Mind-science's lack of progress is also reflected in the abysmal state of psychiatry. In *The Undiscovered Mind*, I argued that pharmaceutical treatments for depression and other mental disorders are not as effective as adherents claim. In retrospect, my critique was too mild. Recent investigations—notably the 2010 book *Anatomy of an Epidemic*, by journalist Robert Whitaker—present evidence that psychiatric medications may in the aggregate hurt patients more than they help.

The funding of mind-science also troubles me. Over the last few decades, psychiatry has begun to resemble a marketing arm of the pharmaceutical industry.[24] Meanwhile, neuroscientists have become increasingly dependent on the U.S. military, which has a keen interest in research that can boost its soldiers' cognitive capacities and degrade those of enemies.

In 2013, President Barack Obama announced that the government was investing more than $100 million a year in a new initiative called BRAIN, for Brain Research through Advancing Innovative Neurotechnologies. The major sponsor of BRAIN is the Defense Advanced Research Projects Agency (DARPA). Neuroscientists have accepted— indeed, eagerly sought—this funding without any significant debate over ethical propriety.

Big Data and the Singularity

Could persistent, precipitous progress in digital technologies lead to breakthroughs in physics, genetics, neuroscience, and other fields? This is the hope of "Big Data" enthusiasts. I'm mightily impressed by some digital technologies. Just now, I used my MacBook Air and Google to learn—in less than a minute!—who coined the term "Big Data" (see the answer in the Notes).[25] My car's GPS navigator, too, seems almost magical.

But the hoopla over Big Data leaves me cold. It reminds me of the rhetoric of chaoplexologists whom I critiqued in *The End of Science*. Like chaoplexologists, Big Data adherents profess an almost religious faith in the power of computers to solve problems that have resisted traditional scientific methods.

Some Big Data manifestos are weirdly anti-intellectual. In 2008 *WIRED* editor-in-chief Chris Anderson prophesied that computers crunching "petabytes" of data would bring about "the end of theory."[26] By uncovering subtle correlations in data, he explained, computers will enable us to predict and manipulate phenomena without having to comprehend them. Science can advance "without coherent models," Anderson wrote, adding, "It's time to ask: What can science learn from Google?"

That same year, brilliant Wall Streeters equipped with the most powerful computer models money can buy failed to foresee one of the worst economic crashes in history. Anderson suggested that Big Data can yield power without comprehension, but Wall Street's computers didn't yield either.

The wackiest prophecy inspired by digital technologies is the Singularity. The term, borrowed from physics, refers to a gigantic leap forward in intelligence—of machines, humans, or some cyborg hybrid—catalyzed by advances in computer science and other fields. Oh, and these super-smart beings will be immortal.

Although I didn't mention the term "Singularity" in *The End of Science*, I presented the basic idea, which was explored decades ago by visionaries like Freeman Dyson and Hans Moravec. I called their speculation "scientific theology," because it didn't even deserve to be called "ironic science."

Modern Singularitarians, notably computer scientist Ray Kurzweil, insist that the Singularity will happen soon, within a generation.[27] They base their predictions on exponential progress in digital technologies but ignore the stagnation in genetics, neuroscience, and other fields that I've highlighted above.

Some very smart, successful people believe in the Singularity. These include the founders of Google, who helped Kurzweil establish a "Singularity University" and in 2013 hired him as their director of engineering. I empathize with the desire of Singularitarians to live forever and boost their IQs 1,000-fold.[28] But the Singularity is a cult for people who have faith in science and technology rather than in God. It is "The Rapture" for nerds.

Progress Toward What Matters Most

I'm often asked what would make me admit I'm wrong about science ending. Sometimes my responses are glib: I'll admit I was wrong when Ray Kurzweil successfully uploads his digitized psyche into a smartphone. Or aliens land in Times Square and announce that the young-earth creationists are right. Or we invent warp-drive spaceships that allow us to squirt down wormholes into parallel universes.

Here is a more serious answer: My book serves up many mini-arguments—for example, that string theory will never be confirmed, and that the mind-body problem will never be solved—that could turn out to be wrong. But nothing can shake my faith in my meta-argument: Science gets things right, it converges on the truth, but it will never give us *absolute* truth. Sooner or later science must bump up against limits.[29]

I see no limits, however, to our ability to create a better world. In my 2012 book *The End of War*, I argue that humanity can achieve permanent peace, not in some distant future, but soon. We don't have to turn into robots, or abolish capitalism or religion, or return to a hunter-gatherer lifestyle. We just have to recognize that war is stupid and immoral and put more effort into nonviolent conflict resolution.

In my classes, I often make my students read President John F. Kennedy's inaugural speech, in which he envisions a world without war, poverty, infectious disease, and tyranny. When I ask if they think Kennedy's vision is plausible or just utopian hogwash, most students go with hogwash. Paradoxically, students who reject my end-of-science argument as too pessimistic are in other respects distressingly gloomy about humanity's prospects.

I point out to my students that humanity has taken huge strides toward every one of the goals mentioned by Kennedy. The world is healthier, wealthier, freer, and more peaceful than it was one hundred years ago—or when Kennedy gave his speech in 1961. Because my college sits on the Hudson River across from Manhattan, I can point out a window and note that the water and air around New York City are much cleaner than during Kennedy's era. I wrap up these classroom riffs by exclaiming, "Things are getting better!"

We may never achieve immortality, or colonize other star systems, or find a unified theory that unlocks all the secrets of existence. But with the help of science, we can create a world in which all people, and not just a lucky elite, flourish. And that is what matters most.

Hoboken, New Jersey, 2015

重启《科学的终结》之争

也许我真的是一个自恋狂吧？所以，在有幸面对那群充满求知欲的纯真青年听众时，我才会不由自主地把自己的邪恶文化基因（meme）强加给他们。

自2005年起，我就职于史蒂文斯理工学院，一所坐落在哈德逊河畔的技术学校，在那里给本科生讲授科学史。在讨论完古希腊"科学"后，我让学生们思考这样一个问题：我们关于宇宙的理论在后人看来有没有可能是完全错误的，就像亚里士多德理论对我们而言一样？

我首先向他们保证并不存在什么标准答案，然后告诉他们我的答案是"不可能"，因为亚里士多德理论的确错了，而我们的理论却是正确的。地球在绕日轨道上运行，而不是相反；世界的构成要素并非土、水、火、气，而是氢、碳以及其他一些元素，这些元素则进一步由夸克和电子构成。

后世的人们对于自然界会了解得更多，还会发明出比智能手机更酷的各种东西，但是，他们关于实在的科学版本，将会与我们的如出一辙，原因有二。首先，我们的版本，正如奈尔·德葛拉司·泰森（Neil deGrasse Tyson）在电视系列片《宇宙》（Cosmos）超赞的重播中所概括的，在许多方面都是正确的；大量新的认识，都只是扩展并填补进了我们现有的宇宙图谱，而不是引发激进的修订。其次，那些一直萦绕不去的重大谜团——譬如说，天地万物由何而来？生命何以肇始？一大块肉何以竟能产生出意识？——也许根本就是无解的。

一言以蔽之，这正是笔者的"科学终结"观点，并且本人至今仍坚信其正确性，一如我在二十年前完成拙著时那样。这足以解释为何笔者会一直不厌其烦地就该论题撰写文章，为何要引导学生们对此加以思考——尽管本人一直希望自己搞错了，并且，若学生们真能把这一悲观预期弃如敝屣，我反而会感到莫名的轻松。

尽管拙著并不讨喜，其文化基因却流布深广，即便那些有关科学之未来的乐观论述，也时常被其侵入。敢于承认"科学终结"说或有可取之处者，直似凤毛麟角；[1]绝大多数论者都拒斥笔者的观点，以一种程式化的时髦套路，直似在表演某种驱邪仪式。[2]

对拙著的含沙射影式批评，通常所采取的都是这样的套路："在20世纪90年代中期，有个叫约翰·霍根的家伙宣称'科学正走向终结'；但看看自那以后我们在认识上取得的所有进步吧，再看看仍有待我们认识的所有未知现象！霍根与19世纪末那些宣称'物理学已经终结'的傻瓜一样，犯了同样的错误！"

拙著《科学的终结》的新版前言，大幅更新了笔者的论据，远远超出了1997年平装本的"跋未尽的终结"的范围，而后者回应的只是首波来袭的批评。在过去的十八年里，科学取得了许多重大进展，从人类基因组计划（human genome project）的顺利完成，到希格斯玻色子的发现。因此，那些只会从字面意义上理解笔者观点的人会觉得其荒谬可笑，也就不足为奇了。

但迄今为止，笔者关于未来将不会再有重大的"启示或革命"的预见，依然坚挺如故——不会再有什么对于自然的洞见，能够像日心说、进化论、量子力学、相对论以及大爆炸理论那样震撼人心。

在某种程度上，与笔者过去在20世纪90年代所揣测的那些相比，今日之科学的状态反而更为恶化了。在拙著《科学的终结》中，我曾预言：在其不断尝试克服限度制约的斗争中，科学家们会变得愈益绝望，愈益倾向于夸张；这一倾向的严峻和广泛程度，已远远超出了笔者曾经的预料。以我三十余年与科学打交道的经历，科学理想与其阴暗面之间的差距总是存在的，这就是人性化的现实，但于今尤烈。

在过去的十年里，计量学家约翰·伊奥尼迪斯（John Ioannidis）曾对期间经同行评议的科学文献进行了归纳分析，结论是："目前发表的研究发现多数都是虚假的。"[3]伊奥尼迪斯把这一问题的产生归咎于科学家之间为了赢得公众瞩目和拨款而展开的日趋激烈的竞争。"许多研究，不是出于对真理的追求，而是出于对其他东西的热衷，"他在2011年的一篇文章里写道，"利益冲突比比皆是，且直接影响到了成果产出。"[4]

根据2012年的一项研究，自1975年以来，生物医学和生命科学领域因学术造假而被撤回的论文数量成十倍地激增。[5] 而大量也许本应被撤掉的论文之所以没被撤掉，是因为其缺陷从来就无人关注。"已公开发表的、正在写作和投稿的科学论文中，存在硬伤的不可胜数，远多于任何人通常所以为的或乐于认可的数字，"这是《经济学人》（*The Economist*）杂志在其2013年的封面文章中所宣称的，该标题是"实验室里的麻烦"。[6]

科学从来就不乏敌手，从宗教极端主义者到全球变暖的异议人士，但最终，就连科学本身都变成了自己的死敌。科学家们正变得越来越像律师或政客，也会时不时地放弃他们的崇高理想，以换取对于财富、权力与声望的赤裸追求。

即便抛开由学术不端行为和研究结果的不可靠所导致的海量问题不论，某些杰出科学家也已变得日渐傲慢，对批评不屑一顾，这与其职业操守极不相称。其中某些人所孜孜以求的，似乎是要把科学从一种追求真理的方法转换成某种意识形态——科学至上主义——任何与世界打交道的非科学方法都将遭受其尖刻的诋毁，其做派令人不得不联想到极端主义。

在其2011年的畅销书《思考，快与慢》（*Thinking, Fast and Slow*）中，心理学家丹尼尔·卡尼曼（Daniel Kahneman）揭示了我们的头脑愚弄我们的种种花招，从而证实了先入之见的确存在。科学家们在痛斥神创论者或气候变化的怀疑分子时，都很乐意引用卡尼曼的发现。但正如卡尼曼本人所强调的：科学家们也不例外，同样会屈从于确认偏见（confirmation bias）。[7]

我对科学仍然充满信心，认为它一定会在道德层面重振雄风，并帮助人类解决各种社会问题，如贫困、气候变化、人口过剩、军国主义，等等。本人虽时常被人指责为悲观论者，但在那些最为重要的事情上，我却是一个乐观主义者。在这篇序文的最后，我会再回到那些最为重要的进展上。但在此之前，请容许我先解释一下，本人的论题怎样容纳科学诸领域的最新进展，包括物理学、生物学、神经科学以及混杂学（我在"混沌"与"复杂性"基础上合成的一个稍嫌刻薄的术语）。

物理学步履蹒跚，宇宙却在加速

三十余年前，在我刚刚涉足科学报道的时候，正是粒子物理学的宏大目标激励了我，今天它却处于如此混乱的状态，以至于我都不忍心再挑剔什么了。

2012年，大型强子对撞机终于给出了有关希格斯玻色子的确凿证据，而关于后者的假说早在20世纪60年代初就已被提出，并认为它可以赋予其他粒子以质量。

希格斯玻色子曾一度被赋予"上帝粒子"的尊号，这当然是带点戏谑色彩的炒作；但其被检测到的意义，相较于物理学的崇高目标而言，却平淡无奇得让人觉得有些滑稽。对于粒子物理学的标准模型（即电磁力与核力的量子描述）而言，希格斯玻色子堪称其压顶石（capstone）；但它却无助于我们去接近物理学的最终目标，就像你即便爬到树上，也不可能使你离月亮更近一样。

半个多世纪以来，物理学家们梦寐以求的一直是发现一种统一理论——有时也被称为"万物至理"，以便完满地解释包括引力在内的所有自然力。他们希望该理论能一揽子解决所有最为深刻的奥秘：宇宙是如何来的？它为什么会采取我们所观察到的形式而不是别的形式？正如爱因斯坦所云：上帝在创造宇宙时是否有别的选择？

在《科学的终结》里，我曾对弦理论——号称最有希望发展成万物至理的备选理论之一——予以严厉批评，因为它所假定的粒子太小了，完全无法被任何可能的实验手段观测到。2002年，我和加来道雄（Michio Kaku）打赌1000美元，说直到2020年都不会有人因为弦理论或任何其他统一理论而获得诺贝尔奖。[8]还有谁想这样赌一把吗？

关于弦理论的研究，有时也被称作M（意指"膜"）理论，在过去的十年里，若说有什么的话，也只有倒退。物理学家们意识到，该理论可以推出近乎无限的版本（据估计量级高达10^{500}），而其中的每一种都预示着一个截然不同的宇宙。

李奥纳特·苏士侃（Leonard Susskind）之类弦的拥趸们，竟厚颜地把上述缺陷重新定义为弦的特色，宣称弦理论所允许的所有宇宙都是实际存在的；在"多元宇宙"的海洋里，我们可观察的宇宙只不过是其中一个微不足道的小泡泡。

若问为什么我们就碰巧生活在这一特定的宇宙中，多元宇宙的狂热信徒们只好引证人择原理。该原理断言：我们身处其中的宇宙，就应该具有我们所观察到的形式，否则的话我们就不可能在这儿观察它。[9]人择原理不过是伪装成解释的重言式命题。

证伪作为区分科学与伪科学的最佳标准，已被人们广泛接受。[10]我在《科学的终结》中曾作过专门介绍的哲学家卡尔·波普尔（Karl Popper），曾经例证过，精神分析理论和马克思主义，因为它们都太过模糊又太易变通，所以借助任何可信的发现，都不可能被证明是错误的，或曰被"证伪"。

面对有关弦与多元宇宙理论无法被证伪的指控，其虔诚信徒如肖恩·卡洛尔（Sean Carroll），如今正大肆宣扬说"证伪的意义被高估了"。还有些声名赫赫的物理学家，尤其是斯蒂芬·霍金（Stephen Hawking）和劳伦斯·克劳斯（Lawrence Krauss），则干脆把哲学贬得一文不值。[11]

物理学家的傲慢自大，淋漓尽致地体现在克劳斯2012年的大作《无中生有的宇宙》（*A Universe from Nothing*）中，该书断言物理学家们已经揭示出为何要有些什么，而不能是一无所有。克劳斯的"答案"不过是毫无新意的臆断，所依据的主要是量子不确定性，而聪明的理论家几乎可以用它来解释一切。而且，克劳斯回避了问题的实质，即他所谓的"无"——原始量子真空，我们的世界据说就是由它孕育而生的——究竟由何而来。[12]他肯定会为写出这段话后悔，因为在一份护封简介中，生物学家理查德·道金斯（Richard Dawkins）把克劳斯比作达尔文。"如果说《物种起源》曾经是生物学给予超自然主义的最致命一击的话，"道金斯鞭辟入里地评论道，"那么，我们似乎也可将《无中生有的宇宙》看作出自宇宙学领域的类似玩意儿。"

像道金斯、克劳斯以及霍金这样的无神论者，其最大愿望无非是把基于信仰的"创世说"置换成基于经验的"创世说"，或者说得更宽泛些，是用理性来反击迷信，对此我完全能够理解。但是，在他们开始为多元宇宙、人择原理等伪科学的歪理邪说辩护时，他们其实是在损毁而不是襄助自己的事业。

自《科学的终结》出版以来，物理学确实曾给我们带来过激动人心的体验，包括过去20年里那些极富戏剧性的科学发现。在20世纪90年代后期，天文学家的视线都聚焦于遥远星系的超新星（可据以测度宇宙的大小与膨胀），发现其膨胀率正变得越来越大。

我并未期望这一令人惊讶的发现会经得起检验，但它的确做到了：宇宙加速现象的发现者们赢得了2011年的诺贝尔奖。早些时候，就曾有物理学家预言：宇宙的加速有可能导致物理学发生革命性转折，就像20世纪初的放射性和其他反常现象一样。但到目前为止，宇宙加速依然还只是大爆炸理论的一个迷人的小转折。

我们人类仍是一如既往的孤独

天文学所涌现的另一项重要成就，是发现为数众多的恒星都有行星环绕。在我刚踏足科学报道这一职业生涯时，天文学家尚未确认任何一颗系外行星的存在。如今，借助高度灵敏的望远镜以及精巧的信号处理方法，他们已经识别出了数以千计的系外行星。在这些数据的基础上进行推算，天文学家估计银河系蕴藏着不下于恒星总数的行星。

但这依然无法解答我们所急于了解的问题：我们人类是孤独的吗？正如笔者在《科学的终结》中所云，我热切地希望外星生命能够在我的有生之年被探测到，因为那可使科学产生意料之外的变化。但考虑到哪怕离我们最近的系外行星与地球间的遥远距离（即便我们最快的宇宙飞船，也要用上千年的时间才能抵达），达成类似发现的机会可以说微乎其微。[13]

科学家们依然搞不清生命在三十亿年前是如何在地球上发生的。2011年《纽约时报》曾刊发了一份关于生命起源研究的报告，其中重点介绍了我在《科学的终结》里曾检讨过的同样的理论，[14] 即定向泛种论（directed panspermia）的沉渣再度泛起，认为是外星人在地球上播下了生命的种子，与智能设计论极端类似。

那些设计婴儿都去哪儿了？

在一些方面，生物学也在曲折的道路上艰难进步着。2003年，就在弗朗西斯·克里克和詹姆斯·沃森解读出DNA双螺旋结构五十周年之后，人类基因组计划胜利竣工（差不多吧），基本上解码了我们人类所有的DNA。这一计划的负责人弗朗西斯·柯林斯（Francis Collins）称，该计划是"深入人类自身的神奇探险之旅"，其结果必将导致医学进步的明显加速。[15]

由于基因测序设备以及相关生物工程技术的大幅提升以及成本的直线下降，这一联邦计划项目在预算内提前完成了。这一进展引来了一堆弹冠相庆的溢美之词，认为我们已经站在了一个全新世界的门槛上，跨过去就可以从根本上理解并调控人类自身，比如说对于传统理论一向束手无策的设计婴儿（designer babies）。

但到目前为止，弗朗西斯·柯林斯所承诺的医学进步依然没有兑现。在过去的二十五年里，研究人员总共实施了约2000例基因治疗方面的临床试验，通过调整患者的DNA来治愈疾患。[16] 在美国，截至笔者写作这篇文章的时候，还没有任何基因疗法被批准用于商业销售（当然，2012年有一项在欧盟获得了批准）。尤其值得关注的是，尽管有了数不胜数的所谓"突破"，癌症的死亡率依然居高不下。[17]

关于基因究竟怎样使我们长成了自己的样子，我们还缺乏基本的了解。行为遗传学试图识别出究竟哪些基因变量决定着人与人之间的差异，比如说，在性格或对精神疾病的易感性方面的差异。该领域已经宣告了无以计数的戏剧性"发现"：同性恋基因、高智商基因、冲动基因、暴力伤害基因、精神分裂症基因，等等。但到目前为止，还没有给出任何一项堪称稳健的发现。[18]

另外一些研究思路同样引人入胜，但对于我们理解基因的工作机理而言，非但没有帮助，反而是越帮越乱。研究人员已经证明，进化与胚胎发育能够以令人惊讶的方式互动（进化发育生物学）；微妙的环境因素会影响基因表达（表观遗传学）；我们的基因组中原本被认为是惰性的部分（垃圾DNA），实际上却很可能有着重要的存在价值。

生物学的困境不禁让人回想起20世纪50年代的物理学，当时，粒子加速器发出了一些奇怪的新粒子，它们很难被传统的量子物理学解释。直到默里·盖尔曼（Murray Gell-Mann）等人提出假说，认为中子和质子都是由三个被称为"夸克"的奇特粒子构成的，混乱的认识才逐渐被澄清。

是否能有某种全新的理念——就像夸克一样新奇大胆——可帮助遗传学澄清其混乱的现状，哪怕是以帮助该领域释放其医疗潜力的方式？但愿如此。生物学是否正面临着深刻的范式转换，就像物理学领域的量子革命一样？

我对此深表怀疑。生物学的基本框架是由新达尔文主义理论再加上以DNA为基础的遗传学构成的，它已经被证明具有足够的弹性。我的猜测是，它会很轻易地吸收掉所有的意外发现，一如量子场论吸收掉夸克。

具有讽刺意味的是，在解释人类行为方面，由于既有研究严重削弱了简单的生物学模型的地位，某些知名生物学家就走向了另一个极端，开始提倡极度决定论、还原论的模型。杰里·科因（Jerry Coyne）断言，自由意志不过是种幻觉；[19] 而理查德·兰厄姆（Richard Wrangham）和爱德华·威尔逊（Edward Wilson）则声称，人类天生就是好战的。[20]

种族主义的智力理论也重新浮出水面。2007年，詹姆斯·沃森（现代遗传学的元老级人物）就曾宣称，黑人问题实际上源自其先天的劣势。[21] 如此看来并非只有物理学家才那么傲慢与不自量力。

神经编码，泛灵论和DARPA

在针对《科学的终结》的最初意见中，最让我有感于心的一条就是，拙著对有关人类心智研究的探讨不够深入；因为与其他科学领域的努力相比，心智研究显然具有产生"启示和革命"的更大潜力。我同意，因此随后出版的两本书所讨论的主题都与心智相关。

在《未知的心灵》（*The Undiscovered Mind*, 1999）一书中，笔者评论了精神病学、神经科学、进化心理学、人工智能，以及其他一些旨在解释心灵并治愈其异常的领域；而《理性的神秘主义》（*Rational Mysticism*, 2003）则探讨了意识的神秘状态，而我本人自懵懂的青春期开始就一直承受着类似神秘状态的困扰。

在过去的十余年里，我也在持续关注神经科学的进展，其中最让我好奇的是破解"神经编码"的努力。[22] 这是一种特殊的算法，或一套规则，据称可以把我们大脑里的电化学冲动转换成知觉、记忆、情感和决断。

神经编码的任何一个解，都会在认识和实践上带来深远的影响。它将有助于解开古老的哲学难题，比如说心—身问题（柏拉图就一直被这一问题困扰）以及自由意志之谜；还有可能使我们对基于大脑的各种顽疾（从精神分裂到阿尔茨海默症）作出更好的理论解释和治疗。

神经编码可以说是科学中最重要的问题，也是最难的问题。就像生命起源研究一样，神经编码研究也催生出了过量的理论，这些理论有着彼此矛盾的观点和假设。虽然说一切生物都共享着显著相似的基因密码，但人类的神经编码却可能不同于其他动物的编码，并且在对新刺激作出反应的过程中不断进化。

神经编码研究所求得的解，可为心智科学提供其迫切需要的统一原则。但就目前而言，心智科学仍停留在哲学家托马斯·库恩所谓的"前科学"状态，在这样那样的时髦理论之间蹒跚前行，甚至就连精神分析和行为主义那样陈腐过时的范式都不乏其拥趸者。

目前的情形变得更加恶劣了。克里斯托弗·科赫（Christof Koch）在过去的几十年里一直是用神经学术语解释意识现象方面的领军人物，也一直是我了解脑科学研究的重要信息源，但近来他却开始支持泛灵论。[23]这一思想可追溯到佛陀和柏拉图那里，认为意识普遍存在于所有物质中，甚至是非生命物质如岩石中。

在《科学的终结》里，我曾杜撰了一个术语"反讽科学"，用以描述那些过于思辨、模糊因而只能把它们当小说看的所谓学术主张，立刻就能联想到超弦和多元宇宙理论。我还曾作出预言：随着传统科学产出的递减，反讽科学的数量将会激增。泛灵论的泛滥，正可代表对笔者预言无懈可击的验证。

心智科学之久无进展，在精神病疗法极度糟糕的现状上也能体现出来。在《未知的心灵》一书中，我曾经指出：对于抑郁之类的精神障碍来说，药物治疗并不像其支持者们所宣称的那样有效。现在回想起来，我的这一批评还是太温和了。最近的调查研究——尤其是新闻工作者罗伯特·惠特克（Robert Whitaker）2010年的《流行病的剖析》（Anatomy of an Epidemic，又译《精神病大流行》）一书所提供的证据显示，精神科药物对病人所造成的伤害，总体而言要远大于其所能带来的治疗效果。

对心智科学研究的资助同样让我感到困扰。在过去的数十年里，精神病学已变得越来越像是一个制药行业的营销部门。[24]与此同时，神经科学家们却变得越来越依赖于美国军方，他们对如何提升己方士兵的认知能力同时抑制敌方士兵的相应能力，似乎有着浓厚的兴趣。

2013年，奥巴马总统宣布，政府正每年斥资上亿美元致力于一个称为"BRAIN"的新计划，旨在"推动神经技术创新，加快大脑研究"；而BRAIN计划的主办方，正是美国国防部高级研究计划局（DARPA）。神经科学家们已接受了——事实上是在热切地寻求着——这份基金资助，就连伦理上是否得当的象征性辩论都没有。

大数据与奇点

那么，数据技术持续的突飞猛进，能否导致物理学、遗传学、神经科学或其他任何领域产生突破呢？这正是热衷于"大数据"的人们所期望的。我本人也会被一些数据技术深深打动，就在刚才，不到一分钟之前，我还在用苹果笔记本电脑和谷歌去查询到底是谁创造了"大数据"这一术语（答案请参阅注释）；[25]我的汽车上安装的GPS导航仪，似乎也很神奇。

但笼罩在大数据上的炒作喧嚣，却只会让我齿冷。这让我想起了拙著《科学的终结》评论过的那些混杂学家的自吹、自赞。大数据的信徒们与混杂学家们一样，对计算机在解决那些为传统科学方法所拒斥的问题上表现出的力量有着某种宗教般的虔诚信仰。

某些大数据的宣言骨子里就是"反智"的。2008年，《连线》（*WIRED*）杂志主编克里斯·安德森（Chris Anderson）就曾警告说，那些永不知疲倦地咀嚼着"PB级"数据的计算机，必将带来"理论的终结"。[26]他解释说，通过发掘数据中的微妙的相关性，计算机使我们能够预言并操纵现象，而不需要任何理解。科学"没了有条理的模型"也能发展，安德森最后画蛇添足地写道，"是时候提出这样的问题了：科学能从谷歌学到些什么？"

就在同一年，才华横溢的华尔街人，装备着用金钱所能买到的最强大的计算机模型，却没能预见到历史上最严重的那一场经济危机。安德森认为大数据无须理解就能带来力量，但华尔街的计算机却没带来两者中任何一点。

数据技术所带来的最为古怪的预言就是奇点理论（the Singularity）。这一借自物理学的术语，意指随着计算机科学以及相关领域的进展，催生出一次巨大的智力飞跃，这里的"智力"可以是机器的、人的，也可以是任何人机混合的。那么，这些超级聪明的存在就将永恒不朽了。

虽然我在《科学的终结》里并未直接提到"奇点"这一术语，但却表达了其基本的理念，即数十年前由愿景家如弗里曼·戴森（Freeman Dyson）、汉斯·莫拉维克（Hans Moravec）之流所探究的那些东西。我把他们的空想研究结论称为"科学神学"，因为它们甚至都不配被称为"反讽的科学"。

现代奇点主义者，如著名计算机科学家雷·库兹韦尔（Ray Kurzweil），坚称奇点很快就会降临，就在不到一代人的时间里。[27]他们作出这样的预言，所依据的自然是数据技术发展的指数增长速度，却忽视了其相关领域停滞不前的事实，如上文所着重讨论的遗传学、神经科学等领域。

不少相当聪明且功成名就的人士，都对奇点理论深信不疑，其中就包括谷歌的创始人，他们帮着库兹韦尔建立了一所"奇点大学"，并在2013年聘请其做谷歌的工程总监。奇点主义者想永远活着，想把自己的智商提升1000倍，这些欲望我都十分理解。[28]但奇点理论却是一种邪教，专为那些信仰科学与技术远甚于信仰上帝的人们设立，是痴迷于计算机技术的呆子们的"救赎"。

最为重要的进步

经常有人问我，究竟怎样才能使我承认自己关于"科学正在终结"的观点错了。大多数时候，我的回答都很漫不经心：要我承认自己错了，等到雷·库兹韦尔把其数字化的灵魂成功上传到智能手机上；或者，等到外星人在时代广场上着陆，并宣布早期地球的创世论者为获胜的一方；要么就是，等到我们发明出曲速飞船，能够冲下虫洞并进入平行宇宙再说。

这里给出的是一个更加严肃的答案：拙著提供了许多具体而微的论点，譬如说，弦理论永远也不可能被确证，心—身问题永远也不可能得到解答，这些最终也许的确会被证明为谬误。但是，任何东西也不可能动摇我对自己的元论点的信心：科学追求对事物的正确认识并且会越来越趋近于真理，但永远也不可能达到绝对真理；早晚有一天，科学会触及其极限。[29]

然而，在我看来，我们创造一个更美好世界的能力是无限的。在我2012年的著作《战争的终结》里，我曾指出：仁爱能成就永久的和平，不是在不久的将来，就是现在。我们并非一定要变成机器人，一定要消除资本主义或宗教信仰，或一定要回归到狩猎—采集的生活方式；我们所必须做的，只是要认识到战争是愚蠢的、不道德的，并把更多的努力投入到非暴力的冲突解决中去。

在我的课堂上，我常常会要求学生们阅读肯尼迪总统的就职演说，他在其中展望了一个没有战争、贫穷、传染性疾病以及暴政的世界。当我问学生们，肯尼迪所描述的世界是合理的还是乌托邦式的废话，大多数学生的回答是后者。让我感到困惑的是，那些拒斥我的科学终结说、认为其过于悲观的学生，却又对人性的预期令人心痛的灰暗。

我提示那些学生，正是人性才使得人类向着肯尼迪所描绘的每一个目标都跨进了一大步。与百年前相比，或者与肯尼迪发表其就职演说的1961年相比，世界已变得更健康、更富足、更自由并且更和平了。因为我执教的大学就坐落在哈德逊河畔，斜对着曼哈顿，所以我又指向一扇窗户，提示他们围绕纽约城的河水与空气，的确比肯尼迪时代干净了许多。我以一声高喊结束了这次课的即兴讨论，"情况已经好转了！"

我们或许永远也不能获得永生，或殖民其他星系，或找到一种能解开关于存在的所有秘密的大一统理论，但是，借助于科学，我们却能创造一个让所有人，而不单单是某些幸运的精英，都能健康幸福的世界。这才是最为重要的。

2015年于新泽西小城霍博肯

Introduction
Searching for *The Answer*

It was in the summer of 1989, during a trip to upstate New York, that I began to think seriously about the possibility that science, pure science, might be over. I had flown to the University of Syracuse to interview Roger Penrose, a British physicist who was a visiting scholar there. Before meeting Penrose, I had struggled through galleys of his dense, difficult book, *The Emperor's New Mind*, which to my astonishment became a best-seller several months later, after being praised in the *New York Times Book Review*.[1] In the book, Penrose cast his eye across the vast panorama of modern science and found it wanting. This knowledge, Penrose asserted, for all its power and richness, could not possibly account for the ultimate mystery of existence, human consciousness.

The key to consciousness, Penrose speculated, might be hidden in the fissure between the two major theories of modern physics: quantum mechanics, which describes electromagnetism and the nuclear forces, and general relativity, Einstein's theory of gravity. Many physicists, beginning with Einstein, had tried and failed to fuse quantum mechanics and general relativity into a single, seamless "unified" theory. In his book, Penrose sketched out what a unified theory might look like and how it might give rise to thought. His scheme, which involved exotic quantum and gravitational effects percolating through the brain, was vague, convoluted, utterly unsupported by evidence from physics or neuroscience. But if it turned out to be in any sense right, it would represent a monumental achievement, a theory that in one stroke would unify physics and solve one of philosophy's most vexing problems, the link between mind and matter. Penrose's ambition alone, I thought, would make him an excellent subject for a profile in *Scientific American*, which employed me as a staff writer.[2]

When I arrived at the airport in Syracuse, Penrose was waiting for me. He was an elfin man, capped with a shock of black hair, who seemed simultaneously distracted and acutely alert. As he drove us back to the Syracuse campus, he kept wondering aloud if he was going in the right direction. He seemed awash in mysteries. I found myself in the disconcerting position of recommending that he take this exit, or make that turn, although I had never been in Syracuse before. In spite of our combined ignorance, we managed to make our way without incident to the building where Penrose worked. On entering Penrose's office we discovered that a colleague had left a brightly colored aerosol can labeled Superstring on his desk. When Penrose pushed the button on the top of the can, a lime green, spaghetti-like strand shot across the room.

Penrose smiled at this little insider's joke. Superstring is the name not only of a child's toy, but also of an extremely small and extremely hypothetical stringlike particle posited by a popular theory of physics. According to the theory, the wriggling of these strings in a 10-dimensional hyperspace generates all the matter and energy in the universe and even space and time. Many of the world's leading physicists felt that superstring theory might turn out to be the unified theory they had sought for so long; some even called it a theory of everything. Penrose was not among the faithful. "It couldn't be right," he told me. "It's just not the way I'd expect the answer to be." I began to realize, as Penrose spoke, that to him "the answer" was more than a mere theory of physics, a way of organizing data and predicting events. He was talking about *The Answer:* the secret of life, the solution to the riddle of the universe.

Penrose is an admitted Platonist. Scientists do not invent the truth; they discover it. Genuine truths exude a beauty, a rightness, a self-evident quality that gives them the power of revelation. Superstring theory did not possess these traits, in Penrose's mind. He conceded that the "suggestion" he set forth in *The Emperor's New Mind*—it did not merit the term *theory* yet, he admitted—was rather ungainly. It might turn out to be wrong, certainly in its details. But he felt sure that it was closer to the truth than was superstring theory. In saying that, I asked, was Penrose implying that one day scientists would find *The Answer* and thus bring their quest to an end?

Unlike some prominent scientists, who seem to equate tentativity with weakness, Penrose actually thinks before he responds, and even as he responds. "I don't think we're close," he said slowly, squinting out his office window, "but it doesn't mean things couldn't move fast at some stage." He cogitated some more. "I guess this is rather suggesting that there is an answer," he continued, "although perhaps that's too pessimistic." This final comment stopped me short. What is so pessimistic, I asked, about a truth seeker thinking that the truth is attainable? "Solving mysteries is a wonderful thing to do," Penrose replied. "And if they were all solved, somehow, that would be rather boring." Then he chuckled, as if struck by the oddness of his own words.[3]

Long after leaving Syracuse, I mulled over Penrose's remarks. Was it possible that science could come to an end? Could scientists, in effect, learn everything there is to know? Could they banish mystery from the universe? It was hard for me to imagine a world without science, and not only because my job depended on it. I had become a science writer in large part because I considered science—pure science, the search for knowledge for its own sake—to be the noblest and most meaningful of human endeavors. We are here to figure out why we are here. What other purpose is worthy of us?

I had not always been so enamored of science. In college, I passed through a phase during which literary criticism struck me as the most thrilling of intellectual endeavors. Late one night, however, after too many cups of coffee, too many hours spent slogging through yet another interpretation of James Joyce's *Ulysses*, I had a crisis of faith. Very smart people had been arguing for decades over the meaning of *Ulysses*. But one of the messages of modern criticism, and of modern literature, was that all texts are "ironic": they have multiple meanings, none of them definitive.[4] *Oedipus Rex, The Inferno*, even the Bible are in a sense "just kidding," not to be taken too literally. Arguments over meaning can never be resolved, since the only true meaning of a text is the text itself. Of course, this message applied to the critics, too. One was left with an infinite regress of interpretations, none of which represented the final word. But everyone still kept arguing! To what end? For each critic to be more clever, more *interesting*, than the rest? It all began to seem pointless.

Although I was an English major, I took at least one course in science or mathematics every semester. Working on a problem in calculus or physics represented a pleasant change of pace from messy humanities assignments; I found great satisfaction in arriving at the correct answer to a problem. The more frustrated I became with the ironic outlook of literature and literary criticism, the more I began to appreciate the crisp, no-nonsense approach of science. Scientists have the ability to pose questions and resolve them in a way that critics, philosophers, historians cannot. Theories are tested experimentally, compared to reality, and those found wanting are rejected. The power of science cannot be denied: it has given us computers and jets, vaccines and thermonuclear bombs, technologies that, for better or worse, have altered the course of history. Science, more than any other mode of knowledge— literary criticism, philosophy, art, religion—yields durable insights into the nature of things. It gets us somewhere. My mini-epiphany led, eventually, to my becoming a science writer. It also left me with this criterion for science: science addresses questions that can be answered, at least in principle, given a reasonable amount of time and resources.

Before my meeting with Penrose, I had taken it for granted that science was open-ended, even infinite. The possibility that scientists might one day find a truth so potent that it would obviate all further investigations had struck me as wishful thinking at best, or as the kind of hyperbole required to sell science (and science books) to the masses. The earnestness, and ambivalence, with which Penrose contemplated the prospect of a final theory forced me to reassess my own views of science's future. Over time, I became obsessed with the issue. What are the limits of science, if any? Is science infinite, or is it as mortal as we are? If the latter, is the end in sight? Is it upon us?

After my original conversation with Penrose, I sought out other scientists who were butting their heads against the limits of knowledge: particle physicists who dreamed of a final theory of matter and energy; cosmologists trying to understand precisely how and even why our universe was created; evolutionary biologists seeking to determine how life began and what laws governed its subsequent unfolding; neuroscientists probing the processes in the brain that give rise to consciousness; explorers of chaos and complexity, who hoped that with computers and new mathematical techniques they could revitalize science. I also spoke to philosophers, including some who allegedly doubted whether science could ever achieve objective, absolute truths. I wrote articles about a number of these scientists and philosophers for *Scientific American*.

When I first thought about writing a book, I envisioned it as a series of portraits, warts and all, of the fascinating truth seekers and truth shunners I have been fortunate enough to interview. I intended to leave it to readers to decide whose forecasts about the future of science made sense and whose did not. After all, who really knew what the ultimate limits of knowledge might be? But gradually, I began to imagine that *I* knew; I convinced myself that one particular scenario was more plausible than all the others. I decided to abandon any pretense of journalistic objectivity and write a book that was overtly judgmental, argumentative, and personal. While still focusing on individual scientists and philosophers, the book would present my views as well. That approach, I felt, would be more in keeping with my conviction that most assertions about the limits of knowledge are, finally, deeply idiosyncratic.

It has become a truism by now that scientists are not mere knowledge-acquisition machines; they are guided by emotion and intuition as well as by cold reason and calculation. Scientists are rarely so human, I have found, so at the mercy of their fears and desires, as when they are confronting the limits of knowledge. The greatest scientists want, above all, to discover truths about nature (in addition to acquiring glory, grants, and tenure and improving the lot of humankind); they want to *know*. They hope, and trust, that the truth is attainable, not merely an ideal or asymptote, which they eternally approach. They also believe, as I do, that the quest for knowledge is by far the noblest and most meaningful of all human activities.

Scientists who harbor this belief are often accused of arrogance. Some *are* arrogant, supremely so. But many others, I have found, are less arrogant than anxious. These are trying times for truth seekers. The scientific enterprise is threatened by technophobes, animal-rights activists, religious fundamentalists, and, most important, stingy politicians. Social, political, and economic constraints will make it more difficult to practice science, and pure science in particular, in the future.

Moreover, science itself, as it advances, keeps imposing limits on its own power. Einstein's theory of special relativity prohibits the transmission of matter or even information at speeds faster than that of light; quantum mechanics dictates that our knowledge of the micro-realm will always be uncertain; chaos theory confirms that even without quantum indeterminacy many phenomena would be impossible to predict; Kurt Gödel's incompleteness theorem denies us the possibility of constructing a complete, consistent mathematical description of reality. And evolutionary biology keeps reminding us that we are animals, designed by natural selection not for discovering deep truths of nature, but for breeding.

Optimists who think they can overcome all these limits must face yet another quandary, perhaps the most disturbing of all. What will scientists do if they succeed in knowing what can be known? What, then, would be the purpose of life? What would be the purpose of humanity? Roger Penrose revealed his anxiety over this dilemma when he called his dream of a final theory pessimistic.

Given these troubling issues, it is no wonder that many scientists whom I interviewed for this book seemed gripped by a profound unease. But their malaise, I will argue, has another, much more immediate cause. *If one believes in science*, one must accept the possibility— even the probability—that the great era of scientific discovery is over. By *science* I mean not applied science, but science at its purest and grandest, the primordial human quest to understand the universe and our place in it. Further research may yield no more great revelations or revolutions, but only incremental, diminishing returns.

The Anxiety of Scientific Influence

In trying to understand the mood of modern scientists, I have found that ideas from literary criticism can serve some purpose after all. In his influential 1973 essay, *The Anxiety of Influence*, Harold Bloom likened the modern poet to Satan in Milton's *Paradise Lost*.[5] Just as Satan fought to assert his individuality by defying the perfection of God, so must the modern poet engage in an Oedipal struggle to define himself or herself in relation to Shakespeare, Dante, and other masters. The effort is ultimately futile, Bloom said, because no poet can hope to approach, let alone surpass, the perfection of such forebears. Modern poets are all essentially tragic figures, latecomers.

Modern scientists, too, are latecomers, and their burden is much heavier than that of poets. Scientists must endure not merely Shakespeare's *King Lear*, but Newton's laws of motion, Darwin's theory of natural selection, and Einstein's theory of general relativity. These theories are not merely beautiful; they are also true, empirically true, in a way that no work of art can be. Most researchers simply concede their inability to supersede what Bloom called "the embarrassments of a tradition grown too wealthy to need anything more."[6] They try to solve what philosopher of science Thomas Kuhn has patronizingly called "puzzles," problems whose solution buttresses the prevailing paradigm. They settle for refining and applying the brilliant, pioneering discoveries of their predecessors. They try to measure the mass of quarks more precisely, or to determine how a given stretch of DNA guides the growth of the embryonic brain. Others become what Bloom derided as a "mere rebel, a childish inverter of conventional moral categories."[7] The rebels denigrate the dominant theories of science as flimsy social fabrications rather than rigorously tested descriptions of nature.

Bloom's "strong poets" accept the perfection of their predecessors and yet strive to transcend it through various subterfuges, including a subtle misreading of the predecessors' work; only by so doing can modern poets break free of the stultifying influence of the past. There are strong scientists, too, those who are seeking to misread and therefore to transcend quantum mechanics or the big bang theory or Darwinian evolution. Roger Penrose is a strong scientist. For the most part, he and others of his ilk have only one option: to pursue science in a speculative, postempirical mode that I call ironic science. Ironic science resembles literary criticism in that it offers points of view, opinions, which are, at best, interesting, which provoke further comment. But it does not converge on the truth. It cannot achieve empirically verifiable surprises that force scientists to make substantial revisions in their basic description of reality.

The most common strategy of the strong scientist is to point to all the shortcomings of current scientific knowledge, to all the questions left unanswered. But the questions tend to be ones that may *never* be definitively answered given the limits of human science. How, exactly, was the universe created? Could our universe be just one of an infinite number of universes? Could quarks and electrons be composed of still smaller particles, ad infinitum? What does quantum mechanics really mean? (Most questions concerning meaning can only be answered ironically, as literary critics know.) Biology has its own slew of insoluble riddles. How, exactly, did life begin on earth? Just how inevitable was life's origin and its subsequent history?

The practitioner of ironic science enjoys one obvious advantage over the strong poet: the appetite of the reading public for scientific "revolutions." As empirical science ossifies, journalists such as myself, who feed society's hunger, will come under more pressure to tout theories that supposedly transcend quantum mechanics or the big bang theory or natural selection. Journalists are, after all, largely responsible for the popular impression that fields such as chaos and complexity represent genuinely new sciences superior to the stodgy old reductionist methods of Newton, Einstein, and Darwin. Journalists, myself included, have also helped Roger Penrose's ideas about consciousness win an audience much larger than they deserve given their poor standing among professional neuroscientists.

Through most of this book, I will examine science as it is practiced today, by humans. (Chapter 2 takes up philosophy.) In the final two chapters I will consider the possibility—advanced by a surprising number of scientists and philosophers—that one day we humans will create intelligent machines that can transcend our puny knowledge. In my favorite version of this scenario, machines transform the entire cosmos into a vast, unified, information-processing network. All matter becomes mind. This proposal is not science, of course, but wishful thinking. It nonetheless raises some interesting questions, questions normally left to theologians. What would an all-powerful cosmic computer do? What would it think about? I can imagine only one possibility. It would try to answer *The Question*, the one that lurks behind all other questions, like an actor playing all the parts of a play: Why is there something rather than nothing? In its effort to find *The Answer* to *The Question*, the universal mind may discover the ultimate limits of knowledge.

寻求"终极答案"

科学——纯科学——是否有可能终结？我对这一问题的严肃思考，始于1989年夏天的一次采访。当时我乘飞机到纽约州北部的锡拉丘兹大学去拜访罗杰·彭罗斯（Roger Penrose），一位正在那里做访问学者的英国物理学家。在采访彭罗斯之前，我是硬着头皮才啃完他那部难解的巨著《皇帝的新脑》，但出乎我的意料，时隔数月，经《纽约时报书评》的宣扬，它竟然成了一本畅销书。[1]彭罗斯在书中全面考察了现代科学，发现它存在着严重的缺失。他断言：现代科学尽管有着强大的威力和丰富的内容，但仍不足以解释存在的终极奥秘，即人的意识问题。

彭罗斯推测，理解意识问题的关键可能就隐藏在现代物理学两大理论之间的裂隙中。一个是量子力学，描述的是电磁学以及粒子相互作用的规律；另一个是广义相对论，即爱因斯坦的引力理论。自爱因斯坦以降，许多物理学家都曾试图把量子力学和广义相对论融汇成一个无内在矛盾的"统一"理论，但都以失败告终。彭罗斯在他的著作中描绘了这种统一理论可能会是什么样子，以及它将给人类思想带来怎样的促进作用。他关于奇异量子和引力效应通过大脑扩散的论述是含混而晦涩的，完全没有什么物理学或神经科学的证据，但一旦在某种程度上被证明是正确的话，将标志着这是一个不朽的成就，它会一举实现物理学的统一，并解决哲学中最让人困扰的意识和物质的关系问题。当时，作为《科学美国人》杂志的专职撰稿人，我认为单凭彭罗斯的这一抱负，就足以使他成为该刊人物专访的合格人选。[2]

抵达锡拉丘兹机场时，彭罗斯正在那里接我。他个头矮小，一头蓬乱的黑发，表现出的神态简直让人无法分清他到底是笨拙还是精明。在驱车返回锡拉丘兹大学校园的路上，他不时地嘀咕着，说不知所走的路线到底对不对，仿佛他正沉浸在某种玄想之中。我很尴尬地发现，尽管自己此前从未来过锡拉丘兹，他却要我来建议是不是要走这个出口，或是不是要在那里转弯，那情景简直就像两个盲人在赶路，居然竟让我俩平安地抵达了彭罗斯工作的楼前。走进他的办公室，就发现在他的办公桌上放着一个色彩艳丽的喷雾玩具盒，那是一位促狭的同事留给他的，上面赫然标着"超弦"（Superstring）的字样。彭罗斯按下盒顶的按钮，便有一束灰绿色的、细面条似的水剂向房间里疾喷而出。

彭罗斯被同伴这个无伤大雅的小把戏逗乐了。超弦不仅是一种儿童玩具的名称，而且是一种流行的物理学理论假设的、极小的、纯属臆测的弦状粒子的名字。根据超弦理论，这些弦在十维超空间中扭曲，产生了宇宙中一切的物质和能量，甚至产生了空间和时间。许多世界著名的物理学家都认为，超弦理论可能会被证明为正是他们寻觅已久的统一理论，有人甚至称之为"万物至理"。彭罗斯却不以为然。"不可能，"他告诉我说，"我所期望的答案绝不会是这个样子。"这时我才开始意识到：对他而言，答案绝不单纯是种物理学理论，一种组织数据和预言事件的方式，他所寻求的是"终极答案"——关于生命的奥秘以及宇宙之谜的答案。

彭罗斯是一位公认的柏拉图主义者，认为科学家不应去发明真理，而要去揭示真理。真正的真理蕴含着美、真实和一种使之具有启示力量的自明品质。他承认自己在《皇帝的新脑》中所提出的见解是十分粗糙的，还够不上"理论"的标准，将来很可能被证明是错误的，尤其在细节上肯定不会完全正确，但可以肯定的是，它比超弦理论更接近真理。我这时插话问道："如此说来，你是否暗示着科学家们有朝一日将会找到'终极答案'，并由此给自己的探索画上句号呢？"

彭罗斯不像某些知名的科学家那样，认为回答问题时迟疑不决是丢面子的事，他在回答之前要思索一段时间，甚至在回答的过程中也是如此。"我不认为我们已经接近真理了，"他凝视着窗外缓缓说道，"但这并不意味着事情不会在某些阶段进展得更快些。"他再度沉思了一会儿："我想这更意味着答案确实存在，尽管这可能让人觉着很沮丧。"最后一句话使我一愣，于是又问："那么，对于真理的追求者来说，认识到真理是可达到的，这有什么可沮丧的呢？""揭示奥秘是一件奇妙的事情，"彭罗斯答道，"如果所有的奥秘都已被解决，这无论怎样说都是让人十分沮丧的。"说到这里，他微微一乐，仿佛被自己古怪的措辞打动了。[3]

离开锡拉丘兹之后的很长一段时间里，我一直在反复思考彭罗斯的话。科学有可能走到尽头吗？科学家们实际上能够认识一切吗？他们能够破解宇宙中的一切神秘现象吗？对我来说，想象一个没有科学的世界是十分困难的，这不仅仅是因为我的职业建立在科学事业之上。我之所以成为一名科学记者，很大程度上是因为我认为科学——纯科学，即仅仅是为了求知的科学——是最崇高、最有意义的人类事业。我们选择了科学，最终是为了理解我们自己，除此之外，还能有什么别的目的呢？

我并非总是这样倾心于科学的。在大学期间，有一段时间我曾认为文学批评是最为振奋人心的智力活动，但后来，当我在某个晚上喝了大量的咖啡，花了大量的时间去啃对詹姆斯·乔伊斯（James Joyce）的《尤利西斯》（Ulysses）的阐释之后，突然陷入了信念危机。睿智的人们已经就《尤利西斯》的意义争论了几十年，但现代文学批评以及现代文学的一个要旨却是：所有的文本都是"反讽的（ironic）"，它们具有多重意义，但没有一种意义是权威性的；[4]《俄狄浦斯王》《地狱篇》甚至《圣经》，在某种意义上说都"只是玩笑"，不能仅仅按字面意义去理解；关于意义的争论永远也不会有结果，因为一种文本唯一的真实意义就是文本自身。当然，这段妙论也适用于批评家们。人们陷入解释的无限回归之中，没有一种解释代表终极的结论，但每个人都仍在争论不休！目的何在？难道仅仅是为了使每个批评家都变得更机智、更有趣吗？于是，所有这些争论在我眼里顿然失去了意义。

尽管我主修的是英语，但我每学期都至少要选修一门科学或数学课。做微积分或物理学习题，意味着愉快地从繁重的人文科学的作业中解脱出来；我在求得一个问题的正确答案的过程中发现了巨大的乐趣。我越是对文学和文学批评的尴尬前景感到灰心，就越是欣赏科学那种简洁而毫不夸饰的方法。科学家提出问题和解决问题的方式，是批评家、哲学家和历史学家们力所难及的。理论必须接受实验的检验，与实际相对照，经不起检验的理论则被剔除。科学的力量是无法否认的，它给我们带来了计算机和喷气式飞机，带来了疫苗和热核炸弹，带来了改变历史进程的技术，不论是福是祸。相对于其他类型的知识，如文学批评、哲学、艺术、宗教等而言，科学能够给出关于事物本质的更为可靠的见解，使我们更有奔头。这种内心的顿悟，引导我最终成了一名科学记者，也形成了我对科学的基本看法：科学至少在原则上处理那些能被解答的问题——当然要提供足够的时间和条件。

在与彭罗斯会晤之前，我理所当然地认为科学是没有尽头的，或者说是无限的。科学家可能在某一天发现一种威力巨大的真理，从而一劳永逸地解决一切有待研究的问题，这种可能性在当时的我看来，最多不过是一厢情愿的幻想，或是向大众推销科学（或科学书籍）时的夸夸其谈。但彭罗斯在思索终极理论可能性时的那种热切而又矛盾的心理，迫使我重估自己关于科学未来的看法。这一问题时时纠缠着我，使我去思索科学的限度（如果存在的话）究竟是什么。科学是无限的，还是如我们的生命一样终有一死？如果是后者，那么科学的末日是否已经在望？末日是否已降临到我们头上？

以采访彭罗斯为开端，我后来又发现了另外一些同样在挑战着知识限度问题的科学家：一心寻求物质和能量的终极理论的粒子物理学家，试图精确理解宇宙怎样产生以及为什么产生的宇宙学家，意欲确定生命怎样发生以及何种规律支配生命发展的进化生物学家，探索着产生意识的大脑内部活动的神经科学家，还有混沌和复杂性的探索者，他们希望能借助计算机和现代数学方法为科学注入新的活力。我也访问了一些哲学家，其中有的怀疑科学是否能不断获得客观的绝对真理。我在《科学美国人》上撰文介绍了许多这类人物。

在我最初萌生写作本书的愿望时，我曾把它设想为一部人物传记集，如实地描述自己有幸采访过的那些各具魅力的人物，不论他们是在追求真理还是在逃避真理。至于哪些人对科学之未来的预测是合理的、哪些人的不合理，我打算把它留给读者自己去判断。毕竟，又有谁真的知道知识的终极限度可能是什么呢？但慢慢地，我开始认为"我知道"，并逐渐相信有一种解释方案比其他的更有说服力。我决定放弃新闻工作客观性的借口，写一本毫不掩饰批判性、论辩性和个人观点的著作，在把焦点仍然聚集在一个个科学家和哲学家的前提下，书中应更多地体现我个人的观点。我觉得这个方案与我的信念是一致的，即几乎所有关于知识的限度的主张都被深深地打上了个人的烙印。

在今天，人们已普遍认识到科学家不仅仅是求解知识的机器，他们也受到激情和直觉的引导，就像他们要受无情的理性和数学计算的约束一样。我发现，面对认识的极限，科学家们更像普通人一样，易受到自己的恐惧和欲望左右。对于那些伟大的科学家来说，第一位的需要是揭示关于自然的真理（另外，当然也需要荣誉、资金和职位，渴望能为更多的人谋福利），他们想"知道"，他们希望——同时也坚信——真理是能够达到的，而不仅仅是作为一种理想，或是一种可无限逼近但永远无法到达的 "渐近线"；他们还像我一样，坚信追求知识是最崇高、最有意义的人类活动。

怀有这一信念的科学家，常常被指责为狂妄自大。事实上也的确有某些科学家狂妄自大，但我发现，更多的科学家与其说狂妄自大，不如说忧心忡忡。真理的追求者们都时光难挨，科学事业正受到来自各方面的威胁：来自那些对技术深怀恐惧的人们、动物保护主义者、宗教极端主义者以及——也是最重要的——吝啬的政客的威胁。社会的、政治的和经济的限制，将使科学事业（尤其是纯科学）在将来的处境更加窘迫。

此外，科学自身在发展的过程中也在不断地给自己的力量套上枷锁。爱因斯坦的狭义相对论，把物质运动甚至信息传递的速度限制在光速范围内；量子力学宣告我们关于微观世界的知识总是不确定的；混沌理论进一步证明，即使不存在量子不确定性，许多现象仍然不可能预测；哥德尔不完备性定理则消除了我们对实在建构一个完备、一致的数学描述系统的可能性；同时，进化生物学在不断地提醒我们：人是动物，自然选择设计出人来，不是为了让人们去揭示自然的深刻真理，而是让人们繁衍后代。

即使那些自认为能克服所有这些局限的乐观主义者，也必然会面临另外的窘境，这可能是所有困境中最恼人的一个：如果科学家们成功地掌握了一切可以掌握的知识，那他们再去做什么呢？到那时，人生的目的又将是什么？人类的目的又将是什么？罗杰·彭罗斯自称他对于终极理论的梦想是悲观的，这充分暴露了他对这种两难处境的焦虑。

本书中我所采访的许多科学家，似乎只要涉及上述沉重的话题，无一不被某种深深的不安所左右，但我认为他们的不安有着另外的更为直接的原因。如果你相信科学，就必须接受这种可能性，甚或已具有几分现实性的可能性，即伟大的科学发现时代已经结束了。这里的科学，并不意味着应用科学，而是指那种最纯粹、最崇高的科学，即希望能理解宇宙、理解人类在宇宙中的位置这类最基本的人类追求。将来的研究已不会产生多少重大的或革命性的新发现了，而只有渐增的收益递减。

对科学影响的焦虑

在试图理解现代科学家们的一般心态时，我发现来自文学批评的思想具有一定的借鉴意义。哈罗德·布鲁姆（Harold Bloom）在1973年发表的颇具影响的著作《影响的焦虑》中，把现代诗人比作是弥尔顿《失乐园》中的撒旦。[5]正如撒旦要通过挑战上帝的完美来维护自己的个性一样，现代诗人也必须致力于一种恋母情结的战斗，以界定他／她自己与莎士比亚、但丁及其他大师的关系。布鲁姆认为这种努力终归是徒劳的，因为没有任何一位诗人能接近这些前辈的高度，更不用说超越他们了。现代诗人作为迟来者（latecomers），实际上都是悲剧性的人物。

现代科学家也是迟来者，并且他们的包袱比诗人的更重。科学家们不仅要承受莎士比亚的《李尔王》，更要承受牛顿的运动定律、达尔文的自然选择理论，以及爱因斯坦的广义相对论。这些理论不仅是美的，而且是真的，被经验所证实了的真，这是任何艺术作品都无可比拟的。面对布鲁姆所谓的"太丰足所以无所求的传统所带来的种种苦恼、惶恐"[6]，许多科学家不得不承认自己的无奈。他们只能在主导"范式"的束缚下，试着去解答被科学哲学家托马斯·库恩（Thomas Kuhn）傲慢地称作"难题"（puzzles）的问题，满足于对前辈们那辉煌的、开创性的发现进行精细的加工和应用。他们试图更精确地测量夸克的质量，或去确定一段特定的DNA如何决定胚脑的发育；另一部分科学家正如布鲁姆所嘲笑的那样，变成了"单纯的叛逆者，幼稚的传统道德范畴颠覆家"[7]，他们把占统治地位的科学理论贬低为脆弱的社会建构产物，而不是在严格检验的基础上建立起来的对自然的描述。

布鲁姆所谓的"强者诗人"（strong poets），承认前辈们登峰造极的成就，但仍然挖空心思地力求超越他们，包括别有用心地误读前辈们的作品，因为只有这样，现代诗人们才能从历史那让人窒息的影响中挣脱出来。也存在着这样的"强者科学家"（strong scientists），他们试图误读并超越量子力学或大爆炸理论或达尔文进化论。罗杰·彭罗斯就是一位强者科学家，他和他的战友们最多也只能有一种选择：以一种思辨的、后实证的（postempirical）方式去追求科学，我称之为反讽的科学（ironic science）。反讽的科学与文学批评的相似之处在于：它所提供的思想、观点，至多是有意义的，能够引发进一步的争论，但它并不趋向真理，不能提供可检验的新奇见解，从而也就不会促使科学家们对描述现实的基本概念做实质性的修改。

强者科学家们最常用的策略是直指当前科学知识的缺陷，指向科学目前尚无法解答而被搁置的所有问题，但因为人类科学局限性的存在，这些问题往往正是那些也许永远无法最终回答的问题。宇宙到底是怎样产生的？我们的宇宙是否只是无限多的宇宙中的一个？夸克和电子是不是由更小的粒子（更更小的粒子……）组成的呢？量子力学的真正意义何在？对于大部分与"意义"（meaning）有关的问题，只能进行反讽式的回答，正如文学批评家所熟知的那样。生物学也有大量自身无法解开的疙瘩：地球上的生命到底是怎样发生的？生命的起源及其随后的发展历史究竟具有怎样的必然性呢？

反讽科学的实践者享有一种"强者诗人"所无法企及的优势，即大众读者们对科学"革命"的渴望。经验科学的停滞，使得像我这样以满足社会需要为天职的新闻工作者面临着更大的压力，去炒作那些估计可能会超越量子力学、大爆炸理论或自然选择论的理论。无论如何，像混沌与复杂性等研究领域，之所以被公众认为是优于牛顿、爱因斯坦和达尔文等僵化的还原主义理论的新学科，与新闻界的炒作是有很大关系的。比如罗杰·彭罗斯关于意识的观点，凭借新闻工作者（包括我自己）的帮助，赢得的注意远大于其所应得到的，而职业神经科学家中支持这一观点的人反而少得可怜。

本书的大部分篇幅将用来考察当今人类正在实践着的科学（第二章考察哲学的问题），在最后的两章中讨论这样一种可能性（赞同这一观点的科学家和哲学家的人数多得出奇），即人类总有一天能够创造出可超越自身的有限认识能力的智能机。关于这点，我所欣赏的设想是：智能机会把整个宇宙转变成一个巨大的、统一的信息加工网络，所有的物质都变成了意识。这一设想当然并不科学，只是一厢情愿的设想，但它仍然引发出一些有趣的问题（这些问题向来属于神学家）：一架全能的超级计算机有什么作用？它会"想"些什么？我只能想象一种可能，它会试图解答"终极问题"（The Question），即潜藏在所有问题背后的那个问题，就像一个演员扮演一出戏剧中的所有角色一样：为什么一定是有些什么，而不是一无所有？或许，在这个"宇宙智慧"为终极问题寻找终极答案的努力中，能够发现知识的终极限度。

目录

/

Contents

The End
of
Science

John Horgan

For Dad

Doubts among the cosmic priesthood.

Andrei Linde's chaotic, fractal, eternally self-reproducing, inflationary universe.

Fred Hoyle, the eternal rebel.

Will cosmology turn into botany?

Chapter Five　The End of Evolutionary Biology

Richard Dawkins, Darwin's greyhound.

Stephen Jay Gould's view of life: shit happens.

Lynn Margulis denounces Gaia.

The organized disorder of Stuart Kauffman.

Stanley Miller ponders the eternally mysterious origin of life.

Chapter Six　The End of Social Science

Edward Wilson's fear of a final theory of sociobiology.

Noam Chomsky on mysteries and puzzles.

The eternal vexation of Clifford Geertz.

Chapter Seven　The End of Neuroscience

Francis Crick, the Mephistopheles of biology, takes on consciousness.

Gerald Edelman postures around the riddle.

John Eccles, the last of the dualists.

Roger Penrose and the quasi-quantum mind.

The counterattack of the mysterians.

Is Daniel Dennett a mysterian?

Marvin Minsky's fear of single-mindedness.

The triumph of materialism.

Chapter Eight　The End of Chaoplexity

What is chaoplexity?

Christopher Langton and the poetry of artificial life.

Per Bak's self-organized criticality.

Chapter One
The End of Progress

In 1989, just a month after my meeting with Roger Penrose in Syracuse, Gustavus Adolphus College in Minnesota held a symposium with the provocative but misleading title, "The End of Science?" The meeting's premise was that *belief in* science—rather than science itself— was coming to an end. As one organizer put it, "There is an increasing feeling that science as a unified, universal, objective endeavor is over."[1] Most of the speakers were philosophers who had challenged the authority of science in one way or another. The meeting's great irony was that one of the scientists who spoke, Gunther Stent, a biologist at the University of California at Berkeley, had for years promulgated a much more dramatic scenario than the one posited by the symposium. Stent had asserted that science itself might be ending, and not because of the skepticism of a few academic sophists. Quite the contrary. Science might be ending because it worked so well.

Stent is hardly a fringe figure. He was a pioneer of molecular biology; he founded the first department dedicated to that field at Berkeley in the 1950s and performed experiments that helped to illuminate the machinery of genetic transmission. Later, after switching from genetics to the study of the brain, he was named chairman of the neurobiology department of the National Academy of Sciences. Stent is also the most astute analyst of the limits of science whom I have encountered (and by astute I mean of course that he articulates my own inchoate premonitions). In the late 1960s, while Berkeley was racked with student protests, he wrote an astonishingly prescient book, now long out of print, called *The Coming of the Golden Age: A View of the End of Progress*. Published in 1969, it contended that science—as well as technology, the arts, and all progressive, cumulative enterprises—was coming to an end.[2]

Most people, Stent acknowledged, consider the notion that science might soon cease to be absurd. How can science possibly be nearing an end when it has been advancing so rapidly throughout this century? Stent turned this inductive argument on its head. Initially, he granted, science advances exponentially through a positive feedback effect; knowledge begets more knowledge, and power begets more power. Stent credited the American historian Henry Adams with having foreseen this aspect of science at the turn of the twentieth century.[3]

Adams's law of acceleration, Stent pointed out, has an interesting corollary. If there are any limits to science, any barriers to further progress, then science may well be moving at unprecedented speed just before it crashes into them. When science seems most muscular, triumphant, potent, that may be when it is nearest death. "Indeed, the dizzy rate at which progress is now proceeding," Stent wrote in *Golden Age*, "makes it seem very likely that progress must come to a stop soon, perhaps in our lifetime, perhaps in a generation or two."[4]

Certain fields of science, Stent argued, are limited simply by the boundedness of their subject matter. No one would consider human anatomy or geography, for example, to be infinite endeavors. Chemistry, too, is bounded. "[T]hough the total number of possible chemical reactions is very great and the variety of reactions they can undergo vast, the goal of chemistry of understanding the principles governing the behavior of such molecules is, like the goal of geography, clearly limited."[5] That goal, arguably, was achieved in the 1930s, when the chemist Linus Pauling showed how all chemical interactions could be understood in terms of quantum mechanics.[6]

In his own field of biology, Stent asserted, the discovery of DNA's twin-corkscrew structure in 1953 and the subsequent deciphering of the genetic code had solved the profound problem of how genetic information is passed on from one generation to the next. Biologists had only three major questions left to explore: how life began, how a single fertilized cell develops into a multicellular organism, and how the central nervous system processes information. When those goals are achieved, Stent said, the basic task of biology, pure biology, will be completed.

Stent acknowledged that biologists could, in principle, continue exploring specific phenomena and applying their knowledge forever. But according to Darwinian theory, science stems not from our desire for truth per se, but from our compulsion to control our environment in order to increase the likelihood that our genes will propagate. When a given field of science begins to yield diminishing practical returns, scientists may have less incentive to pursue their research and society may be less inclined to pay for it.

Moreover, just because biologists complete their empirical investigations, Stent asserted, does not mean that they will have answered all relevant questions. For example, no purely physiological theory can ever really *explain* consciousness, since the "processes responsible for this wholly private experience will be seen to degenerate into seemingly quite ordinary, workaday reactions, no more or less fascinating than those that occur in, say, the liver."[7]

Unlike biology, Stent wrote, the physical sciences seem to be open-ended. Physicists can always attempt to probe more deeply into matter by smashing particles against each other with greater force, and astronomers can always strive to see further into the universe. But in their efforts to gather data from ever-more-remote regimes, physicists will inevitably confront various physical, economic, and even cognitive limits.

Over the course of this century, physics has become more and more difficult to comprehend; it has outrun our Darwinian epistemology, our innate concepts for coping with the world. Stent rejected the old argument that "yesterday's nonsense is today's common sense."[8] Society may be willing to support continued research in physics as long as it has the potential to generate powerful new technologies, such as nuclear weapons and nuclear power. But when physics becomes impractical as well as incomprehensible, Stent predicted, society will surely withdraw its support.

Stent's prognosis for the future was an odd mixture of optimism and pessimism. He predicted that science, before it ends, might help to solve many of civilization's most pressing problems. It could eliminate disease and poverty and provide society with cheap, pollution-free energy, perhaps through the harnessing of fusion reactions. As we gain more dominion over nature, however, we may lose what Nietzsche called our "will to power"; we may become less motivated to pursue further research—especially if such research has little chance of yielding tangible benefits.

As society becomes more affluent and comfortable, fewer young people may choose the increasingly difficult path of science or even of the arts. Many may turn to more hedonistic pursuits, perhaps even abandoning the real world for fantasies induced by drugs or electronic devices feeding directly into the brain. Stent concluded that sooner or later, progress would "stop dead in its tracks," leaving the world in a largely static condition that he called "the new Polynesia." The advent of beatniks and hippies, he surmised, signaled the beginning of the end of progress and the dawn of the new Polynesia. He closed his book with the sardonic comment that "millennia of doing arts and sciences will finally transform the tragicomedy of life into a happening."[9]

A Trip to Berkeley

In the spring of 1992 I traveled to Berkeley to see how Stent thought his predictions had held up over the years.[10] Strolling toward the university from my hotel, I passed what appeared to be the detritus of the sixties: men and women with long gray hair and ragged clothes asking for spare change. Once on the campus, I made my way to the university's biology building, a hulking, concrete complex shadowed by dusty eucalyptus trees. I took an elevator one floor up to Stent's laboratory and found it locked. A few minutes later the elevator door slid open and out walked Stent, a red-faced, sweaty man wearing a yellow bicycle helmet and rolling a dirt-encrusted mountain bike.

Stent had moved to the United States from Germany as a youth, and his gruff voice and attire still bore traces of his origins. He wore wire-rimmed glasses, a blue, short-sleeved shirt with epaulets, dark slacks, and shiny black shoes. He led me through his laboratory, crammed with microscopes, centrifuges, and scientific glassware, to a small office at the rear. The hall outside his office was adorned with photographs and paintings of Buddha. When Stent closed the door of his office behind us, I saw that he had tacked to the door's inner surface a poster from the 1989 meeting at Gustavus Adolphus College. The top half of the poster was covered with the word SCIENCE, written in huge, luridly colored letters. The letters were melting, oozing downward into a pool of Day-Glo protoplasm. Beneath this psychedelic puddle big black letters asked, "The End of Science?"

Stent, at the beginning of our interview, seemed rather suspicious. He asked pointedly if I was following the legal travails of the journalist Janet Malcolm, who had just lost a round in her interminable legal battle with a former profile subject, the psychoanalyst Jeffrey Masson. I mumbled something to the effect that Malcolm's transgressions were too minor to merit any punishment, but that her methods did seem rather careless. If I were writing something critical about a person as obviously volatile as Masson, I told Stent, I would be sure to have all my quotes on tape. (As I spoke, my own tape recorder was silently spinning between us.)

Gradually, Stent relaxed and began to tell me about his life. Born in Berlin to Jewish parents in 1924, he escaped from Germany in 1938 and moved in with a sister living in Chicago. He obtained a doctorate in chemistry at the University of Illinois, but upon reading Erwin Schrödinger's book *What Is Life?* he became entranced by the mystery of genetic transmission. After working at the California Institute of Technology with the eminent biophysicist Max Delbrück, Stent obtained a professorship at Berkeley in 1952. In these early years of molecular biology, Stent said, "none of us knew what we were doing. Then Watson and Crick found the double helix, and within a few weeks we realized we were doing molecular biology."

Stent began pondering the limits of science in the 1960s, partly in reaction to Berkeley's free-speech movement, which challenged the value of Western rationalism and technological progress and other aspects of civilization that Stent held dear. The university appointed him to a committee to "deal with this, to calm things down," by talking to students. Stent sought to fulfill this mandate—and to resolve his own inner conflicts over his role as a scientist—by delivering a series of lectures. These lectures became *The Coming of the Golden Age.*

I told Stent that I could not determine, after finishing *The Coming of the Golden Age,* whether he believed that the new Polynesia, the era of social and intellectual stasis and universal leisure, would be an improvement over our present situation. "I could never decide this!" he exclaimed, looking genuinely distressed. "People called me a pessimist, but I thought I was an optimist." He certainly did not think such a society would be in any sense Utopian. After the horrors wreaked by totalitarian states in this century, he explained, it was no longer possible to take the idea of Utopia seriously.

Stent felt that his predictions had held up reasonably well. Although hippies had vanished (except for the pitiful relics on Berkeley's streets), American culture had become increasingly materialistic and anti-intellectual; hippies had evolved into yuppies. The cold war had ended, although not through the gradual merging of communist and capitalist states that Stent had envisioned. He admitted that he had not anticipated the resurgence, in the wake of the cold war, of long-repressed ethnic conflicts. "I'm very depressed at what's happening in the Balkans," he said. "I didn't think that would happen." Stent was also surprised by the persistence of poverty and of racial conflict in the United States, but he believed these problems would eventually diminish in importance. (Aha, I thought. He was an optimist after all.)

Stent was convinced that science was showing signs of the closure he had predicted in *Golden Age.* Particle physicists were having difficulty convincing society to pay for their increasingly expensive experiments, such as the superconducting supercollider. As for biologists, they still had much to learn about how, say, a fertilized cell is transformed into a complex, multicellular organism, such as an elephant, and about the workings of the brain. "But I think the big picture is basically over," he said. Evolutionary biology in particular "was over when Darwin published *The Origin of Species,"* Stent said. He scoffed at the hope of some evolutionary biologists—notably Edward Wilson of Harvard—that they could remain occupied indefinitely by doing a thorough survey of all life on earth, species by species. Such an enterprise would be a mindless "glass bead game," Stent complained.

He then plunged into a diatribe against environmentalism. It was at heart an antihuman philosophy, one that contributed to the low self-esteem of American youth and poor black children in particular. Alarmed that my favorite Cassandra was revealing himself to be a crank, I changed the subject to consciousness. Did Stent still consider consciousness to be an unsolvable scientific problem, as he had suggested in *Golden Age*? He replied that he thought very highly of Francis Crick, who late in his career had turned his attention to consciousness. If Crick felt that consciousness was scientifically tractable, Stent said, then that possibility must be taken seriously.

Stent was still convinced, though, that a purely physiological explanation of consciousness would not be as comprehensible or as meaningful as most people would like, nor would it help us to solve moral and ethical questions. Stent thought the progress of science might give religion a clearer role in the future rather than eliminate it entirely, as many scientists had once hoped.

When I asked about the possibility that computers might become intelligent and create their own science, Stent snorted in derision. He had a dim view of artificial intelligence, and particularly of its more visionary enthusiasts. Computers may excel at precisely defined tasks such as mathematics and chess, he pointed out, but they still perform abysmally when confronted with the kind of problems—recognizing a face or a voice or walking down a crowded sidewalk—that humans solve effortlessly. "They're full of it," Stent said of Marvin Minsky and others who have predicted that one day we humans will be able to download our personalities into computers. "I wouldn't rule out the possibility that in the twenty-third century you might have an artificial brain," he added, "but it would need experience." One could design a computer to become an expert in restaurants, "but this machine would never know what a steak tastes like."

Stent was similarly skeptical of the claims of investigators of chaos and complexity that with computers and sophisticated mathematics they would be able to transcend the science of the past. In *The Coming of the Golden Age*, Stent discussed the work of one of the pioneers of chaos theory, Benoit Mandelbrot. Beginning in the early 1960s, Mandelbrot showed that many phenomena are intrinsically indeterministic: they exhibit behavior that is unpredictable and apparently random. Scientists can only guess at the causes of individual events and cannot predict them with any accuracy.

Investigators of chaos and complexity were attempting to create effective, comprehensible theories of the same phenomena studied by Mandelbrot, Stent said. He had concluded in *Golden Age* that these indeterministic phenomena would resist scientific analysis, and he saw no reason to change that assessment. Quite the contrary. The work emerging from those fields demonstrated his point that science, when pushed too far, always culminates in incoherence. So Stent did not think that chaos and complexity would bring about the rebirth of science? "No," he replied with a rakish grin. "It's the end of science."

What Science Has Accomplished

We obviously are nowhere near the new Polynesia that Stent envisioned, in part because applied science has not come nearly as far as Stent had hoped (feared?) when he wrote *The Coming of the Golden Age*. But I have come to the conclusion that Stent's prophecy has, in one very important sense, already come to pass. Pure science, the quest for knowledge about what we are and where we came from, has already entered an era of diminishing returns. By far the greatest barrier to future progress in pure science is its past success. Researchers have already mapped out physical reality, ranging from the microrealm of quarks and electrons to the macrorealm of planets, stars, and galaxies. Physicists have shown that all matter is ruled by a few basic forces: gravity, electromagnetism, and the strong and weak nuclear forces.

Scientists have also stitched their knowledge into an impressive, if not terribly detailed, narrative of how we came to be. The universe exploded into existence 15 billion years ago, give or take 5 billion years (astronomers may never agree on an exact figure), and is still expanding outward. Some 4.5 billion years ago, the detritus of an exploding star, a supernova, condensed into our solar system. Sometime during the next few hundred million years, for reasons that may never be known, single-celled organisms bearing an ingenious molecule called DNA emerged on the still-hellish earth. These Adamic microbes gave rise, by means of natural selection, to an extraordinary array of more complex creatures, including *Homo sapiens*.

My guess is that this narrative that scientists have woven from their knowledge, this modern myth of creation, will be as viable 100 or even 1,000 years from now as it is today. Why? Because it is true. Moreover, given how far science has already come, and given the physical, social, and cognitive limits constraining further research, science is unlikely to make any significant additions to the knowledge it has already generated. There will be no great revelations in the future comparable to those bestowed upon us by Darwin or Einstein or Watson and Crick.

The Anticlimax of Immortality

Applied science will continue for a long time to come. Scientists will keep developing versatile new materials; faster and more sophisticated computers; genetic-engineering techniques that make us healthier, stronger, longer-lived; perhaps even fusion reactors that provide cheap energy with few environmental side effects (although given the drastic cutbacks in funding, fusion's prospects now seem dimmer than ever). The question is, will these advances in applied science bring about any surprises, any revolutionary shifts in our basic knowledge? Will they force scientists to revise the map they have drawn of the universe's structure or the narrative they have constructed of our cosmic creation and history? Probably not. Applied science in this century has tended to reinforce rather than to challenge the prevailing theoretical paradigms. Lasers and transistors confirm the power of quantum mechanics, just as genetic engineering bolsters belief in the DNA-based model of evolution.

What constitutes a surprise? Einstein's discovery that time and space, the I beams of reality, are made of rubber was a surprise. So was the observation by astronomers that the universe is expanding, evolving. Quantum mechanics, which unveiled a probabilistic element, a Lucretian swerve, at the bottom of things, was an enormous surprise; God *does* play dice (Einstein's disapproval notwithstanding). The later finding that protons and neutrons are made of smaller particles called quarks was a much lesser surprise, because it merely extended quantum theory to a deeper domain; the foundations of physics remained intact.

Learning that we humans were created not *de novo* by God, but gradually, by the process of natural selection, was a big surprise. Most other aspects of human evolution—those concerning where, when, and how, precisely, *Homo sapiens* evolved—are details. These details may be interesting, but they are not likely to be surprising unless they show that scientists' basic assumptions about evolution are wrong. We may learn, say, that our sudden surge in intelligence was catalyzed by the intervention of alien beings, as in the movie *2001*. That would be a very big surprise. In fact, any proof that life exists—or even once existed—beyond our little planet would constitute a huge surprise. Science, and all human thought, would be reborn. Speculation about the origin of life and its inevitability would be placed on a much more empirical basis.

But how likely is it that we will discover life elsewhere? In retrospect, the space programs of both the United States and the USSR represented elaborate displays of saber rattling rather than the opening of a new frontier for human knowledge. The prospects for space exploration on anything more than a trivial level seem increasingly unlikely. We no longer have the will or the money to indulge in technological muscle flexing for its own sake. Humans, made of flesh and blood, may someday travel to other planets here in our solar system. But unless we find some way to transcend Einstein's prohibition against faster-than-light travel, chances are that we will never even attempt to visit another star, let alone another galaxy. A spaceship that can travel one million miles an hour, a velocity at least one order of magnitude greater than any current technology can attain, would still take almost 3,000 years to reach our nearest stellar neighbor, Alpha Centauri.[11]

The most dramatic advance in applied science I can imagine is immortality. Many scientists are now attempting to identify the precise causes of aging. It is conceivable that if they succeed, scientists may be able to design versions of *Homo sapiens* that can live indefinitely. But immortality, although it would represent a triumph of applied science, would not necessarily change our fundamental knowledge of the universe. We would not have any better idea of why the universe came to be and of what lies beyond its borders than we do now. Moreover, evolutionary biologists suggest that immortality may be impossible to achieve. Natural selection designed us to live long enough to breed and raise our children. As a result, senescence does not stem from any single cause or even a suite of causes; it is woven inextricably into the fabric of our being.[12].

That's What They Thought 100 Years Ago

It is easy to understand why so many people find it hard to believe that science, pure or impure, might be ending. Just a century ago, no one could imagine what the future held in store. Television? Jets? Space stations? Nuclear weapons? Computers? Genetic engineering? It must be as impossible for us to know the future of science—pure or applied— as it would have been for Thomas Aquinas to anticipate Madonna or microwave ovens. There are marvels, utterly unpredictable, lying in wait for us just as there were for our ancestors. We will only fail to seize these treasures if we decide that they do not exist and cease striving to find them. The prophecy can only be self-fulfilling.

This position is often expressed as the that's-what-they-thought-at-the-end-of-the-nineteenth-century argument. The argument goes like this: As the nineteenth century wound down, physicists thought they knew everything. But no sooner had the twentieth century begun, than Einstein and other physicists discovered—invented?—relativity theory and quantum mechanics.

These theories eclipsed Newtonian physics and opened up vast new vistas for modern physics and other branches of science. Moral: Anyone who predicts that science is nearing its end will surely turn out to be as shortsighted as those nineteenth-century physicists.

Those who believe science is finite have a standard retort for this argument: the earliest explorers, because they could not find the edge of the earth, might well have concluded that it was infinite, but they would have been wrong. Moreover, it is by no means a matter of historical record that late-nineteenth-century physicists felt they had wrapped things up. The best evidence for a sense of completion is a speech given in 1894 by Albert Michelson, whose experiments on the velocity of light helped to inspire Einstein's theory of special relativity.

> While it is never safe to say that the future of Physical Science has no marvels even more astonishing than those of the past, it seems probable that most of the grand underlying principles have been firmly established and that further advances are to be sought chiefly in the rigorous application of these principles to all the phenomena which come under our notice. It is here that the science of measurement shows its importance—where quantitative results are more to be desired than qualitative work. An eminent physicist has remarked that the future truths of Physical Science are to be looked for in the sixth place of decimals.[13]

Michelson's remark about the sixth place of decimals has been so widely attributed to Lord Kelvin (after whom the Kelvin, a unit of temperature, is named) that some authors simply credit him with the quote.[14] But historians have found no evidence that Kelvin made such a statement. Moreover, at the time of Michelson's remarks, physicists were vigorously debating fundamental issues, such as the viability of the atomic theory of matter, according to the historian of science Stephen Brush of the University of Maryland. Michelson was so absorbed in his optics experiments, Brush suggested, that he was "oblivious to the violent controversies raging among theorists at the time." The alleged "Victorian calm in physics," Brush concluded, is a "myth."[15]

The Apocryphal Patent Official

Other historians, predictably, disagree with Brush's assessment.[16] Questions concerning the mood of a given era can never be completely resolved, but the view that scientists in the last century were complacent about the state of their field has clearly been exaggerated. Historians have provided a definitive ruling on another anecdote favored by those reluctant to accept that science might be mortal. The story alleges that in the mid-1800s the head of the U.S. Patent Office quit his job and recommended that the office be shut down because there would soon be nothing left to invent.

In 1995, Daniel Koshland, editor of the prestigious journal *Science*, repeated this story in an introduction to a special section on the future of science. In the section, leading scientists offered predictions about what their fields might accomplish over the next 20 years. Koshland, who, like Gunther Stent, is a biologist at the University of California at Berkeley, exulted that his prognosticators "clearly do not agree with that commissioner of patents of yesteryear. Great discoveries with great import for the future of science are in the offing. That we have come so far so fast is not an indication that we have saturated the discovery market, but rather that discoveries will come even faster."[17]

There were two problems with Koshland's essay. First, the contributors to his special section envisioned not "great discoveries" but, for the most part, rather mundane applications of current knowledge, such as better methods for designing drugs, improved tests for genetic disorders, more discerning brain scans, and the like. Some predictions were negative in nature. "Anyone who expects any human-like intelligence from a computer in the next 50 years is doomed to disappointment," proclaimed the physicist and Nobel laureate Philip Anderson.

Second, Koshland's story about the commissioner of patents was apocryphal. In 1940, a scholar named Eber Jeffery examined the patent commissioner anecdote in an article entitled "Nothing Left to Invent," published in the *Journal of the Patent Office Society*.[18] Jeffery traced the story to congressional testimony delivered in 1843 by Henry Ellsworth, then the commissioner of patents. Ellsworth remarked at one point, "The advancement of the arts, from year to year, taxes our credulity and seems to presage the arrival of that period when human improvement must end."

But Ellsworth, far from recommending that his office be shut down, asked for extra funds to cope with the flood of inventions he expected in agriculture, transportation, and communications. Ellsworth did indeed resign two years later, in 1845, but in his resignation letter he made no reference to closing the patent office; rather, he expressed pride at having expanded it. Jeffery concluded that Ellsworth's statement about "that period when human improvement must end" represented "a mere rhetorical flourish intended to emphasize the remarkable strides forward in inventions then current and to be expected in the future." But perhaps Jeffery was not giving Ellsworth enough credit. Ellsworth was, after all, anticipating the argument that Gunther Stent would make more than a century later: the faster science moves, the faster it will reach its ultimate, inevitable limits.

Consider the implications of the alternative position, the one implicitly advanced by Daniel Koshland. He insisted that because science has advanced so rapidly over the past century or so, it can and will continue to do so, possibly forever. But this inductive argument is deeply flawed. Science has only existed for a few hundred years, and its most spectacular achievements have occurred within the last century. Viewed from a historical perspective, the modern era of rapid scientific and technological progress appears to be not a permanent feature of reality, but an aberration, a fluke, a product of a singular convergence of social, intellectual, and political factors.

The Rise and Fall of Progress

I n his 1932 book, *The Idea of Progress*, the historian J. B. Bury stated: "Science has been advancing without interruption during the last three or four hundred years; every new discovery has led to new problems and new methods of solution, and opened up new fields for exploration. Hitherto men of science have not been compelled to halt, they have always found means to advance further. But *what assurance have we that they will not come up against impassable barriers?*" [Italics in the original.][19]

Bury had demonstrated through his own scholarship that the concept of progress was only a few hundred years old at most. From the era of the Roman Empire through the Middle Ages, most truth seekers had a degenerative view of history; the ancient Greeks had achieved the acme of mathematical and scientific knowledge, and civilization had gone downhill from there. Those who followed could only try to recapture some remnant of the wisdom epitomized by Plato and Aristotle. It was such founders of modern, empirical science as Isaac Newton, Francis Bacon, René Descartes, and Gottfried Leibniz who first set forth the idea that humans could systematically acquire and accumulate knowledge through investigations of nature. These ur-scientists believed that the process would be finite, that we could attain complete knowledge of the world and then construct a perfect society, a utopia, based on that knowledge and on Christian precepts. (The new Polynesia!)

Only with the advent of Darwin did certain intellectuals become so enamored with progress that they insisted it might be, or should be, *eternal*. "In the wake of the publication of Darwin's *On the Origin of Species*," Gunther Stent wrote in his 1978 book, *The Paradoxes of Progress*, "the idea of progress was raised to the level of a scientific religion. This optimistic view came to be so widely embraced in the industrialized nations that the claim that progress could presently come to an end is now widely regarded [to be] as outlandish a notion as was in earlier times the claim that the Earth moves around the sun."[20]

It is not surprising that modern nation-states became fervent proponents of the science-is-infinite creed. Science spawned such marvels as The Bomb, nuclear power, jets, radar, computers, and missiles. In 1945 the physicist Vannevar Bush (a distant relative of former presidents George H. W. and George W. Bush) proclaimed in *Science: The Endless Frontier* that science was "a largely unexplored hinterland" and an "essential key" to U.S. military and economic security.[21] Bush's essay served as a blueprint for the construction of the National Science Foundation and other federal organizations that thereafter supported basic research on an unparalleled scale.

The Soviet Union was perhaps even more devoted than its capitalist rival to the concept of scientific and technological progress. The Soviets seemed to have taken their lead from Friedrich Engels, who in *Dialectics of Nature* sought to show off his grasp of Newton's inverse square law of gravity in the following passage.

What Luther's burning of the papal Bull was in the religious field, in the field of natural science was the great work of Copernicus.... But from then on the development of science went forward in great strides, increasing, so to speak, proportionately to the square of the distance in time of its point of departure, as if it wanted to show the world that for the motion of the highest product of organic matter, the human mind, the law of inverse squares holds good, as it does for the motion of inorganic matter.[22]

Science, in the view of Engels, could and would continue striding forward, at an accelerating pace, forever.

Of course, powerful social, political, and economic forces now oppose this vision of boundless scientific and technological progress. The cold war, which was a major impetus for basic research in the United States and the Soviet Union, is over; the United States and the former Soviet republics have much less incentive to build space stations and gigantic accelerators simply to demonstrate their power. Society is also increasingly sensitive to the adverse consequences of science and technology, such as pollution, nuclear contamination, and weapons of mass destruction.

Even political leaders, who have traditionally been the staunchest defenders of the value of scientific progress, have begun voicing antiscience sentiments. The Czech poet and president Václav Havel declared in 1992 that the Soviet Union epitomized and therefore eternally discredited the "cult of objectivity" brought about by science. Havel expressed the hope that It is "the end of the modern era," that the world—and Being as such—is a wholly knowable system governed by a finite number of universal laws that man can grasp and rationally direct for his own benefit.[23]

This disillusionment with science was foreseen early in this century by Oswald Spengler, a German schoolteacher who became the first great prophet of the end of science. In his massive tome, *The Decline of the West*, published in 1918, Spengler argued that science proceeds in a cyclic fashion, with romantic periods of investigation of nature and the invention of new theories giving way to periods of consolidation during which scientific knowledge ossifies. As scientists become more arrogant and less tolerant of other belief systems, notably religious ones, Spengler declared, society will rebel against science and embrace religious fundamentalism and other irrational systems of belief. Spengler predicted that the decline of science and the resurgence of irrationality would begin at the end of the second millennium.[24]

Spengler's analysis was, if anything, too optimistic. His view of science as cyclic implied that science might one day be resurrected and undergo a new period of discovery. Science is not cyclic, however, but linear; we can only discover the periodic table and the expansion of the universe and the structure of DNA once. The biggest obstacle to the resurrection of science—and especially pure science, the quest for knowledge about who we are and where we came from—is science's past success.

No More Endless Horizons

Scientists are understandably loath to state publicly that they have entered an era of diminishing returns. No one wants to be recalled as the equivalent of those allegedly shortsighted physicists of a century ago. There is always the danger, too, that predictions of the demise of science will become self-fulfilling. But Gunther Stent is hardly the only prominent scientist to violate the taboo against such prophecies. In 1971, *Science* published an essay titled "Science: Endless Horizons or Golden Age?" by Bentley Glass, an eminent biologist and the president of the American Association for the Advancement of Science, which publishes *Science*. Glass weighed the two scenarios for science's future posited by Vannevar Bush and Gunther Stent and reluctantly came down on the side of Stent. Not only is science finite, Glass argued, but the end is in sight. "We are like the explorers of a great continent," Glass proclaimed, "who have penetrated to its margins in most points of the compass and have mapped the major mountain chains and rivers. There are still innumerable details to fill in, but the endless horizons no longer exist."[25]

According to Glass, a close reading of Bush's *Endless Frontier* essay suggests that he, too, viewed science as a finite enterprise. Nowhere did Bush specifically state that any fields of science could continue generating new discoveries forever. In fact, Bush described scientific knowledge as an "edifice" whose form "is predestined by the laws of logic and the nature of human reasoning. It is almost as though it already existed." Bush's choice of this metaphor, Glass commented, reveals that he considered scientific knowledge to be finite in extent. Glass proposed that the "bold title" of Bush's essay was "never intended to be taken literally, but supposed merely to imply that from our present viewpoint so much yet remains before us to be discovered that the horizons seem virtually endless."

In 1979, in the *Quarterly Review of Biology*, Glass presented evidence to back up his view that science was approaching a culmination.[26] His analysis of the rate of discoveries in biology showed that they had not kept pace with the exponential increase in researchers and funding. "We have been so impressed by the undeniable acceleration in the rate of magnificent achievements that we have scarcely noticed that we are well into an era of diminishing returns," Glass stated. "That is, more and more scientific effort and expenditure of money must be allocated in order to sustain our progress. Sooner or later this will have to stop, because of the insuperable limits to scientific manpower and expenditure. So rapid has been the growth of science in our own century that we have been deluded into thinking that such a rate of progress can be maintained indefinitely."

When I spoke to him in 1994, Glass confessed that many of his colleagues had been dismayed that he had even broached the subject of science's limits, let alone prophesied its demise.[27] But Glass had felt then, and still felt, that the topic was too important to ignore. Obviously science as a social enterprise has *some* limits, Glass said. If science had continued to grow at the same rate as it did earlier in this century, he pointed out, it would soon have consumed the entire budget of the industrialized world. "I think it's rather evident to everybody that there must be brakes put on the amount of funding for science, pure science." This slowdown, he observed, was evident in the decision of the U.S. Congress in 1993 to terminate the superconducting supercollider, the gargantuan particle accelerator that physicists had hoped would propel them beyond quarks and electrons into a deeper realm of microspace, all for a mere $8 billion.

Even if society were to devote all its resources to research, Glass added, science would one day still reach the point of diminishing returns. Why? Because science *works*; it solves its problems. After all, astronomers have already plumbed the farthest reaches of the universe; they cannot see what, if anything, lies beyond its borders. Moreover, most physicists think that the reduction of matter into smaller and smaller particles will eventually end, or may have already ended for all practical purposes. Even if physicists unearth particles buried beneath quarks and electrons, that knowledge will make little or no difference to biologists, who have learned that the most significant biological processes occur at the molecular level and above. "There's a limit to biology there," Glass explained, "that you don't expect to be able to ever break through, just because of the nature of the constitution of matter and energy."

In biology, Glass said, the great revolutions may be in the past. "It's hard to believe, for me, anyway, that anything as comprehensive and earthshaking as Darwin's view of the evolution of life or Mendel's understanding of the nature of heredity will be easy to come by again. After all, these have been discovered!" Biologists certainly have much to learn, Glass emphasized, about diseases such as cancer and AIDS; about the process whereby a single fertilized cell becomes a complex, multicellular organism; about the relation between brain and mind. "There are going to be new additions to the structure of knowledge. But we have made some of the biggest possible advances. And it's just a question of whether there are any more really big changes in our conceptual universe that are going to be made."

Hard Times Ahead for Physics

In 1992, the monthly journal *Physics Today* published an essay entitled "Hard Times," in which Leo Kadanoff, a prominent physicist at the University of Chicago, painted a bleak picture for the future of physics. "Nothing we do is likely to arrest our decline in numbers, support or social value," Kadanoff declared. "Too much of our base depended on events that are now becoming ancient history: nuclear weapons and radar during World War II, silicon and laser technology thereafter, American optimism and industrial hegemony, socialist belief in rationality as a way of improving the world." Those conditions have largely vanished, Kadanoff contended; both physics and science as a whole are now besieged by environmentalists, animal-rights activists, and others with an antiscientific outlook. "In recent decades, science has had high rewards and has been at the center of social interest and concern. We should not be surprised if this anomaly disappears."[28]

Kadanoff, when I spoke to him over the telephone two years later, sounded even gloomier than he had been in his essay.[29] He laid out his worldview for me with a muffled melancholy, as if he were suffering from an existential head cold. Rather than discussing science's social and political problems, as he had in his article in *Physics Today*, he focused on another obstacle to scientific progress: science's past achievements. The great task of modern science, Kadanoff explained, has been to show that the world conforms to certain basic physical laws. "That is an issue which has been explored at least since the Renaissance and maybe a much longer period of time. For me, that's a settled issue. That is, it seems to me that the world *is* explainable by law." The most fundamental laws of nature are embodied in the theory of general relativity and in the so-called standard model of particle physics, which describes the behavior of the quantum realm with exquisite precision.

Just a half century ago, Kadanoff recalled, many reputable scientists still clung to the romantic doctrine of vitalism, which holds that life springs from some mysterious élan vital that cannot be explained in terms of physical laws. As a result of the findings of molecular biology—beginning with the discovery of the structure of DNA in 1953— "there are relatively few well-educated people" who admit to belief in vitalism, Kadanoff said.

Of course, scientists still have much to learn about how the fundamental laws generate "the richness of the world as we see it." Kadanoff himself is a leader in the field of condensed-matter physics, which studies the behavior not of individual subatomic particles, but of solids or liquids. Kadanoff has also been associated with the field of chaos, which addresses phenomena that unfold in predictably unpredictable ways. Some proponents of chaos—and of a closely related field called complexity—have suggested that with the help of powerful computers and new mathematical methods they will discover truths that surpass those revealed by the "reductionist" science of the past. Kadanoff has his doubts. Studying the consequences of fundamental laws is "in a way less interesting" and "less deep," he said, than showing that the world is lawful. "But now that we know the world is lawful," he added, "we have to go on to other things. And yes, it probably excites the imagination of the average human being less. Maybe with good reason."

Kadanoff pointed out that particle physics has not been terribly exciting lately either. Experiments over the past few decades have merely confirmed existing theories rather than revealing new phenomena requiring new laws; the goal of finding a unified theory of all of nature's forces seems impossibly distant. In fact, no field of science has yielded any truly deep discoveries for a long time, Kadanoff said. "The truth is, there is nothing—there is *nothing*—of the same order of magnitude as the accomplishments of the invention of quantum mechanics or of the double helix or of relativity. Just nothing like that has happened in the last few decades." Is this state of affairs permanent? I asked. Kadanoff was silent for a moment. Then he sighed, as if trying to exhale all his world-weariness. "Once you have proven that the world is lawful," he replied, "to the satisfaction of many human beings, you can't do that again."

Whistling to Keep Our Courage Up

One of the few modern philosophers to devote serious thought to the limits of science is Nicholas Rescher of the University of Pittsburgh. In his 1978 book, *Scientific Progress*, Rescher deplored the fact that Stent, Glass, and other prominent scientists seemed to think that science might be approaching a cul-de-sac. Rescher intended to provide "an antidote to this currently pervasive tendency of thought" by demonstrating that science is at least potentially infinite.[30] But the scenario he sketched out over the course of his book was hardly optimistic. He argued that science, as a fundamentally empirical, experimental discipline, faces economic constraints. As scientists try to extend their theories into more remote domains—seeing further into the universe, deeper into matter—their costs will inevitably escalate and their returns diminish.

"Scientific innovation is going to become more and more difficult as we push out further and further from our home base toward more remote frontiers. If the present perspective is even partly correct, the half-millennium commencing around 1650 will eventually come to be regarded among the great characteristic developmental transformations of human history, with the age of The Science Explosion as unique in its own historical structure as The Bronze Age or The Industrial Revolution or The Population Explosion."[31]

Rescher tacked what he apparently thought was a happy coda onto his depressing scenario: science will never end; it will just go slower and slower and slower, like Zeno's tortoise. Nor should scientists ever conclude that their research must degenerate into the mere filling in of details; it is always *possible* that one of their increasingly expensive experiments will have revolutionary import, comparable to that of quantum mechanics or Darwinian theory.

Bentley Glass, in a review of Rescher's book, called these prescriptions "whistling to keep one's courage up in the face of what, for most practitioners of science, is a bleak and imminent prospect."[32] When I telephoned Rescher in August 1992, he acknowledged that his analysis had been in most respects a grim one. "We can only investigate nature by interacting with it," he said. "To do that we must push into regions never investigated before, regions of higher density, lower temperature, or higher energy. In all these cases we are pushing fundamental limits, and that requires ever more elaborate and expensive apparatuses. So there is a limit imposed on science by the limits of human resources."

But Rescher insisted that "big plums, first-rate discoveries," might— must!—lie ahead. He could not say whence those discoveries might arise. "It's like the jazz musician who was asked where jazz is going, and he said, 'If I knew we'd be there by now.'" Rescher, finally, fell back on the that's-what-they-thought-at-the-end-of-the-last-century argument. The fact that such scientists as Stent, Glass, and Kadanoff seemed to fear that science was drawing to a close, Rescher said, gave him confidence that some marvelous discovery was pending. Rescher, like many other would-be seers, had succumbed to wishful thinking. He admitted that he felt that the end of science would be a tragedy for humanity. If the quest for knowledge ended, what would become of us? What would give our existence meaning?

The Meaning of Francis Bacon's Plus Ultra

The second most common response to the suggestion that science is ending—after "that's what they thought at the end of the last century"— is the old maxim "Answers raise new questions." Kant wrote in *Prolegomena to Any Future Metaphysics* that "every answer given on principles of experience begets a fresh question, which likewise requires its answer and thereby clearly shows the insufficiency of all physical modes of explanation to satisfy reason."[33] But Kant also suggested (anticipating the arguments of Gunther Stent), that the innate structure of our minds constrains both the questions we put to nature and the answers we glean from it.

Of course, science will continue to raise new questions. Most are trivial, in that they concern details that do not affect our basic understanding of nature. Who really cares, except specialists, about the precise mass of the top quark, the existence of which was finally confirmed in 1994 after research costing billions of dollars? Other questions are profound but unanswerable. In fact, the most persistent foil to the *completion* of science—to the attainment of the truly satisfying theory that Roger Penrose and others dream of—is the human ability to invent unanswerable questions. Presented with a purported theory of everything, someone always can and will ask: But how do we know that quarks or even superstrings (in the unlikely event that they are one day shown to exist) are not composed of still smaller entities—ad infinitum? How do we know that the visible universe is not just one of an infinite number of universes? Was our universe necessary or just a cosmic fluke? What about life? Are computers capable of conscious thought? Are amoebas?

No matter how far empirical science goes, our imaginations can always go farther. That is the greatest obstacle to the hopes—and fears—of scientists that we will find *The Answer*, a theory that quenches our curiosity forever. Francis Bacon, one of the founders of modern science, expressed his belief in the vast potential of science with the Latin term *plus ultra*, "more beyond."[34] But *plus ultra* does not apply to science per se, which is a tightly constrained method for examining nature. *Plus ultra* applies, rather, to our imaginations. Although our imaginations are constrained by our evolutionary history, they will always be capable of venturing beyond what we truly know.

Even in the new Polynesia, Gunther Stent suggested, a few persistent souls will keep striving to transcend the received wisdom. Stent called these truth seekers "Faustian" (a term he borrowed from Oswald Spengler). I call them strong scientists (a term I coopted from Harold Bloom's *The Anxiety of Influence)*. By raising questions that science cannot answer, strong scientists can continue the quest for knowledge in the speculative mode that I call ironic science even after empirical science—the kind of science that answers questions—has ended.

The poet John Keats coined the term *negative capability* to describe the ability of certain great poets to remain "in uncertainties, mysteries, doubts, without any irritable reaching after fact and reason." As an example, Keats singled out his fellow poet Samuel Coleridge, who "would let go by a fine isolated verisimilitude caught from the penetralium of mystery, from being incapable of remaining content with half-knowledge."[35] The most important function of ironic science is to serve as humanity's negative capability. Ironic science, by raising unanswerable questions, reminds us that all our knowledge is half-knowledge; it reminds us of how little we know. But ironic science does not make any significant contributions to knowledge itself. Ironic science is thus less akin to science in the traditional sense than to literary criticism—or to philosophy.

Chapter Two
The End of Philosophy

Twentieth-century science has given rise to a marvelous paradox. The same extraordinary progress that has led to predictions that we may soon know everything that can be known has also nurtured doubts that we can know *anything* for certain. When one theory so rapidly succeeds another, how can we ever be sure that any theory is true? In 1987 two British physicists, T. Theocharis and M. Psimopoulos, excoriated this skeptical philosophical position in an essay entitled "Where Science Has Gone Wrong." Published in the British journal *Nature*, the essay blamed the "deep and widespread malaise" in science on philosophers who had attacked the notion that science could achieve objective knowledge. The article printed photographs of four particularly egregious "betrayers of the truth": Karl Popper, Imre Lakatos, Thomas Kuhn, and Paul Feyerabend.[1]

The photographs were grainy, black-and-white shots of the sort that adorn a lurid exposé about a venerable banker who has been caught swindling retirees. These, clearly, were intellectual transgressors of the worst sort. Feyerabend, whom the essayists called "the worst enemy of science," was the most wicked-looking of the bunch. Smirking at the camera over glasses perched on the tip of his nose, he was clearly either anticipating or relishing the perpetration of some diabolical prank. He looked like an intellectual version of Loki, the Norse god of mischief.

The main complaint of Theocharis and Psimopoulos was silly. The skepticism of a few academic philosophers has never represented a serious threat to the massive, well-funded bureaucracy of science. Many scientists, particularly would-be revolutionaries, find the ideas of Popper et al. comforting; if our current knowledge is provisional, there is always the possibility that great revelations lie ahead. Theocharis and Psimopoulos did make one intriguing assertion, however, that the ideas of the skeptics are "flagrantly self-refuting—they negate and destroy themselves." It would be interesting, I thought, to put this argument to the philosophers and see how they responded.

Eventually I had the opportunity to do just that with all the "betrayers of the truth" except for Lakatos, who died in 1974. During my interviews, I also tried to find out whether these philosophers were really as skeptical, as doubtful of science's ability to achieve truth, as some of their own statements implied. I came away convinced that Popper, Kuhn, and Feyerabend each believed very much in science; in fact, their skepticism was motivated by their belief. Their biggest failing, perhaps, was to credit science with more power than it actually has. They feared that science might extinguish our sense of wonder and therefore bring science itself—and all forms of knowledge seeking— to an end. They were trying to protect humanity, scientists included, from the naive faith in science exemplified by such scientists as Theocharis and Psimopoulos.

As science has grown in power and prestige over the past century, too many philosophers have served as science's public relations agents. This trend can be traced to such thinkers as Charles Sanders Peirce, an American who founded the philosophy of pragmatism but could not keep a job or a wife and died penniless and miserable in 1914. Peirce offered this definition of absolute truth: it is whatever scientists say it is when they come to the end of their labors.[2]

Much of philosophy since Peirce has merely elaborated on his view. The dominant philosophy in Europe early in this century was logical positivism, which asserted that we can only know that something is true if it can be logically or empirically demonstrated. The positivists upheld mathematics and science as the supreme sources of truth. Popper, Kuhn, and Feyerabend— each in his own way and for his own reasons—sought to counter this fawning attitude toward science. These philosophers realized that in an age when science is ascendant, the highest calling of philosophy should be to serve as the negative capability of science, to infuse scientists with doubt. Only thus can the human quest for knowledge remain open-ended, potentially infinite; only thus can we remain awestruck before the mystery of the cosmos.

Of the three great skeptics I interviewed, Popper was the first to make his mark.[3] His philosophy stemmed from his effort to distinguish pseudo-science, such as Marxism or astrology or Freudian psychology, from genuine science, such as Einstein's theory of relativity. The latter, Popper decided, was testable; it made predictions about the world that could be empirically checked. The logical positivists had said as much. But Popper denied the positivist assertion that scientists can *prove* a theory through induction, or repeated empirical tests or observations. One never knows if one's observations have been sufficient; the next observation might contradict all that preceded it. Observations can never prove a theory but can only disprove, or falsify it. Popper often bragged that he had "killed" logical positivism with this argument.[4]

Popper expanded his falsification tenet into a philosophy that he called critical rationalism. One scientist ventures a proposal and others try to bat it down with contrary arguments or experimental evidence. Popper viewed criticism, and even conflict, as essential for progress of all kinds. Just as scientists approach the truth through what he calls "conjecture and refutation," so do species evolve through competition and societies through political debate. A "human society without conflict," he once wrote, "would be a society not of friends but of ants."[5] In *The Open Society and Its Enemies*, published in 1945, Popper asserted that politics, even more than science, required the free play of ideas and criticism. Dogmatism inevitably led not to Utopia, as Marxists and fascists alike claimed, but to totalitarian repression.

I began to discern the paradox lurking at the heart of Popper's work—and persona—when, prior to meeting him, I asked other philosophers about him. Queries of this kind usually elicit rather dull, generic praise, but in this case my interlocutors had nothing good to say. They revealed that this man who inveighed against dogmatism was himself almost pathologically dogmatic and demanding of fealty from students. There was an old joke about Popper: *The Open Society and Its Enemies* should have been titled *The Open Society by One of Its Enemies.*

To arrange my interview with Popper, I telephoned the London School of Economics, where he had taught since the late 1940s. A secretary there said that Popper generally worked at his home in Kenley, a suburb south of London, and gave me his number. I called, and a woman with an imperious, German-accented voice answered. Mrs. Mew, assistant to "Sir Karl." Before Sir Karl would see me, I had to send her a sample of my writings. She gave me a reading list that would prepare me for my meeting: a dozen or so books by Sir Karl. Eventually, after several faxes and telephone calls, she set a date. She also provided directions to the train station near Sir Karl's house. When I asked her for directions from the train station to the house, Mrs. Mew assured me that all the cab drivers knew where Sir Karl lived. "He's quite famous."

"Sir Karl Popper's house, please," I said as I climbed into a cab at Kenley station. "Who?" the driver replied. Sir Karl Popper? The famous philosopher? Never heard of him, the driver said. He was familiar with the street on which Popper lived, however, and we found Popper's home—a two-story cottage surrounded by scrupulously trimmed grass and shrubs—with little difficulty.[6]

A tall, handsome woman dressed in dark pants and shirt and with short, dark hair answered the door: Mrs. Mew. She was only slightly less forbidding in person than over the telephone. As she led me into the house, she told me that Sir Karl was quite tired. He had undergone a spate of interviews and congratulations brought on by his 90th birthday the previous month, and he had been working too hard preparing an acceptance speech for the Kyoto Award, known as Japan's Nobel. I should expect to speak to him for only an hour at the most.

I was trying to lower my expectations when Popper made his entrance. He was stooped, equipped with a hearing aid, and surprisingly short; I had assumed that the author of such autocratic prose would be tall. Yet he was

34

as kinetic as a bantamweight boxer. He brandished an article I had written for *Scientific American* about how quantum mechanics was compelling some physicists to abandon the view of physics as a wholly objective enterprise.[7] "I don't believe a word of it," he declared in an Austrian-accented growl. "Subjectivism" has no place in physics, quantum or otherwise. "Physics," he exclaimed, grabbing a book from a table and slamming it down, "is that!" (This from a man who cowrote a book espousing dualism, the notion that ideas and other constructs of the human mind exist independently of the material world.[8])

Once seated, he kept darting away to forage for books or articles that could buttress a point. Striving to dredge a name or date from his memory, he kneaded his temples and gritted his teeth as if in agony. At one point, when the word *mutation* briefly eluded him, he slapped his forehead repeatedly and with alarming force, shouting, "Terms, terms, terms!"

Words poured from him so rapidly and with so much momentum that I began to lose hope that I could ask any of my prepared questions. "I am over 90 and I can still think," he declared, as if suspecting that I doubted it. He tirelessly touted a theory of the origin of life proposed by a former student, Günther Wächtershäuser, a German patent attorney who had a Ph.D. in chemistry.[9] Popper kept emphasizing that he had known all the titans of twentieth-century science: Einstein, Schrödinger, Heisenberg. Popper blamed Bohr, whom he knew "very well," for having introduced subjectivism into physics. Bohr was "a marvelous physicist, one of the greatest of all time, but he was a miserable philosopher, and one couldn't talk to him. He was talking all the time, allowing practically only one or two words to you and then at once cutting in."

As Mrs. Mew turned to leave, Popper abruptly asked her to find one of his books. She disappeared for a few minutes and then returned empty-handed. "Excuse me, Karl, I couldn't find it," she reported. "Unless I have a description, I can't check every bookcase."

"It was actually, I think, on the right of this corner, but I have taken it away maybe..." His voice trailed off. Mrs. Mew somehow rolled her eyes without really rolling them and vanished.

He paused a moment, and I desperately seized the opportunity to ask a question. "I wanted to ask you about..."

"Yes! You should ask me your questions! I have wrongly taken the lead. You can ask me all your questions first."

As I began to question Popper about his views, it became apparent that his skeptical philosophy stemmed from a deeply romantic, idealized view of science. He thus denied the assertion, often made by the logical positivists, that science can ever be reduced to a formal, logical system, in which raw data are methodically converted into truth. A scientific theory, Popper insisted, is an invention, an act of creation as profoundly mysterious as anything in the arts. "The history of science is everywhere speculative," Popper said. "It is a marvelous history. It makes you proud to be a human being." Framing his face in his outstretched hands, Popper intoned, "I believe in the human mind."

For similar reasons, Popper had battled throughout his career against the doctrine of scientific determinism, which he felt was antithetical to human creativity and freedom and thus to science itself. Popper claimed to have realized long before modern chaos theorists that not only quantum systems but even classical, Newtonian ones are inherently unpredictable; he had delivered a lecture on this subject in the 1950s. Waving at the lawn outside the window he said, "There is chaos in every grass."

When I asked Popper if he thought that science was incapable of achieving absolute truth, he exclaimed, "No no!" and shook his head vehemently. He, like the logical positivists before him, believed that a scientific theory could be "absolutely" true. In fact, he had "no doubt" that some current scientific theories were absolutely true (although he refused to say which ones). But he rejected the positivist belief that we can ever *know* that a theory is true. "We must distinguish between truth, which is objective and absolute, and certainty, which is subjective."

If scientists believed too much in their own theories, Popper realized, they might stop seeking truth. And that would be a tragedy, since for Popper truth seeking was what made life worth living. "To search for the truth is a kind of religion," he said, "and I think it is also an ethical belief." Popper's conviction that the search for knowledge must never cease is reflected in the title of his autobiography, *Unended Quest*.

He thus scoffed at the hope of some scientists to achieve a complete theory of nature, one that answers all questions. "Many people think that the problems can be solved; many people think the opposite. I think we have gone very far, but we are much further away. I must show you one passage that bears on this." He shuffled off again and returned with his book *Conjectures and Refutations*. Opening it, he read his own words with reverence: "In our infinite ignorance we are all equal."

Popper also believed that science could never answer questions about the meaning and purpose of the universe. For these reasons he had never completely repudiated religion, although he had long ago abandoned the Lutheranism of his youth. "We know very little, and we should be modest and not pretend we know anything about ultimate questions of this kind."

Yet Popper abhorred those modern philosophers and sociologists who claim that science is incapable of achieving *any* truth and who argue that scientists adhere to theories for cultural and political reasons rather than rational ones. Such critics, Popper charged, resent being viewed as inferior to genuine scientists and are trying to "change their status in the pecking order." I suggested that these critics sought to describe how science *is* practiced, whereas he, Popper, tried to show how it *should* be practiced. Somewhat to my surprise, Popper nodded. "That is a very good statement," he said. "You can't see what science is without having in your head an idea of what science should be." Popper had to agree that scientists often fell short of the ideal he had set for them. "Since scientists got subsidies for their work, science isn't exactly what it should be. This is unavoidable. There is a certain corruption, unfortunately. But I don't talk about that."

Popper then proceeded to talk about it. "Scientists are not as self-critical as they should be," he asserted. "There is a certain wish that you, people like you"—he jabbed a finger at me—"should bring them before the public." He stared at me a moment, then reminded me that he had not sought this interview. "Far from it," he said. "You know that I have really made not only no attempt but have not encouraged you." Popper then plunged into an excruciatingly technical critique— involving triangulation and other arcana—of the big bang theory. "It's always the same," he summed up. "The difficulties are underrated. It is presented in a spirit as if this all has scientific certainty, but scientific certainty doesn't exist."

I asked Popper if he felt that biologists were also too committed to Darwin's theory of natural selection; in the past he had suggested that the theory was tautological and thus pseudo-scientific.[10] "That was perhaps going too far," Popper said, waving his hand dismissively. "I'm not dogmatic about my own views." Suddenly, he pounded the table and exclaimed, "One ought to look for alternative theories! This"—he waved the paper by Günther Wächtershäuser on the origin of life— "is an alternative theory. It seems to be a better theory." That doesn't mean that the theory is true, Popper quickly added. "The origin of life will forever remain untestable, probably," Popper said. Even if scientists create life in a laboratory, he explained, they can never be sure that life actually began in the same way.

It was time to launch my big question. Was his own falsification concept falsifiable? Popper glared at me. Then his expression softened, and he placed his hand on mine. "I don't want to hurt you," he said gently, "but it is a silly question." Peering searchingly into my eyes, he inquired whether one of his critics had persuaded me to ask this question. Yes, I lied. "Exactly," he said, looking pleased.

"The first thing you do in a philosophy seminar when somebody proposes an idea is to say it doesn't satisfy its own criteria. It is one of the most idiotic criticisms one can imagine!" His falsification concept, he said, is a criterion for distinguishing between empirical modes of knowledge, namely, science, and nonempirical ones, such as philosophy. Falsification itself is "decidably unempirical"; it belongs not to science but to philosophy, or "metascience," and it does not even apply to all of science. Popper was admitting, essentially, that his critics were right: falsification is a mere guideline, a rule of thumb, sometimes helpful and sometimes not.

Popper said he had never before responded to the question I had just posed. "I found it too stupid to be answered. You see the difference?" he asked, his voice gentle again. I nodded. The question seemed a bit silly to me, too, I said, but I just thought I should ask. He smiled and squeezed my hand, murmuring, "Yes, very good."

Since Popper seemed so agreeable, I mentioned that one of his former students had accused him of not tolerating criticism of his own ideas. Popper's eyes blazed. "It is completely untrue! I was *happy* when I got criticism! Of course, not when I would answer the criticism, like I have answered it when you gave it to me, and the person would still go on with it. That is the thing which I found uninteresting and would not tolerate." If that happened, Popper would order the student out of his class.

The light in the kitchen was acquiring a ruddy hue when Mrs. Mew stuck her head in the door and informed us that we had been talking for more than three hours. How much longer, she inquired a bit peevishly, did we expect to continue? Perhaps she had better call me a cab? I looked at Popper, who had broken into a bad-boy grin but did appear to be drooping.

I slipped in a final question: Why in his autobiography did Popper say that he was the happiest philosopher he knew? "Most philosophers are really deeply depressed," he replied, "because they can't produce anything worthwhile." Looking pleased with himself, Popper glanced over at Mrs. Mew, who wore an expression of horror. Popper's own smile abruptly faded. "It would be better not to write that," he said, turning back to me. "I have enough enemies, and I better not answer them in this way." He stewed a moment and added, "But it is so."

I asked Mrs. Mew if I could have a copy of the speech Popper was going to deliver at the Kyoto Award ceremony in Japan. "No, not now," she said curtly. "Why not?" Popper inquired. "Karl," she replied, "I've been typing the second lecture nonstop, and I'm a bit..." She sighed. "You know what I mean?" Anyway, she added, she did not have a final version. "What about an uncorrected version?" Popper asked. Mrs. Mew stalked off.

She returned and shoved a copy of Popper's lecture at me. "Have you

got a copy of *Propensities*?" Popper asked her.[11] She pursed her lips and
stomped into the room next door, while Popper explained the book's theme to
me. The lesson of quantum mechanics and even of classical physics, Popper
said, is that nothing is determined, nothing is certain, nothing is completely
predictable; there are only propensities for certain things to occur. "For
example," Popper added, "in this moment there is a certain propensity that
Mrs. Mew may find a copy of my book."

"Oh, please!" Mrs. Mew exclaimed from the next room. She returned, no
longer making any attempt to hide her annoyance. "Sir Karl, Karl, you have
given away the last copy of *Propensities*. Why do you do that?"

"The last copy was given away in your presence," he declared.

"I don't think so," she retorted. "Who was it?"

"I can't remember," he muttered sheepishly.

Outside, a black cab pulled into the driveway. I thanked Popper
and Mrs. Mew for their hospitality and took my leave. As the cab pulled
away, I asked the driver if he knew whose house this was. No, he didn't.
Someone famous, was it? Yes, actually: Sir Karl Popper. Who? Karl
Popper, I replied, one of the greatest philosophers of the twentieth century.
"Is that right?" murmured the driver.

Popper has always been popular among scientists—and with good
reason, since he depicted science as an endless romantic adventure. An
editorial in *Nature* once called Popper, quite justly, "the philosopher *for*
science" [italics added].[12] But Popper's fellow philosophers have been less
kind. His oeuvre, they point out, is rife with contradictions. Popper argued
that science could not be reduced to a method, but his falsification scheme
was just such a method. Moreover, the arguments that he used to kill the
possibility of absolute verification could also be used to kill falsification. If
it is always possible that future observations will contradict a theory, then
it is also possible that future observations may resurrect a theory that has
previously been falsified. It is more reasonable to assume, critics of Popper
have argued, that just as some scientific theories can be falsified, so some can
be confirmed; there is no point in remaining uncertain, after all, that the earth

is round and not flat.

When Popper died in 1994, two years after I met him, the *Economist* hailed him as having been "the best-known and most widely read of living philosophers."[13] It praised, in particular, his insistence on antidogmatism in the political realm. But the obituary also noted that Popper's treatment of induction (the basis of his falsification scheme) had been rejected by later philosophers. "According to his own theories, Popper should have welcomed this fact," the *Economist* noted drily, "but he could not bring himself to do so. The irony is that, here, Popper could not admit he was wrong." Popper's antidogmatism, when applied to science, had become a kind of dogmatism.

Although Popper abhorred psychoanalysis, his own work, finally, may be best understood in psychoanalytic terms. His relationship with authority figures—from scientific giants, such as Bohr, to his assistant, Mrs. Mew—was obviously complex, alternating between defiance and deference. In what is perhaps the single most revealing passage in his autobiography, Popper mentioned that his parents were both Austrian Jews who had converted to Lutheranism. He then argued that the failure of other Jews to assimilate themselves into Germanic culture—and their prominent roles in leftist politics—contributed to the emergence of fascism and state-sponsored anti-Semitism in the 1930s: "anti-Semitism was an evil, to be feared by Jews and non-Jews alike, and... it was the task of all people of Jewish origin to do their best not to provoke it."[14]

The Structure of Thomas Kuhn

"Look," Thomas Kuhn said. The word was weighted with weariness, as if Kuhn was resigned to the fact that I would misinterpret him, but he was still going to try—no doubt in vain—to make his point. Kuhn uttered the word often. "Look," he said again. He leaned his gangly frame and long face forward, and his big lower lip, which ordinarily curled up amiably at the corners, sagged. "For Christ's sake, if I had my choice of having written the book or not having written it, I would choose to have written it. But there have certainly been aspects involving considerable upset about the response to it."

"The book" was *The Structure of Scientific Revolutions*, which may be the most influential treatise ever written on how science does (or does not) proceed. It is notable for having spawned the trendy term *paradigm*. It also fomented the now trite idea that personalities and politics play a large role in science. The book's most profound argument was less obvious: scientists can never truly understand the real world or even each other.[15]

Given this theme, one might think that Kuhn would have expected his own message to be at least partially misunderstood. But when I interviewed Kuhn in his office at the Massachusetts Institute of Technology (of all places) some three decades after the publication of *Structure*, he seemed to be deeply pained by the breadth of misunderstanding of his book. He was particularly upset by claims that he had described science as irrational. "If they had said '*a*rational' I wouldn't have minded at all," he said with no trace of a smile.

Kuhn's fear of compounding the confusion over his work had made him a bit press shy. When I first telephoned him to ask for an interview, he turned me down. "Look. I think not," he said. He revealed that *Scientific American*, my employer, had given *Structure* "the worst review I can remember." (The squib was indeed dismissive; it called Kuhn's argument "much ado about very little." But what did Kuhn expect from a magazine that celebrates science?[16]) Pointing out that I had not been at the magazine then—the review ran in 1964—I begged him to reconsider. Kuhn finally, reluctantly, agreed.

When we at last sat down together in his office, Kuhn expressed nominal discomfort at the notion of delving into the roots of his thought. "One is not one's own historian, let alone one's own psychoanalyst," he warned me. He nonetheless traced his view of science to an epiphany he experienced in 1947, when he was working toward a doctorate in physics at Harvard. While reading Aristotle's *Physics*, Kuhn had become astonished at how "wrong" it was. How could someone who wrote so brilliantly on so many topics be so misguided when it came to physics?

Kuhn was pondering this mystery, staring out his dormitory window

("I can still see the vines and the shade two-thirds of the way down"), when suddenly Aristotle "made sense." Kuhn realized that Aristotle invested basic concepts with different meanings than did modern physicists. Aristotle used the term *motion*, for example, to refer not just to change in position but to change in general—the reddening of the sun as well as its descent toward the horizon. Aristotle's physics, understood on its own terms, was simply different from, rather than inferior to, Newtonian physics.

Kuhn left physics for philosophy, and he struggled for 15 years to transform his epiphany into the theory set forth in *The Structure of Scientific Revolutions*. The keystone of his model was the concept of a paradigm. *Paradigm*, pre-Kuhn, referred merely to an example that serves an educational purpose; *amo, amas, amat*, for instance, is a paradigm for teaching conjugations in Latin. Kuhn used the term to refer to a collection of procedures or ideas that instruct scientists, *implicitly*, what to believe and how to work. Most scientists never question the paradigm. They solve puzzles, problems whose solutions reinforce and extend the scope of the paradigm rather than challenge it. Kuhn called this "mopping up," or "normal science." There are always anomalies, phenomena that the paradigm cannot account for or that even contradict it. Anomalies are often ignored, but if they accumulate they may trigger a revolution (also called a paradigm shift, although not originally by Kuhn), in which scientists abandon the old paradigm for a new one.

Denying the view of science as a continual building process, Kuhn held that a revolution is a destructive as well as a creative act. The proposer of a new paradigm stands on the shoulders of giants (to borrow Newton's phrase) and then bashes them over the head. He or she is often young or new to the field, that is, not fully indoctrinated. Most scientists yield to a new paradigm reluctantly. They often do not understand it, and they have no objective rules by which to judge it. Different paradigms have no common standard for comparison; they are "incommensurable," to use Kuhn's term. Proponents of different paradigms can argue forever without resolving their differences because they invest basic terms—motion, particle, space, time—with different meanings. The conversion of scientists is thus both a subjective and a political process. It may involve sudden, intuitive understanding— like that finally achieved by Kuhn as he pondered Aristotle. Yet scientists often adopt a paradigm simply because it is backed by others with strong reputations or by a majority of the community.

Kuhn's view diverged from Popper's in several important respects. Kuhn (like other critics of Popper) argued that falsification is no more possible than verification; each process implies the existence of absolute standards of evidence, which transcend any individual paradigm. A new paradigm may solve puzzles better than the old one does, and it may yield more practical applications. "But you cannot simply describe the other science as false," Kuhn said. Just because modern physics has spawned computers, nuclear power, and CD players does not mean it is truer, in an absolute sense, than Aristotle's physics. Similarly, Kuhn denied that science is constantly approaching the truth. At the end of *Structure* he asserted that science, like life on earth, does not evolve *toward* anything, but only *away* from something.

Kuhn described himself to me as a "post-Darwinian Kantian." Kant, too, believed that without some sort of *a priori* paradigm the mind cannot impose order on sensory experience. But whereas Kant and Darwin each thought that we are all born with more or less the same innate paradigm, Kuhn argued that our paradigms keep changing as our culture changes. "Different groups, and the same group at different times," Kuhn told me, "can have different experiences and therefore in some sense live in different worlds." Obviously all humans share some responses to experience, simply because of their shared biological heritage, Kuhn added. But whatever is universal in human experience, whatever transcends culture and history, is also "ineffable," beyond the reach of language. Language, Kuhn said, "is not a universal tool. It's not the case that you can say anything in one language that you can say in another."

But isn't mathematics a kind of universal language? I asked. Not really, Kuhn replied, since it has no meaning; it consists of syntactical rules without any semantic content. "There are perfectly good reasons why mathematics can be considered a language, but there is a very good reason why it isn't."

I objected that although Kuhn's view of the limits of language might apply to certain fields with a metaphysical cast, such as quantum mechanics, it did not hold in all cases. For example, the claim of a few biologists that AIDS is not caused by the so-called AIDS virus is either right or wrong; language is not the crucial issue. Kuhn shook his head. "Whenever you get two people interpreting the same data in different ways," he said, "that's metaphysics."

So, were his own ideas true or not? "Look," Kuhn responded with even more weariness than usual; obviously he had heard this question many times before. "I think this way of talking and thinking that I am engaged in opens up a range of possibilities that can be investigated. But it, like any scientific construct, has to be evaluated simply for its utility—for what you can do with it."

But then Kuhn, having set forth his bleak view of the limits of science and indeed of all human discourse, proceeded to complain about the many ways in which his book had been misinterpreted and misused, especially by admirers. "I've often said I'm much fonder of my critics than my fans." He recalled students approaching him to say, "Oh, thank you, Mr. Kuhn, for telling us about paradigms. Now that we know about them we can get rid of them." He insisted that he did not believe that science was *entirely* political, a reflection of the prevailing power structure. "In retrospect, I begin to see why this book fed into that, but boy, was it not meant to, and boy, does it not mean to."

His protests were to no avail. He had a painful memory of sitting in on a seminar and trying to explain that the concepts of truth and falsity are perfectly valid, and even necessary—within a paradigm. "The professor finally looked at me and said, 'Look, you don't know how radical this book is.'" Kuhn was also upset to find that he had become the patron saint of all would-be scientific revolutionaries. "I get a lot of letters saying, 'I've just read your book, and it's transformed my life. I'm trying to start a revolution. Please help me,' and accompanied by a book-length manuscript."

Kuhn declared that, although his book was not intended to be pro-science, he *is* pro-science. It is the rigidity and discipline of science, Kuhn said, that makes it so effective at problem solving. Moreover, science produces "the greatest and most original bursts of creativity" of any human enterprise. Kuhn conceded that he was partly to blame for some of the antiscience interpretations of his model. After all, in *Structure* he did call scientists committed to a paradigm "addicts"; he also compared them to the brainwashed characters in Orwell's *1984*.[17] Kuhn insisted that he did not mean to be condescending by using such terms as *mopping up* or *puzzle solving* to describe what most scientists do. "It was meant to be descriptive." He ruminated a bit. "Maybe I should have said more about the glories that result from puzzle solving, but I thought I was doing that."

As for the word *paradigm*, Kuhn conceded that it had become "hopelessly overused" and was "out of control." Like a virus, the word spread beyond the history and philosophy of science and infected the intellectual community at large, where it came to signify virtually any dominant idea. A 1974 *New Yorker* cartoon captured the phenomenon. "Dynamite, Mr. Gerston!" gushed a woman to a smug-looking man. "You're the first person I ever heard use 'paradigm' in real life." The low point came during the Bush administration, when White House officials introduced an economic plan called "the New Paradigm" (which was really just warmed-over Reaganomics).[18]

Kuhn admitted, again, that the fault was partly his, since in *Structure* he had not defined paradigm as crisply as he might have. At one point paradigm referred to an archetypal experiment, such as Galileo's legendary (and probably apocryphal) dropping of weights from the Leaning Tower of Pisa. Elsewhere the term referred to "the entire constellation of beliefs" that binds a scientific community together. (Kuhn denied, however, that he had defined paradigm in 21 different ways, as one critic claimed.[19]) In a postscript to later editions of *Structure*, Kuhn recommended that *paradigm* be replaced with the term *exemplar*, but it never caught on. He eventually gave up all hope of explaining what he really meant. "If you've got a bear by the tail, there comes a point at which you've got to let it go and stand back," he sighed.

One of the sources of *Structure's* power and persistence is its profound ambiguity; it appeals to relativists and to science worshipers alike. Kuhn acknowledged that "a lot of the success of the book and some of the criticisms are due to its vagueness." (One wonders whether Kuhn's writing style is intended or innate; his speech is as profoundly tangled, as suffused with subjunctives and qualifiers, as is his prose.) *Structure* is clearly a work of literature, and as such it is subject to many interpretations. According to literary theory, Kuhn himself cannot be trusted to provide a definitive account of his own work. Here is one possible interpretation of Kuhn's text, and of Kuhn. Kuhn focused on what science is rather than on what it should be; he had a much more realistic, hard-nosed, psychologically accurate view of science than did Popper. Kuhn realized that, given the power of modern science and the tendency of scientists to believe in theories that have withstood many tests, science may well enter a phase of permanent normalcy, in which no further revolutions, or revelations, are possible.

Kuhn also accepted, as Popper could not, that science might not continue forever, even in a normal state. "There was a beginning to it," Kuhn said. "There are lots of societies that don't have it. It takes very special conditions to support it. Those social conditions are now getting harder to find. Of *course* it could end." Science might even end, Kuhn said, because scientists cannot make any further headway, even given adequate resources.

Kuhn's recognition that science might cease—leaving us with what Charles Sanders Peirce defined as the truth about nature—made it even more imperative for Kuhn than for Popper to challenge science's authority, to deny that science could *ever* arrive at absolute truth. "The one thing I think you shouldn't say is that now we've found out what the world is really like," Kuhn said. "Because that's not what I think the game is about."

Kuhn has tried, throughout his career, to remain true to that original epiphany he experienced in his dormitory at Harvard. During that moment Kuhn saw—he knew!—that reality is ultimately unknowable; any attempt to describe it obscures as much as it illuminates. But Kuhn's insight forced him to take the untenable position that because all scientific theories fall short of absolute, mystical truth, they are all equally untrue; because we cannot discover *The Answer*, we cannot find any answers. His mysticism led him toward a position as absurd as that of the literary sophists who argue that all texts—from *The Tempest* to an ad for a new brand of vodka—are equally meaningless, or meaningful.

At the end of *Structure*, Kuhn briefly raised the question of why some fields of science converge on a paradigm while others, artlike, remain in a state of constant flux. The answer, he implied, was a matter of choice; scientists within certain fields were simply unwilling to commit themselves to a single paradigm. I suspect Kuhn avoided pursuing this issue because he could not abide the answer. Some fields, such as economics and other social sciences, never adhere for long to a single paradigm because they address questions for which no paradigm will suffice. Fields that achieve consensus, or normalcy, to borrow Kuhn's term, do so because their paradigms correspond to something real in nature, something true.

Finding Feyerabend

To say that the ideas of Popper and Kuhn are flawed is not to say that they cannot serve as useful tools for analyzing science. Kuhn's normal-science model accurately describes what most scientists now do: fill in details, solve relatively trivial puzzles that buttress rather than challenge the prevailing paradigm. Popper's falsification criterion can help to distinguish between empirical science and ironic science. But each philosopher, by pushing his ideas too far, by taking them too seriously, ends up in an absurd, self-contradicting position.

How does a skeptic avoid becoming Karl Popper, pounding the table and shouting that he is *not* dogmatic? Or Thomas Kuhn, trying to communicate precisely what he means when he talks about the impossibility of true communication? There is only one way. One must embrace—even revel in—paradox, contradiction, rhetorical excess. One must acknowledge that skepticism is a necessary but impossible exercise. One must become Paul Feyerabend.

Feyerabend's first and still most influential book, *Against Method*, was published in 1975 and has been translated into 16 languages.[20] It argues that philosophy cannot provide a methodology or rationale for science, since there is no rationale to explain. By analyzing such scientific milestones as Galileo's trial before the Vatican and the development of quantum mechanics, Feyerabend sought to show that there is no logic to science; scientists create and adhere to scientific theories for what are ultimately subjective and even irrational reasons. According to Feyerabend, scientists can and must do whatever is necessary to advance. He summed up his anticredo with the phrase "anything goes." Feyerabend once derided Popper's critical rationalism as "a tiny puff of hot air in the positivistic teacup."[21] He agreed with Kuhn on many points, in particular on the incommensurability of scientific theories, but he argued that science is rarely as normal as Kuhn contended. Feyerabend also accused Kuhn—quite rightly—of avoiding the implications of his own view; he remarked, to Kuhn's dismay, that Kuhn's sociopolitical model of scientific change applied nicely to organized crime.[22]

Feyerabend's penchant for posturing made it all too easy to reduce him to a grab bag of outrageous sound bites. He once likened science to voodoo, witchcraft, and astrology. He defended the right of religious fundamentalists to have their version of creation taught alongside Darwin's theory of evolution in public schools.[23] His entry in the 1991 *Who's Who in America* ended with the following remark: "My life has been the result of accidents, not of goals and principles. My intellectual work forms only an insignificant part of it. Love and personal understanding are much more important. Leading intellectuals with their zeal for objectivity kill these personal elements. They are criminals, not the liberators of mankind."

Feyerabend's Dadaesque rhetoric concealed a deadly serious point: the human compulsion to find absolute truths, however noble, too often culminates in tyranny. Feyerabend attacked science not because he truly believed that it had no more claim to truth than did astrology. Quite the contrary. Feyerabend attacked science because he recognized—and was horrified by—its power, its potential to stamp out the diversity of human thought and culture. He objected to scientific certainty for moral and political, rather than for epistemological, reasons.

At the end of his 1987 book, *Farewell to Reason*, Feyerabend revealed just how deep his relativism ran. He addressed an issue that "has enraged many readers and disappointed many friends—my refusal to condemn even an extreme fascism and my suggestion that it should be allowed to thrive."[24] The point was particularly touchy because Feyerabend had served in the German army during World War II. It would be all too easy, Feyerabend argued, to condemn Nazism, but it was that very moral self-righteousness and certitude that made Nazism possible.

By the time I tried to track Feyerabend down in 1992, he had retired from the University of California at Berkeley. No one there knew where he was; colleagues assured me that my efforts to find him would be in vain. At Berkeley, he had had a telephone that allowed him to make calls but not receive them. He would accept invitations to conferences and then fail to show up. By mail, he would invite colleagues to visit him. But when they arrived and knocked on the door of his house in the hills overlooking Berkeley, no one would answer.

Later, while skimming *Isis*, a journal of the history and philosophy of science (titled this way long before the acronym for a terrorist group), I found a short review by Feyerabend of a book of essays. The review displayed Feyerabend's talent for one-liners. In response to a denigrating remark the author had made about religion, Feyerabend retorted, "Prayer may not be very efficient when compared to celestial mechanics, but it surely holds its own vis-à-vis some parts of economics."[26]

I called the editor of *Isis* to ask if he knew how I could contact Feyerabend, and he gave me an address near Zurich, Switzerland. I mailed Feyerabend a fawning letter explaining that I wanted to interview him. To my delight, he responded with a chatty, handwritten note saying an interview would be fine. He divided his time between his home in Switzerland and his wife's place in Rome. He enclosed a telephone number for Rome and a photograph of himself wearing an apron and a big grin and standing before a sink full of dishes. The photograph, he explained, "shows me at my favorite activity, washing dishes for my wife in Rome." In mid-October I received another letter from Feyerabend. "This is to tell you that I should be (93%) in New York during the week from October 25 to November 1 and that we might make an interview then. I'll give you a call as soon as I arrive."

So it happened that one chilly night just a few days before Halloween I met Feyerabend at a luxurious Fifth Avenue apartment. The apartment belonged to a former student who had wisely abandoned philosophy for real estate—apparently with some success. She greeted me and led me into her kitchen, where Feyerabend was sitting at a table sipping a glass of red wine. He thrust himself up from a chair and stood crookedly to greet me, as if he suffered from a stiff back; only then did I remember that Feyerabend had been shot in the back and permanently crippled during World War II.

Feyerabend had the energy and angular face of a leprechaun. When we sat down and began talking, he declaimed, sneered, wheedled, and whispered—depending on his point or plot—while whirling his hands like a conductor. Self-deprecation spiced his hubris. He called himself "lazy" and "a bigmouth." When I asked about his position on a certain point, he winced. "I have no position!" he said. "If you have a position, it is always something screwed down." He twisted an invisible screwdriver into the table. "I have opinions that I defend rather vigorously, and then I find out how silly they are, and I give them up!"

Watching this performance with an indulgent smile was Feyerabend's wife, Grazia Borrini, an Italian physicist whose manner was as calm as Feyerabend's was manic. Borrini had taken Feyerabend's class while pursuing a second degree in public health at Berkeley in 1983; they married six years later. Borrini entered the conversation sporadically, for example, after I asked why Feyerabend thought scientists were so infuriated by his writings.

"I have no idea," he said, the very picture of innocence. "Are they?"

Borrini interjected that *she* had been infuriated when she first heard about Feyerabend's ideas from another physicist. "Someone was taking away from me the keys of the universe," she explained. It was only when she read his books herself that she realized Feyerabend's views were much more subtle and astute than his critics claimed. "This is what I think you should want to

51

write about," Borrini said to me, "the great misunderstanding."

"Oh, forget it, he's not my press agent," Feyerabend said.

Like Popper, Feyerabend had been born and raised in Vienna. He studied acting and opera as a teenager. At the same time, he became intrigued with science after attending lectures by an astronomer. Far from seeing his two passions as irreconcilable, Feyerabend envisioned himself becoming both an opera singer and an astronomer. "I would spend my afternoons practicing singing, and my evenings on the stage, and then late at night I would observe the stars," he said.

Then came the war. Germany occupied Austria in 1938, and in 1942 the 18-year-old Feyerabend enlisted in an officers' school. Although he hoped his training period would outlast the war, he ended up in charge of 3,000 men on the Russian front. While fighting against (actually fleeing from) the Russians in 1945, he was shot in the lower back. "I couldn't get up," Feyerabend recalled, "and I still remember this vision: 'Ah, I shall be in a wheelchair rolling up and down between rows of books.' I was very happy."

He gradually recovered the ability to walk, although only with the help of a cane. Resuming his studies at the University of Vienna after the war, he switched from physics to history, grew bored, returned to physics, grew bored again, and finally settled on philosophy. His talent for advancing absurd positions through sheer cleverness led to a growing suspicion that rhetoric rather than truth is crucial for carrying an argument. "Truth itself is a rhetorical term," Feyerabend asserted. Jutting out his chin he intoned, " 'I am searching for the truth.' Oh boy, what a great person."

Feyerabend studied under Popper at the London School of Economics in 1952 and 1953. There he met Lakatos, another brilliant student of Popper. It was Lakatos who, years later, urged Feyerabend to write *Against Method*. "He was my best friend," Feyerabend said of Lakatos. Feyerabend taught at the University of Bristol until 1959 and then moved to Berkeley, where he befriended Kuhn.

Like Kuhn, Feyerabend denied that he was antiscience. What he *did* claim, first, was that there is no scientific method. "That is exactly how it works in the sciences," Feyerabend said. "You have certain ideas that work, and then some new situation turns up and you try something else. It's opportunism. You need a toolbox full of different kinds of tools. Not only a hammer and pins and nothing else." This is what he meant by his much-maligned phrase "anything goes" (and not, as is commonly thought, that one scientific theory is as good as any other). Restricting science to a particular methodology—even one as loosely defined as Popper's falsification scheme or Kuhn's normal science mode—would destroy it, Feyerabend said.

Feyerabend also objected to the claim that science is superior to other modes of knowledge. He was particularly enraged at the tendency of Western states to foist the products of science—whether the theory of evolution, nuclear power plants, or gigantic particle accelerators—on people against their will. "There is separation between state and church," he complained, "but none between state and science!"

Science "provides fascinating stories about the universe, about the ingredients, about the development, about how life came about, and all this stuff," Feyerabend said. But the prescientific "mythmakers," he emphasized, such as singers, court jesters, and bards, earned their living, whereas most modern scientists are supported by taxpayers. "The public is the patron and should have a say in the matter."

Feyerabend added, "Of course I go to extremes, but not to the extremes people accuse me of, namely, throw out science. Throw out the idea science is *first. That's* all right. It has to be science from case to case." After all, scientists disagree among themselves on many issues. "People should not take it for granted when a scientist says, 'Everybody has to follow this way.'"

If he was not antiscience, I asked, what did he mean by his statement in *Who's Who* that intellectuals are criminals? "I thought so for a long time," Feyerabend responded, "but last year I crossed it out, because there are lots of good intellectuals." He turned to his wife. "I mean, you are an intellectual." "No, I am a physicist," she replied firmly. Feyerabend shrugged. "What does it mean, 'intellectual'? It means people who think about things longer than other people, perhaps. But many of them just ran over other people, saying,

'We have figured it out.'"

Feyerabend noted that many nonindustrialized people had done fine without science. The !Kung bushmen in Africa "survive in surroundings where any Western person would come in and die after a few days," he said. "Now you might say people in this society live much longer, but the question is, what is the quality of life, and that has not been decided."

But didn't Feyerabend realize how annoyed most scientists would be by that kind of statement? Even if the bushmen are happy, they are ignorant, and isn't knowledge better than ignorance? "What's so great about knowledge?" Feyerabend replied. "They are good to each other. They don't beat each other down." People have a perfect right to reject science if they so choose, Feyerabend said.

Did that mean fundamentalist Christians also had the right to have creationism taught alongside the theory of evolution in schools? "I think that 'right' business is a tricky business," Feyerabend responded, "because once somebody has a right they can hit somebody else over the head with that right." He paused. Ideally, he said, children should be exposed to as many different modes of thought as possible so they can choose freely among them. He shifted uneasily in his seat. Sensing an opening, I pointed out that he had not really answered my question about creationism. Feyerabend scowled. "This is a dried-out business. It doesn't interest me very much. Fundamentalism is not the old rich Christian tradition." But American fundamentalists are very powerful, I persisted, and they use the kinds of things Feyerabend says to attack the theory of evolution. "But science has been used to say some people have a low intelligence quotient," he retorted. "So everything is used in many different ways. Science can be used to beat down all sorts of other people."

But shouldn't educators point out that scientific theories are different from religious myths? I asked. "Of course. I would say that science is very popular nowadays," he replied. "But then I have also to let the other side get in as much evidence as possible, because the other side is always given a short presentation." Anyway, so-called primitive people often know far more about their environments, such as the properties of local plants, than do so-called experts. "So to say these people are ignorant is just—*this* is ignorance!"

I unloaded my self-refuting question: Wasn't there something contradictory about the way he used all the techniques of Western rationalism to attack Western rationalism? Feyerabend refused to take the bait. "Well, they are just tools, and tools can be used in any way you see fit," he said mildly. "They can't blame me that I use them." Feyerabend seemed bored, distracted. Although he would not admit it, 1 suspected he was tired of being a radical relativist, of defending the colorful belief systems of the world— astrology, creationism, even fascism!—against the bully of rationalism.

Feyerabend's eyes glittered again, however, when he began talking about a book he was working on. Tentatively titled *The Conquest of Abundance*, it addressed the human passion for reductionism. "All human enterprises," Feyerabend explained, seek to reduce the natural diversity, or "abundance," inherent in reality. "First of all the perceptual system cuts down this abundance or you couldn't survive." Religion, science, politics, and philosophy represent our attempts to compress reality still further. Of course, these attempts to conquer abundance simply create new abundances, new complexities. "Lots of people have been killed, in political wars. I mean, certain opinions are not liked." Feyerabend, I realized, was talking about our quest for *The Answer*, the theory to end all theories.

But *The Answer* will—must—forever remain beyond our grasp, according to Feyerabend. He ridiculed the belief of some scientists that they might someday capture reality in a single theory of everything. "Let them have their belief, if it gives them joy. Let them also give talks about that. 'We touch the infinite!' And some people say"— bored voice—"'Ya ya, he says he touches the infinite.' And some people say"—thrilled voice—"'Ya ya! He says he touches the infinite!' But to tell the little children in school, 'Now that is what the truth is,' that is going much too far."

Any description of reality is necessarily inadequate, Feyerabend said. "You think that this one-day fly, this little bit of nothing, a human being—according to today's cosmology!—can figure it all out? This to me seems so crazy! It cannot possibly be true! What they figured out is one particular response to their actions, and this response gives this universe, and the reality that is behind this is laughing! 'Ha ha! They think they have found me out!'"

A medieval philosopher named Dionysius the Pseudo-Areopagite, Feyerabend said, had argued that to see God directly is to see nothing at all. "This to me makes a lot of sense. I can't explain why. This big thing, out of which everything comes, you don't have the means. Your language has been created by dealing with things, chairs, and a few instruments. And just on this tiny earth!" Feyerabend paused, lost in a kind of exaltation. "God is emanations, you know? And they come down and become more and more material. And down, down at the last emanation, you can see a little trace of it and guess at it."

Surprised by this outburst, I asked Feyerabend if he was religious. "I'm not sure," he replied. He had been raised as a Roman Catholic, and then he became a "vigorous" atheist. "And now my philosophy has taken a completely different shape. It can't just be that the universe— Boom!—you know, and develops. It just doesn't make any sense." Of course, many scientists and philosophers have argued that it is pointless to speculate about the sense, or meaning, or purpose of the universe. "But people ask it, so why not? So all this will be stuffed into this book, and the question of abundance will come out of it, and it will take me a long time."

As I prepared to leave, Feyerabend asked how my wife's birthday party had gone the previous night. (I had told Feyerabend about my wife's birthday in the course of arranging my meeting with him.) Fine, I replied. "You're not drifting apart?" Feyerabend persisted, scrutinizing me. "It wasn't the last birthday you will ever celebrate with her?"

Borrini glared at him, aghast. "Why should it be?"

"I don't know!" Feyerabend exclaimed, throwing his hands up. "Because it happens!" He turned back to me. "How long have you been married?" Three years, I said. "Ah, just the beginning. The bad things will come. Just wait 10 years." Now you really sound like a philosopher, I said. Feyerabend laughed.

He confessed that he had been married and divorced three times before he met Borrini. "Now for the first time I am so happy to be married."

I said that I had heard his marriage to Borrini had made him more easygoing. "Well, this may be two things," Feyerabend replied. "Getting older you don't have the energy not to be easygoing. And she's certainly made a big difference also." He beamed at Borrini, and she beamed back.

Turning to Borrini, I mentioned the photograph that her husband had sent of himself washing dishes, along with the note saying that performing this chore for his wife was the most important thing he did now.

Borrini snorted. "Once in a blue moon," she said.

"What do you mean, once in a blue moon!" Feyerabend bellowed. "Every day I wash dishes!"

"Once in a blue moon," Borrini repeated firmly. I decided to believe the physicist rather than the philosopher.

A little more than a year after my meeting with Feyerabend, the *New York Times* reported, to my dismay, that the "anti-science philosopher" had been killed by a brain tumor.[27] I called Borrini in Zurich to offer my condolences— and, yes, to satisfy my craven journalistic curiosity. She was distraught. It had happened so quickly. Paul had complained of headaches, and then a few months later.... Composing herself, she told me, proudly, that Feyerabend had kept working until the end. Just before he died, he finished a draft of his autobiography. (The book, with the typically Feyerabendian title *Killing Time*, was published in 1995. In the final pages, which Feyerabend wrote in his final days, he concluded that love is all that matters in life.)[28] What about the book on abundance? I asked. No, Paul did not have time to finish that, Borrini murmured.

Recalling Feyerabend's excoriation of the medical profession, I could not resist asking, did her husband seek medical treatment for his tumor? Of course, she replied. He had had "total confidence" in his doctors' diagnosis and had been willing to accept any treatment they recommended; the tumor had simply been detected too late for anything to be done.

Why Philosophy Is So Hard

Theocharis and Psimopoulos, the authors of the essay in *Nature* titled "Where Science Has Gone Wrong," were right after all: the ideas of Popper, Kuhn, and Feyerabend are "self-refuting." All skeptics, finally, fall on their own swords. They become what the critic Harold Bloom derided in *The Anxiety of Influence* as "mere rebels." Their most potent argument against scientific truth is historical: given the rapid turnover of scientific theories over the past century or so, how can we be sure that any current theory will endure? Actually, modern science has been much less revolutionary—and more conservative—than the skeptics, and Kuhn in particular, have suggested. Particle physics rests on the firm foundation of quantum mechanics, and modern genetics bolsters rather than undermines Darwin's theory of evolution. The skeptics' historical arguments are much more devastating when turned against philosophy. If science cannot achieve absolute truth, then what standing should be accorded philosophy, which has exhibited much less ability to resolve its problems? Philosophers themselves have recognized their plight. In *After Philosophy: End or Transformation?*, published in 1987, fourteen prominent philosophers considered whether their discipline had a future. The consensus was philosophical: maybe, maybe not.[29]

One philosopher who has pondered the "chronic lack of progress" of his calling is Colin McGinn, a native of England who has taught at Rutgers University since 1992. McGinn, when I met him in his apartment on Manhattan's upper West Side in August 1994, was disconcertingly youthful. (Of course, I expect all philosophers to have furrowed brows and hairy ears.) He wore jeans, a white T-shirt, and moccasins. He is a compact man, with a defiantly jutting chin and pale blue eyes; he could pass for Anthony Hopkins's younger brother.

When I solicited McGinn's opinion of Popper, Kuhn, and Feyerabend, his mouth curled in distaste. They were "sloppy," "irresponsible"; Kuhn especially was full of "absurd subjectivism and relativism." Few modern philosophers took his views seriously any more. "I don't think that science is provisional in an interesting sense at all," McGinn asserted. *"Some* of it's provisional, but some of it isn't!" Is the periodic table provisional? Or Darwin's theory of natural selection?

Philosophy, on the other hand, does not achieve this kind of resolution, McGinn said. It does not advance in the sense that "you have this problem and you work on it and you solve it, and then you go on to the next problem." Certain philosophical problems have been "clarified"; certain approaches have fallen out of fashion. But the great philosophical questions—What is truth? Does free will exist? How can we know anything?—are as unresolved today as they ever were. That fact should not be surprising, McGinn remarked, since modern philosophy can be defined as the effort to solve problems lying beyond the scope of empirical, scientific inquiry.

McGinn pointed out that many philosophers in this century—notably Ludwig Wittgenstein and the logical positivists—have simply declared that philosophical problems are pseudo-problems, illusions stemming from language or "diseases of thought." Some of these "eliminativists," in order to solve the mind-body problem, have even denied that consciousness exists. That view "can have political consequences that you might not want to accept," McGinn said. "It ends up reducing human beings to nothing. It pushes you toward extreme materialism, toward behaviorism."

McGinn offered a different, and, he suggested, more palatable explanation: the great problems of philosophy are real, but they are beyond our cognitive ability. We can pose them, but we cannot solve them—any more than a rat can solve a differential equation. McGinn said this idea came to him in a late-night epiphany when he still lived in England; only later did he realize that he had encountered a similar idea in the writings of the linguist Noam Chomsky (whose views will be aired in Chapter 6). In his 1993 book, *Problems in Philosophy*, McGinn suggested that perhaps in a million years philosophers would acknowledge that his prediction was correct.[30] Of course, he told me, philosophers probably would cease struggling to achieve the impossible much sooner.

McGinn suspected that science, too, was approaching a cul-de-sac.

"People have great confidence in science and the scientific method," he said, "and it's worked well within its own limits for a few hundred years. But from a larger perspective who's to say that it's going to carry on and conquer everything?" Scientists, like philosophers, are constrained by their cognitive limits. "It's hubris to think we've somehow now got the perfect cognitive instrument in our heads," he said. Moreover, the end of the cold war has removed a major motivation for investment in science, and as the sense of completion in science grows, fewer bright young people will be attracted to scientific careers.

"So it wouldn't surprise me if sometime during the next century people started veering away from studying science as much, except just to learn what they need to know about things, and started to go back into the humanities." In the future, we will look back on science as "a phase, a brilliant phase. People do forget that just 1,000 years ago there was just religious doctrine; that was it." After science has ended, "religion may start to appeal to people again." McGinn, who is a professed atheist, looked rather pleased with himself, and well he might. During our little chat in his airy apartment, with car honks and bus growls and the odor of greasy Chinese food drifting through the window, he had pronounced the impending doom of not just one but two major modes of human knowledge: philosophy and science.

Fearing the Zahir

Of course, philosophy will never really end. It will simply continue in a more overtly ironic, literary mode, like that already practiced by Nietzsche, or Wittgenstein, or Feyerabend. One of my favorite literary philosophers is the Argentinian fabulist Jorge Luis Borges. More than any philosopher I know of, Borges has explored the complex psychological relationship that we have toward the truth. In "The Zahir," Borges told the story of a man who becomes obsessed with a coin he received as change from a storekeeper.[31] The seemingly nondescript coin is a Zahir, an object that is an emblem of all things, of the mystery of existence. A Zahir can be a compass, a tiger, a stone, anything. Once beheld, it cannot be forgotten. It grips the mind of the beholder until all other aspects of reality become insignificant, trivial.

At first, the narrator struggles to free his mind of the Zahir, but he eventually accepts his fate. "I shall pass from thousands of apparitions to one alone: from a very complex dream to a very simple dream. Others will dream that I am mad, and I shall dream of the Zahir. And when everyone dreams of the Zahir day and night, which will be a dream and which a reality, the earth or the Zahir?"[32] The Zahir, of course, is *The Answer*, the secret of life, the theory to end all theories. Popper, Kuhn, and Feyerabend tried to protect us from *The Answer* with doubt and reason, Borges with terror.

Chapter Three
The End of Physics

There are no more dedicated, not to say obsessive, seekers of *The Answer* than modern particle physicists. They want to show that all the complicated things of the world are really just manifestations of one thing. An essence. A force. A loop of energy wriggling in a 10-dimensional hyperspace. A sociobiologist might suspect that a genetic influence lurks behind this reductionist impulse, since it seems to have motivated thinkers since the dawn of civilization. God, after all, was conceived by the same impulse.

Einstein was the first great modern *Answer* seeker. He spent his later years trying to find a theory that would unify quantum mechanics with his theory of gravity, general relativity. To him, the purpose of finding such a theory was to determine whether the universe was inevitable or, as he put it, "whether God had any choice in creating the world." But Einstein, no doubt believing that science made life meaningful, also suggested that no theory could be truly final. He once said of his own theory of relativity, "[It] will have to yield to another one, for reasons which at present we do not yet surmise. I believe that the process of deepening the theory has no limits."[1]

Most of Einstein's contemporaries saw his efforts to unify physics as a product of his dotage and quasi-religious tendencies. But in the 1970s, the dream of unification was revived by several advances. First, physicists showed that just as electricity and magnetism are aspects of a single force, so electromagnetism and the weak nuclear force (which governs certain kinds of nuclear decay) are manifestations of an underlying "electroweak" force.

Researchers also developed a theory for the strong nuclear force, which grips protons and neutrons together in the nuclei of atoms. The theory, called quantum chromodynamics, posits that protons and neutrons are composed of even more elementary particles, called quarks. Together, the electroweak theory and quantum chromodynamics constitute the standard model of particle physics.

Emboldened by this success, workers forged far beyond the standard model in search of a deeper theory. Their guide was a mathematical property called symmetry, which allows the elements of a system to undergo transformations—analogous to rotation or reflection in a mirror—without being fundamentally altered. Symmetry became the *sine qua non* of particle physics. In search of theories with deeper symmetries, theorists began to jump to higher dimensions. Just as an astronaut rising above the two-dimensional plane of the earth can more directly apprehend its global symmetry, so can theorists discern the more subtle symmetries underlying particle interactions by viewing them from a higher-dimensional standpoint.

One of the most persistent problems in particle physics stems from the definition of particles as points. In the same way that division by zero yields an infinite and hence meaningless result, so do calculations involving pointlike particles often culminate in nonsense. In constructing the standard model, physicists were able to sweep these problems under the rug. But Einsteinian gravity, with its distortions of space and time, seemed to demand an even more radical approach.

In the early 1980s, many physicists came to believe that superstring theory represented that approach. The theory replaced pointlike particles with minute loops that eliminated the absurdities arising in calculations. Just as vibrations of violin strings give rise to different notes, so could the vibrations of these strings generate all the forces and particles of the physical realm. Superstrings could also banish one of the bugbears of particle physics: the possibility that there is no ultimate foundation for physical reality but only an endless succession of smaller and smaller particles, nestled inside each other like Russian dolls. According to superstring theory, there is a fundamental scale beyond which all questions concerning space and time become meaningless.

The theory suffers from several problems, however. First, there seem to be countless possible versions, and theorists have no way of knowing which one is correct. Moreover, superstrings are thought to inhabit not only the four dimensions in which we live (the three dimensions of space plus time), but also six extra dimensions that are somehow "compactified," or rolled up into infinitesimal balls, in our universe. Finally, the strings are as small in comparison to a proton as a proton is in comparison to the solar system. They are more distant from us, in a sense, than are the quasars that lurk at the farthest edge of the visible universe. The superconducting supercollider, which was to take physicists much deeper into the microrealm than had any previous accelerator, would have been 54 miles in circumference. To probe the realm superstrings are thought to inhabit, physicists would have to build a particle accelerator 1,000 light-years around. (The entire solar system is only one light-*day* around.) And not even an accelerator that size could allow us to see the extra dimensions where superstrings dance.

Glashow's Gloom

One of the pleasures of being a science writer is feeling superior to run-of-the-mill newshounds. The most primitive species of reporter, to my mind, is the type who tracks down a woman who has watched her only son stabbed to death by a crazed crackhead and asks, "How do you feel?" Yet in the fall of 1993 I found myself stuck with a similar assignment. I was just beginning an article on the future of particle physics when the U.S. Congress killed, once and for all, the superconducting supercollider. (Contractors had already spent more than $2 billion and dug a tunnel in Texas 15 miles long.) Over the next few weeks I had to confront particle physicists who had just seen their brightest hope for the future brutally aborted and ask, "How do you feel?"

The gloomiest place I visited was the department of physics at Harvard

University. The head of the department was Sheldon Glashow, who had shared a Nobel Prize with Steven Weinberg and Abdus Salam for developing the electroweak portion of the standard model. In 1989, Glashow had spoken, along with the biologist Gunther Stent, at the symposium titled "The End of Science?" at Gustavus Adolphus College. Glashow offered a spirited rebuttal of the meeting's "absurd" premise, that philosophical skepticism was eroding belief in science as a "unified, universal, objective endeavor." Does anyone really doubt the existence of the moons of Jupiter, which Galileo discovered centuries ago? Does anyone really doubt the modern theory of disease? "Germs are seen and killed," Glashow declared, "not imagined and unimagined."

Science is "certainly slowing down," Glashow conceded, but not because of attacks from ignorant, antiscientific sophists. His own field of particle physics "is threatened from an entirely different direction: from its very success." The last decade of research has generated countless confirmations of the standard model of particle physics "but has revealed not the slightest flaw, not the tiniest discrepancy.... We have no experimental hint or clue that could guide us to build a more ambitious theory." Glashow added the obligatory hopeful coda: "Nature's road has often seemed impassable, but we have always overcome."[2]

In other statements, Glashow has not hewed to this blithe optimism. He was once a leader in the quest for a unified theory. In the 1970s, he proposed several such theories, although none as ambitious as superstring theory. But with the advent of superstrings, Glashow became disillusioned with the quest for unification. Those working on superstrings and other unified theories were not doing physics at all any more, Glashow contended, because their speculations were so far beyond any possible empirical test. Glashow and a colleague complained in one essay that "contemplation of superstrings may evolve into an activity as remote from conventional particle physics as particle physics is from chemistry, to be conducted at schools of divinity by future equivalents of medieval theologians." They added that "for the first time since the Dark Ages, we can see how our noble search may end, with faith replacing science once again."[3] When particle physics passes beyond the realm of the empirical, Glashow seemed to be suggesting, it may succumb to skepticism, to relativism, after all.

I interviewed Glashow at Harvard in November 1993, shortly after the demise of the superconducting supercollider. His dimly lit office, lined with dark, heavily varnished bookcases and cabinets, was as solemn as a funeral parlor. Glashow himself, a large man chewing restlessly on a cold cigar stub, seemed slightly incongruous there. He had the snowy, tousled hair that seems de rigueur for Nobel laureates in physics, and his glasses were as thick as telescope lenses. Yet one could still detect, beneath the patina of the Harvard professor, the tough, fast-talking New York kid that Glashow had once been.

Glashow was devastated by the death of the supercollider. Physics, he emphasized, cannot proceed on pure thought alone, in spite of what superstring enthusiasts said. Superstring theory "hasn't gotten anywhere despite all the hoopla," he grumbled. More than a century ago some physicists tried to invent unified theories; they failed, of course, because they knew nothing about electrons or protons or neutrons or quantum mechanics. "Now, are we so arrogant as to believe we have all the experimental information we need right now to construct that holy grail of theoretical physics, a unified theory? I think not. I think certainly there are surprises that natural phenomena have in store for us, and we're not going to find them unless we look."

But isn't there much to do in physics besides unification? "Of course there is," Glashow replied sharply. Astrophysics, condensed-matter physics, and even subfields within particle physics are not concerned with unification. "Physics is a very large house filled with interesting puzzles," he said (using Thomas Kuhn's term for problems whose solutions merely reinforce the prevailing paradigm). "Of *course* there will be things done. The question is whether we're getting anywhere toward this holy grail." Glashow believed that physicists would continue to seek "some little interesting tidbit someplace. Something amusing, something new. But it's not the same as the quest as I was fortunate enough to know it in my professional lifetime."

Glashow could not muster much optimism concerning the prospects for his field, given the politics of science funding. He had to admit that particle physics was not terribly useful. "Nobody can make the claim that this kind of research is going to produce a practical device. That would just be a lie. And given the attitude of governments today, the type of research that I fancy doesn't have a very good future."

In that case, could the standard model be the final theory of particle physics? Glashow shook his head. "Too many questions left unanswered," he said. Of course, he added, the standard model would be final in a practical sense if physicists could not forge beyond it with more powerful accelerators. "There will be the standard theory, and that will be the last chapter in the elementary physics story." It is always possible that someone will find a way of generating extremely high energies relatively cheaply. "Maybe someday it will get done. Someday, someday, someday."

The question is, Glashow continued, what will particle physicists do while they are waiting for that someday to come? "I guess the answer is going to be that the [particle-physics] establishment is going to do boring things, futzing around until something becomes available. But they would never admit that it's boring. Nobody will say, 'I do boring things.'" Of course, as the field becomes less interesting and funding dwindles, it will cease attracting new talent. Glashow noted that several promising graduate students had just left Harvard for Wall Street. "Goldman Sachs in particular discovered that theoretical physicists are very good people to have."

The Smartest Physicist of Them All

One reason that superstring theory became so popular in the mid-1980s was that a physicist named Edward Witten decided it represented the best hope for the future of physics. I first saw Witten in the late 1980s when I was eating lunch with another scientist in the cafeteria of the Institute for Advanced Study in Princeton. A man walked by our table holding a tray of food. He had a lantern jaw and a strikingly high forehead, bounded by thick black glasses across the bottom and thick black hair across the top. "Who's that?" I asked my companion. "Oh, that's Ed Witten," my lunchmate replied. "He's a particle physicist."

A year or two later, making idle chitchat between sessions at a physics conference, I asked a number of attendees: Who is the smartest physicist of them all? Several names kept coming up, including the Nobel laureates Steven Weinberg and Murray Gell-Mann. But the name mentioned most often was Witten's. He seemed to evoke a special kind of awe, as though he belonged to a category unto himself. He is often likened to Einstein; one colleague reached even further back for a comparison, suggesting that Witten possessed the greatest mathematical mind since Newton.

Witten may also be the most spectacular practitioner of naive ironic science I have ever encountered. Naive ironic scientists possess an exceptionally strong faith in their scientific speculations, in spite of the fact that those speculations cannot be empirically verified. They believe they do not invent their theories so much as they discover them; these theories exist independently of any cultural or historical context and of any particular efforts to find them.

A naive ironic scientist—like a Texan who thinks everyone but Texans has an accent—does not acknowledge that he or she has adopted any philosophical stance at all (let alone one that might be described as ironic). Such a scientist is just a conduit through which truths pass from the Platonic realm to the world of flesh; background and personality are irrelevant to the scientific work. Thus Witten, when I called him to request an interview, tried to dissuade me from writing about him. He told me that he abhorred journalism focusing on scientists' personalities, and at any rate, many other physicists and mathematicians were much more interesting than he. Witten had been upset by a profile published in the *New York Times Magazine* in 1987, which implied that he had invented superstring theory.[4] Actually, Witten informed me, he had played virtually no role in the creation of superstring theory; he had simply helped to develop it and promote it after it had been discovered.

Every science writer occasionally runs into subjects who genuinely do not want attention from the media, who simply want to be left alone to do their work. What these scientists often fail to realize is that this trait makes them more enticing. Intrigued by Witten's apparently sincere shyness, I persisted in seeking an interview. Witten asked me to send him some things I had written. Stupidly, I included a profile of Thomas Kuhn published in *Scientific American*. Finally, Witten agreed to let me come down to talk to him, but he would give me two hours and not a minute more; I would have to leave at 12 noon, sharp. When I arrived, he immediately began lecturing me on my shoddy journalistic ethics. I had done society a disservice in repeating Thomas Kuhn's view that science is an arational (not irrational) process that does not converge on the truth. "You should be concentrating on serious and substantive contributions to the understanding of science," Witten said. Kuhn's philosophy "isn't taken very seriously except as a debating standard, even by its proponents." Did Kuhn go to a doctor when he was sick? Did he have radial tires on his car? I shrugged and guessed that he probably did. Witten nodded triumphantly. That proved, he declared, that Kuhn believed in science and not in his own relativistic philosophy.

I said that whether or not one agreed with Kuhn's views, they were influential, and provocative, and that one of my aims as a writer was not only to inform readers but also to provoke them. "Aim to report on some of the *truths* that are being discovered, rather than aiming to provoke. *That* should be the aim of a science writer," Witten said sternly. I tried to do both, I replied. "Well, that's a pretty feeble response," Witten said. "Provoking people, or stimulating them intellectually, should be a *by-product* of reporting on some of the truths that are being discovered." This is another mark of the naive ironic scientist: when he or she says "truth," there is never any ironic inflection, no smile; the word is implicitly capitalized. As penance for my journalistic sins, Witten suggested, I should write profiles of five mathematicians in a row; if I did not know which mathematicians were worthy of such attention, Witten would recommend some. (Witten did not realize that he was providing fodder for those who claimed he was less a physicist than a mathematician.)

Since high noon was approaching, I tried to turn the interview toward Witten's career. He refused to answer any "personal" questions, such as what his major had been in college or whether he had considered other careers before becoming a physicist; his history was unimportant. I knew from background reading that although Witten was the son of a physicist and had always enjoyed the subject, he had graduated from Brandeis College in 1971 with a degree in history and plans to become a political journalist. He succeeded in publishing articles in *The New Republic* and the *Nation*. He nonetheless soon decided that he lacked the "common sense" for journalism (or so he told one reporter); he entered Princeton to study physics and obtained his doctorate in 1976.

Witten picked up the story from there. In telling me about his work in physics, he slipped into a highly abstract, impersonal mode of speech. He was reciting, not talking, giving me the history of superstrings, emphasizing not his own role but that of others. He spoke so softly that I worried that my tape recorder wouldn't pick him up over the air conditioner. He paused frequently—for 51 seconds at one point—casting his eyes down and squeezing his lips together like a bashful teenager. He seemed to be striving for the same precision and abstraction in his speech that he achieved in his treatises on superstrings. Now and then—for no reason that I could discern—he broke into convulsive, hiccuping laughter as some private joke flitted through his consciousness.

Witten made a name for himself in the mid-1970s with incisive but fairly conventional papers on quantum chromodynamics and the electroweak force. He learned of superstring theory in 1975, but his initial efforts to understand it were stymied by the "opaque" literature. (Yes, even the smartest person in the world had a hard time understanding superstring theory.) In 1982, however, a review paper by John Schwarz, one of the theory's pioneers, helped Witten to grasp a crucial fact: Superstring theory does not simply allow for the possibility of gravity; it demands that gravity exist. Witten called this realization "the greatest intellectual thrill of my life." Within a few years, any doubts that Witten had had about the theory's potential vanished. "It

70

was clear that if I didn't spend my life concentrating on string theory, I would simply be missing my life's calling," he said. He began publicly proclaiming the theory a "miracle" and predicting that it would "dominate physics for the next 50 years." He also generated a flood of papers on the theory. The 96 articles that Witten produced from 1981 through 1990 were cited by other physicists 12,105 times; no other physicist in the world approached this level of influence.[5]

In his early papers, Witten concentrated on creating a superstring model that was a reasonable facsimile of the real world. But he became increasingly convinced that the best way to achieve that goal was to uncover the theory's "core geometric principles." These principles, he said, might be analogous to the non-Euclidean geometry that Einstein employed to construct his theory of general relativity. Witten's pursuit of these ideas led him deep into topology, which is the study of the fundamental geometric properties of objects, regardless of their particular shape or size. In the eyes of a topologist, a doughnut and a single-handled coffee mug are equivalent, because each has only one hole; one object can be transformed into the other without any tearing. A doughnut and a banana are not equivalent, because one would have to tear the doughnut to mold it into a banana-shaped object. Topologists are particularly fond of determining whether seemingly dissimilar knots can actually be transformed into each other without being cut. In the late 1980s Witten created a technique—which borrowed from both topology and quantum field theory—that allows mathematicians to uncover deep symmetries between hideously tangled, higher-dimensional knots. As a result of his finding, Witten won the 1990 Fields Medal, the most prestigious prize in *mathematics*. Witten called the achievement his "single most satisfying piece of work."

I asked Witten how he responded to the claims of critics that superstring theory is not testable and therefore is not really physics at all. Witten replied that the theory had predicted gravity. "Even though it is, properly speaking, a postprediction, in the sense that the experiment was made before the theory, the fact that gravity is a consequence of string theory, to me, is one of the greatest theoretical insights ever."

He acknowledged, even emphasized, that no one had truly fathomed the theory, and that it might be decades before it yielded a precise description of nature. He would not predict, as others had, that superstring theory might bring about the end of physics. Nevertheless, he was serenely confident that it would eventually yield a profound new understanding of reality. "Good wrong ideas are extremely scarce," he said, "and good wrong ideas that even remotely rival the majesty of string theory have never been seen." When I continued to press Witten on the issue of testability, he grew exasperated. "I don't think I've succeeded in conveying to you its wonder, its incredible consistency, remarkable elegance, and beauty." In other words, superstring theory is too beautiful to be wrong.

Witten then revealed just how strong his faith was. "Generally speaking, all the really great ideas of physics are really spin-offs of string theory," Witten began. "Some of them were discovered first, but I consider that a mere accident of the development on planet earth. On planet earth, they were discovered in this order." Stepping up to his chalkboard, he wrote down general relativity, quantum field theory, superstrings, and supersymmetry (a concept that serves a vital role in superstring theory). "But I don't believe, if there are many civilizations in the universe, that those four ideas were discovered in that order in each civilization." He paused. "I do believe, by the way, that those four ideas were discovered in any advanced civilization."

I could not believe my good fortune. Who's being provocative now? I asked. "I'm not being provocative," Witten retorted. "I'm being provocative in the same way as someone who says the sky is blue is being provocative, if there is a writer somewhere who has said that the sky has pink polka dots."

Particle Aesthetics

In the early 1990s, when superstring theory was still relatively novel, several physicists wrote popular books about its implications. In *Theories of Everything*, the British physicist John Barrow argued that Gödel's incompleteness theorem undermines the very notion of a *complete* theory of nature.[6] Gödel established that any moderately complex system of axioms inevitably raises questions that cannot be answered by the axioms. The implication is that any theory will always have loose ends. Barrow also pointed out that a unified theory of particle physics would not really be a theory of everything, but only a theory of all particles and forces. The theory would have little or nothing to say about phenomena that make our lives meaningful, such as love or beauty.

But Barrow and other analysts at least granted that physicists might achieve a unified theory. That assumption was challenged in *The End of Physics*, written by physicist-turned-journalist David Lindley.[7] Physicists working on superstring theory, Lindley contended, were no longer doing physics because their theories could never be validated by experiments, but only by subjective criteria, such as elegance and beauty. Particle physics, Lindley concluded, was in danger of becoming a branch of aesthetics.

The history of physics supports Lindley's prognosis. Previous theories of physics, however seemingly bizarre, won acceptance among physicists and even the public not because they made sense; rather, they offered predictions that were borne out—often in dramatic fashion— by observations. After all, even Newton's version of gravity violates common sense. How can one thing tug at another across vast spans of space? John Maddox, the editor of *Nature*, once argued that if Newton submitted his theory of gravity to a journal today, it would almost certainly be rejected as too preposterous to believe.[8] Newton's formalism nonetheless provided an astonishingly accurate means of calculating the orbits of planets; it was too effective to deny.

Einstein's theory of general relativity, with its malleable space and time, is even more bizarre. But it became widely accepted as true after observations confirmed his prediction about how gravity would bend light passing around the sun. Likewise, physicists do not believe quantum mechanics because it explains the world, but because it predicts the outcome of experiments with almost miraculous accuracy. Theorists kept predicting new particles and other phenomena, and experiments kept bearing out those predictions.

Superstring theory is on shaky ground indeed if it must rely on aesthetic judgments. The most influential aesthetic principle in science was set forth by the fourteenth-century British philosopher William of Occam. He argued that the best explanation of a given phenomenon is generally the simplest, the one with the fewest assumptions. This principle, called Occam's razor, was the downfall of the Ptolemaic model of the solar system in the Middle Ages. To show that the earth was the center of the solar system, the astronomer Ptolemy was forced to argue that the planets traced elaborately spiraling epicycles around the earth. By assuming that the sun and not the earth was the center of the solar system, later astronomers eventually could dispense with epicycles and replace them with much simpler elliptical orbits.

Ptolemy's epicycles seem utterly reasonable when compared to the undetected—and undetectable—extra dimensions required by superstring theory. No matter how much superstring theorists assure us of the theory's mathematical elegance, the metaphysical baggage it carries with it will prevent it from winning the kind of acceptance—among either physicists or laypeople—that general relativity or the standard model of particle physics have.

Let's give superstring believers the benefit of the doubt, if only for a moment. Let's assume that some future Witten, or even Witten himself, finds an infinitely pliable geometry that accurately describes the behavior of all known forces and particles. In what sense will such a theory explain the world? I have talked to many physicists about superstrings, and none has been able to help me understand what, exactly, a superstring *is*. As far as I can tell, it is neither matter nor energy; it is some kind of mathematical ur-stuff that generates matter and energy and space and time but does not itself correspond to anything in our world.

Good science writers will no doubt make readers think they understand such a theory. Dennis Overbye, in *Lonely Hearts of the Cosmos*, one of the best books ever written on cosmology, imagines God as a cosmic rocker, bringing the universe into being by flailing on his 10-dimensional superstring guitar.[9] (One wonders, is God improvising, or following a score?) The true meaning of superstring theory, of course, is embedded in the theory's austere mathematics. I once heard a professor of literature liken James Joyce's gobbledygookian tome *Finnegans Wake* to the gargoyles atop the cathedral of Notre Dame, built solely for God's amusement. I suspect that if Witten ever finds the theory he so desires, only he—and God, perhaps—will truly appreciate its beauty.

Nightmares of a Final Theory

With his crab-apple cheeks, vaguely Asian eyes, and silver hair still tinged with red, Steven Weinberg resembles a large, dignified elf. He would make an excellent Oberon, king of the fairies in *A Midsummer Night's Dream*. And like a fairy king, Weinberg has demonstrated a powerful affinity for the mysteries of nature, an ability to discern subtle patterns within the froth of data streaming from particle accelerators. In his 1993 book, *Dreams of a Final Theory*, he managed to make reductionism sound romantic. Particle physics is the culmination of an epic quest, "the ancient search for those principles that cannot be explained in terms of deeper principles."[10] The force compelling science, he pointed out, is the simple question, why? That question has led physicists deeper and deeper into the heart of nature. Eventually, he contended, the convergence of explanations down to simpler and simpler principles would culminate in a final theory. Weinberg suspected that superstrings might lead to that ultimate explanation.

Weinberg, like Witten and almost all particle physicists, has a profound faith in the power of physics to achieve absolute truth. But what makes Weinberg such an interesting spokesperson for his tribe is that he, unlike Witten, is acutely aware that his faith is just that, a faith; Weinberg knows that he is speaking with a philosophical accent. If Edward Witten is a philosophically naive scientist, Weinberg is an extremely sophisticated one—too sophisticated, perhaps, for the good of his own field.

I first met Weinberg in New York in March 1993—during the halcyon era before the supercollider was killed—at a dinner held to celebrate the publication of *Dreams of a Final Theory*. He was in an expansive mood, dispensing jokes and anecdotes about famous colleagues and wondering what it would be like to chat with the talk-show host Charlie Rose later that night. Eager to impress the great Nobel laureate, I began name-dropping. I mentioned that Freeman Dyson had recently told me that the whole notion of a final theory was a pipe dream.

Weinberg smiled. The majority of his colleagues, he assured me, believed in a final theory, although many of them preferred to keep that belief private. I dropped another name, Jack Gibbons, whom the newly elected Bill Clinton had just named science advisor. I had recently interviewed Gibbons, I said, and Gibbons had hinted that the United States alone might not be able to afford the supercollider. Weinberg scowled and shook his head, muttering something about society's disturbing lack of appreciation for the intellectual benefits of basic research.

The irony was that Weinberg himself, in *Dreams of a Final Theory*, offered little or no argument as to why society *should* support further research in particle physics. He was careful to acknowledge that neither the superconducting supercollider nor any other earthly accelerator could provide direct confirmation of a final theory; physicists would eventually have to rely on mathematical elegance and consistency as guides. Moreover, a final theory might have no practical value. Weinberg's most extraordinary admission was that a final theory might not reveal the universe to be meaningful in human terms. Quite the contrary. He reiterated an infamous comment made in an earlier book: "The more the universe seems comprehensible, the more it seems pointless."[11] Although the comment had "dogged him ever since," Weinberg refused to back away from it. Instead he elaborated on the remark: "As we have discovered more and more fundamental physical principles they seem to have less and less to do with us."[12] Weinberg seemed to be acknowledging that all our whys would eventually culminate in a because. His vision of the final theory evoked *The Hitchhiker's Guide to the Galaxy* by Douglas Adams. In this science fiction comedy, published in 1980, scientists finally discover the answer to the riddle of the universe, and the answer is... 42. (Adams was practicing philosophy of science in an overtly literary mode.)

The superconducting supercollider was dead and buried when I met Weinberg again in March 1995 at the University of Texas at Austin. His spacious office was cluttered with periodicals that testified to the breadth of his interests, including *Foreign Affairs, Isis, Skeptical Inquirer*, and *American Historical Review*, as well as physics journals. Along one wall ran a chalkboard laced with the obligatory mathematical scribbles. Weinberg spoke with what seemed to be considerable effort. He kept sighing, grimacing, squeezing his eyes shut and rubbing them, even as his deep, sonorous voice rolled forward. He had just eaten lunch and was probably experiencing postprandial fatigue. But I preferred to think he was brooding over the tragic dilemma of particle physicists: they are damned if they achieve a final theory and damned if they don't.

It is a "terrible time for particle physics," Weinberg admitted. "There's never been a time when there's been so little excitement in the sense of experiments suggesting really new ideas or theories being able to make new and qualitatively different kinds of predictions that are then borne out by experiments." With the supercollider dead and plans for other new accelerators in the United States stalled for lack of funding, the prospects for the field were gloomy. Oddly enough, brilliant students were still entering the field, students "better than we deserve, probably," Weinberg added.

Although he shared Witten's belief that physics moves toward absolute truth, Weinberg was acutely aware of the philosophical difficulties of defending this position. He recognized that "the techniques by which we decide on the acceptance of physical theories are extremely subjective." It would always be possible for clever philosophers to make a case that the particle physicists "are just making it up as they go along." (In *Dreams of a Final Theory*, Weinberg even confessed to having a fondness for the writings of the philosophical anarchist Paul Feyerabend.) On the other hand, Weinberg told me, the standard model of particle physics, "whatever the aesthetics were, [has] by now been tested as few theories have been, and it really works. If in fact it was just a social construct it would have fallen apart long before this."

Weinberg realized that physicists would never be able to *prove* a theory final in the same way that mathematicians prove theorems; but if the theory accounted for all experimental data—the masses of all particles, the strengths of all forces—physicists would eventually cease questioning it. "I don't feel I was put here to be sure of anything," Weinberg said. "A lot of philosophy of science going back to the Greeks has been poisoned by the search for certainty, which seems to me a false search. Science is too much fun to sit around wringing our hands because we're not certain about things."

Even as we spoke, Weinberg suggested, someone might be posting the final, correct version of superstring theory on the Internet. "If *she*,"—Weinberg added, with the slightest pause and emphasis on the *she*—"got results that agreed with experiment, then you would say, 'That's it,'" even if researchers could never provide direct evidence of the strings themselves or the extra dimensions they supposedly inhabit; after all, the atomic theory of matter was accepted because it worked, not because experimenters could make pictures of atoms. "I agree strings are much farther away from direct perception than atoms, and atoms are much farther away from direct perception than chairs, but I don't see any philosophical discontinuity there."

There was little conviction in Weinberg's voice. Deep down, he surely knew that superstring theory *did* represent a discontinuity in physics, a leap beyond any conceivable empirical test. Abruptly, he rose and began prowling around the room. He picked up odd objects, fondled them distractedly, put them down, while continuing to speak. He reiterated his belief that a final theory of physics would represent the most fundamental possible achievement of science—the bedrock of all other knowledge. To be sure, some complex phenomena, such as turbulence, or economics, or life, require their own special laws and generalizations. But if you ask why those principles are true, Weinberg added, that question takes you down toward the final theory of physics, on which everything rests. "That's what makes science a hierarchy. And it *is* a hierarchy. It's not just a random net."

Many scientists cannot abide hearing that truth, Weinberg said, but there is no escaping it. "Their final theory is what our final theory explains." If neuroscientists ever explain consciousness, for example, they will explain it in terms of the brain, "and the brain is what it is because of historical accidents and because of universal principles of chemistry and physics." Science will certainly continue after a final theory, perhaps forever, but it will have lost something. "There will be a sense of sadness" about the achievement of a final theory, Weinberg said, since it will bring to a close the great quest for fundamental knowledge.

As Weinberg continued speaking, he seemed to portray the final theory in increasingly negative terms. Asked whether there would ever be such a thing as applied superstring theory, Weinberg grimaced. (In the 1994 book *Hyperspace*, the physicist Michio Kaku foresaw a day when advances in superstring theory would allow us to visit other universes and travel through time.)[13] Weinberg cautioned that "the sands of scientific history are white with the bones of people" who failed to foresee applications of developments in science, but applied superstring theory was "hard to conceive."

Weinberg also doubted that the final theory would resolve all the notorious paradoxes posed by quantum mechanics. "I tend to think these are just puzzles in the way we talk about quantum mechanics," Weinberg said. One way to eliminate these puzzles, he added, would be to adopt the many-worlds interpretation of quantum mechanics. Proposed in the 1950s, this interpretation attempts to explain why the act of observation by a physicist seems to force a particle such as an electron to choose only one path out of the many allowed by quantum mechanics. According to the many-worlds interpretation, the electron actually follows all possible paths, but in separate universes. This explanation does have its troubling aspects, Weinberg conceded. "There may be another parallel time track where John Wilkes Booth missed Lincoln and...." Weinberg paused. "I sort of hope that whole problem will go away, but it may not. That may be just the way the world is."

Is it too much to ask for a final theory to make the world intelligible? Before I could finish the question, Weinberg was nodding. "Yes, it's too much to ask," he replied. The proper language of science is mathematics, he reminded me. A final theory "has to make the universe appear plausible and somehow or other recognizably logical to people who are trained in that language of mathematics, but it may be a long time before that makes sense to other people." Nor will a final theory provide humanity with any guidance in conducting its affairs. "We've learned to absolutely disentangle value judgments from truth judgments," Weinberg said. "I don't see us going back to reconnect them." Science "can certainly help you find out what the consequences of your actions are, but it can't tell you what consequences you ought to wish for. And that seems to me to be an absolute distinction."

Weinberg had little patience for those who suggest that a final theory will reveal the purpose of the universe, or "the mind of God," as Stephen Hawking once put it. Quite the contrary. Weinberg hoped that a final theory would eliminate the wishful thinking, mysticism, and superstition that pervades much of human thought, even among physicists. "As long as we don't know the fundamental rules," he said, "we can hope that we'll find something like a concern for human beings, say, or some guiding divine plan built into the fundamental rules. But when we find out that the fundamental rules of quantum mechanics and some symmetry principles are very impersonal and cold, then it'll have a very demystifying effect. At least that's what I'd like to see."

His face hardening, Weinberg continued: "I certainly would not disagree with people who say that physics in my style or the Newtonian style has produced a certain disenchantment. But if that's the way the world is, it's better we find out. I see it as part of the growing up of our species, just like the child finding out there is no tooth fairy. It's better to find out there is no tooth fairy, even though a world with tooth fairies in it is somehow more delightful."

Weinberg was well aware that many people hungered for a different message from physics. In fact, earlier that day he had heard that Paul Davies, an Australian physicist, had received a million-dollar prize for "advancing public understanding of God or spirituality." Davies had written numerous books, notably *The Mind of God*, suggesting that the laws of physics reveal a plan underlying nature, a plan in which human consciousness may play a central role.[14] After telling me about Davies's prize, Weinberg chuckled mirthlessly. "I was thinking of cabling Davies and saying, 'Do you know of any organization that is willing to offer a million-dollar prize for work showing that there is no divine plan?'"

In *Dreams of a Final Theory*, Weinberg dealt rather harshly with all this talk of divine plans. He raised the embarrassing issue of human suffering. What kind of plan is it that allows the Holocaust, and countless other evils, to happen? What kind of planner? Many physicists, intoxicated by the power of their mathematical theories, have suggested that "God is a geometer." Weinberg retorted, in effect, that he does not see why we should be interested in a God who seems so little interested in us, however good he is at geometry.

I asked Weinberg what gave him the fortitude to sustain such a bleak (and in my view, accurate) vision of the human condition. "I sort of enjoy my tragic view," he replied with a little smile. "After all, which would you rather see, a tragedy or—" he hesitated, his smile fading. "Well, some people would prefer to see a comedy. But.... I think the tragic view adds a certain dimension to life. Anyway, it's the best we have." He stared out his office window, brooding. Fortunately, perhaps, for Weinberg, his view did not include the infamous tower from which a deranged University of Texas student, Charles Whitman, shot 14 people to death in 1966. Weinberg's office overlooks a graceful Gothic church that serves as the university's theological seminary. But Weinberg did not seem to be looking at the church—or, for that matter, at anything else in the material world.

No More Surprises

Even if society musters the will and the money to build larger accelerators and thereby keep particle physics alive—at least temporarily— how likely is it that physicists will learn something as truly new and surprising as, say, quantum mechanics? Not very, according to Hans Bethe. A professor at Cornell University, Bethe won a Nobel Prize in 1967 for his work on the carbon cycle in stellar fusion; in other words, he showed how stars shine. He also headed the theoretical division of the Manhattan Project during World

War II. In that capacity he made what was arguably the most important calculations in the history of the planet. Edward Teller (who, ironically, later became the scientific community's most avid booster of nuclear weapons) had done some calculations suggesting that the fireball from an atomic blast might ignite the earth's atmosphere, triggering a conflagration that would consume the entire world. The scientists studying Teller's conjecture took it very seriously; after all, they were exploring terra incognita. Bethe then examined the problem and made his own calculations. He determined that Teller was wrong; the fireball would not spread.[15]

No one should have to carry out calculations on which the fate of the earth depends. But if someone must, Bethe would be my choice. He exudes wisdom and gravitas. When I inquired whether he had had any lingering doubts, in the moments before the bomb was detonated at Alamogordo, about what would happen, he shook his head. No, he replied. His only concern had been whether the ignition device would work properly. There was not a trace of braggadocio in Bethe's response. He had done the calculations, and he trusted them. (One wonders whether even Edward Witten would entrust the fate of the earth to a prediction based on superstrings.)

When I asked him about the future of his field, Bethe said there were still many open questions in physics, including ones raised by the standard model. Moreover, important discoveries would continue in solid-state physics. But none of these advances would bring about revolutionary changes in the foundations of physics, according to Bethe. As an example, Bethe cited the discovery of so-called high-temperature superconductors, arguably the most exciting advance in physics in decades. These materials, first reported in 1987, conduct electricity with no resistance at relatively high temperatures (which are still well below zero degrees Celsius). "That didn't in any way change our understanding of electric conduction or even of superconductivity," Bethe said. "The basic structure of quantum mechanics, quantum mechanics without relativity, that basic structure is finished." In fact, "the understanding of atoms, molecules, the chemical bond, and so on, that was all complete by 1928." Could there ever be another revolution in physics like the one that accompanied quantum mechanics? "That's very unlikely," Bethe replied in his unsettlingly matter-of-fact way.

Actually, almost all believers in a final theory agree that whatever form it takes, it will still be a *quantum* theory. Steven Weinberg suggested to me that a final theory of physics "might be as far removed from what we now understand as quantum mechanics was from classical mechanics." But he, like Hans Bethe, did not think the final theory would supplant quantum mechanics in any way. "I think we'll be stuck with quantum mechanics," Weinberg said. "So in that sense the development of quantum mechanics may be more revolutionary than anything before or after."

Weinberg's remarks reminded me of an essay published in *Physics Today* in 1990, in which the physicist David Mermin of Cornell University recounted how a certain Professor Mozart (actually Mermin's cranky alter ego) had complained that "particle physics over the last 40 or 50 years has been a disappointment. Who would have expected that in half a century we wouldn't learn anything really profound!" When Mermin asked the fictional professor what he meant, he replied, "All particle physics has taught us about the central mystery is that quantum mechanics still works. Perfectly, as far as anybody can tell. What a letdown!"[16]

John Wheeler and the "It from Bit"

Bethe, Weinberg, and Mermin all seemed to suggest that quantum mechanics is—at least in a qualitative sense—the final theory of physics. Some physicists and philosophers have proposed that if they could only understand quantum mechanics, if they could determine its meaning, they might find *The Answer*. One of the most influential and inventive interpreters of quantum mechanics, and of modern physics in general, is John Archibald Wheeler. Wheeler is the archetypal physics-for-poets physicist. He is famed for his analogies and aphorisms, self-made and coopted. Among the one-liners he bestowed on me when I interviewed him on a warm spring day at Princeton were "If I can't picture it, I can't understand it" (Einstein);

"Unitarianism [Wheeler's nominal religion] is a feather bed to catch falling Christians" (Darwin); "Never run after a bus or woman or cosmological theory, because there'll always be another one in a few minutes" (a friend of Wheeler's at Yale); and "If you haven't found something strange during the day it hasn't been much of a day" (Wheeler).

Wheeler is also renowned for his physical energy. When we left his third-floor office to get some lunch, he spurned the elevator—"elevators are hazardous to your health," he declared—and charged down the stairs. He hooked an arm inside the banister and pivoted at each landing, letting centrifugal force whirl him around the hairpin and down the next flight. "We have contests to see who can take the stairs fastest," he said over a shoulder. Outside, Wheeler marched rather than walked, swinging his fists smartly in rhythm with his stride. He paused only when we reached a door. Invariably he got there first and yanked it open for me. After passing through I paused in reflexive deference— Wheeler was almost 80 at the time—but a moment later he was past me, barreling toward the next doorway.

The metaphor was so obvious I almost suspected Wheeler had it in mind. Wheeler had made a career of racing ahead of other scientists and throwing open doors for them. He had helped gain acceptance—or at least attention—for some of the most outlandish ideas of modern physics, from black holes to multiple-universe theories to quantum mechanics itself. Wheeler might have been dismissed as fun but flaky long ago if he had not had such unassailable credentials. In his early twenties, he had traveled to Denmark to study under Niels Bohr ("because he sees further than any man alive," Wheeler wrote in his application for the fellowship). Bohr was to be the most profound influence on Wheeler's thought. In 1939, Bohr and Wheeler published the first paper that successfully explained nuclear fission in terms of quantum physics.[17]

Wheeler's expertise in nuclear physics led to his involvement in the construction of the first fission-based bomb during World War II and, in the early years of the cold war, the first hydrogen bomb. After the war, Wheeler became one of the leading authorities on general relativity, Einstein's theory of gravity. He coined the term *black hole* in the late 1960s, and he played a major role in convincing astronomers that these bizarre, infinitely dense objects predicted by Einstein's theory might actually exist. I asked Wheeler what characteristic made him able to believe in such fantastical objects, which other physicists came to accept only with great reluctance. "More vividness of imagination," he replied. "There's that phrase of Bohr's I like so much: 'You must be prepared for a surprise, and a very great surprise.'"

Beginning in the 1950s, Wheeler had grown increasingly intrigued by the philosophical implications of quantum physics. The most widely accepted interpretation of quantum mechanics was the so-called orthodox interpretation (although "orthodox" seems an odd descriptor for such a radical worldview). Also called the Copenhagen interpretation, because it was set forth by Wheeler's mentor, Bohr, in a series of speeches in Copenhagen in the late 1920s, it held that subatomic entities such as electrons have no real existence; they exist in a probabilistic limbo of many possible superposed states until forced into a single state by the act of observation. The electrons or photons may act like waves or like particles, depending on how they are experimentally observed.

Wheeler was one of the first prominent physicists to propose that reality might not be wholly physical; in some sense, our cosmos might be a participatory phenomenon, requiring the act of observation—and thus consciousness itself. In the 1960s Wheeler helped to popularize the notorious anthropic principle. It holds, essentially, that the universe must be as it is, because, if it were otherwise, we might not be here to observe it. Wheeler also began to draw his colleagues' attention to some intriguing links between physics and information theory, which was invented in 1948 by the mathematician Claude Shannon. Just as physics builds on an elementary, indivisible entity—namely, the quantum—which is defined by the act of observation, so does information theory. Its quantum is the binary unit, or bit, which is a message representing one of two choices: heads or tails, yes or no, zero or one.

Wheeler became even more deeply convinced of the importance of information after concocting a thought experiment that exposed the

85

strangeness of the quantum world for all to see. Wheeler's delayed-choice experiment is a variation on the classic (but not classical) two-slit experiment, which demonstrates the schizophrenic nature of quantum phenomena. When electrons are aimed at a barrier containing two slits, the electrons act like waves; they go through both slits at once and form what is called an interference pattern, created by the overlapping of the waves, when they strike a detector on the far side of the barrier. If the physicist closes off one slit at a time, however, the electrons pass through the open slit like simple particles and the interference pattern disappears. In the delayed-choice experiment, the experimenter decides whether to leave both slits open or to close one off *after the electrons have already passed through the barrier*—with the same results. The electrons seem to know in advance how the physicist will choose to observe them. This experiment was carried out in the early 1990s and confirmed Wheeler's prediction.

Wheeler accounted for this conundrum with yet another analogy. He likened the job of a physicist to that of someone playing 20 questions in its surprise version. In this variant of the old game, one person leaves the room while the rest of the group—or so the excluded person thinks—selects some person, place, or thing. The single player then reenters the room and tries to guess what the others have in mind by asking a series of questions that can only be answered yes or no. Unbeknownst to the guesser, the group has decided to play a trick. The first person to be queried will think of an object only *after* the questioner asks the question. Each person will do the same, giving a response that is consistent not only with the immediate question but also with all previous questions.

"The word wasn't in the room when I came in even though I thought it was," Wheeler explained. In the same way, the electron, before the physicist chooses how to observe it, is neither a wave nor a particle. It is in some sense unreal; it exists in an indeterminate limbo. "Not until you start asking a question, do you get something," Wheeler said. "The situation cannot declare itself until you've asked your question. But the asking of one question prevents and excludes the asking of another. So if you ask where my great white hope presently lies—and I always find it interesting to ask people what's your great white hope—I'd say it's in the idea that the whole show can be reduced to something similar in a broad sense to this game of 20 questions."

Wheeler has condensed these ideas into a phrase that resembles a Zen koan: "the it from bit." In one of his free-form essays, Wheeler unpacked the phrase as follows: "every it—every particle, every field of force, even the spacetime continuum itself—derives its function, its meaning, its very existence entirely—even if in some contexts indirectly—from the apparatus-elicited answers to yes-or-no questions, binary choices, *bits*."[18]

Inspired by Wheeler, an ever-larger group of researchers—including computer scientists, astronomers, mathematicians, and biologists, as well as physicists—began probing the links between information theory and physics in the late 1980s. Some superstring theorists even joined in, trying to knit together quantum field theory, black holes, and information theory with a skein of strings. Wheeler acknowledged that these ideas were still raw, not yet ready for rigorous testing. He and his fellow explorers were still "trying to get the lay of the land" and "learning how to express things that we already know" in the language of information theory. The effort may lead to a dead end, Wheeler said, or to a powerful new vision of reality, of "the whole show."

Wheeler emphasized that science has many mysteries left to explain. "We live still in the childhood of mankind," he said. "All these horizons are beginning to light up in our day: molecular biology, DNA, cosmology. We're just children looking for answers." He served up another aphorism: "As the island of our knowledge grows, so does the shore of our ignorance." Yet he was also convinced that humans would someday find *The Answer*. In search of a quotation that expressed his faith, he jumped up and pulled down a book on information theory and physics to which he had contributed an essay. After flipping it open, he read: "Surely someday, we can believe, we will grasp the central idea of it all as so simple, so beautiful, so compelling that we will all say to each other, 'Oh, how could it have been otherwise! How could we all have been so blind for so long!'"[19] Wheeler looked up from the book; his expression was beatific. "I don't know whether it will be one year or a decade, but I think we can and will understand. That's the central thing I would like to stand for. We can and will understand."

Many modern scientists, Wheeler noted, shared his faith that humans would one day find *The Answer*. For example, Kurt Gödel, once Wheeler's neighbor in Princeton, believed that *The Answer* might *already* have been discovered. "He thought that maybe among the papers of Leibniz, which in his time had still not been fully smoked out, we would find the—what was the word—the philosopher's key, the magic way to find truth and solve any set of puzzlements." Gödel felt that this key "would give a person who understood it such power that you could only entrust the knowledge of this philosopher's key to people of high moral character."

Yet Wheeler's own mentor, Bohr, apparently doubted whether science or mathematics could achieve such a revelation. Wheeler learned of Bohr's view not from the great man himself, but from his son. After Bohr died, his son told Wheeler that Bohr had felt that the search for the ultimate theory of physics might never reach a satisfying conclusion; as physicists sought to penetrate further into nature, they would face questions of increasing complexity and difficulty that would eventually overwhelm them. "I guess I'm more optimistic than that," Wheeler said. He paused a moment and added, with a rare note of somberness, "But maybe I'm kidding myself."

The irony is that Wheeler's own ideas suggest that a final theory will always be a mirage, that the truth is in some sense imagined rather than objectively apprehended. According to the it from bit, we create not only truth, but even reality itself—the "it"—with the questions we ask. Wheeler's view comes dangerously close to relativism, or worse. In the early 1980s, organizers of the annual meeting of the American Association for the Advancement of Science placed Wheeler on a program with three parapsychologists. Wheeler was furious. At the meeting, he made it clear that he did not share the belief of his cospeakers in psychic phenomena. He passed out a pamphlet that declared, in reference to parapsychology, "Where there's smoke, there's smoke."

Yet Wheeler himself has suggested that there is nothing *but* smoke. "I do take 100 percent seriously the idea that the world is a figment of the imagination," he once remarked to the science writer and physicist Jeremy Bernstein.[20] Wheeler is well aware that this view is, from an empirical viewpoint, unsupportable: Where was mind when the universe was born? And what sustained the universe for the billions of years before we came to be? He nonetheless bravely offers us a lovely, chilling paradox: at the heart of everything is a question, not an answer. When we peer down into the deepest recesses of matter or at the farthest edge of the universe, we see, finally, our own puzzled faces looking back at us.

David Bohm's Implicate Order

Not surprisingly, some other physicist-philosophers have bridled both at Wheeler's views and, more generally, at the Copenhagen interpretation set forth by Bohr. One prominent dissident was David Bohm. Born and raised in Pennsylvania, Bohm left the United States in 1951, at the height of the McCarthy era, after refusing to answer questions from the Un-American Activities Committee about whether he or any of his scientific colleagues (notably Robert Oppenheimer) were communists. After stays in Brazil and Israel, he settled in England in the late 1950s.

By then, Bohm had already begun developing an alternative to the Copenhagen interpretation. Sometimes called the pilot-wave interpretation, it preserves all the predictive power of quantum mechanics but eliminates many of the most bizarre aspects of the orthodox interpretation, such as the schizophrenic character of quanta and their dependence on observers for existence. Since the late 1980s, the pilot-wave theory has attracted increasing attention from physicists and philosophers uncomfortable with the subjectivism and antideterminism of the Copenhagen interpretation.

Paradoxically, Bohm also seemed intent on making physics even *more* philosophical, speculative, holistic. He went much further than Wheeler did in drawing analogies between quantum mechanics and Eastern religion. He developed a philosophy, called the implicate order, that sought to embrace both mystical and scientific knowledge. Bohm's writings on these topics attracted an almost cultlike following; he became a hero to those who hoped to achieve mystical insight through physics. Few scientists combine these two contradictory impulses—the need to clarify reality and to mystify it—in such a dramatic fashion.[21]

In August 1992, I visited Bohm at his home in Edgeware, a suburb north of London. His skin was alarmingly pale, especially in contrast to his purplish lips and dark, wiry hair. His frame, sinking into a large armchair, seemed limp and langorous, but at the same time suffused with nervous energy. He cupped one hand over the top of his head; the other rested on the armrest. His fingers, long and blue veined, with tapered, yellow nails, were splayed. He was recovering, he told me, from a recent heart attack.

Bohm's wife brought us tea and biscuits and then retreated to another part of the house. Bohm spoke haltingly at first, but gradually the words came faster, in a low, urgent monotone. His mouth was apparently dry, because he kept smacking his lips. Occasionally, after making some observation that amused him, he pulled his lips back from his teeth in a semblance of a smile. He also had the disconcerting habit of pausing every few sentences and saying, "Is that clear?" or simply, "Hmmm?" I was often so befuddled, and so hopeless of finding my way to understanding, that I just nodded and smiled. But Bohm could be absolutely, rivetingly clear, too. Later, I learned that he had this same effect on others; like some strange quantum particle, he oscillated in and out of focus.

Bohm said he had begun to question the Copenhagen interpretation in the late 1940s, while writing a book on quantum mechanics. Bohr had rejected the possibility that the probabilistic behavior of quantum systems was actually the result of underlying, deterministic mechanisms, sometimes called hidden variables. Reality was unknowable because it was intrinsically indeterminate, Bohr insisted.

Bohm found this view unacceptable. "The whole idea of science so far has been to say that underlying the phenomenon is some reality which explains things," he explained. "It was not that Bohr denied reality, but he said quantum mechanics implied there was nothing more that could be said about it." Such a view, Bohm decided, reduced quantum mechanics to "a system of formulas that we use to make predictions or to control things technologically. I said, that's not enough. I don't think I would be very interested in science if that were all there was."

In a paper published in 1952, Bohm proposed that particles are indeed particles—and at all times, not just when they are observed. Their behavior is determined by a new, heretofore undetected force, which Bohm called the pilot wave. Any effort to measure these properties precisely would destroy information about them by physically altering the pilot wave. Bohm thus gave the uncertainty principle a purely physical rather than a metaphysical meaning. Bohr had interpreted the uncertainty principle as meaning "not that there is uncertainty, but that there is an inherent ambiguity" in a quantum system, Bohm told me.

Bohm's interpretation did permit, and even highlight, one quantum paradox: nonlocality, the ability of one particle to influence another instantaneously across vast distances. Einstein had drawn attention to non-locality in 1935 in an effort to show that quantum mechanics must be flawed. Together with Boris Podolsky and Nathan Rosen, Einstein proposed a thought experiment—now called the EPR experiment— involving two particles that spring from a common source and fly in opposite directions.[22]

According to the standard model of quantum mechanics, neither particle has a definite position or momentum before it is measured; but by measuring the momentum of one particle, the physicist instantaneously forces the other particle to assume a fixed position—even if it is on the other side of the galaxy. Deriding this effect as "spooky action at a distance," Einstein argued that it violated both common sense and his own theory of special relativity, which prohibits the propagation of effects faster than the speed of light; quantum mechanics must therefore be an incomplete theory. In 1980, however, a group of French physicists carried out a version of the EPR experiment and showed that it did indeed give rise to spooky action. (The reason that the experiment does not violate special relativity is that one cannot exploit non-locality to transmit information.) Bohm never had any doubts about the outcome of the experiment. "It would have been a terrific surprise to find out otherwise," he said.

As Bohm was trying to bring the world into sharper focus through his pilot-wave model, however, he was also arguing that complete clarity was impossible. His ideas were inspired in part by an experiment he saw on television, in which a drop of ink was squeezed onto a cylinder of glycerin. When the cylinder was rotated, the ink diffused through the glycerin in an apparently irreversible fashion; its order seemed to have disintegrated. But when the direction of rotation was reversed, the ink gathered into a drop again.

Upon this simple experiment, Bohm built a worldview called the implicate order. Underlying the apparently chaotic realm of physical appearances—the explicate order—there is always a deeper, hidden, implicate order. Applying this concept to the quantum realm, Bohm proposed that the implicate order is the quantum potential, a field consisting of an infinite number of fluctuating pilot waves. The overlapping of these waves generates what appear to us as particles, which constitute the explicate order. Even such seemingly fundamental concepts as space and time may be merely explicate manifestations of some deeper, implicate order, according to Bohm.

To plumb the implicate order, Bohm said, physicists might need to jettison some basic assumptions about the organization of nature. "Fundamental notions like order and structure condition our thinking unconsciously, and new kinds of theories depend on new kinds of order," he remarked. During the Enlightenment, thinkers such as Newton and Descartes replaced the ancients' organic concept of order with a mechanistic view. Although the advent of relativity and other theories brought about modifications in this order, "the basic idea is still the same," Bohm said, "a mechanical order described by coordinates."

But Bohm, in spite of his own enormous ambitions as a truth seeker, rejected the possibility that scientists could bring their enterprise to an end by reducing all the phenomena of nature to a single phenomenon (such as superstrings). "I think there are no limits to this. People are going to talk about the theory of everything, but that's an assumption, you see, which has no basis. At each level we have something which is taken as appearance and something else is taken as the essence which explains the appearance. But then when we move to another level essence and appearance interchange their rules, right? Is that clear? There's no end to this, you see? The very nature of our knowledge is of that nature, you see. But what underlies it all is unknown and cannot be grasped by thought."

To Bohm, science was "an inexhaustible process." Modern physicists, he pointed out, assume that the forces of nature are the essence of reality. "But why are the forces of nature there? The forces of nature are then taken as the essence. The atoms weren't the essence. Why should these forces be?"

The belief of modern physicists in a final theory could only be self-fulfilling, Bohm said. "If you do that you'll keep away from really questioning deeply." He noted that "if you have fish in a tank and you put a glass barrier in there, the fish keep away from it. And then if you take away the glass barrier they never cross the barrier and they think the whole world is that." He chuckled drily. "So your thought that this is the end could be the barrier to looking further."

Bohm reiterated that "we're not ever going to get a final essence which isn't also the appearance of something." But wasn't that frustrating? I asked. "Well, it depends what you want. You're frustrated if you want to get it all. On the other hand, scientists are going to be frustrated if they get the final answer and then have nothing to do except be technicians, you see." He uttered his dry laugh. So, I said, damned if you do and damned if you don't. "Well, I think you have to look at it differently, you see. One of the reasons for doing science is the extension of perception and not of current knowledge. We are constantly coming into contact with reality better and better."

Science, Bohm continued, is sure to evolve in totally unexpected ways. He expressed the hope that future scientists would be less dependent on mathematics for modeling reality and would draw on new sources of metaphor and analogy. "We have an assumption now that's getting stronger and stronger that mathematics is the only way to deal with reality," Bohm said. "Because it's worked so well for a while we've assumed that it has to be that way."

Like many other scientific visionaries, Bohm expected that science and art would someday merge. "This division of art and science is temporary," he observed. "It didn't exist in the past, and there's no reason why it should go on in the future." Just as art consists not simply of works of art but of an "attitude, the artistic spirit," so does science consist not in the accumulation of knowledge but in the creation of fresh modes of perception. "The ability to perceive or think differently is more important than the knowledge gained," Bohm explained. There was something poignant about Bohm's hope that science might become more artlike. Most physicists have objected to his pilot-wave interpretation on aesthetic grounds: it is too ugly to be right.

Bohm, trying to convince me once and for all of the impossibility of final knowledge, offered the following argument: "Anything known has to be determined by its limits. And that's not just quantitative but qualitative. The theory is this and not that. Now it's consistent to propose that there is the unlimited. You have to notice that if you say there is the unlimited, it cannot be different, because then the unlimited will limit the limited, by saying that the limited is not the unlimited, right? The unlimited must include the limited. We have to say, from the unlimited the limited arises, in a creative process; that's consistent. Therefore we say that no matter how far we go there is the unlimited. It seems that no matter how far you go, somebody will come up with another point you have to answer. And I don't see how you could ever settle that."

At this moment, to my relief, Bohm's wife entered the room and asked if we wanted more tea. As she refilled my cup, I pointed out a book on Tibetan mysticism on a shelf behind Bohm. When I asked Bohm if he had been influenced by such writings, he nodded. He had been a friend and student of the Indian mystic Krishnamurti, who died in 1986. Krishnamurti was one of the first modern Indian sages to try to show westerners how to achieve enlightenment. Was Krishnamurti himself enlightened? "In some ways, yes," Bohm replied. "His basic thing was to go into thought, to get to the end of it, completely, and thought would become a different kind of consciousness." Of course, one could never truly plumb one's own mind, Bohm said. Any attempt to examine one's own thought changes it—just as the measurement of an electron alters its course. There could be no final mystical knowledge, Bohm seemed to be implying, any more than there could be a final theory of physics.

Was Krishnamurti a happy person? Bohm looked puzzled at my question. "That's hard to say," he replied eventually. "He was unhappy at times, but I think he was pretty happy overall. The thing is not about happiness, really." Bohm frowned, as if realizing the import of what he had just said.

In *Science, Order, and Creativity*, cowritten with F. David Peat, Bohm insisted on the importance of "playfulness" in science, and in life.[23] But Bohm himself, both in his writings and in person, was anything but playful. For him, this was not a game, this truth seeking; it was a dreadful, impossible, but necessary task. Bohm was desperate to know, to discover the secret of everything, either through physics or through meditation, through mystical knowledge. And yet he insisted that reality was unknowable—because, I believe, he was repelled by the thought of finality. He recognized that any truth, no matter how initially wondrous, eventually ossifies into a dead, inanimate thing that does not reveal the absolute but conceals it. Bohm craved not truth but revelation, perpetual revelation. As a result, he was doomed to perpetual doubt.

I finally said good-bye to Bohm and his wife and departed. Outside, a light rain was falling. I walked up the path to the street and glanced back at the Bohm's house, a modest, whitewashed cottage on a street of modest, whitewashed cottages. Bohm died of a heart attack two months later.[24]

Feynman's Gloomy Prophecy

In *The Character of Physical Law*, Richard Feynman, who won a Nobel Prize in 1965 for devising a quantum version of electromagnetism, offered a rather dark prophecy about the future of physics:

We are very lucky to live in an age in which we are still making discoveries. It is like the discovery of America—you only discover it once. The age in which we live is the age in which we are discovering the fundamental laws of nature, and that day will never come again. It is very exciting, it is marvelous, but this excitement will have to go. Of course in the future there will be other interests. There will be the interest of the connection of one level of phenomena to another—phenomena in biology and so on, or, if you are talking about exploration, exploring other planets, but there will not be the same things we are doing now.[25]

After the fundamental laws are discovered, physics will succumb to second-rate thinkers, that is, philosophers. "The philosophers who are always on the outside making stupid remarks will be able to close in, because we cannot push them away by saying, 'If you were right we would be able to guess all the laws,' because when the laws are all there they will have an explanation for them.... There will be a degeneration of ideas, just like the degeneration that great explorers feel is occurring when tourists begin moving in on a new territory."[26]

Feynman's vision was uncannily on target. He erred only in thinking that it would be millennia, not decades, before the philosophers closed in. I saw the future of physics in 1992 when I attended a symposium at Columbia University in which philosophers and physicists discussed the meaning of quantum mechanics.[27] The symposium demonstrated that more than 60 years after quantum mechanics was invented, its meaning remained, to put it politely, elusive. In the lectures, one could hear echoes of Wheeler's it from bit approach, and Bohm's pilot-wave hypothesis, and the many-worlds model favored by Steven Weinberg and others. But for the most part each speaker seemed to have arrived at a private understanding of quantum mechanics, couched in idiosyncratic language; no one seemed to understand, let alone agree with, anyone else. The bickering brought to mind what Bohr once said of quantum mechanics: "If you think you understand it, that only shows you don't know the first thing about it."[28]

Of course, the apparent disarray could have stemmed entirely from my own ignorance. But when I revealed my impression of confusion and dissonance to one of the attendees, he reassured me that my perception was accurate. "It's a mess," he said of the conference (and, by implication, the whole business of interpreting quantum mechanics). The problem, he noted, arose because, for the most part, the different interpretations of quantum mechanics cannot be empirically distinguished from one another; philosophers and physicists favor one interpretation over another for aesthetic and philosophical—that is, subjective—reasons.

This is the fate of physics. The vast majority of physicists, those employed in industry and even academia, will continue to apply the knowledge they already have in hand—inventing more versatile lasers and superconductors and computing devices—without worrying about any underlying philosophical issues.[29] A few diehards dedicated to truth rather than practicality will practice physics in a nonempirical, ironic mode, plumbing the magical realm of superstrings and other esoterica and fretting about the meaning of quantum mechanics. The conferences of these ironic physicists, whose disputes cannot be experimentally resolved, will become more and more like those of that bastion of literary criticism, the Modern Language Association.

Chapter Four
The End of Cosmology

In 1990 I traveled to a remote resort in the mountains of northern Sweden to attend a symposium entitled "The Birth and Early Evolution of Our Universe." When I arrived, I found that about 30 particle physicists and astronomers from around the world—the United States, Europe, the Soviet Union, and Japan— were there. I had come to the meeting in part to meet Stephen Hawking. The compelling symbolism of his plight—powerful brain in a paralyzed body—had helped to make him one of the best-known scientists in the world.

Hawking's condition, when I met him, was worse than I had expected. He sat in a semifetal position, hunch shouldered, slack jawed, and painfully frail, his head tipped to one side, in a wheelchair loaded with batteries and computers. As far as I could tell, he could move only his left forefinger. With it he laboriously selected letters, words, or sentences from a menu on his computer screen. A voice synthesizer uttered the words in an incongruously deep, authoritative voice— reminiscent of the cyborg hero of *Robocop*. Hawking seemed, for the most part, more amused than distressed by his plight. His purple-lipped, Mick Jagger mouth often curled up at one corner in a kind of smirk.

Hawking was scheduled to give a talk on quantum cosmology, a field he had helped to create. Quantum cosmology assumes that at very small scales, quantum uncertainty causes not merely matter and energy, but the very fabric of space and time, to flicker between different states. These space-time fluctuations might give rise to wormholes, which could link one region of space-time with another one very far away, or to "baby universes." Hawking had stored the hour-long speech, titled "The Alpha Parameters of Wormholes," in his computer; he had merely to tap a key to prompt his voice synthesizer to read it, sentence by sentence.

In his eerie cyber-voice, Hawking discussed whether we might someday be able to slip into a wormhole in our galaxy and a moment later pop out the other end in a galaxy far, far away. Probably not, he concluded, because quantum effects would scramble our constituent particles beyond recognition. (Hawking's argument implied that "warp drive," the faster-than-light transport depicted in *Star Trek*, was impossible.) He wrapped up his lecture with a digression on superstring theory. Although all we see around us is the "mini-superspace" that we call space-time, "we are really living in the infinite dimensional super-space of string theory."[1]

My response to Hawking was ambivalent. He was, on the one hand, a heroic figure. Trapped in a crippled, helpless body, he could still imagine realities with infinite degrees of freedom. On the other hand, what he was saying struck me as being utterly preposterous. Wormholes? Baby universes? Infinite dimensional superspace of string theory? This seemed more science fiction than science.

I had more or less the same reaction to the entire conference. A few talks— those in which astronomers discussed what they had gleaned from probing the cosmos with telescopes and other instruments— were firmly grounded in reality. This was empirical science. But many of the presentations addressed issues hopelessly divorced from reality, from any possible empirical test. What was the universe like when it was the size of a basketball, or a pea, or a proton, or a superstring? What is the effect on our universe of all the other universes linked to it by wormholes? There was something both grand and ludicrous about grown men (no women were present) bickering over such issues.

Over the course of the meeting, I struggled to quell that instinctive feeling of preposterousness, with some success. I reminded myself that these were terribly smart people, "the greatest geniuses in the world," as a local Swedish newspaper had put it. They would not waste their time on trivial pursuits. I therefore did my best, in writing later about the ideas of Hawking and other cosmologists, to make them sound plausible, to instill awe and comprehension instead of skepticism and confusion in readers. That is the job of the science writer, after all.

But sometimes the clearest science writing is the most dishonest. My initial reaction to Hawking and others at the conference was, to some extent, appropriate. Much of modern cosmology, particularly those aspects inspired by unified theories of particle physics and other esoteric ideas, *is* preposterous. Or, rather, it is ironic science, science that is not experimentally testable or resolvable even in principle and therefore is not science in the strict sense at all. Its primary function is to keep us awestruck before the mystery of the cosmos.

The irony is that Hawking was the first prominent physicist of his generation to predict that physics might soon achieve a complete, unified theory of nature and thus bring about its own demise. He offered up this prophecy in 1980, just after he had been named the Lucasian Professor of Mathematics at the University of Cambridge; Newton had held this chair some 300 years earlier. (Few observers noted that at the end of his speech, titled "Is the End of Theoretical Physics in Sight?," Hawking suggested that computers, given their accelerated evolution, might soon surpass their human creators in intelligence and achieve the final theory on their own.)[2] Hawking spelled out his prophecy in more detail in *A Brief History of Time*. The attainment of a final theory, he declared in the book's closing sentence, might help us to "know the mind of God."[3] The phrase suggested that a final theory would bequeath us a mystical revelation in whose glow we could bask for the rest of time.

But earlier in the book, in discussing what he called the no-boundary proposal, Hawking offered a very different view of what a final theory might accomplish. The no-boundary proposal addressed the age-old questions: What was there before the big bang? What exists beyond the borders of our universe? According to the no-boundary proposal, the entire history of the universe, all of space and all of time, forms a kind of four-dimensional sphere: space-time. Talking about the beginning or end of the universe is thus as meaningless as talking about the beginning or end of a sphere. Physics, too, Hawking conjectured, might form a perfect, seamless whole after it is unified; there might be only one fully consistent unified theory capable of generating space-time as we know it. God might not have had any choice in creating the universe.

"What place, then, for a creator?" Hawking asked.[4] There is *no* place, was his reply; a final theory would exclude God from the universe, and with him all mystery. Like Steven Weinberg, Hawking hoped to rout mysticism, vitalism, creationism from one of their last refuges, the origin of the universe. According to one biographer, Hawking and his wife, Jane, separated in 1990 in part because she, as a devout Christian, had become increasingly offended by his atheism.[5]

After the publication of *Brief History*, several other books took up the question of whether physics could achieve a complete and final theory, one that would answer all questions and thereby bring physics to an end. Those who claimed that no such theory was possible tended to fall back on Gödel's theorem and other esoterica. But throughout his career, Hawking has demonstrated that there is a much more basic obstacle to a theory of everything. Physicists can never eradicate mystery from the universe, they can never find *The Answer*, as long as there are physicists with imaginations as florid as Hawking's.

I suspect that Hawking—who may be less a truth seeker than an artist, an illusionist, a cosmic joker—knew all along that finding and empirically validating a unified theory would be extremely difficult, even impossible. His declaration that physics was on the verge of finding *The Answer* may well have been an ironic statement, less an assertion than a provocation. In 1994, he admitted as much when he told an interviewer that physics might never achieve a final theory after all.[6] Hawking is a master practitioner of ironic physics and cosmology.

Cosmology's Great Surprises

The most remarkable fact about modern cosmology is that it is not *all* ironic. Cosmology has given us several genuine, irrefutable surprises. Early in this century, the Milky Way, the island of stars within which our own sun nestles, was thought to constitute the entire universe. Then astronomers realized that minute smudges of light called nebulae, thought to be mere clouds of gas within the Milky Way, were actually islands of stars. The Milky Way was only one of a vast number of galaxies in a universe that was much, much larger than anyone had imagined. That finding represented an enormous, empirical, irrevocable surprise, one that even the most hardcore relativist would be hard-pressed to deny. To paraphrase Sheldon Glashow, galaxies are not imagined and unimagined; they exist.

Another great surprise was to come. Astronomers discovered that the glow of galaxies was invariably shifted toward the red end of the visible spectrum. Apparently the galaxies were hurtling away from the earth and from each other, and this recessional velocity caused the light to undergo a Doppler shift (the same shift that makes an ambulance siren deepen in pitch as it races away from a listener). The red-shift evidence supported a theory, based on Einstein's theory of relativity, that the universe had begun in an explosion that was ongoing.

In the 1950s, theorists predicted that the fiery birth of the universe billions of years ago should have left an afterglow in the form of faint microwaves. In 1964 two radio engineers at Bell Laboratories stumbled upon the so-called cosmic background radiation. Physicists had also proposed that the fireball of creation would have served as a nuclear furnace in which hydrogen fused into helium and other light elements. Careful observations over the past few decades have shown that the abundances of light elements in the Milky Way and in other galaxies precisely match theoretical predictions.

David Schramm of Fermilab and the University of Chicago likes to call these three lines of evidence—the red shift of galaxies, the microwave background, and the abundance of elements—the pillars on which the big bang theory stands. Schramm is a big, barrel-chested, ebullient man, a pilot, mountain climber, and former collegiate champion in Greco-Roman wrestling. He is an indefatigable booster of the big bang—and of his own role in refining the calculations of light-element abundances. After I arrived at the symposium in Sweden, Schramm sat me down and went over the evidence for the big bang in great detail. "The big bang is in *fantastic* shape," he said. "We have the basic framework. We just need to fill in the gaps."

Schramm acknowledged that some of those gaps were rather large. Theorists cannot determine precisely how the hot plasma of the early universe condensed into stars and galaxies. Observations have suggested that the visible, starry stuff that astronomers can see through their telescopes is not massive enough to keep galaxies from flying apart; some invisible, or dark, matter must be binding the galaxies together. All the matter that we can see, in other words, may be just foam on a deep, dark sea.

Another question concerns what cosmologists like to call "largescale structure." In the early days of cosmology, galaxies seemed to be scattered more or less evenly throughout the universe. But as observations improved, astronomers found that galaxies tend to huddle together in clusters surrounded by gigantic voids. Finally, there is the question of how the universe behaved in the so-called quantum gravity era, when the cosmos was so small and hot that all the forces of the universe were thought to be unified. These were the issues that dominated discussion during the Nobel symposium in Sweden. But none of these problems, Schramm insisted, threatened the basic framework of the big bang. "Just because you can't predict tornadoes," he said, "doesn't mean the earth isn't round."[7]

Schramm delivered much the same message to his fellow cosmologists at the Nobel symposium. He kept proclaiming that cosmology was in a "golden age." His chamber of commerce enthusiasm seemed to grate on some of his colleagues; after all, one does not become a cosmologist to fill in the details left by the pioneers. After Schramm's umpteenth "golden age" proclamation, one physicist snapped that you cannot know an age is golden when you are *in* that age but only in retrospect. Schramm jokes proliferated. One colleague speculated that the stocky physicist might represent the solution to the dark matter problem. Another proposed that Schramm be employed as a plug to prevent our universe from being sucked down a wormhole.

Toward the end of the meeting in Sweden, Hawking, Schramm, and all the other cosmologists piled into a bus and drove to a nearby village to hear a concert. When they entered the Lutheran church where the concert was to be held, it was already filled. The orchestra, a motley assortment of flaxen-haired youths and wizened elders clutching violins, clarinets, and other instruments, was already seated at the front of the church. Their neighbors jammed the balconies and the seats at the rear of the church. As the scientists filed down the center aisle to the pews at the front reserved for them, Hawking leading the way in his motorized wheelchair, the townspeople started to clap, tentatively at first, then passionately, for almost a full minute. The symbolism was perfect: for this moment, at least, in this place and for these people, science had usurped the role of religion as the source of truth about the universe.

Doubts had infiltrated the scientific priesthood, however. In the moments before the concert began, I overheard a conversation between David Schramm and Neil Turok, a young British physicist. Turok confided to Schramm that he was so concerned about the intractability of questions related to dark matter and the distribution of galaxies that he was thinking of quitting cosmology and entering another field. "Who says we have any right to understand the universe, anyway?" Turok asked plaintively. Schramm shook his big head. The basic framework of cosmology, the big bang theory, was absolutely sound, he whispered insistently, as the orchestra began warming up; cosmologists just needed to tie up a few loose ends. "Things will sort themselves out," Schramm said.

Turok seemed to find Schramm's words comforting, but he probably should have been alarmed. What if Schramm was right? What if cosmologists already had, in the big bang theory, the major answer to the puzzle of the universe? What if all that remained was tying up loose ends, those that could be tied up? Given this possibility, it is no wonder that "strong" scientists such as Hawking have vaulted past the big bang theory into postempirical science. What else can someone so creative and ambitious do?

The Russian Magician

One of Stephen Hawking's few rivals as a practitioner of ironic cosmology is Andrei Linde, a Russian physicist who emigrated to Switzerland in 1988 and to the United States two years later. Linde, too, was at the Nobel symposium in Sweden, and his antics were among the highlights of the meeting. After imbibing a drink or two at an outdoor cocktail party, Linde snapped a rock in half with a karate chop. He stood on his hands and then flipped himself backward and landed on his feet. He pulled a box of wooden matches out of his pocket and placed two of them, forming a cross, on his hand. While Linde kept his hand—at least seemingly—perfectly still, the top match trembled and hopped as if jerked by an invisible string. The trick maddened his colleagues. Before long, matches and curses were flying every which way as a dozen or so of the world's most prominent cosmologists sought in vain to duplicate Linde's feat. When they demanded to know how Linde did it, he smiled and growled, "Ees kvantum fluctuation."

Linde is even more renowned for his theoretical sleights of hand. In the early 1980s he helped to win acceptance for one of the more extravagant ideas to emerge from particle physics: inflation. The invention of inflation (the term *discovery* is not appropriate here) is generally credited to Alan Guth of MIT, but Linde helped to refine the theory and win its acceptance. Guth and Linde proposed that very early in the history of our universe—at $T = 10^{-43}$ seconds, to be precise, when the cosmos was allegedly much smaller than a proton—gravity might briefly have become a repulsive rather than an attractive force. As a result, the universe supposedly passed through a tremendous, exponential growth spurt before settling down to its current, much more leisurely rate of expansion.

Guth and Linde based their idea on untested—and almost certainly untestable—unified theories of particle physics. Cosmologists nonetheless fell in love with inflation, because it could explain some nagging problems raised by the standard big bang model. First, why does the universe appear more or less the same in all directions? The answer is that just as blowing up a balloon smooths out its wrinkles, so would the exponential expansion of the universe render it relatively smooth. Conversely, inflation also explains why the universe is not a *completely* homogeneous consommé of radiation, but contains lumps of matter in the form of stars and galaxies. Quantum mechanics suggests that even empty space is brimming with energy; this energy constantly fluctuates, like waves dancing on the surface of a windblown lake. According to inflation, the peaks generated by these quantum fluctuations in the very early universe could have become large enough, after being inflated, to serve as the gravitational seeds from which stars and galaxies would grow.

Inflation has some startling implications, one of which is that everything we can see through our telescopes represents an infinitesimal fraction of the vastly larger realm created during inflation. But Linde did not stop there. Even that prodigious universe, he contended, is only one of an infinite number spawned by inflation. Inflation, once it begins, can never end; it has created not only our universe—the galaxy-emblazoned realm we ponder through our telescopes—but also countless others. This megaverse has what is known as a fractal structure: the big universes sprout little universes, which sprout still smaller ones, and so on. Linde called his model the chaotic, fractal, eternally self-reproducing, inflationary universe.[8]

For someone so publicly playful and inventive, Linde can be surprisingly dour. I glimpsed this side of his character when I visited him at Stanford University, where he and his wife, Renata Kallosh, who is also a theoretical physicist, began working in 1990. When I arrived at the gray, cubist home they were renting, Linde gave me a perfunctory tour. In the backyard we encountered Kallosh, who was rooting happily in a flower bed. "Look, Andrei!" she cried, pointing to a nest filled with cheeping birds on a tree branch above her. Linde, his pallor and squint betraying his unfamiliarity with sunlight, merely nodded. When I asked if he found California relaxing, he muttered, "Maybe too relaxing."

As Linde recounted his life story, it became apparent that anxiety, even depression, had played a significant role in motivating him. At various stages of his career, he would despair of having any insight into the nature of things—just before achieving a breakthrough. Linde had stumbled on the basic concept of inflation in the late 1970s, while in Moscow, but decided that the idea was too flawed to pursue. His interest was revived by Alan Guth's proposal that inflation could explain several puzzling features of the universe, such as its smoothness, but Guth's version, too, was flawed. After pondering the problem so obsessively that he developed an ulcer, Linde showed how Guth's model could be adjusted to eliminate the technical problems.

But even this model of inflation depended on features of unified theories that were, Linde felt, suspect. Eventually—after falling into a gloom so deep that he had difficulty getting out of bed—he determined that inflation could result from much more generic quantum processes first proposed by John Wheeler. According to Wheeler, if one had a microscope trillions upon trillions of times more powerful than any in existence, one would see space and time fluctuating wildly because of quantum uncertainty. Linde contended that what Wheeler called "space-time foam" would inevitably give rise to the conditions necessary for inflation.

Inflation is a self-exhausting process; the expansion of space quickly causes the energy driving inflation to dissipate. But Linde argued that once inflation begins, it will always continue somewhere—again, because of quantum uncertainty. (A handy feature, this quantum uncertainty.) New universes are sprouting into existence at this very moment. Some immediately collapse back into themselves. Others inflate so fast that matter never has a chance to coalesce. And some, like ours, settle down to an expansion leisurely enough for gravity to mold matter into galaxies, stars, and planets.

Linde sometimes compared this supercosmos to an infinite sea. Viewed up close, the sea conveys an impression of dynamism and change, of waves heaving up and down. We humans, because we live within one of these heaving waves, think that the entire universe is expanding. But if we could rise above the sea's surface, we would realize that our expanding cosmos is just a tiny, insignificant, local feature of an infinite, eternal ocean. In a way, Linde contended, the old steady-state theory of Fred Hoyle (which I discuss later in this chapter) was right; when seen from a God-like perspective, the supercosmos exhibits a kind of equilibrium.

Linde was hardly the first physicist to posit the existence of other universes. But whereas most theorists treat other universes as mathematical abstractions, and slightly embarrassing ones at that, Linde delighted in speculating about their properties. Elaborating on his self-reproducing universe theory, for example, Linde borrowed from the language of genetics. Each universe created by inflation gives birth to still other "baby universes." Some of these offspring retain the "genes" of their predecessors and evolve into similar universes with similar laws of nature—and perhaps similar inhabitants. Invoking the anthropic principle, Linde proposed that some cosmic version of natural selection might favor the perpetuation of universes that are likely to produce intelligent life. "The fact that somewhere else there is life like ours is to me almost certain," he said. "But we can never know this."

Like Alan Guth and several other cosmologists, Linde liked to speculate on the feasibility of creating an inflationary universe in a laboratory. But only Linde asked: Why would one *want* to create another universe? What purpose would it serve? Once some cosmic engineer created a new universe, it would instantaneously dissociate itself from its parent at faster-than-light speeds, according to Linde's calculations. No further communication would be possible.

On the other hand, Linde surmised, perhaps the engineer could manipulate the seed of preinflationary stuff in such a way that it would evolve into a universe with particular dimensions, physical laws, and constants of nature. In that way, the engineer could impress a message of some sort onto the very structure of the new universe. In fact, Linde suggested, our own universe might have been created by beings in another universe, and physicists such as Linde, in their fumbling attempts to unravel the laws of nature, might actually be decoding a message from our cosmic parents.

Linde presented these ideas warily, watching my reaction. Only at the end, perhaps taking satisfaction in my gaping mouth, did he permit himself a little smile. His smile faded, however, when I wondered what the message imbedded in our universe might be. "It seems," he said wistfully, "that we are not quite grown up enough to know." Linde looked even more glum when I asked if he ever worried that all his work might be—I struggled to find the right word—bullshit.

"In my moments of depression I feel myself a complete idiot," he replied. "What I am playing with are some very primitive toys." He added that he tried not to become too attached to his own ideas. "Sometimes the models are quite strange, and if you take them too seriously then you are in danger to be trapped. I would say this is similar to running over very thin ice on the lake. If you just are running very fast, you may not sink, and you may go a large distance. If you stay and think if you are running in the right direction, you may go down."

Linde seemed to be saying that his goal as a physicist was not to achieve resolution, to arrive at *The Answer* or even simply at *an* answer, but to keep moving, to keep skating. Linde feared the thought of finality. His self-reproducing universe theory makes sense in this light: if the universe is infinite and eternal, then so is science, the quest for knowledge. But even a physics confined to this universe, Linde suggested, cannot be close to resolution. "For example, you do not include consciousness. Physics studies matter, and consciousness is not matter." Linde agreed with John Wheeler that reality must be in some sense a participatory phenomenon. "Before you make measurement, there is no universe, nothing you can call objective reality," Linde said.

Linde, like Wheeler and David Bohm, seemed to be tormented by mystical yearnings that physics alone could never satisfy. "There is some limit to rational knowledge," he said. "One way to study the irrational is to jump into it and just meditate. The other is to study the boundaries of the irrational with the tools of rationality." Linde chose the latter route, because physics offered a way "not to say total nonsense" about the workings of the world. But sometimes, he confessed, "I'm depressed when I think I will die like a physicist."

The Deflation of Inflation

The fact that Linde has earned so much respect—he was courted by
several U.S. universities before choosing Stanford—bears witness
both to his rhetorical skills and to cosmologists' hunger for new ideas.
Nonetheless, by the early 1990s, inflation and many of the other exotic ideas
that had emerged from particle physics in the previous decade had begun
losing support from mainstream cosmologists. Even David Schramm, who
had been quite bullish on inflation when I met him in Sweden, had his doubts
when I spoke to him several years later. "I *like* inflation," Schramm said, but
it can never be thoroughly verified because it does not generate any unique
predictions, predictions that cannot be explained in some other way. "You
won't see that for inflation," Schramm continued, "whereas for the big bang
itself you do see that. The beautiful, cosmic microwave background and the
light-element abundances tell you, 'This is it.' There's no other way of getting
these observations."

Schramm acknowledged that as cosmologists venture further back toward
the beginning of time, their theories become more speculative. Cosmology
needs a unified theory of particle physics to describe processes in the very
early universe, but validating a unified theory may be extremely difficult.
"Even if somebody comes up with a really beautiful theory, like superstring
theory, there's not any way it can be tested. So you're not really doing the
scientific method, where you make predictions and then check it. There's not
that experimental check going on. It's more just mathematical consistency."

Could the field end up being like the interpretation of quantum
mechanics, where the standards are primarily aesthetic? "That's a real
problem I have with it," Schramm replied, "that unless one comes up with
tests, we are into the more philosophical rather than physics area. The tests
have to give the universe as we observe it, but that's more of a post-diction
rather than a pre-diction." It is always possible that theoretical explorations

of black holes, superstrings, Wheeler's it from bit, and other exotica might yield some sort of breakthough. "But until someone comes up with definitive tests," Schramm said, "or we're lucky enough to find a black hole that we can probe in a careful way, we're not going to have that kind of 'Eureka' where you're really confident you know the answer."

Realizing the significance of what he was saying, Schramm suddenly switched back to his customary boosterish, public relations mode. The fact that cosmologists are having so much difficulty advancing beyond the big bang model is a *good* sign, he insisted, falling back on an all-too-familiar argument. "For example, at the turn of the century, physicists were saying most of physics is solved. There are a few nagging little problems, but it's basically solved. And we found that was certainly not the case. In fact, what we find is, that usually is the clue that there's another big step coming. Just when you think the end is in sight, you find that's a wormhole into a whole new perspective into the universe. And I think that may be what's going to happen, that we're zooming in and starting to see things. We'll see certain nagging problems that we haven't been able to solve. And I expect the solution to those problems will lead to a whole new rich and exciting area. The enterprise is not going to die."[9]

But what if cosmology *has* passed its peak, in the sense that it is unlikely to deliver any more empirical surprises as profound as the big bang theory itself? Cosmologists are lucky to know anything with certainty, according to Howard Georgi, a particle physicist at Harvard University. "I think you have to regard cosmology as a historical science, like evolutionary biology," said Georgi, a cherub-faced man with a cheerfully sardonic manner. "You're trying to look at the present-day universe and extrapolate back, which is an interesting but dangerous thing to do, because there may have been accidents that had big effects. And they try very hard to understand what kinds of things can be accidental and what features are robust. But I find it difficult to understand those arguments well enough to really be convinced." Georgi suggested that cosmologists might acquire some much-needed humility by reading the books of the evolutionary biologist Stephen Jay Gould, who discusses the potential pitfalls of reconstructing the past based on our knowledge of the present (see Chapter 5).

Georgi chuckled, perhaps recognizing the improbability of any cosmologist's taking his advice. Like Sheldon Glashow, whose office was just down the hall, Georgi had once been a leader in the search for a unified theory of physics. And like Glashow, Georgi eventually denounced superstring theory and other candidates for a unified theory as nontestable and thus nonscientific. The fate of particle physics and cosmology, Georgi noted, are to some extent intertwined. Cosmologists hope that a unified theory will help them understand the universe's origins more clearly. Conversely, some particle physicists hope that in lieu of terrestrial experiments, they can find confirmation of their theories by peering through telescopes at the edge of the universe. "That strikes me as a bit of a push," Georgi remarked mildly, "but what can I say?" When I asked him about quantum cosmology, the field explored by Hawking, Linde, and others, Georgi smiled mischievously. "A simple particle physicist like myself has trouble in those uncharted waters," he said. He found papers on quantum cosmology, with all their talk of wormholes and time travel and baby universes, "quite amusing. It's like reading Genesis." As for inflation, it is "a wonderful sort of scientific myth, which is *at least as* good as any other creation myth I've ever heard."[10]

The Maverick of Mavericks

There will always be those who reject not only inflation, baby universes, and other highly speculative hypotheses, but the big bang theory itself. The dean of big bang bashers is Fred Hoyle, a British astronomer and physicist. A selective reading of Hoyle's résumé might make him appear the quintessential insider. He studied at the University of Cambridge under the Nobel laureate Paul Dirac, who correctly predicted the existence of antimatter. Hoyle became a lecturer at Cambridge in 1945, and in the 1950s he helped to show how stars forge the heavy elements of which planets and people are made. Hoyle founded the prestigious Institute of Astronomy at Cambridge in the early 1960s and served as its first director. For these and other achievements he was knighted in 1972. Yes, Hoyle is Sir Fred. Yet Hoyle's stubborn refusal to accept the big bang theory—and his adherence to fringe ideas in other fields—made him an outlaw in the field he had helped to create.[11]

Since 1988, Hoyle has lived in a high-rise apartment building in Bournemouth, a town on England's southern coast. When I visited him there, his wife, Barbara, let me in and took me into the living room, where I found Hoyle sitting in a chair watching a cricket match on television. He rose and shook my hand without taking his eyes off the match. His wife, gently admonishing him for his rudeness, went over to the TV and turned it off. Only then did Hoyle, as if waking from a spell, turn his full attention to me.

I expected Hoyle to be odd and embittered, but he was, for the most part, all too amiable. With his pug nose, jutting jaw, and penchant for slang— colleagues were "chaps" and a bogus theory a "bust flush"—he exuded a kind of blue-collar integrity and geniality. He seemed to revel in the role of outsider. "When I was young the old regarded me as an outrageous young fellow, and now that I'm old the young regard me as an outrageous old fellow." He chuckled. "I should say that nothing would embarrass me more than if I were to be viewed as someone who is repeating what he has been saying year after year," as many astronomers do. "What I would be worried about is somebody coming along and saying, 'What you've been saying is technically not sound.' That would worry me." (Actually, Hoyle has been accused of both repetitiveness and technical errors.[12])

Hoyle had a knack for sounding reasonable—for example, when arguing that the seeds of life must have come to our planet from outer space. The spontaneous generation of life on the earth, Hoyle once remarked, would have been as likely as the assemblage of a 747 aircraft by a tornado passing through a junkyard. Elaborating on this point during our interview, Hoyle pointed out that asteroid impacts rendered the earth uninhabitable until at least 3.8 billion years ago and that cellular life had almost certainly appeared by 3.7 billion years ago. If one thought of the entire 4.5-billion-year history of the planet as a 24-hour day, Hoyle elaborated, then life appeared in about half an hour. "You've got to discover DNA; you've got to make thousands of enzymes in that half an hour," he explained. "And you've got to do it in a very hostile situation. So I find when you put all this together it doesn't add up to a very attractive situation." As Hoyle spoke, I found myself nodding in agreement. Yes, of course life could not have originated here. What could be more obvious? Only later did I realize that according to Hoyle's timetable, apes were transmogrified into humans some 20 seconds ago, and modern civilization sprang into existence in less than 1/10 second. Improbable, perhaps, but it happened.

Hoyle had first started thinking seriously about the origin of the universe shortly after World War II, during long discussions with two other physicists, Thomas Gold and Hermann Bondi. "Bondi had a relative somewhere—he seemed to have relatives everywhere—and one sent him a case of rum," Hoyle recalled. While imbibing Bondi's liquor, the three physicists turned to a perennial puzzle of the young and intoxicated: How did we come to be?

The finding that all galaxies in the cosmos are receding from one another had already convinced many astronomers that the universe had exploded into being at a specific time in the past and was still expanding. Hoyle's fundamental objection to this model was philosophical. It did not make sense to talk about the creation of the universe unless one already had space and time for the universe to be created in. "You lose the universality of the laws of physics," Hoyle explained to me. "Physics is no longer." The only alternative to this absurdity, Hoyle decided, was that space and time must have always existed. He, Gold, and Bondi thus invented the steady-state theory, which posits that the universe is infinite both in space and time and constantly generates new matter through some still-unknown mechanism.

Hoyle stopped promoting the steady-state theory after the discovery of the microwave background radiation in the early 1960s seemed to provide conclusive evidence for the big bang. But his old doubts resurfaced in the 1980s, as he watched cosmologists struggle to explain the formation of galaxies and other puzzles. "I began to get the sense that there was something seriously wrong"—not only with new concepts such as inflation and dark matter, he said, but with the big bang itself. "I'm a great believer that if you have a correct theory, you show a lot of positive results. It seems to me that they'd gone on for 20 years, by 1985, and there wasn't much to show for it. And that couldn't be the case if it was right."

Hoyle thus resurrected the steady-state theory in a new and improved form. Rather than one big bang, he said, many little bangs occurred in preexisting space and time. These little bangs were responsible for light elements and the red shifts of galaxies. As for the cosmic microwave background, Hoyle's best guess was that it is radiation emitted by some sort of metallic interstellar dust. Hoyle acknowledged that his "quasi–steady state theory," which in effect replaced one big miracle with many little ones, was far from perfect. But he insisted that recent versions of the big bang theory, which posit the existence of inflation, dark matter, and other exotica, are much more deeply flawed. "It's like medieval theology," he exclaimed in a rare flash of anger.

The longer Hoyle spoke, however, the more I began to wonder just how sincere his doubts about the big bang were. In some of his statements, he revealed a proprietary fondness for the theory. One of the great ironies of modern science is that Hoyle coined the term *big bang* in 1950 while he was doing a series of radio lectures on astronomy. Hoyle told me that he meant not to disparage the theory, as many accounts have suggested, but merely to describe it. At the time, he recalled, astronomers often referred to the theory as "the Friedman cosmology," after a physicist who showed how Einstein's theory of relativity gave rise to an expanding universe.

"That was poison," Hoyle declared. "You had to have something vivid. So I thought up the big bang. If I had patented it, copyrighted it...." he mused. The August 1993 *Sky and Telescope* magazine initiated a contest to rename the theory. After mulling over thousands of suggestions, the judges announced they could find none worthy of supplanting *big bang*.[13] Hoyle said he was not surprised. "Words are like harpoons," he commented. "Once they go in, they are very hard to pull out."

Hoyle also seemed obsessed with how close he had come to discovering the cosmic microwave background. It was 1963, and during an astronomy conference Hoyle fell into a conversation with Robert Dicke, a physicist from Princeton who was planning to search for the cosmic microwaves predicted by the big bang model. Dicke told Hoyle that he expected the microwaves to be about 20 degrees above absolute zero, which is what most theorists were predicting. Hoyle informed Dicke that in 1941, a Canadian radio astronomer named Andrew McKellar had found interstellar gas radiating microwaves at 3 degrees, not 20.

To Hoyle's everlasting regret, neither he nor Dicke, during their conversation, spelled out the implication of McKellar's finding: that the microwave background might be 3 degrees. "We just sat there drinking coffee," Hoyle remembered, his voice rising. "If either of us had said, 'Maybe it *is* 3 degrees,' we'd have gone straightaway and checked it, and then we'd have had it in 1963." A year later, just before Dicke turned on his microwave experiment, Arno Penzias and Robert Wilson of Bell Laboratories discovered the 3-degree microwave radiation, an achievement for which they later received the Nobel Prize. "I've always felt that was one of the worst misses of my life," Hoyle sighed, shaking his head slowly.

Why should Hoyle care that he had nearly discovered a phenomenon that he now derided as spurious? I think Hoyle, like many mavericks, once hoped to be a member of the inner circle of science, draped in honor and glory. He went far toward realizing that goal. But in 1972, officials at Cambridge forced Hoyle to resign from his post as director of the Institute for Astronomy—for political rather than scientific reasons. Hoyle and his wife left Cambridge for a cottage on an isolated moor in northern England, where they lived for 15 years before moving to Bournemouth. During this period, Hoyle's anti-authoritarianism, which had always served him so well, became less creative than reactionary. He degenerated into what Harold Bloom derided as a "mere rebel," although he still dreamed of what might have been.

Hoyle also seemed to suffer from another problem. The task of the scientist is to find patterns in nature. There is always the danger that one will see patterns where there are none. Hoyle, in the latter part of his career, seemed to have succumbed to this pitfall. He saw patterns—or, rather, conspiracies—both in the structure of the cosmos and among those scientists who rejected his radical views. Hoyle's mind-set is most evident in his views on biology. Since the early 1970s he has argued that the universe is pervaded by viruses, bacteria, and other organisms. (Hoyle first broached this possibility in 1957 in *The Black Cloud*, which remains the best known of his many science fiction novels.) These space-faring microbes supposedly provided the seeds for life on earth and spurred evolution thereafter; natural selection played little or no role in creating the diversity of life.[14] Hoyle has also asserted that epidemics of influenza, whooping cough, and other diseases are triggered when the earth passes through clouds of pathogens.

Discussing the biomedical establishment's continued belief in the more conventional, person-to-person mode of disease transmission, Hoyle glowered. "They don't look at those data and say, "Well, it's wrong,' and stop teaching it. They just go on doping out the same rubbish. And that's why if you go to the hospital and there's something wrong with you, you'll be lucky if they cure it." But if space is swarming with organisms, I asked, why haven't they been detected? Oh, but they probably were, Hoyle assured me. He suspected that U.S. experiments on high-altitude balloons and other platforms had turned up evidence of life in space in the 1960s, but that officials had hushed it up. Why? Perhaps for reasons related to national security, Hoyle suggested, or because the results contradicted received wisdom. "Science today is locked into paradigms," he intoned solemnly. "Every avenue is blocked by beliefs that are wrong, and if you try to get anything published by a journal today, you will run against a paradigm and the editors will turn it down."

Hoyle emphasized that, contrary to certain reports, he did not believe the AIDS virus came from outer space. It "is such a strange virus I have to believe it's a laboratory product," he said. Was Hoyle implying that the pathogen might have been produced by a biological warfare program that went awry? "Yes, that's my feeling," he replied.

Hoyle also suspected that life and indeed the entire universe must be unfolding according to some cosmic plan. The universe is an "obvious fix,"

Hoyle said. "There are too many things that look accidental which are not." When I asked if Hoyle thought some supernatural intelligence was guiding things, he nodded gravely. "That's the way I look on God. It is a fix, but how it's being fixed I don't know." Of course, many of Hoyle's colleagues— and probably a majority of the public at large— share his view that the universe is, must be, a conspiracy. And perhaps it is. Who knows? But his assertion that scientists would deliberately suppress evidence of space-faring microbes or of legitimate flaws in the big bang theory reveals a fundamental misunderstanding of his colleagues. Most scientists *yearn* for such revolutionary discoveries.

The Sun Principle

Hoyle's eccentricities aside, future observations may prove his skepticism toward the big bang to be at least partially prescient. Astronomers may find that the cosmic microwave background stems not from the flash of the big bang, but from some more mundane source, such as dust in our own Milky Way. The nucleosynthesis evidence, too, may not hold up as well as Schramm and other big bang boosters have claimed. But even if one knocks these two pillars out from under the big bang, the theory can still stand on the red-shift evidence, which even Hoyle agrees provides proof that the universe is expanding.

The big bang theory does for astronomy what Darwin's theory of natural selection did for biology: it provides cohesion, sense, meaning, a unifying narrative. That is not to say that the theory can—or will ever—explain all phenomena. Cosmology, in spite of its close conjunction with particle physics, the most painstakingly precise of sciences, is far from being precise itself. That fact has been demonstrated by the persistent inability of astronomers to agree on a value for the Hubble constant, which is a measure of the size, age, and rate of expansion of the universe. To derive the Hubble constant, one must measure the breadth of the red shift of galaxies and their distance from the earth. The former measurement is straightforward, but the latter is horrendously complicated. Astronomers cannot assume that the apparent brightness of a galaxy is proportional to its distance; the galaxy might be nearby, or it might simply be intrinsically bright. Some astronomers insist that the universe is 10 billion years old or even younger; others are equally sure that it cannot be less than 20 billion years old.[15]

The debate over the Hubble constant offers an obvious lesson: even when performing a seemingly straightforward calculation, cosmologists must make various assumptions that can influence their results; they must interpret their data, just as evolutionary biologists and historians do. One should thus take with a large grain of salt any claims based on high precision (such as Schramm's assertion that nucleosynthesis calculations agree with theoretical predictions to five decimal points).

More detailed observations of our cosmos will not necessarily resolve questions about the Hubble constant or other issues. Consider: the most mysterious of all stars is our own sun. No one really knows, for example, what causes sunspots or why their numbers wax and wane over periods of roughly a decade. Our ability to describe the universe with simple, elegant models stems in large part from our lack of data, our ignorance. The more clearly we can see the universe in all its glorious detail, the more difficult it will be for us to explain with a simple theory how it came to be that way. Students of human history are well aware of this paradox, but cosmologists may have a hard time accepting it.

This sun principle suggests that many of the more exotic suppositions of cosmology are due for a fall. As recently as the early 1970s black holes were still considered theoretical curiosities, not to be taken seriously. (Einstein himself thought black holes were "a blemish to be removed from his theory by a better mathematical formulation," according to Freeman Dyson.[16]) Gradually, as a result of the proselytizing of John Wheeler and others, they have come to be accepted as real objects. Many theorists are now convinced that almost all galaxies, including our own, harbor gigantic black holes at their cores. The reason for this acceptance is that no one can imagine a better way to explain the violent swirling of matter at the center of galaxies.

These arguments depend on our ignorance. Astronomers should ask themselves this: If they could somehow be whisked to the center of the Andromeda galaxy or our own Milky Way, what would they find there? Would they find something that resembles the black holes described by current theory, or would they encounter something entirely different, something that no one imagined or could have imagined? The sun principle suggests that the latter outcome is more probable. We humans may never see directly into the dust-obscured heart of our own galaxy, let alone into any other galaxy, but we may learn enough to raise doubts about the black hole hypothesis. We may learn enough to learn, once again, how little we know.

The same is true of cosmology in general. We have learned one astounding, basic fact about the universe. We know that the universe is expanding, and may have been for 10 to 20 billion years, just as evolutionary biologists know that all life evolved from a common ancestor through natural selection. But cosmologists are as unlikely to transcend that basic understanding as evolutionary biologists are to leap beyond Darwinism. David Schramm was right. In the future, the late 1980s and early 1990s will be remembered as the golden age of cosmology, when the field achieved a perfect balance between knowledge and ignorance. As more data flood in in years to come, cosmology may become more like botany, a vast collection of empirical facts only loosely bound by theory.

The End of Discovery

Finally, scientists do not enjoy an infinite ability to discover interesting new things about the universe. Martin Harwit, an astrophysicist and historian of science who until 1995 directed the Smithsonian Institution's National Air and Space Museum in Washington, D.C., made this point in his 1981 book, *Cosmic Discovery:*

The history of most efforts at discovery follows a common pattern, whether we consider the discovery of varieties of insects, the exploration of the oceans for continents and islands, or the search for oil reserves in the ground. There is an initial accelerating rise in the discovery rate as increasing numbers of explorers become attracted. New ideas and tools are brought to bear on the search, and the pace of discovery quickens. Soon, however, the number of discoveries remaining to be made dwindles, and the rate of discovery declines despite the high efficiency of the methods developed. The search is approaching an end. An occasional, previously overlooked feature can be found or a particular rare species encountered; but the rate of discovery begins to decline quickly and then diminishes to a trickle. Interest drops, researchers leave the field, and there is virtually no further activity.[17]

Unlike more experimental fields of science, Harwit pointed out, astronomy is an essentially passive activity. We can only detect celestial phenomena by means of information falling to us from the sky, mostly in the form of electromagnetic radiation. Harwit made various guesses about improvements in mature observational technologies, such as optical telescopes, as well as others that were still nascent, such as gravitational-wave detectors. He presented a graph estimating the rate at which new cosmic discoveries had been made in the past and would be made in the future. The graph was a bell-shaped curve peaking sharply at the year 2000. By that year, we will have discovered roughly half of all the phenomena we can discover, according to Harwit. We will have discovered roughly 90 percent of the accessible phenomena by the year 2200, and the rest will trickle in at a decreasing rate over the next few millennia.

Of course, Harwit conceded, various developments could accelerate or arrest this schedule. "Political factors might dictate that astronomy receive less support in the future. A war might slow the search to a virtual halt, though the postwar era, if there was one, could provide astronomers with discarded military equipment that would again accelerate the discovery rate."[18] Every cloud has a silver lining.

Ironic cosmology will continue, of course, as long as we have poets as imaginative and ambitious as Hawking, Linde, Wheeler, and, yes, Hoyle. Their visions are both humbling, in that they show the limited scope of our empirical knowledge, and exhilarating, since they also testify to the limitlessness of human imagination. At its best, ironic cosmology can keep us awestruck. But it is not science.

John Donne could have been speaking for Hawking, and for all of us, when he wrote: "My thoughts reach all, comprehend all. Inexplicable mystery; I their Creator am in a close prison, in a sick bed, anywhere, and any one of my Creatures, my thoughts, is with the Sunne and beyond the Sunne, overtakes the Sunne, and overgoes the Sunne in one pace, one steppe, everywhere."[19] Let that serve as the epitaph of cosmology.

Chapter Five
The End of Evolutionary Biology

No other field of science is as burdened by its past as is evolutionary biology. It reeks of what the literary critic Harold Bloom called the anxiety of influence. The discipline of evolutionary biology can be defined to a large degree as the ongoing attempt of Darwin's intellectual descendants to come to terms with his overwhelming influence. Darwin based his theory of natural selection, the central component of his vision, on two observations. First, plants and animals usually produce more offspring than their environment can sustain. Second, these offspring differ slightly from their parents and from each other. Darwin concluded that each organism, in its struggle to survive long enough to reproduce, competes either directly or indirectly with others of its species. Chance plays a role in the survival of any individual organism, but nature will favor, or select, those organisms whose variations make them slightly more fit, that is, more likely to survive long enough to reproduce and pass on those adaptive variations to their offspring.

Darwin could only guess what gives rise to the all-important variations between generations. *On the Origin of Species*, first published in 1859, mentioned a proposal set forth by the French biologist Jean-Baptiste Lamarck, that organisms could pass on not only inherited but also acquired characteristics to their heirs. For example, the constant craning of a giraffe to reach leaves high in a tree would alter its sperm or egg so that its offspring would be born with longer necks. But Darwin was clearly uncomfortable with the idea that adaptation is self-directed. He preferred to think that variations between generations are random, and that only under the pressure of natural selection do they become adaptive and lead to evolution.[1]

Unbeknownst to Darwin, during his lifetime an Austrian monk named Gregor Mendel was conducting experiments that would help refute Lamarck's theory and vindicate Darwin's intuition. Mendel was the first scientist to recognize that natural forms can be subdivided into discrete traits, which are transmitted from one generation to the next by what Mendel termed *hereditary particles* and are now called genes. Genes prevent the blending of traits and thereby preserve them. The recombination of genes that takes place during sexual reproduction, together with occasional genetic mistakes, or mutations, provides the variety needed for natural selection to work its magic.

Mendel's 1868 paper on breeding pea plants went largely unnoticed by the scientific community until the turn of the century. Even then, Mendelian genetics was not immediately reconciled with Darwin's ideas. Some early geneticists felt that genetic mutation and sexual recombination might guide evolution along certain paths independently of natural selection. But in the 1930s and 1940s, Ernst Mayr of Harvard University and other evolutionary biologists fused Darwin's ideas with genetics into a powerful restatement of his theory, called the new synthesis, which affirmed that natural selection is the primary architect of biological form and diversity.

The discovery in 1953 of the structure of DNA—which serves as the blueprint from which all organisms are constructed—confirmed Darwin's intuition that all life is related, descended from a common source. Watson and Crick's finding also revealed the source of both continuity and variation that makes natural selection possible. In addition, molecular biology suggested that all biological phenomena could be explained in mechanical, physical terms.

That conclusion was by no means foregone, according to Gunther Stent. In *The Coming of the Golden Age*, he noted that prior to the unraveling of DNA's structure, some prominent scientists felt that the conventional methods and assumptions of science would prove inadequate for understanding heredity and other basic biological questions. The physicist Niels Bohr was the major proponent of this view. He contended that just as physicists had to cope with an uncertainty principle in trying to understand the behavior of an electron, so would biologists face a fundamental limitation when they tried to probe living organisms too deeply:

> there must remain an uncertainty as regards the physical condition to which [the organism is] subjected, and the idea suggests itself that the minimal freedom we must allow the organism in this respect is just large enough to hide its ultimate secrets from us. On this view, the existence of life must be considered as a starting point in biology, in a similar way as the quantum of action, which appears as an irrational element from the point of view of classical mechanical physics, taken together with the existence of the elementary particles, forms the foundation of quantum mechanics.[2]

Stent accused Bohr of trying to revive the old, discredited concept of vitalism, which holds that life stems from a mysterious essence or force that cannot be reduced to a physical process. But Bohr's vitalist vision has not been borne out. In fact, molecular biology has proved one of Bohr's own dicta, that science, when it is most successful, reduces mysteries to trivialities (although not necessarily comprehensible ones).[3]

What can an ambitious young biologist do to make his or her mark in the post-Darwin, post-DNA era? One alternative is to become more Darwinian than Darwin, to accept Darwinian theory as a supreme insight into nature, one that cannot be transcended. That is the route taken by the arch-clarifier and reductionist Richard Dawkins of the University of Oxford. He has honed Darwinism into a fearsome weapon, one with which he obliterates any ideas that challenge his resolutely materialistic, nonmystical view of life. He seems to view the persistence of creationism and other anti-Darwinian ideas as a personal affront.

I met Dawkins at a gathering convened by his literary agent in Manhattan.[4] He is an icily handsome man, with predatory eyes, a knife-thin nose, and incongruously rosy cheeks. He wore what appeared to be an expensive, custom-made suit. When he held out his finely veined hands to make a point, they quivered slightly. It was the tremor not of a nervous man, but of a finely tuned, high-performance competitor in the war of ideas: Darwin's greyhound.

As in his books, Dawkins in person exuded a supreme self-assurance. His statements often seemed to have an implied preamble: "As any fool can see...." An unapologetic atheist, Dawkins announced that he was not the sort of scientist who thought science and religion addressed separate issues and thus could easily coexist. Most religions, he contended, hold that God is responsible for the design and purpose evident in life. Dawkins was determined to stamp out this point of view. "All purpose comes ultimately from natural selection," he said. "This is the credo that I want to put forward."

Dawkins then spent some 45 minutes setting forth his ultrareductionist version of evolution. He suggested that we think of genes as little bits of software that have only one goal: to make more copies of themselves. Carnations, cheetahs, and all living things are just elaborate vehicles that these "copy-me programs" have created to help them reproduce. Culture, too, is based on copy-me programs, which Dawkins called memes. Dawkins asked us to imagine a book with the message: Believe this book and make your children believe it or when you die you will all go to a very unpleasant place called hell. "That's a very effective piece of copy-me code. Nobody is foolish enough to just accept the injunction, 'Believe this and tell your children to believe it.' You have to be a little more subtle and dress it up in some more elaborate way. And of course we know what I'm talking about." Of course. Christianity, like all religions, is an extremely successful chain letter. What could make more sense?

Dawkins then fielded questions from the audience, a motley assortment of journalists, educators, book editors, and other quasi-intellectuals. One listener was John Perry Barlow, a former Whole Earth hippie and occasional lyricist for the Grateful Dead who had mutated into a New Age cyber-prophet. Barlow, a bearish man with a red bandanna tied around his throat, asked

Dawkins a long question having something to do with where information *really* exists.

Dawkins's eyes narrowed, and his nostrils flared ever so slightly as they caught the scent of woolly-headedness. Sorry, he said, but he did not understand the question. Barlow spoke for another minute or so. "I feel you are trying to get at something which interests you but doesn't interest me," Dawkins said and scanned the room for another questioner. Suddenly, the room seemed several degrees chillier.

Later, during a discussion about extraterrestrial life, Dawkins set forth his belief that natural selection is a cosmic principle; wherever life is found, natural selection has been at work. He cautioned that life cannot be too common in the universe, because thus far we have found no evidence of life on other planets in the solar system or elsewhere in the cosmos. Barlow bravely broke in to suggest that our inability to detect alien life-forms may stem from our perceptual inadequacies. "We don't know who discovered water," Barlow added meaningfully, "but we can be pretty sure it wasn't fish." Dawkins turned his level gaze on Barlow. "So you mean we're looking at them all the time," Dawkins asked, "but we don't see them?" Barlow nodded. "Yessss," Dawkins sighed, as if exhaling all hope of enlightening the unutterably stupid world.

Dawkins can be equally harsh with his fellow biologists, those who have dared to challenge the basic paradigm of Darwinism. He has argued, with devastating persuasiveness, that all attempts to modify or transcend Darwin in any significant way are flawed. He opened his 1986 book *The Blind Watchmaker* with the following proclamation: "Our existence once presented the greatest of all mysteries, but.... it is a mystery no longer because it is solved. Darwin and Wallace solved it, though we shall continue to add footnotes to their solution for a while yet."[5]

"There's always an element of rhetoric in those things," Dawkins replied when I asked him later about the footnotes remark. "On the other hand, it's a legitimate piece of rhetoric," in that Darwin did solve "the mystery of how life came into existence and how life has the beauty, the adaptiveness, the complexity it has." Dawkins agreed with Gunther Stent that all the great advances in biology since Darwin—Mendel's demonstration that genes come in discrete packages, Watson and Crick's discovery of the double-helical structure of DNA— buttressed rather than undermined Darwin's basic idea.

Molecular biology has recently revealed that the process whereby DNA interacts with RNA and proteins is more complicated than previously thought, but the basic paradigm of genetics—DNA-based genetic transmission—is in no danger of collapsing. "What would be a serious reversal," Dawkins said, "would be if you could take a whole organism, a zebra on the Serengeti Plain, and allow it to acquire some characteristic, like learning a new route to the water hole, and have that backward encoded into the genome. Now if anything like that were to happen I really would eat my hat."

There are still some rather large biological mysteries left, such as the origin of life, of sex, and of human consciousness. Developmental biology—which seeks to show how a single fertilized cell becomes a salamander or an evangelist—also raises important issues. "We certainly need to know how that works, and it's going to be very, very complicated." But Dawkins insisted that developmental biology, like molecular genetics before it, would simply fill in more details within the Darwinian paradigm.

Dawkins was "fed up" with those intellectuals who argued that science alone could not answer ultimate questions about existence. "They think science is too arrogant and that there are certain questions that science has no business to ask, that traditionally have been of interest to religious people. As though *they* had any answers. It's one thing to say it's very difficult to know how the universe began, what initiated the big bang, what consciousness is. But if science has difficulty explaining something, there sure as hell is no one else who is going to explain it." Dawkins quoted, with great gusto, a remark by the great British biologist Peter Medawar that some people "'enjoy wallowing in a nonthreatening squalor of incomprehension.' I want to understand," Dawkins added fiercely, "and understanding means to me scientific understanding."

I asked Dawkins why he thought his message—that Darwin basically told us all we know and all we need to know about life—met with resistance not only from creationists or New Agers or philosophical sophists, but even from obviously competent biologists. "It may be I don't get the point across with sufficient clarity," he replied. But the opposite, of course, is more likely to be true. Dawkins gets his point across with utter clarity, so much so that he leaves no room for mystery, meaning, purpose—or for great scientific revelations beyond the one that Darwin himself gave us.

Gould's Contingency Plan

Naturally, some modern biologists bridle at the notion that they are merely adding footnotes to Darwin's magnum opus. One of Darwin's strongest (in Bloom's sense) descendants is Stephen Jay Gould of Harvard University. Gould has sought to resist the influence of Darwin by denigrating his theory's power, by arguing that it doesn't explain that much. Gould began staking out his philosophical position in the 1960s by attacking the venerable doctrine of uniformitarianism, which holds that the geophysical forces that shaped the earth and life have been more or less constant.[6]

In 1972, Gould and Niles Eldredge of the American Museum of Natural History in New York extended this critique of uniformitarianism to biological evolution by introducing the theory of punctuated equilibrium (also called punk eek or, by critics of Gould and Eldredge, evolution by jerks).[7] New species are only rarely created through the gradual, linear evolution that Darwin described, Gould and Eldredge argued. Rather, speciation is a relatively rapid event that occurs when a group of organisms veers away from its stable parent population and embarks on its own genetic course. Speciation must depend not on the kind of adaptive processes described by Darwin (and Dawkins), but on much more particular, complex, contingent factors.

In his subsequent writings, Gould has hammered away relentlessly at ideas that he claims are implicit in many interpretations of Darwinian theory: progress and inevitability. Evolution does not demonstrate any coherent direction, according to Gould, nor are any of its products—such as *Homo sapiens*—in any sense inevitable; replay the "tape of life" a million times, and this peculiar simian with the oversized brain might never come to be. Gould has also attacked genetic determinism wherever he has found it, whether in pseudo-scientific claims about race and intelligence or in much more respectable theories related to sociobiology. Gould packages his skepticism in a prose rich with references to culture high and low and suffused with an acute awareness of its own existence as a cultural artifact. He has been stunningly successful; almost all his books have been bestsellers, and he is one of the most widely quoted scientists in the world.[8]

Before meeting Gould, I was curious about several, seemingly contradictory aspects of his thought. I wondered, for example, just how deep his skepticism, and his aversion to progress, ran. Did he believe, with Thomas Kuhn, that science itself has not demonstrated any coherent progress? Is the course of science as meandering, as aimless, as that of life? Then how did Gould avoid the contradictions that Kuhn fell prey to? Moreover, some critics—and Gould's success ensured that he had legions—accused him of being a crypto-Marxist. But Marx espoused a highly deterministic, progressivist view of history that seemed antithetical to Gould's.

I also wondered whether Gould was backing down on the issue of punctuated equilibrium. In the headline of their original 1972 paper, Gould and Eldredge boldly called punk eek an "alternative" to Darwin's gradualism that might someday supersede it. In the headline of a retrospective essay published in *Nature* in 1993, called "Punctuated Equilibrium Comes of Age," Gould and Eldredge suggested that their hypothesis might be "a useful extension" or "complement" to Darwin's basic model. Punk meek, Gould and Eldredge concluded the 1993 essay with a spasm of disarming honesty. They noted that their theory was only one of many modern scientific ideas emphasizing randomness and discontinuity rather than order and progress. "Punctuated equilibrium, seen in this light, is only paleontology's contribution to a *Zeitgeist*, and *Zeitgeists*, as (literally) transient ghosts of time, should never be trusted. Thus, in developing punctuated equilibrium, we have either been toadies and panderers to fashion, and therefore destined to history's ashheap, or we had a spark of insight about nature's constitution. Only the punctuated and unpredictable future will tell."[9]

I suspected that this uncharacteristic modesty could be traced to events that had transpired in the late 1970s, when journalists were touting punctuated equilibrium as a "revolutionary" advance beyond Darwin. Inevitably, creationists seized on punk eek as proof that the theory of evolution was not universally accepted. Some biologists blamed Gould and Eldredge for having encouraged such claims with their rhetoric. In 1981 Gould tried to set the record straight by testifying at a trial held in Arkansas over whether creationism should be taught alongside evolution in schools. Gould was forced to admit, in effect, that punctuated equilibrium was not a truly revolutionary theory; it was a rather minor technical matter, a squabble among experts.

Gould is disarmingly ordinary-looking. He is short and plump; his face, too, is chubby, adorned with a button nose and graying Charlie Chaplin mustache. When I met him, he was wearing wrinkled khaki pants and an oxford shirt; he looked like the archetypal rumpled, absent-minded professor. But the illusion of ordinariness vanished as soon as Gould opened his mouth. When discussing scientific issues, he talked in a rapid-fire murmur, laying out even the most complex, technical argument with an ease that hinted at much vaster knowledge held in reserve. He decorated his speech, like his writings, with quotations, which he invariably prefaced by saying, "Of course you know the famous remark of...." As he spoke, he often appeared distracted, as if he were not paying attention to his own words. I had the impression that mere speech was not enough to engage him fully; the higher-level programs of his mind roamed ahead, conducting reconnaissance, trying to anticipate possible objections to his discourse, searching for new lines of argument, analogies, quotations. I had the sense that, no matter where I was, Gould was way ahead of me.

Gould acknowledged that his approach to evolutionary biology had been inspired in part by Kuhn's *Structure of Scientific Revolutions*, which he read shortly after it was published in 1962. The book helped Gould believe that he, a young man from "a lower-middle-class family in Queens where nobody had gone to college," might be able to make an important contribution to science. It also led Gould to reject the "inductivist, ameliorative, progressive, add-a-fact-at-a-time-don't-theorize-'til-you're-old model of doing science."

I asked Gould if he believed, as Kuhn did, that science did not advance toward the truth. Shaking his head adamantly, Gould denied that Kuhn held such a position. "I know him, obviously," Gould said. Although Kuhn was the "intellectual father" of the social constructivists and relativists, he nonetheless believed that "there's an objective world out there," Gould asserted; Kuhn felt that this objective world is in *some* sense very hard to define, but he certainly acknowledged that "we have a better sense of what it is now than we did centuries ago."

So was Gould, who had strived so ceaselessly to banish the notion of progress from evolutionary biology, a believer in scientific progress? "Oh sure," he said mildly. "I think all scientists are." No real scientist could possibly be a true cultural relativist, Gould elaborated, because science is too boring. "The day-to-day work of science is *intensely* boring. You've got to clean the mouse cages and titrate your solutions. And you gotta clean your petri dishes." No scientist could endure such tedium unless he or she thought it would lead to "greater empirical adequacy." Gould added, again in reference to Kuhn, that "some people who have large-scale ideas will often express them in an almost strangely exaggerated way just to accentuate the point." (Musing over this remark later, I had to wonder: Was Gould offering a backhanded apology for his own rhetorical excesses?)

Gould glided just as easily past my queries on Marx. He acknowledged that he found some of Marx's proposals quite attractive. For example, Marx's view that ideas are socially embedded and change through conflict, through the clash of theses and antitheses, "is actually a very sensible and interesting theory of change," Gould remarked. "You move by negating the previous one, and then you negate the first negation, and you don't go back to the first one. You've actually moved on somewhere else. I think all of that is quite interesting." Marx's view of social change and revolution, in which "you accumulate small insults to the system until the system itself breaks," is also quite compatible with punctuated equilibrium.

I hardly had to ask the next question: Was Gould, or had he ever been, a Marxist? "You just remember what Marx said," Gould replied before my mouth closed.

No intellectual, Gould explained, wants to identify himself too closely with any "ism," especially one so capacious. Gould also disliked Marx's ideas about progress. Darwin, while "too eminent a Victorian to dispense with progress totally," was much more critical of Victorian concepts of progress than Marx had been.

On the other hand, Gould, the adamant antiprogressivist, did not rule out the possibility that culture could display some sort of progress. "Because social inheritance is Lamarckian, there is more of a theoretical basis for belief in progress in culture. It gets derailed all the time by war, *et cetera*, and therefore it becomes contingent. But at least because anything we invent is passed directly to an offspring, there is that possibility of directional accumulation."

When, finally, I asked Gould about punctuated equilibrium, he defended it in spirited fashion. The real significance of the idea, he said, is that "you can't explain [speciation] at the level of the adaptive struggle of the individuals in Darwinian, conventional Darwinian, terms." The trends can only be accounted for by mechanisms operating at the level of species. "You get trends because some species speciate more often, because some species live longer than others," he said. "Since the causes of the birth and death of species are quite different from the causes of the birth and death of organisms, it is a different kind of theory. That's what's interesting. That's where the new theory was in punk eek."

Gould refused to admit that he was in any sense backing down on the issue of punctuated equilibrium or conceding Darwin's supremacy. When I asked him about the switch from "alternative" in his original 1972 paper to "complement" in the 1993 *Nature* retrospective, he exclaimed, "I didn't write that!" Gould accused John Maddox, the editor of *Nature*, of inserting "complement" into the paper's headline without checking with him or with Eldredge. "I'm mad at him about that," Gould fumed. But then Gould proceeded to argue that alternative and complement do not really have such different meanings.

"Look, by saying it's an alternative, that doesn't mean that the old kind of gradualism doesn't exist. See, that's another thing I think people miss. The world is full of alternatives, right? I mean we've got men and women, which are alternative states of gender in *Homo sapiens*. I mean if you claim something is an alternative, that doesn't mean it operates exclusively. Gradualism had pretty complete hegemony before we wrote. Here's an alternative to test. I think punctuated equilibrium has an overwhelmingly dominant frequency in the fossil record, which means gradualism exists but it's not really important in the overall pattern of things."

As Gould continued speaking, I began to doubt whether he was really interested in resolving debates over punctuated equilibrium or other issues. When I asked him if he thought biology could ever achieve a final theory, he grimaced. Biologists who hold such beliefs are "naive inductivists," he said. "They actually think that once we sequence the human genome, well, we'll have it!" Even some paleontologists, he admitted, probably think "if we keep at it long enough we really will know the basic features of the history of life and then we'll have it." Gould disagreed. Darwin "had the answer right about the basic interrelationships of organisms, but to me that's only a beginning. It's not over; it's started."

So what did Gould consider the outstanding issues for evolutionary biology? "Oh, there are so many I don't know where to start." He noted that theorists still had to determine the "full panoply of causes" underlying evolution, from molecules on up to large populations of organisms. Then there were "all these contingencies," such as the asteroid impacts that are thought to have caused mass extinctions. "So I would say causes, strengths of causes, levels of causes, and contingency." Gould mused a moment. "That's not a bad formulation," he said, whereupon he took a little notebook from his shirt pocket and scribbled in it.

Then Gould cheerfully rattled off all the reasons that science would *never* answer all these questions. As a historical science, evolutionary biology can offer only retrospective explanations and not predictions, and sometimes it can offer nothing at all because it lacks sufficient data. "If you're missing the evidence of antecedent sequences, then you can't do it at all," he said. "That's why I think we'll never know the origins of language. Because it's not a question of theory; it's a question of contingent history."

Gould also agreed with Gunther Stent that the human brain, created for survival in preindustrial society, is simply not capable of solving certain questions. Research has shown that humans are inept at handling problems that involve probabilities and the interactions of complex variables—such as nature and nurture. "People do not understand that if both genes and culture interact—of *course* they do—you can't then say it's 20 percent genes and 80 percent environment. You can't do that. It's not meaningful. The emergent property is the emergent property and that's all you can ever say about it." Gould was not one of those who invested life or the mind with mystical properties, however. "I'm an old-fashioned materialist," he said. "I think the mind arises from the complexities of neural organization, which we don't really understand very well."

To my surprise, Gould then plunged into a rumination on infinity and eternity. "These are two things that we can't comprehend," he said. "And yet theory almost demands that we deal with it. It's probably because we're not thinking about them right. Infinity is a paradox within Cartesian space, right? When I was eight or nine I used to say, 'Well, there's a brick wall out there.' Well, what's beyond the brick wall? But that's Cartesian space, and even if space is curved you still can't help thinking what's beyond the curve, even if that's not the right way of thinking about it. Maybe all of that's just wrong! Maybe it's a universe of fractal expansions! I don't know what it is. Maybe there are ways in which this universe is structured we just can't think about." Gould doubted whether scientists in any discipline could achieve a final theory, given their tendency to sort things according to preconceived concepts. "I really wonder whether any claim for a final theory isn't just reflecting the way in which we conceptualize them."

Was it possible, given all these limits, that biology, and even science as a whole, might simply go as far as it could and then come to an end? Gould shook his head. "People thought science was ending in 1900, and since then we've got plate tectonics, the genetic basis for life. Why should it stop?" And anyway, Gould added, our theories might reflect our own limitations as truth seekers rather than the true nature of reality. Before I could respond, Gould had already leaped ahead of me. "Of course, if those limits are intrinsic, then science will be complete within the limits. Yeah, yeah. Okay, that's a fair argument. I don't think it's right, but I can understand the structure of it."

Moreover, there may still be great conceptual revolutions in biology's future, Gould argued. "The evolution of life on this planet may turn out to be a very small part of *the* phenomenon of life." Life elsewhere, he elaborated, may very well not conform to Darwinian principles, as Richard Dawkins believed; in fact, the discovery of extraterrestrial life might falsify the claim of Dawkins that Darwin reigns not only here on little earth but throughout the cosmos.

So Gould believes that life exists elsewhere in the universe? I asked. "Don't you?" he retorted. I replied that I thought the issue was entirely a matter of opinion. Gould winced in annoyance; for once, perhaps, he had been caught off guard. Yes, of *course* the existence of life elsewhere is a matter of opinion, he said, but one can still engage in informed speculation. Life seems to have emerged rather readily here on earth, since the oldest rocks that *could* show evidence of life *do* show such evidence. Moreover, "the immensity of the universe and the improbability of absolute uniqueness of any part of it leads to the immense probability that there is some kind of life all over. But we don't know that. Of *course* we don't know that, and I know it's not philosophically incoherent to claim the other."

The key to understanding Gould may be not his alleged Marxism, or liberalism, or anti-authoritarianism, but his fear of his own field's potential for closure. By liberating evolutionary biology from Darwin—and from science as a whole, science defined as the search for universal laws— he has sought to make the quest for knowledge open-ended, even infinite. Gould is far too sophisticated to deny, as some clumsy relativists do, that the fundamental laws uncovered by science exist. Instead, he contends, very persuasively, that the laws do not have much explanatory power; they leave many questions unanswered. He is an extremely adept practitioner of ironic science in its negative capability mode. His view of life can nonetheless be summed up by the old bumper-sticker slogan "Shit happens."

Gould, of course, puts it more elegantly. He noted during our interview that many scientists do not consider history, which deals with particulars and contingency, a part of science. "I think that's a false taxonomy. History is a different type of science." Gould admitted that he found the fuzziness of history, its resistance to straightforward analysis, exhilarating. "I love it! That's because I'm a historian at heart." By transforming evolutionary biology into history—an intrinsically interpretive, ironic discipline, like literary criticism—Gould makes it more amenable for someone with his considerable rhetorical skills. If the history of life is a bottomless quarry of largely random events, he can keep mining it, verbally cherishing one odd fact after another, without ever fearing that his efforts have become trivial or redundant. Whereas most scientists seek to discern the signal underlying nature, Gould keeps drawing attention to the noise. Punctuated equilibrium is not really a theory at all; it is a description of noise.

Gould's great bugaboo is lack of originality. Darwin anticipated the basic concept of punctuated equilibrium in *On the Origin of Species:* "Many species once formed never undergo any further change ... and the periods during which species have undergone further modification, though long as measured by years, have probably been short in comparison with the periods during which they retain the same form."[10] Ernst Mayr, Gould's colleague at Harvard, proposed in the 1940s that species could appear so rapidly—through the geographic isolation of small populations, for example—that they would leave no trace of the intermediate steps in the fossil record.

Richard Dawkins can find little of value in Gould's oeuvre. To be sure, Dawkins said, speciation may sometimes or even often occur in rapid bursts. But so what? "The important thing is that you have gradualistic selection going on even if that gradualistic selection is telescoped into brief periods around about the time of speciation," Dawkins commented. "So I don't see it as an important point. I see it as an interesting wrinkle on the neo-Darwinian theory."

Dawkins also belittled Gould's insistence that there was no inevitability to the appearance of humans or any other form of intelligent life on earth. "I agree with him on that!" Dawkins said. "So, I think, does everybody else! That's my point! He's tilting at windmills!" Life was single celled for almost three billion years, Dawkins pointed out, and it could well have continued for another three billion without giving rise to multicellular organisms. "So, yes, of *course* there is no inevitability."

Was it possible, I asked Dawkins, that in the long run Gould's view of evolutionary biology would prevail? After all, Dawkins had proposed that the fundamental questions of biology might well be finite, whereas the historical issues Gould addressed were virtually infinite. "If you mean that once all the interesting questions have been solved all you've got to do is find the details," Dawkins replied drily, "I suppose that's got to be true." On the other hand, he added, biologists can never be sure which biological principles have truly universal significance "'til we've been to a few other planets that have life." Dawkins was acknowledging, implicitly, that the deepest questions of biology— To what extent is life on earth inevitable? Is Darwinism a universal or merely terrestrial law?—will never be truly, empirically answered as long as we have only one form of life to study.

The Gaia Heresy

Just as Darwin has absorbed Gould's proposals with barely a burp, so he has metabolized the ideas of another would-be strong scientist, Lynn Margulis of the University of Massachusetts at Amherst. Margulis has challenged what she calls the ultra-Darwinian orthodoxy with several ideas. The first, and most successful, was the concept of symbiosis. Darwin and his heirs had emphasized the role of competition between individuals and species in evolution. In the 1960s, Margulis began arguing that symbiosis had been an equally important factor— and perhaps more important—in the evolution of life. One of the greatest mysteries in evolutionary biology concerns the evolution of prokaryotes, cells that lack a nucleus and are the simplest of all organisms, into eukaryotes, cells that have nuclei. All multicellular organisms, including we humans, consist of eukaryotic cells.

Margulis proposed that eukaryotes may have emerged when one prokaryote absorbed another, smaller one, which became the nucleus. She suggested that such cells should be considered not as individual organisms but as composites. After Margulis was able to provide examples of symbiotic relationships among living microorganisms, she gradually won support for her views on the role of symbiosis in early evolution. She did not stop there, however. Like Gould and Eldredge, she argued that conventional Darwinian mechanisms could not account for the stops and starts observed in the fossil record. Symbiosis, she suggested, could explain why species appear so suddenly and why they persist so long without changing.[11]

Margulis's emphasis on symbiosis led naturally to a much more radical idea: Gaia. The concept and term (Gaia was the Greek goddess of the earth) were originally proposed in 1972 by James Lovelock, a self-employed British chemist and inventor who is perhaps even more iconoclastic than Margulis. Gaia comes in many guises, but the basic idea is that the biota, the sum of all life on earth, is locked in a symbiotic relationship with the environment, which includes the atmosphere, the seas, and other aspects of the earth's surface. The biota chemically regulates the environment in such a way as to promote its own survival. Margulis was immediately taken with the idea, and she and Lovelock have collaborated in promulgating it ever since.[12]

I met Margulis in May 1994 in the first-class lounge of New York's Pennsylvania Station, where she was waiting for a train. She resembled an aging tomboy: she had short hair and ruddy skin, and she wore a striped, short-sleeved shirt and khaki pants. She dutifully played the radical, at first. She ridiculed the suggestion of Dawkins and other ultra-Darwinians that evolutionary biology might be nearing completion. *"They're* finished," Margulis declared, "but that's just a small blip in the twentieth century history of biology rather than a full-fledged and valid science."

She emphasized that she had no problem with the basic premise of Darwinism. "Evolution no doubt occurs, and it's been seen to occur, and it's occurring now. Everyone who's scientific minded agrees with that. The question is, how does it occur? And that's where everyone parts company." Ultra-Darwinians, by focusing on the gene as the unit of selection, had failed to explain how speciation occurs; only a much broader theory that incorporated symbiosis and higher-level selection could account for the diversity of the fossil record and of life today, according to Margulis.

Symbiosis, she added, also allows a kind of Lamarckianism, or inheritance of acquired characteristics. Through symbiosis, one organism can genetically absorb or infiltrate another and thereby become more fit. For example, if a translucent fungus absorbs an alga that can perform photosynthesis, the fungus may acquire the capacity for photosynthesis, too, and pass it on to its offspring. Margulis said that Lamarck had been unfairly cast as the goat of evolutionary biology. "We have this British-French business. Darwin's all right and Lamarck is bad. It's really terrible." Margulis noted that symbiogenesis, the creation of new species through symbiosis, was not really an original idea. The concept was first proposed late in the last century, and it has been resurrected many times since then.

Before meeting Margulis, I had read a draft of a book she was writing with her son, Dorion Sagan, called *What Is Life?* The book was an amalgam of philosophy, science, and lyric tributes to "life: the eternal enigma." It argued, in effect, for a new holistic approach to biology, in which the animist beliefs of the ancients would be fused with the mechanistic views of post-Newton, post-Darwin science.[13] Margulis conceded that the book was aimed less at advancing testable, scientific assertions than at encouraging a new philosophical outlook among biologists. But the only difference between her and biologists like Dawkins, she insisted, was that she admitted her philosophical outlook instead of pretending that she didn't have one. "Scientists are no cleaner with respect to being untouched by culture than anyone else."

Did that mean she did not believe that science could achieve absolute truth? Margulis pondered the question a moment. She noted that science derives its power and persuasiveness from the fact that its assertions can be checked against the real world—unlike the assertions of religion, art, and other modes of knowledge. "But I don't think that's the same as saying there's absolute truth. I don't think there's absolute truth, and if there is, I don't think any person has it."

But then, perhaps realizing how close she was edging toward relativism—toward being what Harold Bloom had called a mere rebel— Margulis took pains to steer herself back toward the scientific mainstream. She said that, although she was often considered a feminist, she was not and resented being typecast as one. She conceded that, in comparison to such concepts as "survival of the fittest" and "nature red in tooth and claw," Gaia and symbiosis might *seem* feminine. "There is that cultural overtone, but I consider that just a complete distortion."

She rejected the notion—often associated with Gaia—that the earth is in some sense a living organism. "The earth is *obviously* not a live organism," Margulis said, "because no single living organism cycles its waste. That's *so* anthropomorphic, *so* misleading." James Lovelock encouraged this metaphor, she claimed, because he thought it would aid the cause of environmentalism, and because it suited his own quasi-spiritual leanings. "He says it's an okay metaphor because it's better than the old one. I think it's bad because it's just getting the scientists mad at you, because you're encouraging irrationality." (Actually, Love-lock, too, has reportedly voiced doubts about some of his earlier claims concerning Gaia and has even thought of renouncing the term.[14])

Both Gould and Dawkins have ridiculed Gaia as pseudo-science, poetry posing as a theory. But Margulis is, in at least one sense, much more hard-nosed, more of a positivist, than they are. Gould and Dawkins resorted to speculation about extraterrestrial life in order to buttress their views of life on earth. Margulis scoffed at these tactics. Any proposals concerning the existence of life elsewhere in the universe—or its Darwinian or non-Darwinian nature—are sheer speculation, she said. "You have no constraints on the answer to that, whether it's a frequent or infrequent thing. So I don't see how people can have strong opinions on that. Let me put it this way: opinions aren't science. There's no scientific basis! It's just opinion!"

She remembered that in the early 1970s she had received a call from the director Steven Spielberg, who was in the process of writing the movie *ET*. Spielberg asked Margulis if she thought it was likely or even possible that an extraterrestrial would have two hands, each with five fingers. "I said, 'You're making a movie! Just make it fun! What the hell do you care! Don't try to confuse yourself that it's science!'"

Toward the end of our interview, I asked Margulis if she minded always being referred to as a provocateur or gadfly, or as someone who was "fruitfully wrong," as one scientist had put it.[15] She pressed her lips together, brooding over the question. "It's kind of dismissive, not serious," she replied. "I mean, you wouldn't do this to a serious scientist, would you?" She stared at me, and I finally realized her question was not rhetorical; she really wanted an answer. I agreed that the descriptions seemed somewhat condescending.

"Yeah, that's right," she mused. Such criticism did not bother her, she insisted. "Anyone who makes this kind of *ad hominem* criticism exposes himself, doesn't he? I mean, if their argument is just based on provocative adjectives about me rather than the substance of the issue, then...." Her voice trailed off. Like so many strong scientists, Margulis cannot help but yearn, now and then, to be simply a respected member of the status quo.

Kauffman's Passion for Order

Perhaps the most ambitious and radical modern challenger of Darwin is Stuart Kauffman, a biochemist at the Santa Fe Institute, headquarters of the ultratrendy field of complexity (see Chapter 8). In the 1960s, when he was still a graduate student, Kauffman began to suspect that Darwin's theory of evolution was seriously flawed, in that it could not account for the seemingly miraculous ability of life to appear and then to perpetuate itself in such marvelous ways. After all, the second law of thermodynamics decrees that everything in the universe is drifting inexorably toward "heat death," or universal blandness.

Kauffman tested his ideas by simulating the interaction of various abstract agents—supposedly representative of chemical and biological substances—on computers. He came to several conclusions. One was that when a system of simple chemicals reaches a certain level of complexity, it undergoes a dramatic transition, akin to the phase change that occurs when liquid water freezes. The molecules begin spontaneously combining to create larger molecules of increasing complexity and catalytic capability. Kauffman argued that this process of self-organization or autocatalysis, rather than the fortuitous formation of a molecule with the ability to replicate and evolve, led to life.

Kauffman's other hypothesis was even farther reaching, in that it challenged perhaps the central tenet of biology: natural selection. According to Kauffman, complex arrays of interacting genes subject to random mutations do not evolve randomly. Instead, they tend to converge toward a relatively small number of patterns, or attractors, to use the term favored by chaos theorists. In his 709-page 1993 book, *The Origins of Order: Self-Organization and Selection in Evolution*, Kauffman contended that this ordering principle, which he sometimes called antichaos, may have played a larger role than natural selection in guiding the evolution of life, particularly as life grew in complexity.[16]

Kauffman, when I first met him during a visit to Santa Fe in May 1994, had a broad, deeply tanned face and wavy gray hair thinning at the crown. He wore standard-issue Santa Fe apparel: denim shirt, khakis, hiking boots. He seemed at once tremulous and vulnerable and supremely confident. His speech, like the improvisations of a jazz musician, was short on melody, long on digression. Like a salesperson trying to establish bonhomie, he kept calling me John. He clearly enjoyed talking about philosophy. Over the course of our conversation he presented minilectures not only on his antichaos theories but also on the limits of reductionism, the difficulty of falsifying theories, and the social context of scientific facts.

Early in our conversation, Kauffman recalled an article I had written for *Scientific American* on the origin of life.[17] In the article I had quoted Kauffman saying, of his own origin-of-life theory, "I'm sure I'm right." Kauffman told me he had been embarrassed to see this quotation in print; he had vowed to avoid such hubris in the future. Somewhat to my regret, Kauffman kept his word, for the most part. During our conversation, he took great pains to qualify his assertions: "Now I'm not going to say that I'm sure I'm right, John, but..."

Kauffman had just finished a book, *At Home in the Universe*, that spelled out the implications of his theories on biological evolution. "The line that I take in my Oxford book, John, that I think is right—I mean, I think it's right in the sense that it's all perfectly plausible; *much* to be shown experimentally. But at least in terms of mathematical models, here you have a body of models that says the emergence of life might be a natural phenomenon, in the sense that, given a sufficiently complicated set of reacting molecules, you'd expect to crystallize autocatalytic subsets. So if that view is right, as I told you with abundant enthusiasm a couple of years ago"—big smile—"then we're not incredibly improbable accidents." In fact, life almost certainly exists elsewhere in the universe, he added. "And therefore we are at home in the universe in a different way than we would be if life were this incredibly improbable event that happened on one planet and one planet only because it was so improbable that you wouldn't expect it to happen at all."

Kauffman put the same spin on his theory about how networks of genes tend to settle into certain recurring patterns. "Again, suppose that I'm right," Kauffman said; then much of the order displayed by biological systems results not from "the hard-won success of natural selection" but from these pervasive order-generating effects. "The whole point of it is that it's spontaneous order. It's order for free, okay? Once again, if that view is right, then not only do we have to modify Darwinism to account for it, but we understand something about the emergence and order of life in a different way."

His computer simulations offer another, more sobering message, Kauffman said. Just as the addition of a single grain of sand to a large sandpile can trigger avalanches down its sides, so can a change in the fitness of one species cause a sudden change in the fitness of all the other species in the ecosystem, which can culminate in an avalanche of extinctions. "To say it metaphorically, the best adaptation each of us achieves may unleash an avalanche that leads to our ultimate demise, okay? Because we're all playing the game together and sending out ripples into the system we mutually create. Now *that* bespeaks humility." Kauffman credited the sandpile analogy to Per Bak, a physicist associated with the Santa Fe Institute who had developed a theory called self-organized criticality (which is discussed in Chapter 8).

When I brought up my concern that many scientists, and those at the Santa Fe Institute in particular, seemed to confuse computer simulations with reality, Kauffman nodded. "I agree with you. That personally bothers me a lot," he replied. Viewing some simulations, he said, "I cannot tell where the boundary is between talking about *the world*—I mean, everything out there—and really neat computer games and art forms and toys." When *he* did computer simulations, however, he was always "trying to figure out how something in the world works, or almost always. Sometimes I'm busy just trying to find things that just seem interesting, and I wonder if they apply. But I don't think you're doing science unless what you're doing winds up fitting something out there in the world, demonstrably. And that ultimately means being testable."

His model of genetic networks "makes all *kinds* of predictions" that will probably be tested within the next 15 or 20 years, Kauffman said. "It's testable with some caveats. When you've got a system with 100,000 components, and you can't take the system apart in detail—yet—what are the appropriate testable consequences? They're going to have to be statistical consequences, right?"

How could his origin-of-life theory be tested? "You could be asking two different questions," Kauffman replied. In one sense, the question concerns the way in which life actually arose on earth some four billion years ago. Kauffman did not know whether his theory, or any theory, could satisfactorily address this historical question. On the other hand, one might test his theory by trying to create autocatalytic sets in the laboratory. "Tell you what. We'll make a deal. Whether I do it or somebody else does it, if somebody makes collectively autocatalytic sets of molecules with the phase transitions and reaction graphs, you owe me a dinner, okay?"

There are parallels between Kauffman and other challengers of the status quo in evolutionary biology. First, Kauffman's ideas have historical precedents, just as punctuated equilibrium and symbiosis do. Kant, Goethe, and other pre-Darwinian thinkers speculated that general mathematical principles or rules might underlie the patterns of nature. Even after Darwin, many biologists remained convinced of the existence of some order-generating force in addition to natural selection that counteracts the universal drift toward thermodynamic sameness and gives rise to biological order. Twentieth-century proponents of this viewpoint, which is sometimes called rational morphology, include D'Arcy Wentworth Thompson, William Bateson, and, most recently, Brian Goodwin.[18]

Moreover, Kauffman seems to be motivated at least as much by philosophical conviction about how things must be as by scientific curiosity about how things really are. Gould stresses the importance of chance, contingency, in shaping evolution. Margulis eschews the reductionism of neo-Darwinism for a more holistic approach. Kauffman, similarly, feels that accident alone cannot have created life; our cosmos must harbor some fundamental order-generating tendency.

Finally, Kauffman, like Gould and Margulis, has struggled to define his relationship to Darwin. In his interview with me he said he viewed anti-chaos as a complement to Darwinian natural selection. At other times, he has proclaimed that antichaos is the primary factor in evolution and that natural selection's role has been minor or nonexistent. Kauffman's continued ambivalence over this issue was starkly revealed in a typeset draft of *At Home in the Universe*, which he gave me in the spring of 1995. On the book's first page, Kauffman proclaimed that Darwinism was "wrong," but he had crossed out "wrong" and replaced it with "incomplete." Kauffman went back to "wrong" in the galleys of his book, released several months later. What did the final, published version say? "Incomplete."

Kauffman has one powerful ally in Gould, who proclaimed on the cover of Kauffman's *Origins of Order* that it would become "a landmark and a classic as we grope towards a more comprehensive and satisfying theory of evolution." This is a strange alliance. Whereas Kauffman has argued that the laws of complexity he discerns in his computer simulations have lent the evolution of life a certain inevitability, Gould has devoted his career to arguing that virtually *nothing* in the history of life was inevitable. In his conversation with me, Gould also specifically repudiated the proposal that the history of life unfolded according to mathematical laws. "It's a very deep position," Gould said, "but I also think it's very deeply wrong." What Gould and Kauffman do have in common is that each has challenged the assertion of Richard Dawkins and other hard-core Darwinians that evolutionary theory has already more or less explained the history of life. In providing blurbs for Kauffman's books, Gould shows that he adheres to the old maxim "The enemy of my enemy is my friend."

For the most part, Kauffman has had little success at winning a following for his ideas. Perhaps the major problem is that his theories are statistical in nature, as he himself admits. But one cannot confirm a statistical prediction about the probability of life's origin and its subsequent evolution when one only has one data point—terrestrial life—to consider. One of the harshest assessments of Kauffman's work comes from John Maynard Smith, a British evolutionary biologist who, like Dawkins, is renowned for his sharp tongue and who pioneered the use of mathematics in evolutionary biology. Kauffman once studied under Maynard Smith, and he has spent countless hours trying to convince his former mentor of the importance of his work—

apparently in vain. In a public debate in 1995, Maynard Smith said of self-organized criticality, the sandpile model advanced by Per Bak and embraced by Kauffman, "I just find the whole enterprise contemptible." Maynard Smith told Kauffman later, over beers, that he did not find Kauffman's approach to biology interesting.[19] To a practitioner of ironic science such as Kauffman, there can be no crueler insult.

Kauffman is at his most eloquent and persuasive when he is in his critical mode. He has implied that the evolutionary theory promulgated by biologists such as Dawkins is cold and mechanical; it does not do justice to the majesty and mystery of life. Kauffman is right; there *is* something unsatisfying, tautological, about Darwinian theory, even when it is explicated by a rhetorician as skilled as Dawkins. But Dawkins, at least, distinguishes between living and nonliving things. Kauffman seems to see all phenomena, from bacteria to galaxies, as manifestations of abstract mathematical forms that undergo endless permutations. He is a mathematical aesthete. His vision is similar to that of the particle physicists who call God a geometer. Kauffman has suggested that his vision of life is more meaningful and comforting than that of Dawkins. But most of us, I suspect, can identify more with Dawkins's pushy little replicators than we can with Kauffman's Boolean functions in N-space. Where is the meaning and comfort in these abstractions?

The Conservatism of Science

One constant threat to the status quo in science—and in all human endeavors—is the desire of new generations to make their mark on the world. Society also has an insatiable appetite for the new. These twin phenomena are largely responsible for the rapid turnover of styles in the arts, where change for its own sake is embraced. Science is hardly immune to these influences. Gould's punctuated equilibrium, Margulis's Gaia, and Kauffman's antichaos have all had their minute of fame. But it is much more difficult to achieve lasting change in science than in the arts, for obvious reasons.

Science's success stems in large part from its conservatism, its insistence on high standards of effectiveness. Quantum mechanics and general relativity were as new, as surprising, as anyone could ask for. But they were believed ultimately not because they imparted an intellectual thrill but because they were effective: they accurately predicted the outcome of experiments. Old theories are old for good reason. They are robust, flexible. They have an uncanny correspondence to reality. They may even be True.

Would-be revolutionaries face another problem. The scientific culture was once much smaller and therefore more susceptible to rapid change. Now it has become a vast intellectual, social, and political bureaucracy, with inertia to match. Stuart Kauffman, during one of our conversations, compared the conservatism of science to that of biological evolution, in which history severely constrains change. Not only science, but many other systems of ideas—and particularly those with important social consequences—tend to "stabilize and freeze in" over time, Kauffman noted. "Think of the evolution of standard operating procedure on ships, or aircraft carriers," he said. "It's an incredibly conservative process. If you wandered in and tried to design, *ab initio* [from scratch], procedures on an aircraft carrier, I mean, you'd just blow it to shit!"

Kauffman leaned toward me. "This is really interesting," he said. "Take the law, okay? British common law has evolved for what, 1,200 years? There is this enormous corpus with a whole bunch of concepts about what constitutes reasonable behavior. It would be really hard to change all that! I wonder if you could show that as a web of concepts matures in any area—in all these cases we're making maps to make our way in the world—I wonder if you could show that somehow the center of it got more and more resistant to change." Oddly enough, Kauffman was presenting an excellent argument why his own radical theories about the origin of life and of biological order would probably never be accepted. If any scientific idea has proved its ability to overcome all challengers, it is Darwin's theory of evolution.

The Mysterious Origin of Life

If I were a creationist, I would cease attacking the theory of evolution—which is so well supported by the fossil record—and focus instead on the origin of life. This is by far the weakest strut of the chassis of modern biology. The origin of life is a science writer's dream. It abounds with exotic scientists and exotic theories, which are never entirely abandoned or accepted, but merely go in and out of fashion.[20]

One of the most diligent and respected origin-of-life researchers is Stanley Miller. He was a 23-year-old graduate student in 1953 when he sought to recreate the origin of life in a laboratory. He filled a sealed glass apparatus with a few liters of methane, ammonia, and hydrogen (representing the atmosphere) and some water (the ocean). A spark-discharge device zapped the gases with simulated lightning, while a heating coil kept the waters bubbling. Within a few days, the water and gases were stained with a reddish goo. On analyzing the substance, Miller found to his delight that it was rich in amino acids. These organic compounds are the building blocks of proteins, the basic stuff of life.

Miller's results seemed to provide stunning evidence that life could arise from what the British chemist J. B. S. Haldane had called the "primordial soup." Pundits speculated that scientists, like Mary Shelley's Dr. Frankenstein, would shortly conjure up living organisms in their laboratories and thereby demonstrate in detail how genesis unfolded. It hasn't worked out that way. In fact, almost 40 years after his original experiment, Miller told me that solving the riddle of the origin of life had turned out to be more difficult than he or anyone else had envisioned. He recalled one prediction, made shortly after his experiment, that within 25 years scientists would "surely" know how life began. "Well, 25 years have come and gone," Miller said drily.

After his 1953 experiment, Miller dedicated himself to the search for the secret of life. He developed a reputation as both a rigorous experimentalist and a bit of a curmudgeon, someone who is quick to criticize what he feels is shoddy work. When I met Miller in his office at the University of California at San Diego, where he is a professor of biochemistry, he fretted that his field still had a reputation as a fringe discipline, not worthy of serious pursuit. "Some work is better than others. The stuff that is awful does tend to drag it down. I tend to get very upset about that. People do good work, and then you see this garbage attract attention." Miller seemed unimpressed with any of the current proposals on the origin of life, referring to them as "nonsense" or "paper chemistry." He was so contemptuous of some hypotheses that, when I asked his opinion of them, he merely shook his head, sighed deeply, and snickered—as if overcome by the folly of humanity. Stuart Kauffman's theory of autocatalysis fell into this category. "Running equations through a computer does not constitute an experiment," Miller sniffed.

Miller acknowledged that scientists may never know precisely where and when life emerged. "We're trying to discuss a historical event, which is very different from the usual kind of science, and so criteria and methods are very different," he remarked. But when I suggested that Miller sounded pessimistic about the prospects for discovering life's secret, he looked appalled. Pessimistic? Certainly not! He was optimistic!

One day, he vowed, scientists would discover the self-replicating molecule that had triggered the great saga of evolution. Just as the discovery of the microwave afterglow of the big bang legitimized cosmology, so would the discovery of the first genetic material legitimize Miller's field. "It would take off like a rocket," Miller muttered through clenched teeth. Would such a discovery be immediately self-apparent? Miller nodded. "It will be in the nature of something that will make you say, 'Jesus, there it is. How could you have overlooked this for so long?' And everybody will be totally convinced."

When Miller performed his landmark experiment in 1953, most scientists still shared Darwin's belief that proteins were the likeliest candidates for self-reproducing molecules, since proteins were thought to be capable of reproducing and organizing themselves. After the discovery that DNA is the basis for genetic transmission and for protein synthesis, many researchers began to favor nucleic acids over proteins as the ur-molecules. But there was a major hitch in this scenario. DNA can make neither proteins nor copies of itself without the help of catalytic proteins called enzymes. This fact turned the origin of life into a classic chicken-or-egg problem: which came first, proteins or DNA?

In *The Coming of the Golden Age*, Gunther Stent, prescient as always, suggested that this conundrum could be solved if researchers found a self-replicating molecule that could act as its own catalyst.[21] In the early 1980s, researchers identified just such a molecule: ribonucleic acid, or RNA, a single-strand molecule that serves as DNA's helpmate in manufacturing proteins. Experiments revealed that certain types of RNA could act as their own enzymes, snipping themselves in two and splicing themselves back together again. If RNA could act as an enzyme then it might also be able to replicate itself without help from proteins. RNA could serve as both gene and catalyst, egg and chicken.

But the so-called RNA-world hypothesis suffers from several problems. RNA and its components are difficult to synthesize under the best of circumstances, in a laboratory, let alone under plausible prebiotic conditions. Once RNA is synthesized, it can make new copies of itself only with a great deal of chemical coaxing from the scientist. The origin of life "has to happen under easy conditions, not ones that are very special," Miller said. He is convinced that some simpler—and possibly quite dissimilar—molecule must have paved the way for RNA.

Lynn Margulis, for one, doubts whether investigations of the origin of life will yield the kind of simple, self-validating answer of which Miller dreams. "I think that may be true of the cause of cancer but not of the origin of life," Margulis said. Life, she pointed out, emerged under complex environmental conditions. "You have day and night, winter and summer, changes in temperature, changes in dryness. These things are historical accumulations. Biochemical systems are effectively historical accumulations. So I don't think there is ever going to be a packaged recipe for life: add water and mix and get life. It's not a single-step process. It's a cumulative process that involves a lot of changes." The smallest bacterium, she noted, "is so much more like people than Stanley Miller's mixtures of chemicals, because it already has these system properties. So to go from a bacterium to people is less of a step than to go from a mixture of amino acids to that bacterium."

Francis Crick wrote in his book *Life Itself* that "the origin of life appears to be almost a miracle, so many are the conditions which would have to be satisfied to get it going."[22] (Crick, it should be noted, is an agnostic leaning toward atheism.) Crick proposed that aliens visiting the earth in a spacecraft billions of years ago may have deliberately seeded it with microbes.

Perhaps Stanley Miller's hope will be fulfilled: scientists will find some clever chemical or combination of chemicals that can reproduce, mutate, and evolve under plausible prebiotic conditions. The discovery would be sure to launch a new era of applied chemistry. (The vast majority of researchers focus on this goal, rather than on the elucidation of life's origin.) But given our lack of knowledge about the conditions under which life began, any theory of life's origin based on such a finding would always be subject to doubts. Miller has faith that biologists will know the answer to the riddle of life's origin when they see it. But his belief rests on the premise that the answer will be plausible, if only retrospectively. Who said the origin of life on earth was plausible? Life might have emerged from a freakish convergence of improbable and even unimaginable events.

Moreover, the discovery of a plausible ur-molecule, when or if it happens, is unlikely to tell us what we really want to know: Was life on earth inevitable or a freak occurrence? Has it happened elsewhere or only in this lonely, lonely spot? These questions can only be resolved if we discover life beyond the earth. Society seems increasingly reluctant to underwrite such investigations. In 1993, Congress shut down NASA's SETI (Search for Extraterrestrial Intelligence) program, which scanned the heavens for radio signals generated by other civilizations. The dream of a manned mission to Mars, the most likely site for extraterrestrial life in the solar system, has been indefinitely deferred.

Even so, scientists may find evidence of life beyond the earth tomorrow. Such a discovery would transform all of science and philosophy and human thought. Stephen Jay Gould and Richard Dawkins might be able to settle their argument over whether natural selection is a cosmic or merely terrestrial phenomenon (although each would doubtless find ample evidence for his point of view). Stuart Kauffman might be able to determine whether the laws he discerns in his computer simulations prevail in the real world. If the extraterrestrials are intelligent enough to have developed their own science, Edward Witten may learn whether superstring theory really is the inevitable culmination of any search for the fundamental rules governing reality. Science fiction will become fact. The *New York Times* will resemble the *Weekly World News*, one of the supermarket tabloids that prints "photographs" of presidents hobnobbing with aliens. One can always hope.

Chapter Six
The End of Social Science

Everything would have been easy for Edward O. Wilson if he had just stuck to ants. Ants lured him into biology when he was a boy growing up in Alabama, and they remain his greatest source of inspiration. He has written stacks of papers and several books on the tiny creatures. Ant colonies line Wilson's office at Harvard University's Museum of Comparative Zoology. Showing them off to me, he was as proud and excited as a 10-year-old child. When I asked Wilson if he had exhausted the topic of ants yet, he exulted, "We're only just beginning!" He had recently embarked on a survey of *Pheidole*, one of the most abundant genera in the animal kingdom. *Pheidole* is thought to include more than 2,000 species of ants, most of which have never been described or even named. "I guess with that same urge that makes men in their middle age decide that at last they are going to row across the Atlantic in a rowboat or join a group to climb K2, I decided that I would take on *Pheidole*," Wilson said.[1]

Wilson was a leader in the effort to conserve the earth's biological diversity, and his grand goal was to make *Pheidole* a benchmark of sorts for biologists seeking to monitor the biodiversity of different regions. Drawing on Harvard's collection of ants, the largest in the world, Wilson was generating a set of painstaking pencil drawings of each species of *Pheidole* along with descriptions of its behavior and ecology. "It probably looks crushingly dull to you," Wilson apologized as he flipped through his drawings of *Pheidole* species (which were actually compellingly monstrous). "To me it's one of the most satisfying activities imaginable." He confessed that, when he peered through his microscope at a previously unknown species, he had "the sensation of maybe looking upon—I don't want to get too poetic—of looking upon the face of creation." A single ant was enough to render Wilson awestruck before the universe.

I first detected a martial spirit glinting through Wilson's boyish exuberance when we walked over to the ant farm sprawling across a counter in his office. These are leaf-cutter ants, Wilson explained, which range from South America as far north as Louisiana. The scrawny little specimens scurrying across the surface of the spongelike nest are the workers; the soldiers lurk within. Wilson pulled a plug from the top of the nest and blew into the hole. An instant later several bulked-up behemoths boiled to the surface, BB-sized heads tossing, mandibles agape. "They can cut through shoe leather," Wilson remarked, a bit too admiringly. "If you tried to dig into a leaf-cutter nest, they would gradually dispatch you, like a Chinese torture, by a thousand cuts." He chuckled.

Wilson's pugnacity—innate or inculcated?—emerged more clearly later on, when he discussed the continued reluctance of American society to confront the role played by genes in shaping human behavior. "This country is so seized by our civic religion, egalitarianism, that it just averts its gaze from anything that would seem to detract from that central ethic we have that everybody is equal, that perfect societies can be built with the good will of people." As he delivered this sermon, Wilson's long-boned, Yankee-farmer face, usually so genial, became as stony as a Puritan preacher's.

There are two—at least two—Edward Wilsons. One is the poet of social insects and the passionate defender of all the earth's biodiversity. The other is a fiercely ambitious, competitive man struggling with his sense that he is a latecomer, that his field is more or less complete. Wilson's response to the anxiety of influence was quite different from that of Gould, Margulis, Kauffman, and others who have wrestled with Darwin. For all their differences, they responded to Darwin's dominance by arguing that Darwinian theory has limited explanatory power, that evolution is much more complicated than Darwin and his modern heirs have suggested. Wilson took the opposite course. He sought to extend Darwinism, to show that it could explain more than anyone—even Richard Dawkins—had thought.

Wilson's role as the prophet of sociobiology can be traced back to a crisis of faith he suffered in the late 1950s, just after his arrival at Harvard. Although he was already one of the world's authorities on social insects, he began to brood over the apparent insignificance—at least in the eyes of other scientists—of his field of research. Molecular biologists, exhilarated by their discovery of the structure of DNA, the basis of genetic transmission, had begun questioning the value of studying whole organisms, such as ants. Wilson once recalled that James Watson, who was then at Harvard and still flushed with the excitement of having discovered the double helix, "radiated contempt" for evolutionary biology, which he saw as a glorified version of stamp collecting rendered obsolete by molecular biology.[2]

Wilson responded to this challenge by broadening his outlook, by seeking the rules of behavior governing not only ants but all social animals. That effort culminated in *Sociobiology: The New Synthesis*. Published in 1975, the book was for the most part a magisterial survey of nonhuman social animals, from ants and termites to antelope and baboons. Drawing on ethology, population genetics, and other disciplines, Wilson showed how mating behavior and division of labor could be understood as adaptive responses to evolutionary pressure.

Only in the last chapter did Wilson turn to humans. He drew attention to the obvious fact that sociology, the study of human social behavior, was badly in need of a unifying theory. "There have been attempts at system building but.... they came to little. Much of what passes for theory in sociology today is really labeling of phenomena and concepts, in the expected manner of natural history. Process is difficult to analyze because the fundamental units are elusive, perhaps nonexistent. Syntheses commonly consist of the tedious cross-referencing of differing sets of definitions and metaphors erected by the more imaginative thinkers."[3]

Sociology would only become a truly scientific discipline, Wilson argued, if it submitted to the Darwinian paradigm. He noted, for example, that warfare, xenophobia, the dominance of males, and even our occasional spurts of altruism could all be understood as adaptive behaviors stemming from our primordial compulsion to propagate our genes. In the future, Wilson predicted, further advances in evolutionary theory as well as in genetics and neuroscience would enable sociobiology to account for a wide variety of human behavior; sociobiology would eventually subsume not only sociology, but also psychology, anthropology, and all the "soft" social sciences.

The book received generally favorable reviews. Yet some scientists, including Wilson's colleague Stephen Jay Gould, rebuked Wilson for suggesting that the human condition was somehow *inevitable*. Wilson's views, critics argued, represented an updated version of social Darwinism; that notorious Victorian-era doctrine, by conflating what is with what should be, had provided a scientific justification for racism, sexism, and imperialism. The attacks on Wilson peaked in 1978 at a meeting of the American Association for the Advancement of Science. A member of a radical group called the International Committee Against Racism dumped a pitcher of water on Wilson's head while shouting, "You're all wet."[4]

Undeterred, Wilson went on to write two books on human sociobiology with Charles Lumsden, a physicist at the University of Toronto: *Genes, Mind, and Culture* (1981) and *Promethean Fire* (1983). Wilson and Lumsden conceded in the latter book "the sheer difficulty of creating an accurate portrayal of genetic and cultural interaction." But they declared that the way to cope with this difficulty was not to continue "the honored tradition of social theory written as literary criticism," but to create a rigorous mathematical theory of the interaction between genes and culture. "The theory we wished to build," Wilson and Lumsden wrote, "would contain a system of linked abstract processes expressed as far as possible in the form of explicit mathematical structures that translate the processes back to the real world of sensory experience."[5]

The books that Wilson wrote with Lumsden were not as well received as *Sociobiology* had been. One critic recently called their view of human nature "grimly mechanistic" and "simplistic."[6] During our interview, Wilson was nonetheless more bullish than ever about the prospects for sociobiology. While granting that support for his proposals was very slim in the 1970s, he insisted that "a lot more evidence exists today" that many human traits— ranging from homosexuality to shyness—have a genetic basis; advances in medical genetics have also made genetic explanations of human behavior more acceptable to scientists and to the public. Human sociobiology has flourished not only in Europe, which has a Sociobiological Society, but also in the United States, Wilson said; although many scientists in the United States shun the term *sociobiology* because of its political connotations, disciplines with names such as biocultural studies, evolutionary psychology, and Darwinian studies of human behavior are all "sprigs" growing from the trunk of sociobiology.[7]

Wilson was still convinced that sociobiology would eventually subsume not only the social sciences but also philosophy. He was writing a book, tentatively titled *Natural Philosophy*, about how findings from sociobiology would help to resolve political and moral issues. He intended to argue that religious tenets can and should be "empirically tested" and rejected if they are incompatible with scientific truths. He suggested, for example, that the Catholic church might examine whether its prohibition against abortion— a dogma that contributes to overpopulation—conflicts with the larger moral goal of preserving all the earth's biodiversity. As Wilson spoke, I recalled one colleague's comment that Wilson combined great intelligence and learnedness with, paradoxically, a kind of naïveté, almost an innocence.

Even those evolutionary biologists who admire Wilson's efforts to lay the foundations for a detailed theory of human nature doubt whether such an endeavor can succeed. Dawkins, for example, loathed the "knee-jerk hostility" toward sociobiology displayed by Stephen Jay Gould and other left-leaning scientists. "I think Wilson was shabbily treated, not least by his colleagues at Harvard," Dawkins said. "And so if there's an opportunity to be counted, I would stand up and be counted with Wilson." Yet Dawkins was not as confident as Wilson seemed to be that "the messiness of human life" could be completely understood in scientific terms. Science is not intended to explain "highly complex systems arising out of lots and lots of details," Dawkins elaborated. "Explaining sociology would be rather like using science to explain or to predict the exact course of a molecule of water as it goes over Niagara Falls. You couldn't do it, but that doesn't mean that there's anything *fundamentally* difficult about it. It's just very, very complicated."

I suspect that Wilson himself may doubt whether sociobiology will ever become as all-powerful as he once believed. At the end of *Sociobiology*, he implied that the field would eventually culminate in a complete, final theory of human nature. "To maintain the species indefinitely," Wilson wrote, "we are compelled to drive toward total knowledge, right down to the levels of the neuron and the gene. When we have progressed enough to explain ourselves in these mechanistic terms, and the social sciences come to full flower, the result might be hard to accept." He closed with a quote from Camus: "in a universe divested of illusions and lights, man feels an alien, a stranger. His exile is without remedy since he is deprived of the memory of a lost home or the hope of a promised land."[8]

When I recalled this gloomy coda, Wilson admitted that he had finished *Sociobiology* in a slight depression. "I thought after a period of time, as we knew more and more about where we came from and why we do what we do, in precise terms, that it would reduce—what's the word I'm looking for—our exalted self-image, and our hope for indefinite growth in the future." Wilson also believed that such a theory would bring about the end of biology, the discipline that had given meaning to his own life. "But then I talked myself out of that," he said. Wilson decided that the human mind, which has been and is still being shaped by the complex interaction between culture and genes, represented an endless frontier for science. "I saw that here was an immense unmapped area of science and human history which we would take forever to explore," he recalled. "That made me feel much more cheerful." Wilson resolved his depression by acknowledging, in effect, that his critics were right: science cannot explain all the vagaries of human thought and culture. There can be no *complete* theory of human nature, one that resolves all the questions we have about ourselves.

Just how revolutionary is sociobiology? Not very, according to Wilson himself. For all his creativity and ambition, Wilson is a rather conventional Darwinian. That became clear when I asked him about a concept called biophilia, which holds that the human affinity for nature, or at least certain aspects of it, is innate, a product of natural selection. Biophilia represents Wilson's effort to find common ground between his two great passions, sociobiology and biodiversity. Wilson wrote a monograph on biophilia, published in 1984, and later edited a collection of essays on the topic by himself and others. During my conversation with Wilson, I made the mistake of remarking that biophilia reminded me of Gaia, because each idea evokes an altruism that embraces all of life rather than just one's kin or even one's species.

"Actually, not," Wilson replied, so sharply that I was taken aback. Biophilia does not posit the existence of "some phosphorescent altruism in the air," Wilson scoffed. "I take a very strong mechanistic view of where human nature came from," he said. "Our concern for other organisms is very much a product of Darwinian natural selection." Biophilia evolved, Wilson continued, not for the benefit of all life, but for the benefit of individual humans. "My view is pretty strictly anthropocentric, because what I see and understand, from all that I know of evolution, supports that view and not the other."

I asked Wilson whether he agreed with his Harvard colleague Ernst Mayr that modern biology had been reduced to addressing puzzles whose solution would merely reinforce the prevailing paradigm of neo-Darwinism.[9] Wilson smirked. "Fix the constants to the next decimal point," he said, alluding to the quote that helped to create the legend of the complacent nineteenth-century physicists. "Yeah, we've heard that." But having gently mocked Mayr's view of completion, Wilson went on to agree with it. "We are not about to dethrone evolution by natural selection, or our basic understanding of speciation," Wilson said. "So I, too, am skeptical that we are going to go through any revolutionary changes of how evolution works or how diversification works or how biodiversity is created, at the species level." There is much to learn about embryonic development, about the interaction between human biology and culture, about ecologies and other complex systems. But the basic rules of biology, Wilson asserted, are "beginning to fall pretty much, in my judgment, permanently into place. How evolution works, the algorithm, the machine, what drives it."

What Wilson might have added is that the chilling moral and philosophical implications of Darwinian theory were spelled out long ago. In his 1871 book, *The Descent of Man*, Darwin noted that if humans had evolved as bees had, "there can be hardly a doubt that our unmarried females would, like the worker-bees, think it a sacred duty to kill their brothers, and mothers would strive to kill their fertile daughters; and no one would think of interfering."[10] In other words, we humans are animals, and natural selection has shaped not only our bodies but our very beliefs, our fundamental sense of right and wrong. One dismayed Victorian reviewer of *Descent* fretted in the *Edinburgh Review*, "If these views be true, a revolution in thought is imminent, which will shake society to its very foundations by destroying the sanctity of the conscience and the religious sense."[11] That revolution happened long ago. Before the end of the nineteenth century, Nietzsche had proclaimed that there were no divine underpinnings to human morality: God is dead. We did not need sociobiology to tell us that.

A Few Words from Noam Chomsky

One of the more intriguing critics of sociobiology and other Darwinian approaches to social science is Noam Chomsky, who is both a linguist and one of America's most uncompromising social critics. The first time I saw Chomsky in the flesh he was giving a talk on the practices of modern labor unions. He was wiry, with the slight hunch of a chronic reader. He wore steel-rimmed glasses, sneakers, chinos, and an open-necked shirt. But for the lines in his face and the gray in his longish hair, he could have passed for a college student, albeit one who would rather discuss Hegel than guzzle beer at fraternity parties.

Chomsky's main message was that union leaders were more concerned with maintaining their own power than with representing workers. His audience? Union leaders. During the question-and-answer period they reacted, as one might expect, with defensiveness and even hostility. But Chomsky met their arguments with such serene, unshakable conviction—and such a relentless barrage of *facts*—that before long the targets of his criticism were nodding in agreement: yes, perhaps they *were* selling out to their capitalist overlords.

When I later expressed surprise to Chomsky at the harshness of his lecture, he informed me that he was not interested in "giving people A's for being right." He opposed all authoritarian systems. Of course, he usually focused his wrath not on labor unions, which have lost much of their power, but on the U.S. government, industry, and the media. He called the United States a "terrorist superpower" and the media its "propaganda agent." He told me that if the *New York Times*, one of his favorite targets, started reviewing his books on politics, it would be a sign to him that he was doing something wrong. He summed up his world view as "whatever the establishment is, I'm against it."

I said I found it ironic that his political views were so antiestablishment, given that in linguistics he *is* the establishment. "No I'm not," he snapped. His voice, which ordinarily is hypnotically calm—even when he is eviscerating someone—suddenly had an edge. "My position in linguistics is a minority position, and it always has been." He insisted that he was "almost totally incapable of learning languages" and that, in fact, he was not even a professional linguist. MIT had only hired him and given him tenure, he suggested, because it really did not know or care much about the humanities; it simply needed to fill a slot.[12]

I provide this background for its cautionary value. Chomsky is one of the most contrarian intellectuals I have met (rivaled only by the anarchic philosopher Paul Feyerabend). He is compelled to put all authority figures in their places, even himself. He exemplifies the anxiety of self-influence. One should thus take all of Chomsky's pronouncements with a grain of salt. In spite of his denials, Chomsky is the most important linguist who has ever lived. "It is hardly an exaggeration to say that there is no major theoretical issue in linguistics today that is debated in terms other than those in which he has chosen to define it," declares the *Encyclopaedia Britannica*.[13] Chomsky's position in the history of ideas has been likened to that of Descartes and Darwin.[14] When Chomsky was in graduate school in the 1950s, linguistics— and all the social sciences—was dominated by behaviorism, which hewed to John Locke's notion that the mind begins as a *tabula rasa*, a blank slate that is inscribed upon by experience. Chomsky challenged this approach. He contended that children could not possibly learn language solely through induction, or trial and error, as behaviorists believed. Some fundamental principles of language—a kind of universal grammar— must be embedded in our brains. Chomsky's theories, which he first set forth in his 1957 book *Syntactic Structures*, helped to rout behaviorism once and for all and paved the way for a more Kantian, genetically oriented view of human language and cognition.[15]

Edward Wilson and other scientists who attempt to explain human nature in genetic terms are all, in a sense, indebted to Chomsky. But Chomsky has never been comfortable with Darwinian accounts of human behavior. He accepts that natural selection may have played *some* role in the evolution of language and other human attributes. But given the enormous gap between human language and the relatively simple communication systems of other animals, and given our fragmentary knowledge of the past, science can tell us little about how language evolved. Just because language is adaptive now, Chomsky elaborates, does not mean that it arose in response to selection pressures. Language may have been an incidental by-product of a spurt in intelligence that only later was coopted for various uses. The same may be true of other properties of the human mind. Darwinian social science, Chomsky has complained, is not a real science at all but "a philosophy of mind with a little bit of science thrown in." The problem, according to Chomsky, is that "Darwinian theory is so loose it can incorporate anything they discover."[16]

Chomsky's evolutionary perspective has, if anything, convinced him that we may have only a limited ability to understand nature, human or inhuman. He rejects the notion—popular among many scientists—that evolution shaped the brain into a general-purpose learning and problem-solving machine. Chomsky believes, as Gunther Stent and Colin McGinn do, that the innate structure of our minds imposes limits on our understanding. (Stent and McGinn arrived at this conclusion in part because of Chomsky's research.)

Chomsky divides scientific questions into problems, which are at least potentially answerable, and mysteries, which are not. Before the seventeenth century, Chomsky explained to me, when science did not really exist in the modern sense, almost all questions appeared to be mysteries. Then Newton, Descartes, and others began posing questions and solving them with the methods that spawned modern science. Some of those investigations have led to "spectacular progress," but many others have proved fruitless. Scientists have made absolutely no progress, for example, investigating such issues as consciousness and free will. "We don't even have bad ideas," Chomsky said.

All animals, he argued, have cognitive abilities shaped by their evolutionary histories. A rat, for example, may learn to navigate a maze that requires it to turn left at every second fork but not one that requires it to turn left at every fork corresponding to a prime number. If one believes that humans are animals—and not "angels," Chomsky added sarcastically—then we, too, are subject to these biological constraints. Our language capacity allows us to formulate questions and resolve them in ways that rats cannot, but ultimately we, too, face mysteries as absolute as that faced by the rat in a prime-number maze. We are limited in our ability to ask questions as well. Chomsky thus rejected the possibility that physicists or other scientists could attain a theory of everything; at best, physicists can only create a "theory of what they know how to formulate."

In his own field of linguistics "there's a lot of understanding now about how human languages are more or less cast in the same mold, what the principles are that unify them and so on." But many of the most profound issues raised by language remain impenetrable. Descartes, for instance, struggled to comprehend the human ability to use language in endlessly creative ways. "We're facing the same blank wall Descartes did" on that issue, Chomsky said.

In his 1988 book, *Language and Problems of Knowledge*, Chomsky suggested that our verbal creativity may prove more fruitful than our scientific skills for addressing many questions about human nature. "It is quite possible—overwhelmingly probable, one might guess—that we will always learn more about human life and human personality from novels than from scientific psychology," he wrote. "The science-forming capacity is only one facet of our mental endowment. We use it where we can but are not restricted to it, fortunately."[17]

The success of science, Chomsky proposed to me, stems from "a kind of chance convergence of the truth about the world and the structure of our cognitive space. And it *is* a chance convergence because evolution didn't design us to do this; there's no pressure on differential reproduction that led to the capacity to solve problems in quantum theory. We had it. It's just there for the same reason that most other things are there: for some reason that nobody understands."

Modern science has stretched the cognitive capacity of humans to the breaking point, according to Chomsky. In the nineteenth century, any well-educated person could grasp contemporary physics, but in the twentieth century "you've got to be some kind of freak." That was my opening. Does the increasing difficulty of science, I asked, imply that science might be approaching its limits? Might science, defined as the search for *comprehensible* regularities or patterns in nature, be ending? Abruptly, Chomsky took back everything he had just said. "Science is hard, I would agree with that. But when you talk to young children, they want to understand nature. It's driven out of them. It's driven out of them by boring teaching and by an educational system that tells them they're too stupid to do it." Suddenly it was the establishment, not our innate limitations, that had brought science to its current impasse.

Chomsky insisted that "there are major questions for the natural sciences which we can formulate and that are within our grasp, and that's an exciting prospect." For example, scientists still must show— and almost certainly will show—how fertilized cells grow into complex organisms and how the human brain generates language. There is still plenty of science left to do, "plenty of physics, plenty of biology, plenty of chemistry."

In denying the implication of his own ideas, Chomsky may have been exhibiting just another odd spasm of self-defiance. But I suspect he was really succumbing to wishful thinking. Like so many other scientists, he cannot imagine a world without science. I once asked Chomsky which work he found more satisfying, his political activism or his linguistic research. He seemed surprised that I needed to ask. Obviously, he replied, he spoke out against injustice merely out of a sense of duty; he took no intellectual pleasure from it. If the world's problems were suddenly to disappear, he would happily, joyfully, devote himself to the pursuit of knowledge for its own sake.

The Antiprogress of Clifford Geertz

Practitioners of ironic science can be divided into two types: naïfs, who believe or at least hope they are discovering objective truths about nature (the superstring theorist Edward Witten is the archetypal example), and sophisticates, who realize that they are, in fact, practicing something more akin to art or literary criticism than to conventional science. There is no better example of a sophisticated ironic scientist than the anthropologist Clifford Geertz. He is simultaneously a scientist and a philosopher of science; his work is one long comment on itself. If Stephen Jay Gould serves as the negative capability of evolutionary biology, Geertz does the same for social science. Geertz has helped to fulfill Gunther Stent's prophecy in *The Coming of the Golden Age* that the social sciences "may long remain the ambiguous, impressionistic disciplines that they are at present."[18]

I first encountered Geertz's writings in college, when I took a class on literary criticism and the instructor assigned Geertz's 1973 essay "Thick Description: Toward an Interpretive Theory of Culture."[19] The essay's basic message was that an anthropologist cannot portray a culture by merely "recording the facts." He or she must interpret phenomena, must try to guess what they *mean*. Consider the blink of an eye, Geertz wrote (crediting the example to the British philosopher Gilbert Ryle). The blink may represent an involuntary twitch stemming from a neurological disorder, or from fatigue, or from nervousness. Or it may be a wink, an intentional signal, with many possible meanings. A culture consists of a virtually infinite number of such messages, or signs, and the anthropologist's task is to interpret them. Ideally, the anthropologist's interpretation of a culture should be as complex and richly imagined as the culture itself. But just as literary critics never can hope to establish, once and for all, what *Hamlet* means, so anthropologists must eschew all hope of discovering absolute truths. "Anthropology, or at least interpretive anthropology, is a science whose progress is marked less by a perfection of consensus than by a refinement of debate," Geertz wrote. "What gets better is the precision with which we vex each other."[20] The point of his brand of science, Geertz realized, is not to bring discourse to a close, but to perpetuate it in ever-more-interesting ways.

In later writings, Geertz likened anthropology not only to literary criticism but also to literature. Ethnography involves "telling stories, making pictures, concocting symbolisms and deploying tropes," Geertz wrote, just as literature does. He called anthropology "faction," or "imaginative writing about real people in real places at real times."[21] (Of course, substituting art for literary criticism hardly represents a radical step for one such as Geertz, since to most postmodernists a text is a text is a text.)

Geertz displayed his own talents as a faction writer in "Deep Play: Notes on the Balinese Cockfight." The first sentence of the 1972 essay established his anything-but-straightforward style: "Early in April of 1958, my wife and I arrived, malarial and diffident, in a Balinese village we intended, as anthropologists, to study."[22] (Geertz's prose has been likened to that of both Marcel Proust and Henry James. Geertz told me he was flattered by the former comparison, but thought the latter was probably closer to the mark.)

The opening section of the essay described how the young couple gained the confidence of the normally aloof Balinese. Geertz, his wife, and a group of villagers had been watching a cockfight when police raided the scene. The American couple fled along with their Balinese neighbors. Impressed that the scientists had not sought privileged treatment from the police, the villagers accepted them.

Having thus established his credentials as an insider, Geertz proceeded to depict and then to analyze the Balinese obsession with cockfighting. He eventually concluded that the bloody sport—in which roosters armed with razor-sharp spurs fight to the death—mirrored and thus exorcized the Balinese people's fear of the dark forces underlying their superficially calm society. Like *King Lear* or *Crime and Punishment*, cockfighting "catches up these themes—death, masculinity, rage, pride, loss, beneficence, chance—and [orders] them into an encompassing structure."[23]

Geertz is a shambling bear of a man, with shaggy, whitening hair and beard. When I first interviewed him one drizzly spring day at the Institute for Advanced Study in Princeton, he fidgeted incessantly, pulling on his ear, pawing his cheek, slouching down in his chair and abruptly drawing himself up.[24] Now and then, as he listened to me pose a question, he pulled the top of his sweater up over the tip of his nose, like a bandit trying to conceal his identity. His discourse, too, was elusive. It mimicked his writing: all stops and starts, headlong assertions punctuated by countless qualifications, and suffused with hypertrophic self-awareness.

Geertz was determined to correct what he felt was a common misimpression, that he was a universal skeptic who did not believe that science could achieve any durable truths. Some fields, Geertz said, notably physics, obviously do have the capability to arrive at the truth. He also stressed that, contrary to what I might have heard, he did not consider anthropology to be merely an art form, devoid of any empirical content, and thus not a legitimate field of science. Anthropology is "empirical, responsive to evidence, it theorizes," Geertz said, and practitioners can sometimes achieve a nonabsolute falsification of ideas. Hence it is a science, one that can achieve progress of a kind.

On the other hand, "nothing in anthropology has anything like the status of the harder parts of the hard sciences, and I don't think it ever will," Geertz said. "Some of the assumptions that [anthropologists] made about how easy it is to understand all this and what you need to do in order to do that are no longer really—well, nobody believes them anymore." He laughed. "It doesn't mean it's impossible to know anybody, either, or to do anthropological work. I don't think that at all. But it's not easy."

In modern anthropology, disagreement rather than consensus is the norm. "Things get more and more complicated, but they don't converge to a single point. They spread out and disperse in a very complex way. So I don't see everything heading toward some grand integration. I see it as much more pluralistic and differentiated."

As Geertz continued speaking, it seemed that the progress he envisioned was a kind of antiprogress, in which anthropologists would eliminate, one by one, all the assumptions that could make consensus possible; firm beliefs would dwindle, and doubts multiply. He noted that few anthropologists still believed that they could extract universal truths about humanity from the study of so-called "primitive" tribes, those supposedly existing in a pristine state, uncorrupted by modern culture; neither could anthropologists pretend to be purely objective data gatherers, free of biases and preconceptions.

Geertz found laughable the predictions of Edward Wilson that the social sciences could eventually be rendered as rigorous as physics by grounding them in evolutionary theory, genetics, and neuroscience. Would-be revolutionaries have always come forward with some grand idea that would unify the social sciences, Geertz recalled. Before sociobiology there was general systems theory, and cybernetics, and Marxism. "The notion that someone is going to come along and revolutionize everything overnight is a kind of academicians' disease," Geertz said.

At the Institute for Advanced Study, Geertz was occasionally approached by physicists or mathematicians who had developed highly mathematical models of racial relations and other sociological problems. "But they don't know anything about what goes on in the inner cities!" Geertz exclaimed. "They just have a mathematical model!" Physicists, he grumbled, would never stand for a theory of physics that lacked an empirical foundation. "But somehow or other social science doesn't count. And if you want to have a general theory of war and peace, all you have to do is to sit down and write an equation without having any knowledge of history or people."

Geertz was painfully aware that the introspective, literary style of science he had promulgated also had its pitfalls. It could lead to excessive self-consciousness, or "epistemological hypochondria" on the part of the practitioner. This trend, which Geertz dubbed "I-witnessing," had produced some interesting works but also some abysmal ones. Some anthropologists, Geertz noted, had become so determined to expose all their potential biases, ideological or otherwise, that their writings resembled confessionals, revealing far more about the author than about the putative subject.

Geertz had recently revisited two regions he had studied early in his career, one in Morocco and the other in Indonesia. Both places had changed, drastically; he had also changed. As a result, he became even *more* aware of just how hard it is for the anthropologist to discern truths that transcend their time, place, context. "I always felt it could end in total failure," he said. "I'm still reasonably optimistic, in the sense that I think it's doable, doable as long as you don't claim too much for it. Am I pessimistic? No, but I am chastened." Anthropology is not the only field grappling with questions about its own limitations, Geertz pointed out. "I do sense the same mood in all kinds of fields"—even in particle physics, he said, which seems to be reaching the limits of empirical verification. "The kind of simple self-confidence in science that there once was doesn't seem to me to be so pervasive. Which doesn't mean everybody is giving up hope and wringing their hands in anguish and so on. But it is extraordinarily difficult."

At the time of our meeting in Princeton, Geertz was writing a book about his excursions into his past. When the book was published in 1995, its title neatly summarized Geertz's anxious attitude: *After the Fact*. Geertz peeled apart the title's multiple meanings in the book's final paragraph: Scientists like him are, of course, chasing after facts, but they can only capture facts, if at all, retrospectively; by the time they reach some understanding of what has taken place, the world has moved on, inscrutable as ever.

The phrase also referred, Geertz concluded, to the "post-positivist critique of empirical realism, the move away from simple correspondence theories of truth and knowledge which makes of the very term 'fact' a delicate matter. There is not much assurance or sense of closure, not even much of a sense of knowing what it is one precisely *is* after, in so indefinite a quest, amid such various people, over such a diversity of times. But it is an excellent way, interesting, dismaying, useful, and amusing, to spend a life."[25] Ironic social science may not get us anywhere, but at least it can give us something to do, forever if we like.

Chapter Seven
The End of Neuroscience

Mind, not space, is science's final frontier. Even the most avid believers in the power of science to solve its problems consider the mind a potentially endless source of questions. The problem of mind can be approached in many ways. There is the historical dimension: How, why, did *Homo sapiens* become so smart? Darwin provided a general answer long ago: natural selection favored hominids who could use tools, anticipate the actions of potential competitors, organize into hunting parties, share information through language, and adapt to changing circumstances. Together with modern genetics, Darwinian theory has much to say about the structure of our minds and thus about our sexual and social behavior (although not as much as Edward Wilson and other sociobiologists might like).

But modern neuroscientists are interested less in how and why our minds evolved, in a historical sense, than in how they are structured and work right now. The distinction is similar to the one that can be made between cosmology, which seeks to explain the origins and subsequent evolution of matter, and particle physics, which addresses the structure of matter as we find it here in the present. One discipline is historical and thus necessarily tentative, speculative, and open-ended. The other is, by comparison, much more empirical, precise, and amenable to resolution and finality.

Even if neuroscientists restricted their studies to the mature rather than embryonic brain, the questions would be legion. How do we learn, remember, see, smell, taste, and hear? Most researchers would say these problems, although profoundly difficult, are tractable; scientists will solve them by reverse engineering our neural circuitry. Consciousness, our subjective sense of awareness, has always seemed to be a different sort of puzzle, not physical but metaphysical. Through most of this century, consciousness was not considered a proper subject for scientific investigation. Although behaviorism had died, its legacy lived on in the reluctance of scientists to consider subjective phenomena, and consciousness in particular.

That attitude changed when Francis Crick turned his attention to the problem. Crick is one of the most ruthless reductionists in the history of science. After he and James Watson unraveled the twin-corkscrew structure of DNA in 1953, Crick went on to show how genetic information is encoded into DNA. These achievements gave Darwin's theory of evolution and Mendel's theory of heredity the hard empirical base they had lacked. In the mid-1970s, Crick moved from Cambridge, England, where he had spent most of his career, to the Salk Institute for Biological Research, a cubist fortress overlooking the Pacific Ocean just north of San Diego, California. He worked on developmental biology and the origin of life before finally turning his attention to the most elusive and inescapable of all phenomena: consciousness. Only Nixon could go to China. And only Francis Crick could make consciousness a legitimate subject for science.[1]

In 1990, Crick and a young collaborator, Christof Koch, a German-born neuroscientist at the California Institute of Technology, proclaimed in *Seminars in the Neurosciences* that it was time to make consciousness a subject of empirical investigation. They asserted that one could not hope to achieve true understanding of consciousness or any other mental phenomena by treating the brain as a black box, that is, an object whose internal structure is unknown and even irrelevant. Only by examining neurons and the interactions between them could scientists accumulate the kind of unambiguous knowledge needed to create truly scientific models of consciousness, models analogous to those that explain heredity in terms of DNA.[2]

Crick and Koch rejected the belief of many of their colleagues that consciousness could not be defined, let alone studied. Consciousness, they argued, was synonymous with awareness, and all forms of awareness—whether directed toward objects in the world or highly abstract, internal concepts—seem to involve the same underlying mechanism, one that combines attention with short-term memory. (Crick and Koch credited William James with inventing this definition.) Crick and Koch urged investigators to focus on visual awareness as a synecdoche for consciousness, since the visual system had been so well mapped. If researchers could find the neural mechanisms underlying this function, they might be able to unravel more complex and subtle phenomena, such as self-consciousness, that might be unique to humans (and thus much more difficult to study at the neural level). Crick and Koch had done the seemingly impossible: they had transformed consciousness from a philosophical mystery to an empirical problem. A theory of consciousness would represent the apogee—the culmination—of neuroscience.

Legend has it that some students of the arch-behaviorist B. F. Skinner, on being exposed firsthand to his relentlessly mechanistic view of human nature, fell into an existential despair. I recalled this factoid on meeting Crick in his huge, airy office at the Salk Institute. He was not a gloomy or dour man. Quite the contrary. Clad in sandals, corn-colored slacks, and a gaudy Hawaiian shirt, he was almost preternaturally jolly. His eyes and mouth curled up at the corners in a perpetual wicked grin. His bushy white eyebrows flared out and up like horns. His ruddy face flushed even darker when he laughed, which he did often and with gusto. Crick seemed particularly cheery when he was skewering some product of wishful and fuzzy thinking, such as my vain hope that we humans have free will.[3]

Even an act as apparently simple as seeing, Crick informed me in his crisp, Henry Higgins accent, actually involves vast amounts of neural activity. "The same could be said as to how you make a move, say, picking up a pen," he continued, plucking a ballpoint from his desk and waving it before me. "A lot of computation goes on preparing you for that movement. What you're aware of is a decision, but you're not aware of what makes you do the decision. It seems free to you, but it's the result of things you're not aware of." I frowned, and Crick chuckled.

Trying to help me understand what he and Koch meant by attention—which was the crucial component of their definition of consciousness—Crick emphasized that it involved more than the simple processing of information. To demonstrate this point, he handed me a sheet of paper imprinted with a familiar black-and-white pattern: I saw a white vase on a black background one moment and two silhouetted human profiles the next. Although the visual input to my brain remained constant, Crick pointed out, what I was aware of—or paying attention to—kept changing. What change in the brain corresponded to this change in attention? If neuroscientists could answer that question, Crick said, they might go far toward solving the mystery of consciousness.

Crick and Koch had actually offered a tentative answer to this question in their 1990 paper on consciousness. Their hypothesis was based on evidence that when the visual cortex is responding to stimulation, certain groups of neurons fire extremely rapidly and in synchrony. These oscillating neurons, Crick explained to me, might correspond to aspects of the scene to which attention is being directed. If one envisions the brain as a vast, muttering crowd of neurons, the oscillating neurons are like a group of people who suddenly start singing the same song. Going back to the vase-profiles figure, one set of neurons sings "vase" and then another sings "faces."

The oscillation theory (which has been advanced independently by other neuroscientists) has its weaknesses, as Crick was quick to admit. "I think it's a good, brave, first attempt," he said, "but I have my doubts that it will turn out to be right." He observed that he and Watson had only succeeded in discovering the double helix after numerous false starts. "Exploratory research is really like working in a fog. You don't know where you're going. You're just groping. Then people learn about it afterwards and think how straightforward it was." Crick was nonetheless confident the issue would be resolved not by arguing over psychological concepts and definitions, but by doing "lots of experiments. That's what science is about."

Neurons must be the basis for any model of the mind, Crick told me.

Psychologists have treated the brain as a black box, which can be understood in terms merely of inputs and outputs rather than of internal mechanisms. "Well, you can do that if the black box is simple enough, but if the black box is complicated the chances of you getting the right answer are rather small," Crick said. "It's just the same as in genetics. We had to know about genes, and what genes did. But to pin it down, we had to get down to the nitty-gritty and find the molecules and things involved."

Crick gloated that he was in a perfect position to promote consciousness as a scientific problem. "I don't have to get grants," he said, because he had an endowed chair at the Salk Institute. "The main reason I do it is because I find the problem fascinating, and I feel I've earned the right to do what I like." Crick did not expect researchers to solve these problems overnight. "What I want to stress is that the problem is important and has been too long neglected."

While talking to Crick, I could not help but think of the famous first line of *The Double Helix*, James Watson's memoir about how he and Crick had deciphered DNA's structure: "I have never seen Francis Crick in a modest mood."[4] Some historical revisionism is in order here. Crick is often modest. During our conversation, he expressed doubts about his oscillation theory of consciousness; he said parts of a book he was writing on the brain were "dreadful" and needed rewriting. When I asked Crick how he interpreted Watson's quip, he laughed. What Watson meant, Crick suggested, was not that he was immodest, but that he was "full of confidence and enthusiasm and things like that." If he was also a tad bumptious at times, and critical of others, well, that was because he wanted so badly to get to the bottom of things. "I can be patient for about 20 minutes," he said, "but that's it."

Crick's analysis of himself, like his analysis of most things, seemed on target. He has the perfect personality for a scientist, an empirical scientist, the kind who answers questions, who gets us somewhere. He is, or appears to be, singularly free of self-doubt, wishful thinking, and attachments to his own theories. His immodesty, such as it is, comes simply from wanting to know how things work, regardless of the consequences. He cannot tolerate obfuscation or wishful thinking or untestable speculation, the hallmarks of ironic science. He is also eager to share his knowledge, to make things as clear as possible. This trait is not as common among prominent scientists as one might expect.

In his autobiography, Crick revealed that as a youth and would-be scientist he had worried that by the time he grew up everything would already be discovered. "I confided my fears to my mother, who reassured me. 'Don't worry, Ducky,' she said. 'There will be plenty left for you to find out.'"[5] Recalling this passage, I asked Crick if he thought there would *always* be plenty left for scientists to find out. It all depends on how one defines science, he replied. Physicists might soon determine the fundamental rules of nature, but they could then apply that knowledge by inventing new things forever. Biology seems to have an even longer future. Some biological structures— such as the brain— are so complex that they may resist elucidation for some time. Other puzzles, especially historical ones, such as the origin of life, may never be fully answered simply because the available data are insufficient. "There are enormous numbers of interesting problems" in biology, Crick said. "There's enough to keep us busy at least through our grandchildren's time." On the other hand, Crick agreed with Richard Dawkins that biologists already had a good general understanding of the processes underlying evolution.

As Crick escorted me out of his office, we passed a table bearing a thick stack of paper. It was a draft of Crick's book on the brain, entitled *The Astonishing Hypothesis*. Would I like to read his opening paragraph? Sure, I said. "The Astonishing Hypothesis," the book began, "is that 'You,' your joys and your sorrows, your memories and your ambitions, your sense of personal identity and free will, are in fact no more than the behavior of a vast assembly of nerve cells and their associated molecules. As Lewis Caroll's Alice might have phrased it, 'You're nothing but a pack of neurons.'"[6] I looked at Crick. He was grinning from ear to ear.

I was talking to Crick on the telephone several weeks later, checking the facts of an article I had written about him, when he asked me for some advice. He confessed that his editor wasn't thrilled with the title *The Astonishing*

Hypothesis; she didn't think the view that "we are nothing but a pack of neurons" was all that astonishing. What did I think? I told Crick I had to agree; his view of the mind was, after all, just old-fashioned reductionism and materialism. I suggested that *The Depressing Hypothesis* might be a more fitting title, but it might repel would-be readers. The title didn't matter that much anyway, I added, since the book would sell on the strength of Crick's name.

Crick absorbed all this with his usual good humor. When his book appeared in 1994, it was still called *The Astonishing Hypothesis*. However, Crick, or, more likely, his editor, had added a subtitle: *The Scientific Search for the Soul.* I had to smile when I saw it; Crick was obviously trying not to find the soul—that is, some spiritual essence that exists independently of our fleshy selves—but to eliminate the possibility that there was one. His DNA discovery had gone far toward eradicating vitalism, and now he hoped to stamp out any last vestiges of that romantic worldview through his work on consciousness.

Gerald Edelman Postures around the Riddle

One of the premises of Crick's approach to consciousness is that no theory of mind advanced to date has much value. Yet at least one prominent scientist, a Nobel laureate no less, claims to have gone far toward solving the problem of consciousness: Gerald Edelman. His career, like Crick's, has been eclectic, and highly successful. While still a graduate student, Edelman helped to determine the structure of immunoglobulins, proteins that are crucial to the body's immune response. In 1972 he shared a Nobel Prize for that work. Edelman moved on to developmental biology, the study of how a single fertilized cell becomes a full-fledged organism. He found a class of proteins, called cell-adhesion molecules, thought to play an important role in embryonic development.

All this was merely prelude, however, to Edelman's grand project of creating a theory of mind. Edelman has set forth his theory in four books:

Neural Darwinism, Topobiology, The Remembered Present, and *Bright Air, Brilliant Fire*.[7] The gist of the theory is that just as environmental stresses select the fittest members of a species, so inputs to the brain select groups of neurons—corresponding to useful memories, for example—by strengthening the connections between them.

Edelman's tumescent ambition and personality have made him an alluring subject for journalists. A *New Yorker* profile called him "a dervish of motion, energy and raw intellect" who was "as much Henny Youngman as Einstein"; it mentioned that detractors consider him "an empire-building egomaniac."[8] In a *New York Times Magazine* cover story in 1988, Edelman conferred divine powers upon himself. Discussing his work in immunology, he said that "before I came to it, there was darkness—afterwards there was light." He called a robot based on his neural model his "creature" and said: "I can only observe it, like God. I look down on its world."[9]

I experienced Edelman's self-regard firsthand when I visited him at Rockefeller University in June 1992. (Sometime later, Edelman left Rockefeller to head his own laboratory at the Scripps Institute in La Jolla, California, just down the road from Crick.) Edelman is a large man. Clad in a dark, broad-shouldered suit, he exuded a kind of menacing elegance and geniality. As in his books, he kept interrupting his scientific discourse to dispense stories, jokes, or aphorisms whose relevance was often obscure. The digressions seemed intended to demonstrate that Edelman represented the ideal intellectual, both cerebral and earthy, learned and worldly; no mere experimentalist, he.

Explaining how he became interested in the mind, Edelman said: "I'm very excited by dark and romantic and open issues of science. I'm not averse to working on details, but pretty much only in the service of trying to address this issue of closure." Edelman wanted to find the answer to great questions. His Nobel Prize–winning research on antibody structure had transformed immunology into "more or less a closed science"; the central question, which concerned how the immune system responds to invaders, was resolved. He and others helped to show that self-recognition happens through a process known as selection: the immune system has innumerable different antibodies, and the presence of foreign antigens spurs the body to accelerate the production of, or select, antibodies specific to that antigen and to suppress the production of other antibodies.

Edelman's search for open questions led him inexorably to the development and operation of the brain. He realized that a theory of the human mind would represent the ultimate closure for science, for then science could account for its own origin. Consider superstring theory, Edelman said. Could it explain the existence of Edward Witten? Obviously not. Most theories of physics relegate issues involving the mind to "philosophy or sheer speculation," Edelman noted. "You read that section of my book where Max Planck says we'll never get this mystery of the universe because we are the mystery? And Woody Allen said if I had my life to live over again I'd live in a delicatessen?"

Describing his approach to the mind, Edelman sounded, at first, as resolutely empirical as Crick. The mind, Edelman emphasized, can only be understood from a biological standpoint, not through physics or computer science or other approaches that ignore the structure of the brain. "We will not have a deeply satisfactory brain theory unless we have a deeply satisfactory theory of neural anatomy, okay? It's as simple as that." To be sure, "functionalists," such as the artificial-intelligence maven Marvin Minsky, say they can build an intelligent being without paying attention to anatomy. "My answer is, 'When you show me, fine.'"

But as Edelman continued speaking, it became clear that, unlike Crick, he viewed the brain through the filter of his idiosyncratic obsessions and ambitions. He seemed to think that all his insights were totally original; no one had truly seen the brain before he had turned his attention to it. He recalled that when he had started studying the brain, or, rather, brains, he was immediately struck by their variability. "It seemed to me very curious that the people who worked in neuroscience always talked about brains as if they were identical," he said. "When you look at papers everybody talked about it as if it were a replicable machine. But when you actually look in depth, at every level— and there are amazing numbers of levels—the thing that really hits you is the diversity." Even identical twins, he remarked, show great differences in the organization of their neurons. These differences, far from being insignificant noise, are profoundly important. "It's quite scary," Edelman said. "That's something you just can't get around."

The vast variability and complexity of the brain may be related to a problem with which philosophers from Kant to Wittgenstein had wrestled: how do we categorize things? Wittgenstein, Edelman elaborated, highlighted the troublesome nature of categories by pointing out that different games often had nothing in common except the fact that they were games. "Typical Wittgenstein," Edelman mused. "There is a kind of ostentation in his modesty. I don't know what that is. He provokes you and it's very powerful. It's ambiguous, sometimes, and it's not cute. It's riddle, it's posturing around the riddle."

A little girl playing hopscotch, chess players, and Swedish sailors doing naval exercises are all playing games, Edelman continued. To most observers, these phenomena seem to have little or nothing to do with each other, and yet they are all members of the set of possible games. "This defines what is known in the business as a polymorphous set. It's a very hard thing. It means a set defined by neither necessary nor sufficient conditions. I can show you pictures of it in *Neural Darwinism.*" Edelman grabbed the book off his table and flipped through it until he found an illustration of two sets of geometrical forms that represented polymorphous sets. He then pushed the book away and transfixed me once again. "I'm *astonished* that people don't sit and put these things together," Edelman said.

Edelman, of course, did put these things together: the polymorphous diversity of the brain allows it to respond to the polymorphous diversity of nature. The brain's diversity is not irrelevant noise, but is "the very basis on which selection is going to be made, when encountering an unknown set of physical correspondences in the world. All right? Well that's very promising. Let's go one step further. Could the *unit* of selection be the neuron?" No, because the neuron is too binary, inflexible; it is either on, firing, or off, dormant. But *groups* of linked, interacting neurons could do the job. These groups compete with each other in an effort to create effective representations, or maps, of the infinite variety of stimuli entering from the world. Groups that form successful maps grow still stronger, while other groups wither.

Edelman continued asking and answering his own questions. He spoke slowly, portentously, as if trying to physically impress his words on my brain. How do these groups of linked neurons solve the problem of polymorphous sets that troubled Wittgenstein? Through reentry. What is reentry? "Reentry is the ongoing recursive signaling between mapped areas," Edelman said, "so that you map maps by massively parallel reciprocal connections. It's *not* feedback, which is between two wires, in which I have a definite function, instruction—sine wave in, amplified sine wave out." He was grim, almost angry, as if I had suddenly become the symbol of all his puny-minded, envious critics, who said that reentry *was* merely feedback.

He paused a moment, as if to collect himself, and began again, speaking loudly, slowly, pausing between words, like a tourist trying to make a presumably dim native understand him. Contrary to what his critics said, his model was unique; it had nothing in common with neural networks, he said, lacing the term *neural networks* with scorn. To gain his trust—and because it was true—I confessed that I had always found neural networks difficult to grasp. (Neural networks consist of artificial neurons, or switches, linked by connections of varying strengths.) Edelman smiled triumphantly. "Neural nets involve the stretching of a metaphor," he said. "There is this yawning gap, and you say, 'Is it me, or am I missing something?'" *His* model, he assured me, did not suffer from that problem.

I began asking another question about reentry, but Edelman held up his hand. It was time, he said, to tell me about his latest creature, *Darwin 4*. The best way to validate his theory would be to observe the behavior of neurons in a living animal, which was of course impossible. The only solution, Edelman said, was to construct an automaton that embodied the principles of neural Darwinism. Edelman and his coworkers had built four robots, each named *Darwin*, each more sophisticated than the last. Indeed, *Darwin 4*, Edelman assured me, was not a robot at all but a "real creature." It was "the first nonliving thing that truly learns, okay?"

Again he paused, and I felt his evangelical fervor washing over me. He seemed to be trying to build a sense of drama, as if he were pulling back a succession of veils, each of which concealed a deeper mystery. "Let's go take a look," he said. We headed out of his office and down the hall. He opened the door to a room containing a huge, humming mainframe computer. This, Edelman assured me, was *Darwin 4*'s "brain." Then we walked to another room, where the creature itself awaited us. A pile of machinery on wheels, it sat on a plywood stage littered with blue and red blocks. Perhaps sensing my disappointment—real robots will always disappoint anyone who has seen *Star Wars*—Edelman reiterated that *Darwin 4* "looks like a robot but it's not."

Edelman pointed out the "snout," a bar tipped with a light-sensitive sensor and a magnetic gripper. A television monitor mounted on one wall flashed some patterns that represented, Edelman informed me, the state of *Darwin's* brain. "When it does find an object it will poke up to it, it'll grab it, and then it will get good or bad values.... That will alter the diffuse relationships and synaptology of these things, which are brain maps"—he pointed at the television monitor—"that weaken or strengthen synapses that alter how muscles move."

Edelman stared at *Darwin 4*, which remained stubbornly immobile. "Uh, it takes a fair amount of time," he said, adding that "the amount of computation involved is hair-raising." Finally the robot stirred, to Edelman's evident relief, and began rolling slowly around the platform, nudging blocks, leaving blue ones, picking up red ones with its magnetic snout and taking them over to a big box that Edelman called "home."

Edelman gave me a running commentary. "Uh oh, it just moved its eye. It just found an object. It picked up an object. Now it's going to search for home."

What is its end goal? I asked. "It *has* no end goals," Edelman reminded me with a frown. "We have given it *values*. Blue is bad, red is good." Values are general and thus better suited to helping us cope with the polymorphous world than are goals, which are much more specific. When he was a teenager, Edelman elaborated, he desired Marilyn Monroe, but Marilyn Monroe was not his *goal*. He possessed *values* that led him to desire certain feminine properties, which Marilyn Monroe happened to exemplify.

Brutally repressing an upwelling image of Edelman and Marilyn Monroe, I asked how this robot differed from all the others built by scientists over the past few decades, many of which were capable of feats at least as impressive as those achieved by *Darwin 4*. The difference, Edelman replied, his jaw setting, was that *Darwin 4* possessed values or instincts, whereas other robots needed specific instructions to accomplish any task. But don't all neural networks, I asked, eschew specific instructions for general learning programs? Edelman frowned. "But all of those, you have to exclusively define the input and the output. That's the big distinction. Isn't this correct, Julio?" He turned to a dour, young postdoc who had joined us and was listening to our conversation silently.

After a moment's hesitation, Julio nodded. Edelman, with a broad smile, noted that most artificial-intelligence designers tried to program knowledge in from the top down with explicit instructions for every situation, instead of having knowledge arise naturally from values. Take a dog, he said. Hunting dogs acquire their knowledge from a few basic instincts. "That is more efficacious than any bunch of Harvard boys writing a program for swamps!" Edelman guffawed and glanced at Julio, who joined in uneasily.

But *Darwin 4* is still a computer, a robot, with a limited repertoire of responses to the world, I persisted; Edelman was using language metaphorically when he called it a "creature" with a "brain." As I spoke, Edelman muttered, "Yup, all right, all right," while nodding rapidly. If a computer, he said, is defined as something driven by algorithms, or effective procedures, then *Darwin 4* is *not* a computer. True, computer scientists might program robots to do what *Darwin 4* does. But they would just be faking biological behavior, whereas *Darwin 4*'s behavior is authentically biological. If some random electronic glitch scrambles a line of code in his creature, Edelman informed me, "it'll just correct like a wounded organism and it'll go around again. I do that for the other one and it'll drop dead in its tracks."

Rather than pointing out that all neural networks and many conventional computer programs have this capability, I asked Edelman about the complaints of some scientists that they simply did not understand his theories. Most genuinely new scientific concepts, he replied, must overcome such resistance. He had invited those who had accused him of obscurity—notably Gunther Stent, whose complaints about Edelman's incomprehensibility were quoted in the *New York Times* Magazine—to visit him so he could explain his work in person. (Stent had reached his decision about Edelman's work after sitting next to him during a trans-Atlantic flight.) No one had accepted Edelman's offer. "The opacity, I believe, is in the reception, not the transmission," Edelman said.

By this time, Edelman was no longer making an attempt to hide his irritation. When I asked about his relationship with Francis Crick, Edelman abruptly announced that he had to attend an important meeting. He would leave me in the capable hands of Julio. "I have a very long-term relationship with Francis, and that's not something one can answer—Boom! Boom!—on the way out the door. Or, as Groucho Marx said, 'Leave, and never darken my towels again!'" With that, he departed on a wave of hollow laughter.

Edelman has admirers, but most dwell on the fringes of neuroscience. His most prominent fan is the neurologist Oliver Sacks, whose beautifully written accounts of his dealings with brain-damaged patients have set the standard for literary—that is, ironic—neuroscience. Francis Crick spoke for many of his fellow neuroscientists when he accused Edelman of hiding "presentable" but not terribly original ideas behind a "smoke screen of jargon." Edelman's Darwinian terminology, Crick added, has less to do with any real analogies to Darwinian evolution than with rhetorical grandiosity. Crick suggested that Edelman's theory be renamed "neural Edelmanism." "The trouble with Jerry," Crick said, is that "he tends to produce slogans and sort of waves them about without really paying attention to what other people are saying. So it's really too much hype, is what one is complaining about."[10]

The philosopher Daniel Dennett of Tufts University remained unimpressed after visiting Edelman's laboratory. In a review of Edelman's *Bright Air, Brilliant Fire*, Dennett argued that Edelman had merely presented rather crude versions of old ideas. Edelman's denials nothwithstanding, his model *was* a neural network, and reentry *was* feedback, according to Dennett. Edelman also "misunderstands the philosophical issues he addresses at an elementary level," Dennett asserted. Edelman may profess scorn for those who think the brain is a computer, but his use of a robot to "prove" his theory shows that he holds the same belief, Dennett explained.[11]

Some critics accuse Edelman of deliberately trying to take credit for others' ideas by wrapping them in his own idiosyncratic terminology. My own, somewhat more charitable, interpretation is that Edelman has the brain of an empiricist and the heart of a romantic. He seemed to acknowledge as much, in his typically oblique way, when I asked him if he thought science was in principle finite or infinite. "I don't know what that *means*," he replied. "I know what it means when I say that a series in mathematics is finite or infinite. But I don't know what it means to say that science is infinite. Example, okay? I'll quote Wallace Stevens: *Opus Posthumus*. 'In the very long run even the truth doesn't matter. The risk is taken.'" The *search* for truth is what counts, Edelman seemed to be implying, not the truth itself.

Edelman added that Einstein, when asked whether science was exhausted, reportedly answered, "Possibly, but what's the use of describing a Beethoven symphony in terms of air-pressure waves?" Einstein, Edelman explained, was referring to the fact that physics alone could not address questions related to value, meaning, and other subjective phenomena. One might respond by asking: What is the use of describing a Beethoven symphony in terms of reentrant neural loops? How does the substitution of neurons for air-pressure waves or atoms or any physical phenomenon do justice to the magic and mystery of the mind? Edelman cannot accept, as Francis Crick does, that we are "nothing but a pack of neurons." Edelman therefore obfuscates his basic neural theory—infusing it with terms and concepts borrowed from evolutionary biology, immunology, and philosophy—to lend it added grandeur, resonance, mystique. He is like a novelist who risks obscurity—even seeks it—in the hope of achieving a deeper truth. He is a practitioner of ironic neuroscience, one who, unfortunately, lacks the requisite rhetorical skills.

Quantum Dualism

There is one issue on which Crick, Edelman, and indeed almost all neuroscientists agree: the properties of the mind do not depend in any crucial way on quantum mechanics. Physicists, philosophers, and others have speculated about links between quantum mechanics and consciousness since at least the 1930s, when some philosophically inclined physicists began to argue that the act of measurement—and hence consciousness itself—played a vital role in determining the outcome of experiments involving quantum effects. Such theories have involved little more than hand waving, and proponents invariably have ulterior philosophical or even religious motives. Crick's partner Christof Koch summed up the quantum-consciousness thesis in a syllogism: Quantum mechanics is mysterious, and consciousness is mysterious. Q.E.D.: quantum mechanics and consciousness must be related.[12]

One vigorous advocate of a quantum theory of consciousness is John Eccles, a British neuroscientist who won a Nobel Prize in 1963 for his studies of neural transmission. Eccles is perhaps the most prominent modern scientist to espouse dualism, which holds that the mind exists independently of its physical substrate. He and Karl Popper authored a book defending dualism, called *The Self and Its Brain*, published in 1977. They rejected physical determinism in favor of free will: the mind could choose between different thoughts and courses of action undertaken by the brain and body.[13]

The most common objection to dualism is that it violates the conservation of energy: how can the mind, if it has no physical existence, initiate physical changes in the brain? Together with the German physicist Friedrich Beck, Eccles has provided the following answer: The brain's nerve cells fire when charged molecules, or ions, accumulate at a synapse, causing it to release neurotransmitters. But the presence of a given number of ions at a synapse does not always trigger the firing of a neuron. The reason, according to Eccles, is that for at least an instant, the ions exist in a quantum superposition of states; in some states the neuron discharges and in others it does not.

The mind exerts its influence over the brain by "deciding" which neurons will fire and which will not. As long as probability is conserved throughout the brain, this exercise of free will does not violate conservation of energy. "We have no proof of any of this," Eccles cheerfully acknowledged after explaining his theory to me in a telephone interview. He nonetheless called the hypothesis "a tremendous advance" that would inspire a resurgence of dualism. Materialism and all its vile progeny—logical positivism, behaviorism, identity theory (which equates states of mind with physical states of the brain)—"are finished," Eccles declared.

Eccles was frank—too frank for his own good—about his motive for turning to quantum mechanics in explaining the mind's properties. He was a "religious person," he told me, who rejected "cheap materialism." He believed that "the very nature of the mind is the same as the nature of life. It's a divine creation." Eccles also insisted that "we're only at the beginning of discovering the mystery of existence." Could we ever plumb that mystery, I asked, and thereby bring science to an end? "I don't think so," he replied. He paused and added, heatedly, "I don't *want* it to end. The only important thing is to go on." He agreed with his fellow dualist and falsificationist Karl Popper that we must and will keep "discovering and discovering and discovering. And thinking. And we must not claim to have the last word on anything."

What Roger Penrose Really Wants

Roger Penrose is somewhat better at obscuring his ulterior motives—perhaps because he perceives them only dimly himself. Penrose first made his reputation as an authority on black holes and other exotica of physics. He also invented Penrose tiles, simple geometric forms that when fit together generate infinitely varied, quasi-periodic patterns. Since 1989 he has been renowned for the arguments set forth in *The Emperor's New Mind*. The book's main purpose was to refute the claim of artificial-intelligence proponents that computers could replicate all the attributes of humans, including consciousness.

The key to Penrose's argument is Gödel's incompleteness theorem. The theorem states that any consistent system of axioms beyond a certain basic level of complexity yields statements that can be neither proved nor disproved with those axioms; hence the system is always incomplete. According to Penrose, the theorem implies that no "computable" model—that is, neither classical physics, computer science, nor neuroscience as presently construed—can replicate the mind's creative, or, rather, intuitive powers. The mind must derive its power from some more subtle phenomenon, probably one related to quantum mechanics.

Three years after my first meeting with Penrose in Syracuse—which had triggered my interest in the limits of science—I visited him at his home base, the University of Oxford. Penrose said he was working on a sequel to *The Emperor's New Mind,* which would spell out his theory in greater detail. He was more convinced than ever, he said, that he was on the right track with his theory of quasi-quantum consciousness. "One is going out on a limb here, but I feel quite strongly about those things. I can't see any way out."[14]

I noted that some physicists had begun thinking about how exotic quantum effects, such as superposition, might be harnessed for performing computations that classical computers could not achieve. If these quantum computers proved feasible, would Penrose grant that they might be capable of thinking? Penrose shook his head. A computer capable of thought, he said, would have to rely on mechanisms related not to quantum mechanics in its present form but to a deeper theory not yet discovered. What he was *really* arguing against in *The Emperor's New Mind*, Penrose confided, was the assumption that the mystery of consciousness, or of reality in general, could be explained by the current laws of physics. "I'm saying this is wrong," he announced. "The laws that govern the behavior of the world are, I believe, much more subtle than that."

Contemporary physics simply does not make sense, he elaborated. Quantum mechanics, in particular, *has* to be flawed, because it is so glaringly inconsistent with ordinary, macroscopic reality. How can electrons act like particles in one experiment and waves in another? How can they be in two places at the same time? There must be some deeper theory that eliminates the paradoxes of quantum mechanics and its disconcertingly subjective elements. "Ultimately our theory has to accommodate subjectivism, but I wouldn't like to see the theory itself being a subjective theory." In other words, the theory should allow for the existence of minds but should not *require* it.

Neither superstring theory—which is after all a quantum theory— nor any other current candidate for a unified theory has the qualities that Penrose feels are necessary. "If there is going to be such a kind of total theory of physics, in some sense it couldn't conceivably be the character of any theory I've seen," he said. Such a theory would need a "kind of compelling naturalism." In other words, the theory would have to make sense.

Yet Penrose was just as conflicted as he had been in Syracuse over whether physics would achieve a truly complete theory. Gödel's theorem, he said, suggests that there will always be open questions in physics, as in mathematics. "Even if you could find the end of how the physical world actually was as a mathematical structure," Penrose remarked, "there would still be no end to the subject, if you like, just because there's no end to mathematics." He spoke very deliberately, much more so than when we first discussed the topic in 1989. He had clearly given the issue more thought.

I recalled Richard Feynman's comparison of physics to chess: once we learn the basic rules, we can still explore their consequences forever. "Yes, it's not dissimilar to that viewpoint," Penrose said. Did that mean he thought it would be possible to know the fundamental rules, if not all the consequences of those rules? "I suppose in my more optimistic moods I believe that." He added heatedly, "I'm certainly not one of these people who thinks there's no end to our physical understanding of the world." In Syracuse, Penrose had said it was pessimistic to believe in *The Answer;* now he felt this view was optimistic.

Penrose said he was pleased, for the most part, with the reception that had been given to his ideas; most of his critics had been polite, at least. One exception was Marvin Minsky. Penrose had had an unpleasant encounter with Minsky at a meeting in Canada where they both lectured. At Minsky's insistence, Penrose spoke first. Minsky then rose to offer a rebuttal. After announcing that wearing a jacket "implies you're a gentleman," he took his jacket off, exclaiming, "Well, I don't feel like a gentleman!" He proceeded to attack *The Emperor's New Mind* with arguments that were, according to Penrose, silly. Recalling this scene, Penrose seemed still mystified, and pained. I marveled again, as I had when I first met him, at the contrast between Penrose's mildness of manner and the audacity of his intellectual views.

In 1994, two years after my meeting with Penrose in Oxford, his book *Shadows of the Mind* was published. In *The Emperor's New Mind*, Penrose had been rather vague about where quasi-quantum effects might work their magic. In *Shadows* he hazarded a guess: microtubules, minute tunnels of protein that serve as a kind of skeleton for most cells, including neurons. Penrose's hypothesis was based on a claim by Stuart Hameroff, an anesthesiologist at the University of Arizona, that anesthesia inhibits the movement of electrons in microtubules. Erecting a mighty theoretical edifice on this frail claim, Penrose conjectured that microtubules perform nondeterministic, quasi-quantum computations that somehow give rise to consciousness. Each neuron is thus not a simple switch, but a complex computer in its own right.

Penrose's microtubule theory could not help but be anticlimactic. In his first book, he built up an air of suspense, anticipation, and mystery, as does a director of a horror movie who offers only tantalizing glimpses of the monster. When Penrose finally unveiled his monster, it looked like an overweight actor wearing a cheap rubber suit, complete with flapping fins. Some skeptics have responded, not unexpectedly, with ridicule rather than with awe. They have noted that microtubules are found in almost all cells, not just in neurons. Does that mean that our livers are conscious? What about our big toes? What about paramecia? (Penrose's partner, Stuart Hameroff, when I put this question to him in April 1994, replied, "I'm not going to contend that a paramecium is conscious, but it does show pretty intelligent behavior.")

Penrose can also be countered with Crick's argument against free will. Because Penrose, through mere introspection, cannot retrace the computational logic of his perception of a mathematical truth, he insists that the perceptions must stem from some mysterious, non-computational phenomenon. But as Crick pointed out, just because we are not aware of the neural processes leading up to a decision does not mean that those processes did not occur. Proponents of artificial intelligence rebut Penrose's Gödel argument by contending that one can always design a computer to broaden its own base of axioms to solve a new problem; in fact, such learning algorithms are rather common (although they are still extremely crude in comparison to the human mind).[15]

Some critics of Penrose have accused him of being a vitalist, someone who secretly hopes that the mystery of the mind will not yield to science. But if Penrose was a vitalist, he would have kept his ideas vague and untestable. He would never have unveiled the microtubule monster. Penrose is a true scientist; he wants to *know*. He sincerely believes that our *present* understanding of reality is incomplete, logically flawed, and, well, mysterious. He is looking for a key, an insight, some clever quasi-quantum trick that will make everything suddenly clear. He is looking for *The Answer*. He has made the great mistake of thinking that physics should render the world completely intelligible and meaningful. Steven Weinberg could have told him that physics lacks that capacity.

Attack of the Mysterians

Penrose, although he pushed a theory of consciousness far beyond the horizon of current science, at least held out the hope that the theory could one day be reached. But some philosophers have questioned whether *any* purely materialistic model—involving conventional neural processes or the exotic, nondeterministic mechanisms envisioned by Penrose—can *really* account for consciousness. The philosopher Owen Flanagan named these doubters "the new mysterians," after the sixties rock group Question Mark and the Mysterians, which performed the hit song "96 Tears." (Flanagan himself is not a mysterian, but a down-to-earth materialist.)[16]

The philosopher Thomas Nagel offered one of the clearest expressions of the mysterian viewpoint (an oxymoronic undertaking?) in his famous 1974 essay "What Is It Like to Be a Bat?" Nagel assumed that subjective experience is a fundamental attribute of humans and many higher-level animals, such as bats. "No doubt it occurs in countless forms totally unimaginable to us, on other planets in other solar systems throughout the universe," Nagel wrote. "But no matter how the form may vary, the fact that an organism has conscious experience *at all* means, basically, that there is something it is like to *be* that organism."[17] Nagel argued that no matter how much we learn about the physiology of bats, we cannot *really* know what it is like to be one, because science cannot penetrate the realm of subjective experience.

Nagel is what one might call a weak mysterian: he holds out the possibility that philosophy and/or science might one day reveal a natural way to bridge the gap between our materialistic theories and subjective experience. Colin McGinn is a strong mysterian. McGinn is the same philosopher who believes that most major philosophical questions are unsolvable because they are beyond our cognitive abilities (see Chapter 2). Just as rats have cognitive limitations, so do humans, and one of our limitations is that we cannot solve the mind-body problem. McGinn considers his position on the mind-body problem—that it is unsolvable—the logical conclusion of Nagel's analysis in "What Is It Like to Be a Bat?" McGinn defends his viewpoint as superior to what he calls the "eliminativist" position, which attempts to show that the mind-body problem is not really a problem at all.

It is quite possible, McGinn said, for scientists to invent a theory of the mind that can predict the outcome of experiments with great precision and yield a wealth of medical benefits. But an effective theory is not necessarily a comprehensible one. "There's no real reason why part of our mind can't develop a formalism with these remarkable predictive properties, but we can't make sense of the formalism in terms of the part of our mind which understands things. So it could be in the case of consciousness that we come up with a theory which is analogous to quantum theory in this respect, a theory which is actually a good theory of consciousness, but we wouldn't be able to interpret it, or to understand it."[18]

This kind of talk infuriates Daniel Dennett. A philosopher at Tufts University, Dennett is a big, hoary-bearded man with a look of perpetual, twinkly amusement—Santa Claus on a diet. Dennett exemplifies what McGinn calls the eliminativist position. In his 1992 book, *Consciousness Explained*, Dennett contended that consciousness—and our sense that we possess a unified self—was an illusion arising out of the interaction of many different "subprograms" run on the brain's hardware.[19] Dennett, when I asked him about McGinn's mysterian argument, called it ridiculous. He denigrated McGinn's comparison of humans to rats. Unlike humans, Dennett asserted, rats cannot conceive of scientific questions, so of *course* they cannot solve them. Dennett suspected that McGinn and other mysterians "don't want consciousness to fall to science. They like the idea that this is off-limits to science. Nothing else could explain why they welcome such slipshod arguments."

Dennett tried another strategy, one that struck me as oddly Platonic for such an avowed materialist—and also dangerous for a writer. He recalled that Borges, in his story "The Library of Babel," imagined an infinite library of all possible statements, those that have been, will be, and could be, from the most nonsensical to the most sublime. Surely somewhere in the Library of Babel is a perfectly stated resolution of the mind-body problem, Dennett said. Dennett served up this argument with such confidence that I suspected he believed that the Library of Babel really exists.

Dennett granted that neuroscience might never produce a theory of consciousness that satisfied everyone. "We can't explain *anything* to everyone's satisfaction," he said. After all, many people are dissatisfied with science's explanations of, say, photosynthesis or biological reproduction. But "the sense of mystery is gone from photosynthesis or reproduction," Dennett said, "and I think in the end we will have a similar account of consciousness."

Abruptly, Dennett tacked in a completely different direction. "There's a curious paradox looming" in modern science, he said. "One of the very trends that makes science proceed so rapidly these days is a trend that leads science away from human understanding. When you switch from trying to model things with elegant equations to doing massive computer simulations, you may end up with a model that exquisitely models nature, the phenomena you're interested in, but you don't understand the model. That is, you don't understand it the way you understood models in the old days."

A computer program that accurately modeled the human brain, Dennett noted, might be as inscrutable to us as the brain itself. "Software systems are already at the very edge of human comprehensibility," he observed. "Even a system like the Internet is absolutely trivial compared to a brain, and yet it's been patched and built on so much that nobody really understands how it works or whether it will go on working. And thus, as you start using software-writing programs and software-debugging programs and code that heals itself, you create new artifacts that have a life of their own. And they become objects that are no longer within the epistemological hegemony of their makers. And so that's going to be sort of like the speed of light. It's going to be a barrier against which science is going to keep butting its head forever."

Astonishingly, Dennett was implying that he, too, had mysterian inclinations. He thought a theory of the mind, although it might be highly effective and have great predictive power, was unlikely to be intelligible to mere humans. The only hope humans have of comprehending their own complexity may be to cease being human. "Anybody who has the motivation or talent," he said, "will be able in effect to merge with these big software systems." Dennett was referring to the possibility, advanced by some artificial-intelligence enthusiasts, that one day we humans will be able to abandon our mortal, fleshy selves and become machines. "I think that's logically possible," Dennett added. "I'm not sure how plausible it is. It's a coherent future. I think it's not self-contradictory." But Dennett seemed to doubt whether even superintelligent machines would ever fully comprehend themselves. Trying to know themselves, the machines would have to become still more complicated; they would thus be caught in a spiral of ever-increasing complexity, chasing their own tails for all eternity.

How Do I Know You're Conscious?

In the spring of 1994, I witnessed a remarkable collision of the philosophical and scientific worldviews at a meeting called "Toward a Scientific Basis for Consciousness," held at the University of Arizona.[20] On the first day

David Chalmers, a long-haired Australian philosopher who bears an uncanny resemblance to the subject of Thomas Gains-borough's famous painting *Blue Boy*, set forth the mysterian viewpoint in forceful terms. Studying neurons, he declared, cannot reveal why the impingement of sound waves on our ears gives rise to our subjective *experience* of Beethoven's Fifth Symphony. All physical theories, Chalmers said, describe only functions—such as memory, attention, intention, introspection—correlating to specific physical processes in the brain. But none of these theories can explain why the performance of these functions is accompanied by subjective experience. After all, one can certainly imagine a world of androids that resemble humans in every respect—except that they do not have a conscious experience of the world. However much they learn about the brain, neuroscientists cannot bridge that "explanatory gap" between the physical and subjective realms with a strictly physical theory, according to Chalmers.

Up to this point, Chalmers had expressed the basic mysterian viewpoint, the same one associated with Thomas Nagel and Colin McGinn. But then Chalmers proclaimed that although science could not solve the mind-body problem, philosophy still might. Chalmers thought he had found a possible solution: scientists should assume that information is as essential a property of reality as matter and energy. Chalmers's theory was similar to the it from bit concept of John Wheeler—in fact, Chalmers acknowledged his debt to Wheeler—and it suffered from the same fatal flaw. The concept of information does not make sense unless there is an information processor—whether an amoeba or a particle physicist—that gathers information and acts on it. Matter and energy were present at the dawn of creation, but life was not, as far as we know. How, then, can information be as fundamental as matter and energy? Nevertheless, Chalmers's ideas struck a chord among his audience. They thronged around him after his speech, telling him how much they had enjoyed his message.[21]

At least one listener was displeased: Christof Koch, Francis Crick's collaborator. That night Koch, a tall, rangy man wearing red cowboy boots, tracked Chalmers down at a cocktail party for the conferees and chastized him for his speech. It is precisely because philosophical approaches to consciousness have all failed that scientists must focus on the brain, Koch declared in his rapid-fire, German-accented voice as rubberneckers gathered. Chalmers's information-based theory of consciousness, Koch continued, like all philosophical ideas, was untestable and therefore useless. "Why don't you just say that when you have a brain the Holy Ghost comes down and makes you conscious!" Koch exclaimed. Such a theory was unnecessarily complicated, Chalmers responded drily, and it would not accord with his own subjective experience. "But how do I know your subjective experience is the same as mine?" Koch sputtered. "How do I even know you're conscious?"

Koch had brought up the embarrassing problem of solipsism, which lies at the heart of the mysterian position. No person really *knows* that any other being, human or inhuman, has a subjective experience of the world. By raising this ancient philosophical conundrum, Koch, like Dennett, was revealing himself to be a mysterian. Koch admitted as much to me later. All science can do, he asserted, is provide a detailed map of the physical processes that correlate with different subjective states. But science cannot truly "solve" the mind-body problem. No empirical, neurological theory can explain why mental functions are accompanied by specific subjective states. "I don't see how any science can explain that," Koch said. For the same reason, Koch doubted that science would ever provide a definitive answer to the question of whether machines can become conscious and have subjective experiences. "The debate might not be resolved ever," he told me, adding gratuitously, "How do I even know you are conscious?"

Even Francis Crick, although he was more optimistic than Koch, had to acknowledge that the solution to consciousness might not be intuitively comprehensible. "I don't think it will be a commonsense answer that we get when we understand the brain," Crick said. After all, natural selection cobbles organisms together not according to any logical plan, but with various gimmicks and tricks, with whatever works. Crick went on to suggest that the mysteries of the mind might not yield as readily as those of heredity. The mind "is a much more complicated system" than the genome, he remarked, and theories of the mind would probably have more limited explanatory power.

Holding up his pen, Crick explained that scientists should be able to determine which neural activity correlated with my perception of the pen. "But if you were to ask, 'Do you see red and blue the same way I see red and blue?' well, that's something you can't communicate to me. So I don't think we'll be able to explain everything that we're conscious of."

Just because the mind stems from deterministic processes, Crick continued, does not mean that scientists will be able to predict all its meanderings; they may be chaotic and thus unpredictable. "There may be other limitations in the brain. Who knows? I don't think you can look too far ahead." Crick doubted that quantum phenomena played a crucial role in consciousness, as Roger Penrose had suggested. On the other hand, Crick added, some neural equivalent of Heisenberg's uncertainty principle might restrict our ability to trace the brain's activity in minute detail, and the processes underlying consciousness might be as paradoxical and difficult for us to grasp as quantum mechanics is. "Remember," Crick elaborated, "our brains have evolved to deal with everyday matters when we were hunter-gatherers, and before that when we were monkeys." Yes, that was the point of Colin McGinn, and Chomsky, and Stent.

The Many Minds of Marvin Minsky

The most unlikely mysterian of all is Marvin Minsky. Minsky was one of the founders of artificial intelligence (AI), which holds that the brain is nothing more than a very complicated machine whose properties can be duplicated with computers. Before I visited him at MIT, colleagues warned me that he might be cranky, even hostile. If I did not want the interview cut short, I should not ask him too directly about the falling fortunes of artificial intelligence or of his own particular theories of the mind. One former associate pleaded with me not to take advantage of Minsky's penchant for outrageous utterances. "Ask him if he means it, and if he doesn't say it three times you shouldn't use it," the ex-colleague urged.

When I met Minsky, he was rather edgy, but the condition seemed congenital rather than acquired. He fidgeted ceaselessly, blinking, waggling his foot, pushing things about his desk. Unlike most scientific celebrities, he gave the impression of conceiving ideas and tropes from scratch rather than retrieving them whole from memory. He was often, but not always, incisive. "I'm rambling here," he muttered after a riff on how to verify models of the mind collapsed in a heap of sentence fragments.[22]

Even his physical appearance had an improvisational air. His large, round head seemed entirely bald, but was actually fringed by hairs as transparent as optical fibers. He wore a braided belt that supported, in addition to his pants, a belly pack and a tiny holster containing pliers with retractable jaws. With his paunch and vaguely Asian features, he resembled Buddha—Buddha reincarnated as a hyperactive hacker.

Minsky seemed unable, or unwilling, to inhabit any emotion for long. Early on, as predicted, he lived up to his reputation as a curmudgeon, and arch-reductionist. He expressed contempt for those who doubted whether computers could be conscious. Consciousness is a trivial issue, he said. "I've solved it, and I don't understand why people don't listen." Consciousness is merely a type of short-term memory, a "low-grade system for keeping records." Computer programs such as LISP, which have features that allow their processing steps to be retraced, are "extremely conscious," more so than humans, with their pitifully shallow memory banks.

Minsky called Roger Penrose a "coward" who could not accept his own physicality, and he derided Gerald Edelman's reentrant-loops hypothesis as warmed-over feedback theory. Minsky even snubbed MIT's own Artificial Intelligence Laboratory, which he had founded and where we happened to be meeting. "I don't consider this to be a serious research institution at the moment," he announced.

When we wandered through the lab looking for a lecture on a chess-playing computer, however, a metamorphosis occurred. "Isn't the chess meeting supposed to be here?" Minsky queried a group of researchers chatting in a lounge. "That was yesterday," someone replied. After asking a few questions about the talk, Minsky spun tales about the history of chess-playing programs. This minilecture evolved into a reminiscence of Minsky's friend Isaac Asimov, who had just died. Minsky recounted how Asimov—who had popularized the term *robot* and explored its metaphysical implications in his science fiction—always refused Minsky's invitations to see the robots being built at MIT out of fear that his imagination "would be weighed down by this boring realism."

One lounger, noticing that he and Minsky wore the same pliers, yanked his instrument from its holster and with a flick of his wrist snapped the retractable jaws into place. "*En garde,*" he said. Minsky, grinning, drew his weapon, and he and his challenger whipped their pliers repeatedly at each other, like punks practicing their switchblade technique. Minsky expounded on both the versatility and—an important point for him—the limitations of the pliers; his pair pinched him during certain maneuvers. "Can you take it apart with itself?" someone asked. Minsky and his colleagues shared an insiders' laugh at this reference to a fundamental problem in robotics.

Later, returning to Minsky's office, we encountered a young, extremely pregnant Korean woman. She was a doctoral candidate and was scheduled for an oral exam the next day. "Are you nervous?" asked Minsky. "A little," she replied. "You shouldn't be," he said, and gently pressed his forehead against hers, as if seeking to infuse her with his strength. I realized, watching this scene, that there were many Minskys.

But of course there would be. Multiplicity is central to Minsky's view of the mind. In his book *The Society of Mind* he contended that brains contain many different, highly specialized structures that evolved in order to solve different problems.[23] "We have many layers of networks of learning machines," he explained to me, "each of which has evolved to correct bugs or to adapt the other agencies to the problems of thinking." It is thus unlikely that the brain can be reduced to a particular set of principles or axioms, "because we're dealing with a real world instead of a mathematical one that is defined by axioms."

If AI had not lived up to its early promise, Minsky said, that was because modern researchers had succumbed to "physics envy"—the desire to reduce the intricacies of the brain to simple formulas. Such researchers, to Minsky's annoyance, had failed to heed his message that the mind has many different methods for coping with even a single, relatively simple problem. For example, someone whose television set fails to work will probably first consider the problem to be purely physical. He or she will check to see whether the television is properly programmed or whether the cord is plugged in. If that fails, the person may try to have the machine repaired, thus turning the problem from a physical one to a social one—how to find someone who can repair the television quickly and cheaply.

"That's one lesson I can't get across to these people," Minsky said of his fellow AI workers. "It seems to me that the problem the brain has more or less solved is how to organize different methods into working when the individual methods fail pretty often." The only theorist other than he who truly grasped the mind's complexity, Minsky asserted, was dead. "Freud has the best theories so far, next to mine, of what it takes to make a mind."

As Minsky continued speaking, his emphasis on multiplicity took on a metaphysical and even moral cast. He blamed the problems of his field—and of science in general—on what he called "the investment principle," which he defined as the tendency of humans to keep doing something that they have learned to do well rather than to move on to new problems. Repetition, or, rather, single-mindedness, seemed to hold a kind of horror for Minsky. "If there's something you like very much," he asserted, "then you should regard this not as you feeling good but as a kind of brain cancer, because it means that some small part of your mind has figured out how to turn off all the other things."

The reason Minsky had mastered so many skills during his career—he is an adept in mathematics, philosophy, physics, neuroscience, robotics, and computer science and has written several science fiction novels— was that he had learned to enjoy the "feeling of awkwardness" triggered by having to learn something new. "It's so thrilling not to be able to do something. It's such a rare experience to treasure. It won't last."

Minsky was a child prodigy in music, too, but eventually he decided that music was a soporific. "I think the reason people like music is to *suppress* thought—the wrong kinds of thought—not to produce it." Minsky still occasionally found himself composing "Bach-like things"—an electric piano crowded his office—but he tried to resist the impulse. "I had to kill the musician at some point," he said. "It comes back every now and then, and I hit it."

Minsky had no patience for those who claim the mind is too subtle to understand. "Look, before Pasteur people said, 'Life is different. You can't explain it mechanically.' It's just the same thing." But a final theory of the mind, Minsky emphasized, would be much more complex than a final theory of physics—which Minsky also believed was attainable. All of particle physics might be condensed to a page of equations, Minsky said, but to describe all the components of the mind would require much more space. After all, consider how long it would take precisely to describe an automobile, or even a single spark plug. "It would take a fair-sized book to explain how they welded and sintered the spline to the ceramic without it leaking when it starts."

Minsky said the truth of a model of mind could be demonstrated in several ways. First, a machine based on the model's principles should be able to mimic human development. "The machine ought to be able to start as a baby and grow up by seeing movies and playing with things." Moreover, as imaging technology improves, scientists should be able to determine whether the neural processes in living humans corroborate the model. "It seems to me that it's perfectly reasonable that once you get a [brain] scanner that had one angstrom [one ten-billionth of a meter] resolution, then you could see every neuron in someone's brain. You watch this for 1,000 years and you say, 'Well, we know exactly what happens whenever this person says *blue*. And people check this out for generations and the theory is sound. Nothing goes wrong, and that's the end of it."

If humans achieve a final theory of the mind, I asked, what frontiers will be left for science to explore? "Why are you asking me this question?" Minsky retorted. The concern that scientists will run out of things to do is pitiful, he said. "There's *plenty* to do." We humans may well be approaching our limits as scientists, but we will someday create machines much smarter than we that can continue doing science. But that would be machine science, not human science, I said. "You're a racist, in other words," Minsky said, his great domed forehead purpling. I scanned his face for signs of irony, but found none. "I think the important thing for us is to grow," Minsky continued, "not to remain in our own present stupid state." We humans, he added, are just "dressed up chimpanzees." Our task is not to preserve present conditions but to evolve, to create beings better, more intelligent than we.

But Minsky, surprisingly, was hard-pressed to say precisely what kinds of questions these brilliant machines might be interested in. Echoing Daniel Dennett, Minsky suggested, rather halfheartedly, that machines might try to comprehend themselves as they evolved into ever-more-complex entities. He seemed more enthusiastic discussing the possibilities of converting human personalities into computer programs that could then be downloaded into machines. Minsky saw downloading as a way to indulge in pursuits that he would ordinarily consider too dangerous, such as taking LSD or indulging in religious faith.

Minsky confessed that he would love to know what Yo-Yo Ma, the great cellist, felt like when playing a concerto, but Minsky doubted whether such an experience would be possible. To share Yo-Yo Ma's experience, Minsky explained, he would have to possess all Yo-Yo Ma's memories, he would have to *become* Yo-Yo Ma. But in becoming Yo-Yo Ma, Minsky suspected, he would cease to be Minsky.

This was an extraordinary admission for Minsky to make. Like literary critics who claim that the only true interpretation of a text is the text itself, Minsky was implying that our humanness is irreducible; any attempt to convert an individual into an abstract mathematical program—a string of ones and zeros that could be downloaded onto a disk and transferred from one machine to another or combined with another program representing another person—might well destroy that individual's essence. In his own oblique way, Minsky was suggesting that the how-do-I-know-you're-conscious problem is insurmountable. If no two personalities could ever be fully fused, then downloading, too, might be impossible. In fact, the whole premise of artificial intelligence, if intelligence is defined in human terms, might be flawed.

Minsky, for all his reputation as a rabid reductionist, is actually an *anti*reductionist. He is even more of a romantic, in his way, than is Roger Penrose. Penrose holds out the hope that the mind can be reduced to a single quasi-quantum trick. Minsky insists that no such reduction is possible, because multiplicity is the essence of the mind, of all minds, those of humans and machines alike. Minsky's revulsion toward single-mindedness, simplicity, reflects not just a scientific judgment, I think, but something deeper. Minsky, like Paul Feyerabend and David Bohm and other great romantics, seems to fear *The Answer*, the revelation to end all revelations. Fortunately for Minsky, no such revelation is likely to emerge from neuroscience, since any useful theory of the mind will probably be hideously complex, as he recognizes. Unfortunately for Minsky, it also seems unlikely, given this complexity, that he or even his grandchildren will witness the birth of machines with human attributes. If and when we do construct intelligent, autonomous machines, they will surely be aliens, as unlike us as a 747 is unlike a sparrow. And we could never be *sure* that they were conscious, any more than any of us knows that anyone else is conscious.

Did Bacon Solve the Consciousness Problem?

The conquest of consciousness will take time. The brain is marvelously complicated. But is it infinitely complicated? Given the rate at which neuroscientists are learning about it, within a few decades they may have a highly effective map of the brain, one that correlates specific neural processes to specific mental functions—including consciousness as defined by Crick and Koch. This knowledge may yield many practical benefits, such as treatments for mental illness and information-processing tricks that can be transferred to computers. In *The Coming of the Golden Age*, Gunther Stent proposed that advances in neuroscience might one day give us great power over our own selves. We might be able to "direct specific electrical inputs into the brain. These inputs can then be made to generate synthetically sensations, feelings, and emotions.... Mortal men will soon live like gods without sorrow of heart and remote from grief, as long as their pleasure centers are properly wired."[24]

But Stent, anticipating the mysterian arguments of Nagel, McGinn, and others, also wrote, "the brain may not be capable, in the last analysis, of providing an explanation of itself."[25] Scientists and philosophers will still strive to accomplish the impossible. They will ensure that neuroscience continues in a postempirical, ironic mode in which practitioners argue about the meaning of their physical models, much as physicists argue about the meaning of quantum mechanics. Every now and then a particularly evocative interpretation, set forth by some latter-day Freud steeped in neural and cybernetic knowledge, may attract a vast following and threaten to become the final theory of mind. Neomysterians will then sally forth and point out the theory's inevitable shortcomings. Can it provide a truly satisfying explanation of dreams, or mystical experience? Can it tell us whether amoebas are conscious, or computers?

One could argue that consciousness was "solved" as soon as someone decided that it was an epiphenomenon of the material world. Crick's blunt materialism echoes that of the British philosopher Gilbert Ryle, who coined the phrase "ghost in the machine" in the 1930s to ridicule dualism.[26] Ryle pointed out that dualism—which held that the mind was a separate phenomenon, independent of its physical substrate and capable of exerting influence over it—violated conservation of energy and thus all of physics. Mind is a property of matter, according to Ryle, and only by tracing the intricate meanderings of matter in the brain can one "explain" consciousness.

Ryle was not the first to propose this materialistic paradigm, which is at once so empowering and deflating. Four centuries ago, Francis Bacon urged the philosophers of his day to cease trying to show how the universe evolved from thought and to begin considering how thought evolved from the universe.[27] Here, arguably, Bacon anticipated modern explanations of consciousness within the context of the theory of evolution and, more generally, of the materialist paradigm. The scientific conquest of consciousness will be the ultimate anticlimax, yet another demonstration of Niels Bohr's dictum that science's job is to reduce all mysteries to trivialities. But human science will not, cannot, solve the how-do-I-know-you're-conscious problem. There may be only one way to solve it: to make all minds one mind.

Chapter Eight
The End of Chaoplexity

I miss the Reagan era. Ronald Reagan made moral and political choices so easy. What he liked, I disliked. Star Wars, for example. Formally known as the Strategic Defense Initiative, it was Reagan's plan to build a shield in space that would protect the United States from the nuclear missiles of the Soviet Union. Of the many stories I wrote about Star Wars, the one I am most embarrassed by now involved Gottfried Mayer-Kress, a physicist at, of all places, the Los Alamos National Laboratory, the cradle of the atomic bomb. Mayer-Kress had constructed a simulation of the arms race between the Soviet Union and the United States that employed "chaotic" mathematics. His simulation suggested that Star Wars would destabilize relations between the superpowers and possibly lead to a catastrophe, that is, nuclear war. Because I approved of Mayer-Kress's conclusions—and because his place of employment added a nice touch of irony—I wrote up an admiring report of his work. Of course, if Mayer-Kress's simulation had suggested that Star Wars was a good idea, I would have dismissed his work as the nonsense that it obviously was. Star Wars could well have destabilized relations between the superpowers, but did we need some computer model to tell us that?

I don't mean to beat up on Mayer-Kress. He meant well. (In 1993, several years after I wrote about Mayer-Kress's Star Wars research, I saw a press release from the University of Illinois, where he was then employed, announcing that his computer simulations had suggested solutions to the conflicts in Bosnia and Somalia.[1]) His work is just one of the more blatant examples of over-reaching by someone in the field of chaoplexity (pronounced kay-oh-*plex-ity*). By *chaoplexity* I mean both *chaos* and its close relative *complexity*. Each term, and chaos in particular, has been defined in specific, distinct ways by specific individuals. But each has also been defined in so many overlapping ways by so many different scientists and journalists that the terms have become virtually synonymous, if not meaningless.

The field of chaoplexity emerged as a full-blown pop-culture phenomenon with the publication in 1987 of *Chaos: Making a New Science*, by former *New York Times* reporter James Gleick. After Gleick's masterful book became a best-seller, scores of journalists and scientists sought to duplicate his success by writing similar books on similar topics.[2] There are two, somewhat contradictory aspects to the chaoplexity message. One is that many phenomena are nonlinear and hence inherently unpredictable, because arbitrarily tiny influences can have enormous, unforeseeable consequences. Edward Lorenz, a meteorologist at MIT and a pioneer of chaoplexity, called this phenomenon the butterfly effect, because it meant that a butterfly fluttering in Iowa could, in principle, trigger an avalanche of effects culminating in a monsoon in Indonesia. Because we can never possess more than approximate knowledge of a weather system, our ability to predict its behavior is severely limited.

This insight is hardly new. Henri Poincaré warned at the turn of the century that "small differences in the initial conditions produce very great ones in the final phenomena. A small error in the former will produce an enormous error in the latter. Prediction becomes impossible."[3] Investigators of chaoplexity— whom I will call chaoplexologists— also like to emphasize that many phenomena in nature are "emergent"; they exhibit properties that cannot be predicted or understood simply by examining the system's parts. Emergence, too, is a hoary idea, related to holism, vitalism, and other antireductionist creeds that date back to the last century at least. Certainly Darwin did not think that natural selection could be derived from Newtonian mechanics.

So much for the negative side of the chaoplexity message. The positive side goes as follows: The advent of computers and of sophisticated nonlinear mathematical techniques will help modern scientists understand chaotic, complex, emergent phenomena that have resisted analysis by the reductionist methods of the past. The blurb on the back of Heinz Pagels's *The Dreams of Reason*, one of the best books on the "new sciences of complexity," put it this way: "Just as the telescope opened up the universe and the microscope revealed the secrets of the microcosm, the computer is now opening an exciting new window on the nature of reality. Through its capacity to process what is too complex for the unaided mind, the computer enables us for the first time to simulate reality, to create models of complex systems like large molecules, chaotic systems, neural nets, the human body and brain, and patterns of evolution and population growth."[4]

This hope stems in large part from the observation that simple sets of mathematical instructions, when carried out by a computer, can yield fantastically complicated and yet strangely ordered effects. John von Neumann may have been the first scientist to recognize this capability of computers. In the 1950s, he invented the cellular automaton, which in its simplest form is a screen divided into a grid of cells, or squares. A set of rules relates the color, or state, of each cell to the state of its immediate neighbors. A change in the state of a single cell can trigger a cascade of changes throughout the entire system. "Life," created in the early 1970s by the British mathematician John Conway, remains one of the most celebrated of cellular automatons. Whereas most cellular automatons eventually settle into predictable, periodic behavior, Life generates an infinite variety of patterns—including cartoonlike objects that seem to be engaged in inscrutable missions. Inspired by Conway's strange computer world, a number of scientists began using cellular automatons to model various physical and biological processes.

Another product of computer science that seized the imagination of the scientific community was the Mandelbrot set. The set is named after Benoit Mandelbrot, an applied mathematician at IBM who is one of the protagonists of Gleick's book *Chaos* (and whose work on indeterministic phenomena led Gunther Stent to conclude that the social sciences would never amount to much). Mandelbrot invented fractals, mathematical objects displaying what is known as fractional dimensionality: they are fuzzier than a line but never quite fill a plane. Fractals also display patterns that keep recurring at finer and finer scales. After coining the term *fractal*, Mandelbrot pointed out that many real-world phenomena—notably clouds, snowflakes, coastlines, stock-market fluctuations, and trees—have fractal-like properties.

The Mandelbrot set, too, is a fractal. The set corresponds to a simple mathematical function that is repeatedly iterated; one solves the function and plugs the answer back into it and solves it again, ad infinitum. When plotted by a computer, the numbers generated by the function cluster into a now-famous shape, which has been likened to a tumor-ridden heart, a badly burned chicken, and a warty figure eight lying on its side. When one magnifies the set with a computer, one finds that its borders do not form crisp lines, but shimmer like flames. Repeated magnification of the borders plunges the viewer into a bottomless phantasmagoria of baroque imagery. Certain patterns, such as the basic heartlike shape, keep recurring, but always with subtle variations.

The Mandelbrot set, which has been called "the most complex object in mathematics," has become a kind of laboratory in which mathematicians can test ideas about the behavior of nonlinear (or chaotic, or complex) systems. But what relevance do those findings have to the real world? In his 1977 magnum opus, *The Fractal Geometry of Nature*, Mandelbrot warned that it was one thing to observe a fractal pattern in nature and quite another to determine the *cause* of that pattern. Although exploring the consequences of self-similarity yielded "extraordinary surprises, helping me to understand the fabric of nature," Mandelbrot said, his attempts to unravel the causes of self-similarity "had few charms."[5]

Mandelbrot seemed to be alluding to the seductive syllogism that underlies chaoplexity. The syllogism is this: There are simple sets of mathematical rules that when followed by a computer give rise to extremely complicated patterns, patterns that never quite repeat themselves. The natural world also contains many extremely complicated patterns that never quite repeat themselves. Conclusion: Simple rules underlie many extremely complicated phenomena in the world. With the help of powerful computers, chaoplexologists can root out those rules.

Of course, simple rules *do* underlie nature, rules embodied in quantum mechanics, general relativity, natural selection, and Mendelian genetics. But chaoplexologists insist that much more powerful rules remain to be found.

The 31 Flavors of Complexity

Blue and red dots skittered across a computer screen. But these were not just colored dots. These were agents, simulated people, doing the things that real people do: foraging for food, seeking mates, competing and cooperating with each other. At least, that's what Joshua Epstein, the creator of this computer simulation, claimed. Epstein, a sociologist from the Brookings Institution, was showing his simulation to me and two other journalists at the Santa Fe Institute, where Epstein was a visiting fellow. The institute, founded in the mid-1980s, quickly became the headquarters of complexity, the self-proclaimed successor to chaos as the new science that would transcend the stodgy reductionism of Newton, Darwin, and Einstein.

As my colleagues and I watched Epstein's colored dots and listened to his even more colorful interpretation of their movements, we offered polite murmurs of interest. But behind his back we exchanged jaded smiles. None of us took this kind of thing very seriously. We all understood, implicitly, that this was ironic science. Epstein himself, when pressed, acknowledged that his model was not predictive in any way; he called it a "laboratory," a "tool," a "neural prosthesis" for exploring ideas about the evolution of societies. (These were all favorite terms of Santa Fe'ers.) But during public presentations of his work, Epstein had also claimed that simulations such as his would revolutionize the social sciences, helping to solve their most intractable problems.[6]

Another believer in the power of computers is John Holland, a computer scientist with joint appointments at the University of Michigan and the Santa Fe Institute. Holland was the inventor of genetic algorithms, which are segments of computer code that can rearrange themselves to produce a new program that can solve a problem more efficiently. According to Holland, the algorithms are, in effect, evolving, just as the genes of living organisms evolve in response to the pressure of natural selection.

Holland has proposed that it may be possible to construct a "unified theory of complex adaptive systems" based on mathematical techniques such as those embodied in his genetic algorithms. He spelled out his vision in a 1993 lecture:

Many of our most troubling long-range problems—trade imbalances, sustainability, AIDS, genetic defects, mental health, computer viruses— center on certain systems of extraordinary complexity. The systems that host these problems—economies, ecologies, immune systems, embryos, nervous systems, computer networks—appear to be as diverse as the problems. Despite appearances, however, the systems do share significant characteristics, so much so that we group them under a single classification at the Santa Fe Institute, calling them *complex adaptive systems* (cas). This is more than terminology. It signals our intuition that there are general principles that govern all cas behavior, principles that point to ways of solving the attendant problems. Much of our work is aimed at turning this intuition into fact.[7]

The ambition revealed by this statement is breathtaking. Chaoplexologists often ridicule particle physicists for their hubris, for thinking they can achieve a theory of everything. But actually particle physicists are rather modest in their ambition: they merely hope that they can wrap up the forces of nature in one tidy package and perhaps illuminate the origin of the universe. Few are so bold as to propose that their unified theory will yield both *truth* (that is, insight into nature) and *happiness* (solutions to our worldly problems), as Holland and others have proposed. And Holland is considered one of the more modest scientists associated with the field of complexity.

But can scientists achieve a unified theory of complexity if they cannot agree what, precisely, complexity is? Students of complexity have struggled, with little success, to distinguish themselves from those who study chaos. According to University of Maryland physicist James Yorke, chaos refers to a restricted set of phenomena that evolve in predictably unpredictable ways—demonstrating sensitivity to initial conditions, aperiodic behavior, the recurrence of certain patterns at different spatial and temporal scales, and so on. (Yorke ought to know, since he coined the term *chaos* for a paper published in 1975.) Complexity seems to refer to "anything you want," according to Yorke.[8]

One widely touted definition of complexity involves "the edge of chaos." This picturesque phrase was incorporated in the subtitles of two books by journalists published in 1992: *Complexity: Life at the Edge of Chaos*, by Roger Lewin, and *Complexity: The Emerging Science at the Edge of Order and Chaos*, by M. Mitchell Waldrop.[9] (The authors no doubt intended the phrase to evoke the style as well as the substance of the field.) The basic idea of the edge of chaos is that nothing novel can emerge from systems with high degrees of order and stability, such as crystals; on the other hand, completely chaotic, or aperiodic, systems, such as turbulent fluids or heated gases, are *too* formless. Truly complex things—amoebas, bond traders, and the like— happen at the border between rigid order and randomness.

Most popular accounts credit this idea to the Santa Fe researchers Norman Packard and Christopher Langton. Packard, whose experience as a leading figure in chaos theory educated him in the importance of idea packaging, coined the all-important phrase "edge of chaos" in the late 1980s. In experiments with cellular automatons, he and Langton concluded that a system's computational potential—that is, its ability to store and process information—peaks in a regime between highly periodic and chaotic behavior. But two other researchers at the Santa Fe Institute, Melanie Mitchell and James Crutchfield, have reported that their own computer experiments do not support the conclusions of Packard and Langton. They also questioned whether "anything like a drive toward universal-computational capabilities is an important force in the evolution of biological organisms."[10] Although a few Santa Fe'ers still employ the phrase "edge of chaos" (notably Stuart Kauffman, whose work was described in Chapter 5), most now disavow it.

Many other definitions of complexity have been proposed—at least 31, according to a list compiled in the early 1990s by physicist Seth Lloyd of the Massachusetts Institute of Technology, who is also associated with the Santa Fe Institute.[11] The definitions typically draw on thermodynamics, information theory, and computer science and involve concepts such as entropy, randomness, and information— which themselves have proved to be notoriously slippery terms. All definitions of complexity have drawbacks. For example, algorithmic information theory, proposed by the IBM mathematician Gregory Chaitin (among others), holds that the complexity of a system can be represented by the shortest computer program describing it. But according to this criterion, a text created by a team of typing monkeys is more complex—because it is more random and therefore less compressible—than *Finnegans Wake*.

Such problems highlight the awkward fact that complexity exists, in some murky sense, in the eye of a beholder (like pornography, for instance).[12] Researchers have at times debated whether complexity has become so meaningless that it should be abandoned, but they have invariably concluded that the term has too much public relations value. Santa Fe'ers often employ "interesting" as a synonym for "complex." But what government agency would supply funds for research on a "unified theory of interesting things"?

The Poetry of Artificial Life

Members of the Santa Fe Institute may not agree on what they are studying, but they concur on how they should study it: with computers. Christopher Langton embodies the faith in computers that gave rise to the chaos and complexity movements. He has proposed that simulations of life run on a computer are alive—not sort of, or in a sense, or metaphorically, but actually. Langton is the founding father of artificial life, a subfield of chaoplexity that has attracted much attention in its own right. Langton has helped organize several conferences on artificial life—the first held at Los Alamos in 1987—attended by biologists, computer scientists, and mathematicians who share his affinity for computer animation.[13]

Artificial life is an outgrowth of artificial intelligence, a field that preceded it by several decades. Whereas artificial-intelligence researchers seek to understand the mind better by simulating it on a computer, proponents of artificial life hope to gain insights into a broad range of biological phenomena through their simulations. And just as artificial intelligence has generated more portentous rhetoric than tangible results, so has artificial life. As Langton stated in an essay introducing the inaugural issue of the quarterly journal *Artificial Life* in 1994:

Artificial life will teach us much about biology—much that we could not have learned by studying the natural products of biology alone—but artificial life will ultimately reach beyond biology, into a realm we do not yet have a name for, but which must include culture and our technology in an extended view of nature. I don't want to paint a rosy picture of the future of artificial life. It will not solve all our problems. Indeed, it may well add to them.... Perhaps the simplest way to emphasize this point is by merely pointing out that Mary Shelley's prophetic story of Dr. Frankenstein can no longer be considered science fiction.[14]

Even before I met Langton, I felt I knew him. He played a prominent role in several popular journalistic treatments of chaoplexity. And no wonder. He is the archetypal hip young scientist: simultaneously intense and mellow, longhaired, given to wearing jeans, leather vests, hiking boots, Indian jewelry. He also comes equipped with a marvelous life story; its centerpiece is a hang-gliding accident that led to a coma and an epiphany. (Those who want to hear the tale can read Waldrop's *Complexity*, Lewin's *Complexity*, or Steven Levy's *Artificial Life*.[15])

When I finally encountered Langton in the flesh in Santa Fe in May 1994, we decided to talk while eating lunch at one of his favorite restaurants. Langton's car—hadn't I read about it in one of the books about him?—was a beat-up old compact filled with miscellanea, from audiotapes and pliers to plastic containers of hot sauce, all covered with a layer of beige desert dust. As we drove to the restaurant, Langton dutifully banged the chaoplexity boilerplate. Most scientists from Newton on have studied systems exhibiting stability, periodicity, and equilibrium, but he and other researchers at Santa Fe wanted to understand the "transient regimes" underlying many biological phenomena. After all, he said, "once you reach the equilibrium point for a living organism, you're dead."

He grinned. It had begun raining, and he turned on his wipers. The windshield quickly went from translucent to opaque, as the wipers smeared dirt across the glass. Langton, peering through an unsmeared corner of the glass, continued talking, seemingly unperturbed by the windshield's metaphorical message. Science, he said, had obviously made enormous progress by breaking things up into pieces and studying those pieces. But that methodology provided only limited understanding of higher-level phenomena, which were created to a large extent through historical accidents. One could transcend those limitations, however, through a synthetic methodology, in which the basic components of existence were put together in new ways in computers to explore what might have happened or could have happened.

"You end up with a much larger set" of possibilities, Langton said. "You can then probe the set not just of existing chemical compounds but of possible chemical compounds. And it's only really within that ground of the possible chemical compounds that you're going to see any regularity. The regularity is there but you can't see it in the very small set of things that nature initially provided you with." With computers, biologists can explore the role of chance by simulating the beginning of life on earth, altering the conditions and observing the consequences. "So part of what artificial life is all about, and part of the broader scheme that I just call synthetic biology in general, is probing beyond, pushing beyond the envelope of what occurred naturally." In this way, Langton suggested, artificial life might reveal which aspects of our history were inevitable and which were merely contingent.

In the restaurant, as he chewed on chicken fajitas, Langton confirmed that, yes, he really did adhere to the view known as "strong a-life," which holds that computer simulations of living things are themselves alive. He described himself as a functionalist, who believed life was characterized by what it did rather than by what it was made of. If a programmer created molecule-like structures that, following certain laws, spontaneously organized themselves into entities that could seemingly eat, reproduce, and evolve, Langton would consider those entities to be alive—"even if they're in a computer."

Langton said his belief had moral consequences. "I like to think that if I saw somebody sitting next to me at a computer terminal who is *torturing* these creatures, you know, sending them to some digital equivalent of hell, or rewarding only a select few who spelled out his name on the screen, I would try to get this guy some psychological help!"

I told Langton that he seemed to be conflating metaphor, or analogy, with reality. "What I'm trying to do, actually, is something a little more seditious than that," Langton replied, smiling. He wanted people to realize that life might be a process that could be implemented by any number of arrangements of matter, including the ebb and flow of electrons in a computer. "At some level the actual physical realization is irrelevant to the functional properties," he said. "Of course there are differences," he added. "There are going to be differences if there's a different material base. But are the differences fundamental to the property of being alive or not?"

Langton did not support the claim—commonly made by artificial-intelligence enthusiasts—that computer simulations can also have subjective experiences. "This is why I like artificial life more than AI," he said. Unlike most biological phenomena, subjective states cannot be reduced to mechanical functions. "No mechanical explanation you could possibly give is going to give you that explanation for this sense of awareness, of I-ness, of my being here now." Langton, in other words, was a mysterian, someone who believed that an explanation of consciousness was beyond the reach of science. He finally conceded that the question of whether computer simulations are *really* alive was also, ultimately, a philosophical, and therefore unresolvable, issue. "But for artificial life to do its job and help broaden the empirical database for biological science and for a theory of biology, they don't have to solve that problem. Biologists have never really had to solve that."

The longer Langton spoke, the more he seemed to acknowledge— and even welcome the fact—that artificial life would never be the basis for a truly empirical science. Artificial-life simulations, he said, "force me to look over my shoulder about the assumptions I make about the real world." In other words, the simulations can enhance our negative capability; they can serve to challenge rather than to support theories about reality. Moreover, scientists studying artificial life might have to settle for something less than the "complete understanding" they derived from the old, reductionist methods. "For certain categories in nature there won't be anything more we can do by way of an explanation than to be able to say, 'Well, here's the history.'"

Then he confessed that such an outcome would suit him fine; he hoped the universe was in some fundamental sense "irrational." "Rationality is very much connected with the tradition in science for the last 300 years, when you're going to end up with some sort of understandable explanation of something. And I would be disappointed if that were the case."

Langton complained that he was frustrated with the linearity of scientific language. "There's a reason for poetry," he said. "Poetry is a very nonlinear use of language, where the meaning is more than just the sum of the parts. And science requires that it be nothing more than the sum of the parts. And just the fact that there's stuff to explain out there that's more than the sum of the parts means that the traditional approach, just characterizing the parts and the relations, is not going to be adequate for capturing the essence of many systems that you would like to be able to do. That's not to say that there isn't a way to do it in a more scientific way than poetry, but I just have the feeling that culturally there's going to be more of something like poetry in the future of science."

The Limits of Simulation

In February 1994, the journal *Science* published a paper, "Verification, Validation, and Confirmation of Numerical Models in the Earth Sciences," that addressed the problems posed by computer simulations. The remarkably postmodern article was written by Naomi Oreskes, a historian and geophysicist at Dartmouth College; Kenneth Belitz, a geophysicist also at Dartmouth; and Kristin Shrader Frechette, a philosopher at the University of South Florida. Although they focused on geophysical modeling, their warnings were really applicable to numerical models of all kinds (as they acknowledged in a letter printed by *Science* some weeks later).[16]

The authors observed that numerical models were becoming increasingly influential in debates over global warming, the depletion of oil reserves, the suitability of nuclear-waste sites, and other issues. Their paper was meant to serve as a warning that "verification and validation of numerical models of natural systems is impossible." The only propositions that can be verified—that is, proved true—are those dealing in pure logic or mathematics. Such systems are closed, in that all their components are based on axioms that are true by definition. Two plus two equals four by common agreement, not because the equation corresponds to some external reality. Natural systems are always open, Oreskes and her colleagues pointed out; our knowledge of them is always incomplete, approximate, at best, and we can never be sure we are not overlooking some relevant factors.

"What we call data," they explained, "are inference-laden signifiers of natural phenomena to which we have incomplete access. Many inferences and assumptions can be justified on the basis of experience (and some uncertainties can be estimated), but the degree to which our assumptions hold in any new study can never be established a priori. The embedded assumptions thus render the system open." In other words, our models are always idealizations, approximations, guesses.

The authors emphasized that when a simulation accurately mimics or even predicts the behavior of a real phenomenon, the model is still not verified. One can never be sure whether a match stems from some genuine correspondence between the model and reality or is coincidental. Moreover, it is always possible that other models, based on different assumptions, could yield the same results.

Oreskes and her coauthers noted that the philosopher Nancy Cartwright called numerical models "a work of fiction." They continued,

> While not necessarily accepting her viewpoint, we might ponder this aspect of it: A model, like a novel, may resonate with nature, but it is not a "real" thing. Like a novel, a model may be convincing—it may ring true if it is consistent with our experience of the natural world. But just as we may wonder how much the characters of a novel are drawn from real life and how much is artifice, we might ask the same of a model: How much is based on observation and measurement of accessible phenomena, how much is based on informed judgement, and how much is convenience?... [We] must admit that a model may confirm our biases and support incorrect intuitions. Therefore, models are most useful when they are used to challenge existing formulations, rather than to validate or verify them.

Numerical models work better in some cases than in others. They work particularly well in astronomy and particle physics, because the relevant objects and forces conform to their mathematical definitions so precisely. Moreover, mathematics helps physicists define what is otherwise undefinable. A quark is a purely mathematical construct. It has no meaning apart from its mathematical definition. The properties of quarks—charm, color, strangeness—are mathematical properties that have no analogue in the macroscopic world we inhabit. Mathematical theories are less compelling when applied to more concrete, complex phenomena, such as anything in the biological realm. As the evolutionary biologist Ernst Mayr has pointed out, each organism is unique; each one also changes from moment to moment.[17] That is why mathematical models of biological systems generally have less predictive power than physics does. We should be equally wary of their ability to yield truths about nature.

The Self-Organized Criticality of Per Bak

This kind of "wishy-washy" philosophical skepticism irks Per Bak. Bak, a Danish physicist who came to the United States in the 1970s, is like a parody of Harold Bloom's strong poet. He is a tall, stout man, at once owlish and pugnacious, who bristles with opinions. Trying to convince me of the superiority of complexity to other modes of science, he scoffed at the suggestion that particle physicists could uncover the secret of existence by probing ever-smaller scales of matter. "The secret doesn't come from going deeper and deeper into the system," Bak asserted with a distinct Danish accent. "It comes from going the *other* direction."[18]

Particle physics is dead, Bak proclaimed, killed by its own success. Most particle physicists, he noted, "think they're still doing science when they're really just cleaning up the mess after the party." The same was true of solid-state physics, the field in which Bak began his career. The fact that thousands of physicists were working on high-temperature superconductivity—for the most part in vain—showed how desiccated the field had become: "There's very little meat and many animals who want to eat it." As for chaos (which Bak defined in the same narrow way that James Yorke did), physicists had come to a basic understanding of the processes underlying chaotic behavior by 1985, two years before Gleick's book *Chaos* was published. "That's how things go!" Bak barked. "Once something reaches the masses, it's already done." (Complexity, of course, is the exception to Bak's rule.)

Bak had nothing but contempt for scientists who were content merely to refine and extend the work of the pioneers. "There's no need for that! We don't need the cleanup team here!" Fortunately, Bak said, many mysterious phenomena continue to resist scientific understanding: the evolution of species, human cognition, economics. "What these things have in common is that they are very large things with many degrees of freedom. They are what we call complex systems. And there will be a revolution in science. These things will be made into hard sciences in the next years in the same way that

222

[particle] physics and solid-state physics were made hard sciences in the last 20 years." Bak rejected the "pseudophilosophical, pessimistic, wishy-washy" view that these problems are simply too difficult for our puny human brains. "If I thought that was true I wouldn't be doing these things!" Bak exclaimed. "We should be optimistic, concrete, and then we can go on. And I'm sure that science will look totally different 50 years from now than it does today."

In the late 1980s Bak and two colleagues proposed what quickly became a leading candidate for a unified theory of complexity: self-organized criticality. His paradigmatic system is a sandpile. As one adds sand to the top of the pile, it approaches what Bak calls the critical state, in which even a single additional grain of sand dropped on the top of the pile can trigger an avalanche down the pile's sides. If one plots the size and frequency of the avalanches occurring in this critical state, the results conform to what is known as a power law: the frequency of avalanches is inversely proportional to a power of their size.

Bak credited the chaos pioneer Benoit Mandelbrot with having pointed out that earthquakes, stock-market fluctuations, the extinction of species, and many other phenomena displayed the same pattern of power-law behavior. (In other words, the phenomena that Bak defined as complex were also all chaotic.) "Since economics, geophysics, astronomy, biology have these singular features there *must* be a theory here," Bak said. He hoped his theory might explain why small earthquakes are common and large ones uncommon, why species persist for millions of years and then vanish, why stock markets crash. "We can't explain everything about everything, but something about everything."

Bak thought models such as his would eventually revolutionize economics. "Traditional economics is not a real science. It's a mathematical discipline where they talk about perfect markets and perfect rationality and perfect equilibrium." This approach was a "grotesque approximation" that could not explain real-world economic behavior. "Any real person who works on Wall Street and observes what happens knows that fluctuations come from chain reactions in the system. It comes from the coupling between the various agents: bank traders, customers, thieves, robbers, governments, economies, whatever. Traditional economics has no description of this phenomenon."

Can mathematical theories provide meaningful insights into cultural phenomena? Bak groaned at the question. "I don't understand what meaning is," he said. "In science there is no meaning to anything. It just observes and describes. It doesn't ask the atom why it's going left when it's subjected to a magnetic field. So social scientists should go out and observe the behavior of people and then figure out what the consequences of that are for society."

Bak acknowledged that such theories offered statistical descriptions rather than specific predictions. "The whole idea is we cannot predict. But nevertheless we can understand those systems that we cannot predict. We can understand *why* they cannot be predicted." That, after all, was what thermodynamics and quantum mechanics, which are also probabilistic theories, had accomplished. The models should be specific enough to be falsifiable, but not too specific, Bak said. "I think it's a losing game to make very specific and detailed models. That doesn't give any insight." That would be mere engineering, Bak sniffed.

When I asked whether he thought researchers would someday converge on a single, true theory of complex systems, Bak seemed to lose a bit of his nerve. "It is a much more fluid situation," he said. He doubted whether scientists would ever achieve a simple, unique theory of the brain, for instance. But they might find "some principles, hopefully not too many, that govern the behavior of the brain." He mused for a moment, then added, "I think it is a much more long-term thing than, say, chaos theory."

Bak also feared that the federal government's growing antipathy toward pure science and its increased emphasis on practical applications might hinder progress in complexity studies. It was increasingly difficult to pursue science for its own sake; science had to be useful. Most scientists were forced to do "deadly boring stuff that cannot be of any real interest." His own primary employer, Brookhaven National Laboratory, was making people do *"horrifying things, incredible* garbage." Even Bak, as fiercely optimistic as he was, had to acknowledge the dismal plight of modern science.

Self-organized criticality has been touted by, among others, Al Gore. In his 1992 best-seller, *Earth in the Balance*, Gore revealed that self-organized criticality had helped him to understand not only the sensitivity of the environment to potential disruptions but also "change in my own life."[19] Stuart Kauffman has found affinities between self-organized criticality, the edge of chaos, and the laws of complexity he has glimpsed in his computer simulations of biological evolution. But other researchers have complained that Bak's model does not even provide a very good description of his paradigmatic system: a sandpile. Experiments by physicists at the University of Chicago have shown that sandpiles behave in many different ways, depending on the size and shape of the grains; few sandpiles display the power-law behavior predicted by Bak.[20] Moreover, Bak's model may be too general and statistical in nature really to illuminate any of the systems it describes. After all, many phenomena can be described by a so-called Gaussian curve, more commonly known as a bell curve. But few scientists would claim, for example, that human intelligence scores and the apparent luminosity of galaxies must derive from common mechanisms.

Self-organized criticality is not really a theory at all. Like punctuated equilibrium, self-organized criticality is merely a description, one of many, of the random fluctuations, the noise, permeating nature. By Bak's own admission, his model can generate neither specific predictions about nature nor meaningful insights. What good is it, then?

Cybernetics and Other Catastrophes

History abounds with failed attempts to create a mathematical theory that explains and predicts a broad range of phenomena, including social ones. In the seventeenth century Leibniz fantasized about a system of logic so compelling that it could resolve not only all mathematical questions but also philosophical, moral, and political ones.[21] Leibniz's dream has persisted even in the century of doubt. Since World War II scientists have become temporarily infatuated with at least three such theories: cybernetics, information theory, and catastrophe theory.

Cybernetics was created largely by one person, Norbert Wiener, a mathematician at the Massachusetts Institute of Technology. The subtitle of his 1948 book, *Cybernetics*, revealed his ambition: *Control and Communication in the Animal and the Machine*.[22] Wiener based his neologism on the Greek term *kubernetes*, or steersman. He proclaimed that it should be possible to create a single, overarching theory that could explain the operation not only of machines but also of all biological phenomena, from single-celled organisms up through the economies of nation-states. All these entities process and act on information; they all employ such mechanisms as positive and negative feedback and filters to distinguish signals from noise.

By the 1960s cybernetics had lost its luster. The eminent electrical engineer John R. Pierce noted drily in 1961 that "in this country the word cybernetics has been used most extensively in the press and in popular and semiliterary, if not semiliterate, magazines."[23] Cybernetics still has a following in isolated enclaves, notably Russia (which during the Soviet era was highly receptive to the fantasy of society as a machine that could be fine-tuned by following the precepts of cybernetics). Wiener's influence persists in U.S. pop culture if not within science itself: we owe words such as *cyberspace, cyberpunk*, and *cyborg* to Wiener.

Closely related to cybernetics is information theory, which Claude Shannon, a mathematician at Bell Laboratories, spawned in 1948 with a two-part paper titled "A Mathematical Theory of Communication."[24] Shannon's great achievement was to invent a mathematical definition of information based on the concept of entropy in thermodynamics. Unlike cybernetics, information theory continues to thrive—within the niche for which it was intended. Shannon's theory was designed to improve the transmission of information over a telephone or telegraph line subject to electrical interference, or noise. The theory still serves as the theoretical foundation for coding, compression, encryption, and other aspects of information processing.

By the 1960s information theory had infected other disciplines outside communications, including linguistics, psychology, economics, biology, and even the arts. (For example, various sages tried to concoct formulas relating the quality of music to its information content.) Although information theory is enjoying a renaissance in physics as a result of the influence of John Wheeler (the it from bit) and others, it has yet to contribute to physics in any concrete way. Shannon himself doubted whether certain applications of his theory would come to much. "Somehow people think it can tell you things about meaning," he once said to me, "but it can't and wasn't intended to."[25]

Perhaps the most oversold metatheory was the appropriately named catastrophe theory, invented by the French mathematician René Thom in the 1960s. Thom developed the theory as a purely mathematical formalism, but he and others began to claim that it could provide insights into a broad range of phenomena that displayed discontinuous behavior. Thom's magnum opus was his 1972 book, *Structural Stability and Morphogenesis*, which received awestruck reviews in Europe and the United States. A reviewer in the *Times* of London declared that "it is impossible to give a brief description of the impact of this book. In one sense the only book with which it can be compared is Newton's *Principia*. Both lay out a new conceptual framework for the understanding of nature, and equally both go on to unbounded speculation."[26]

Thom's equations revealed how a seemingly ordered system could undergo abrupt, "catastrophic" shifts from one state to another. Thom and his followers suggested that these equations could help to explain not only such purely physical events as earthquakes, but also biological and social phenomena, such as the emergence of life, the metamorphosis of a caterpillar into a butterfly, and the collapse of civilizations. By the late 1970s, the counterattack had begun. Two mathematicians declared in *Nature* that catastrophe theory "is one of many attempts to deduce the world by thought alone." They called that "an appealing dream, but a dream that cannot come true." Other critics charged that Thom's work "provides no new information about anything" and is "exaggerated, not wholly honest."[27]

Chaos, as defined by James Yorke, exhibited this same boom-bust cycle. By 1991, at least one pioneer of chaos theory, the French mathematician David Ruelle, had begun to wonder whether his field had passed its peak. Ruelle invented the concept of strange attractors, mathematical objects that have fractal properties and can be used to describe the behavior of systems that never settle into a periodic pattern. In his book *Chance and Chaos*, Ruelle noted that chaos "has been invaded by swarms of people who are attracted by success, rather than the ideas involved. And this changes the intellectual atmosphere for the worse.... The physics of chaos, in spite of frequent triumphant announcements of 'novel' breakthroughs, has had a declining output of interesting discoveries. Hopefully, when the craze is over, a sober appraisal of the difficulties of the subject will result in a new wave of high-quality results."[28]

"More Is Different"

Even some researchers associated with the Santa Fe Institute seem to doubt that science can achieve the kind of transcendent, unified theory of complex phenomena that John Holland, Per Bak, and Stuart Kauffman all dream of. One skeptic is Philip Anderson, a notoriously hard-nosed physicist who won a Nobel Prize in 1977 for his work on superconductivity and other antics of condensed matter and was one of the founders of the Santa Fe Institute. Anderson was a pioneer of antireductionism. In "More Is Different," an essay published in *Science* in 1972, Anderson contended that particle physics, and indeed all reductionist approaches, have a limited ability to explain the world. Reality has a hierarchical structure, Anderson argued, with each level independent, to some degree, of the levels above and below. "At each stage, entirely new laws, concepts and generalizations are necessary, requiring inspiration and creativity to just as great a degree as in the previous one," Anderson noted. "Psychology is not applied biology, nor is biology applied chemistry."[29]

"More Is Different" became a rallying cry for the chaos and complexity movements. Ironically, the principle suggests that these so-called anti-reductionist efforts may never culminate in a unified theory of complex, chaotic systems, one that illuminates everything from immune systems to economies, as chaoplexologists such as Bak have suggested. (The principle also suggests that the attempt of Roger Penrose to explain the mind in terms of quasi-quantum mechanics was misguided.) Anderson acknowledged as much when I visited him at Princeton University, his home base. "I don't think there is a theory of everything," he said. "I think that there are basic principles that have very wide generality," such as quantum mechanics, statistical mechanics, thermodynamics, and symmetry breaking. "But you mustn't give in to the temptation that when you have a good general principle at one level that it's going to work at all levels." (Of quantum mechanics, Anderson said, "There seems to me to be no possible modification of that in the foreseeable future.") Anderson agreed with the evolutionary biologist Steven Jay Gould that life is shaped less by deterministic laws than by contingent, unpredictable circumstances. "I guess the prejudice I'm trying to express is a prejudice in favor of natural history," Anderson elaborated.

Anderson did not share the faith of some of his colleagues at the Santa Fe Institute in the power of computer models to illuminate complex systems. "Since I know a little bit about global economic models," he explained, "I know they don't work! I always wonder whether global climate models or oceanic circulation models or things like that are as full of phony statistics and phony measurements." Making simulations more detailed and realistic is not necessarily the solution, Anderson noted. It is possible for a computer to simulate the phase transition of a liquid into a glass, for example, "but have you learned anything? Do you understand it any better than you did before? Why not just take a piece of glass and say it's going through the glass transition? Why do you have to look at a computer go through the glass transition? So that's the *reductio ad absurdum*. At some point the computer's not telling you what the system itself is doing."

And yet, I said, there seemed to be this abiding faith among some of his colleagues that they would someday find a theory that would dispel all mysteries. "Yeah," Anderson said, shaking his head. Abruptly, he threw his arms in the air and cried out, like a born-again parishioner: "I've finally seen the light! I understand everything!" He lowered his arms and smiled ruefully. "You *never* understand everything," he said. "When one understands everything, one has gone crazy."

The Quark Master Rules Out "Something Else"

An even more improbable leader of the Santa Fe Institute is Murray Gell-Mann. Gell-Mann is a master reductionist. He won a Nobel Prize in 1969 for discovering a unifying order beneath the alarmingly diverse particles streaming from accelerators. He called his particle-classification system the Eight-fold way, after the Buddhist road to wisdom. (The name was meant to be a joke, Gell-Mann has often pointed out, not an indication that he was one of these flaky New Age types who thought physics and Eastern mysticism had something in common.) He showed the same flair for discerning unity in complexity—and for coining terms—when he proposed that neutrons, protons, and a host of other shorter-lived particles were all made of triplets of more fundamental particles called quarks. Gell-Mann's quark theory has been amply demonstrated in accelerators, and it remains a cornerstone of the standard model of particle physics.

Gell-Mann is fond of recalling how he found the neologism *quark* while perusing James Joyce's *Finnegans Wake*. (The passage reads, "Three quarks for Muster Mark!") This anecdote serves notice that Gell-Mann's intellect is far too powerful and restless to be satisfied by particle physics alone. According to a "personal statement" that he distributes to reporters, his interests include not only particle physics and modern literature, but also cosmology, nuclear arms–control policy, natural history, human history, population growth, sustainable human development, archaeology, and the evolution of language. Gell-Mann seems to have at least some familiarity with most of the major languages of the world and with many dialects; he enjoys telling people about the etymology and correct native pronunciation of their names. He was one of the first major scientists to climb aboard the complexity bandwagon. He helped to found the Santa Fe Institute, and he became its first full-time professor in 1993. (He had spent almost 40 years before that as a professor at the California Institute of Technology.)

Gell-Mann is unquestionably one of this century's most brilliant scientists. (His literary agent, John Brockman, said that Gell-Mann "has five brains, and each one is smarter than yours."[30]) He may also be one of the most annoying. Virtually everyone who knows Gell-Mann has a story about his compulsion to tout his own talents and to belittle those of others. He displayed this trait almost immediately after we first met in 1991 in a New York City restaurant, several hours before he was to catch a flight to California. Gell-Mann is a small man with large black glasses, short white hair, and a skeptical squint. I had barely sat down when he began to tell me—as I set out my tape recorder and yellow pad—that science writers were "ignoramuses" and a "terrible breed" who invariably got things wrong; only scientists were really qualified to present their work to the masses. As time went on, I felt less offended, since it became clear that Gell-Mann held most of his scientific colleagues in contempt as well. After a series of particularly demeaning comments about some of his fellow physicists, Gell-Mann added, "I don't want to be quoted insulting people. It's not nice. Some of these people are my friends."

To stretch out the interview, I had arranged for a limousine to take us to the airport together. Once there, I accompanied Gell-Mann as he checked his bags and went to the first-class lounge. He began to fret that he did not have enough money to take a taxi home after he arrived in California. (Gell-Mann had not yet moved permanently to Santa Fe.) If I could lend him some money, he would write me a check. I gave him $40. As Gell-Mann handed me a check, he suggested that I consider not cashing it, since his signature would probably be quite valuable someday. (I cashed the check, but I kept a photocopy.)[31]

My suspicion is that Gell-Mann doubts that his colleagues in Santa Fe will discover anything truly profound, anything approaching, say, Gell-Mann's own quark theory. However, if a miracle occurs, and the chaoplexologists somehow manage to accomplish something important, Gell-Mann wants to be able to share the glory. His career would therefore encompass the whole range of modern science, from particle physics on up to chaos and complexity.

For a putative leader of chaoplexity, Gell-Mann espouses a world-view remarkably similar to that of the arch-reductionist Steven Weinberg—although of course Gell-Mann does not put it that way. "I have no idea what Weinberg said in his book," Gell-Mann said when I asked him during an interview in Santa Fe in 1995 if he agreed with what Weinberg had said in *Dreams of a Final Theory* about reductionism. "But if you read *mine* you saw what I said about it." Gell-Mann then went on to repeat some of the major themes of his 1994 book, *The Quark and the Jaguar*.[32] To Gell-Mann, science forms a hierarchy. At the top are those theories that apply everywhere in the known universe, such as the second law of thermodynamics and his own quark theory. Other laws, such as those related to genetic transmission, apply only here on earth, and the phenomena they describe entail a great deal of randomness and historical circumstance.

"With biological evolution we see a gigantic amount of history enters, huge numbers of accidents that could have gone different ways and produced different life-forms than we have on the earth, constrained of course by selection pressures. Then we get to human beings and the characteristics of human beings are determined by huge amounts of history. But still, there's clear determination from the fundamental laws and from history, or fundamental laws and specific circumstances."

Gell-Mann's reductionist predilections can be seen in his attempts to get his colleagues at the Santa Fe Institute to substitute his own neologism, *plectics*, for complexity. "The word is based on the Indo-European word *plec*, which is the basis of both simplicity and complexity. So in plectics we try to understand the relation between the simple and the complex, and in particular how we get from the simple fundamental laws that govern the behavior of all matter to the complex fabric that we see around us," he said. "We're trying to make theories of how this process works, in general and also special cases, and how those special cases relate to the general situation." (Unlike *quark*, *plectics* has never caught on. I have never heard anyone other than Gell-Mann use the term—except to deride Gell-Mann's fondness for it.)

Gell-Mann rejected the possibility that his colleagues would discover a single theory that embraced all complex adaptive systems. "There are huge differences among these systems, based on silicon, based on protoplasm, and so on. It's not the same." I asked Gell-Mann if he agreed with the "More Is Different" principle set forth by his colleague Philip Anderson. "I have no idea what he said," Gell-Mann replied disdainfully. I explained Anderson's idea that reductionist theories have limited explanatory power; one cannot go back up the chain of explanation from particle physics to biology. "You can! You can!" Gell-Mann exclaimed. "Did you read what I wrote about this? I devoted two or three chapters to this!"

Gell-Mann said that in principle one can go back up the chain of explanation, but in practice one often cannot, because biological phenomena stem from so many random, historical, contingent circumstances. That is not to say that biological phenomena are ruled by some mysterious laws of their own that act independently of the laws of physics. The whole point of the doctrine of emergence is that "we don't need *something else* in order to get *something else*," Gell-Mann said. "And when you look at the world that way, it just falls into place! You're not tortured by these strange questions any more!"

Gell-Mann thus rejected the possibility—raised by Stuart Kauffman and others—that there might be a still-undiscovered law of nature that explains why the universe has generated so much order in spite of the supposedly universal drift toward disorder decreed by the second law of thermodynamics. This issue, too, was settled, Gell-Mann replied. The universe began in a wound-up state far from thermal equilibrium. As the universe winds down, entropy increases, on average, throughout the system, but there can be many local violations of that tendency. "It's a tendency, and there are lots and lots of eddies in that process," he said. "That's *very* different from saying complexity increases! The envelope of complexity grows, expands. It's obvious from these other considerations it doesn't need another new law, however!"

The universe does create what Gell-Mann calls frozen accidents— galaxies, stars, planets, stones, trees—complex structures that serve as a foundation for the emergence of still more complex structures. "As a general rule, more complex life-forms emerge, more complex computer programs, more complex astronomical objects emerge in the course of nonadaptive stellar and galactic evolution and so on. But! If we look very very very far into the future, maybe it won't be true any more!" Eons from now the era of complexity could end, and the universe could degenerate into "photons and neutrinos and junk like that and not a lot of individuality." The second law would get us after all.

"What I'm trying to oppose is a certain tendency toward obscurantism and mystification," Gell-Mann continued. He emphasized that there was much to be understood about complex systems; that was why he helped to found the Santa Fe Institute. "There's a huge amount of wonderful research going on. What I say is that there is no evidence that we need—I don't know how else to say it—*something else*!" Gell-Mann, as he spoke, wore a huge sardonic grin, as if he could scarcely contain his amusement at the foolishness of those who might disagree with him.

Gell-Mann noted that "the last refuge of the obscurantists and mystifiers is self-awareness, consciousness." Humans are obviously more intelligent and self-aware than other animals, but they are not qualitatively different. "*Again*, it's a phenomenon that appears at a certain level of complexity and presumably emerges from the fundamental laws plus an awful lot of historical circumstances. Roger Penrose has written two foolish books based on the long-discredited fallacy that Gödel's theorem has something to do with consciousness requiring"— pause—"*something else.*"

If scientists did discover a new fundamental law, Gell-Mann said, they would do so by forging further into the microrealm, in the direction of superstring theory. Gell-Mann felt that superstring theory would probably be confirmed as the final, fundamental theory of physics early in the next millennium. But would such a far-fetched theory, with all its extra dimensions, ever really be accepted? I asked. Gell-Mann stared at me as if I had expressed a belief in reincarnation. "You're looking at science in this weird way, as if it were a matter of an opinion poll," he said. "The world is a certain way, and opinion polls have nothing to do with it! They do exert pressures on the scientific enterprise, but the ultimate selection pressure comes from comparison with the world." What about quantum mechanics? Would we be stuck with its strangeness? "I don't think there's anything strange about it! It's just quantum mechanics! Acting like quantum mechanics! That's all it does!" To Gell-Mann, the world made perfect sense. He already had *The Answer*.

Is science finite or infinite? For once, Gell-Mann did not have a prepackaged answer. "That's a very difficult question," he replied soberly. "I can't say." His view of how complexity emerges from fundamental laws, he said, "still leaves open the question of whether the whole scientific enterprise is open-ended. After all, the scientific enterprise can also concern itself with all kinds of details." Details.

One of the things that makes Gell-Mann so insufferable is that he is almost always right. His assertion that Kauffman, Bak, Penrose, and others will fail to find *something else* just beyond the horizon of current science— something that can explain better than current science can the mystery of life and of human consciousness and of existence itself—will probably prove to be correct. Gell-Mann may err—dare one say it?—only in thinking that superstring theory, with all its extra dimensions and its infinitesimal loops, will ever become an accepted part of the foundation of physics.

Ilya Prigogine and the End of Certainty

In 1994, Arturo Escobar, an anthropologist at Smith College, wrote an essay in the journal *Current Anthropology* about some of the new concepts and metaphors emerging from modern science and technology. He noted that chaos and complexity offered different visions of the world than did traditional science; they emphasized "fluidity, multiplicity, plurality, connectedness, segmentarity, heterogeneity, resilience; not 'science' but knowledges of the concrete and the local, not laws but knowledge of the problems and the self-organizing dynamics of nonorganic, organic and social phenomena." Note the quotation marks around the word *science*.[33]

It is not only postmodernists such as Escobar who view chaos and complexity as what I would call ironic endeavors. The artificial-life maven Christopher Langton clearly was putting forward this same idea when he foresaw more "poetry" in the future of science. Langton's ideas, in turn, echoed those espoused much earlier by the chemist Ilya Prigogine. In 1977, Prigogine won a Nobel Prize for studies of so-called dissipative systems, unusual mixtures of chemicals that never achieve equilibrium but keep fluctuating among multiple states. Upon these experiments, Prigogine, who fluctuates between institutes that he founded at the Free University in Belgium and the University of Texas at Austin, constructed a tower of ideas about self-organization, emergence, the links between order and disorder—in short, chaoplexity.

Prigogine's great obsession is time. For decades he complained that physics was not paying sufficient heed to the obvious fact that time proceeds only in one direction. In the early 1990s, Prigogine announced that he had forged a new theory of physics, one that finally did justice to the irreversible nature of reality. The probabilistic theory supposedly eliminated the philosophical paradoxes that had plagued quantum mechanics and reconciled it with classical mechanics, nonlinear dynamics, and thermodynamics. As a bonus, Prigogine declared, the theory would help to bridge the chasm between the sciences and the humanities and bring about the "reenchantment" of nature.

Prigogine has his fans, at least among nonscientists. Futurist Alvin Toffler, in the forward to Prigogine's 1984 book, *Order out of Chaos* (a bestseller in Europe), likened Prigogine to Newton and prophesied that the science of the Third Wave future would be Prigoginian.[34] But scientists familiar with Prigogine's work—including the many younger practitioners of chaos and complexity who have clearly borrowed his ideas and rhetoric—have little or nothing good to say about him. They accuse him of being arrogant and self-aggrandizing. They claim that he has made little or no concrete contribution to science; that he has merely recreated experiments by others and waxed philosophical about them; and that he had won a Nobel Prize for less cause than any other recipient.

These charges may well be true. But Prigogine may also have earned his colleagues' enmity by revealing the dirty secret of late-twentieth-century science, that it is, in a sense, digging its own grave. In *Order out of Chaos*, cowritten with Isabelle Stengers, Prigogine pointed out that the major discoveries of science in this century have proscribed the limits of science. "Demonstrations of impossibility, whether in relativity, quantum mechanics, or thermodynamics, have shown us that nature cannot be described 'from the outside,' as if by a spectator," Prigogine and Stengers stated. Modern science, with its probabilistic descriptions, also "leads to a kind of 'opacity' as compared to the transparency of classical thought."[35]

I met Prigogine in Austin in March 1995, a day after his return from a stint in Belgium. He showed no signs of jet lag. At age 78, he was exceedingly alert and energetic. Although short and compact, he possessed a regal bearing; he seemed not arrogant so much as calmly accepting of his own greatness. As I reviewed the issues I hoped to discuss with him, he nodded and muttered, "Yays, yays," with some impatience; he was eager, I soon realized, to begin enlightening me about the nature of things.

Shortly after we sat down, two researchers at the center joined us. A secretary told me later that they had been instructed to interrupt Prigogine if he proved too overwhelming for me and did not allow me to ask any questions. They did not fulfill this task. Prigogine, once started, was unstoppable. Words, sentences, paragraphs issued from him in a steady, implacable stream. His accent was almost parodically thick—it reminded me of Inspector Clouseau in *The Pink Panther*—and yet I had no difficulty understanding him.

He briefly recounted his youth. He was born in Russia in 1917, during the revolution, and his bourgeois family soon fled to Belgium. His interests were eclectic: he played the piano, studied literature, art, philosophy—and science, of course. He suspected that the turbulent setting of his youth had provoked his career-long fascination with time. "I may have been impressed with the fact that science had so little to say about time, about history, evolution, and that perhaps brought me to the problem of thermodynamics. Because in thermodynamics, the main quantity is entropy, and entropy means just evolution."

In the 1940s, Prigogine proposed that the increase of entropy decreed by
the second law of thermodynamics need not always create disorder; in some
systems, such as the churning chemical cells he studied in his laboratory,
entropic drift could generate striking patterns. He also began to realize that
"structure is rooted in the direction of time, irreversible time, and the arrow
of time is a very important element in the structure of the universe. Now
this already was bringing me in a sense in conflict with great physicists like
Einstein, who was saying that time is an illusion."

Most physicists, according to Prigogine, think irreversibility is an illusion
stemming from the limits of their observations. "Now this I could never
believe, because in a sense that seems to indicate that our measurements, or
our approximations, introduce irreversibility in a time-reversible universe!"
Prigogine exclaimed. "We are not the father of time. We are the children of
time. We come from evolution. What we have to do is to include evolutionary
patterns in our descriptions. What we need is a Darwinian view of physics, an
evolutionary view of physics, a biological view of physics."

Prigogine and his colleagues had been developing just such a physics.
As a result of this new model, Prigogine told me, physics would be reborn,
contrary to the pessimistic predictions of reductionists such as Steven
Weinberg (who worked in the same building as Prigogine). The new physics
might also heal the great rift between science, which had always depicted
nature as the outcome of deterministic laws, and the humanities, which
emphasized human freedom and responsibility. "You cannot on one side
believe that you are part of an automaton and on the other hand believe in
humanism," Prigogine declared.

Of course, he emphasized, this unification was metaphorical rather than
literal; it would not by any means help science to solve all its problems.
"One should not exaggerate and dream about a unified theory which will
include politics and economics and the immune system and physics and
chemistry," Prigogine said, in a rebuke to researchers at the Santa Fe Institute
and elsewhere who dreamed of just such a theory. "One should not think
that progress in chemical nonequilibrium reactions will give you the key

for human politics. Of course not! Of course not! But still, it brings in a unified element. It brings in the element of bifurcation, it brings in the idea of historical dimension, it brings in the idea of evolutionary patterns, which indeed you find on all levels. And in this sense it is a unifying element of our view of the universe."

Prigogine's secretary poked her head in at the door to remind him that she had made reservations for lunch at the faculty club at noon. After her third reminder—at five minutes after noon—Prigogine ended his peroration with a flourish and announced that it was time to go to lunch. At the faculty lounge, Prigogine and I were joined by a dozen or so other researchers employed at his center: Prigoginians. We assembled at a long rectangular table. Prigogine sat in the middle of one side, like Jesus at the Last Supper, and I sat beside him, like Judas, listening, along with everyone else, as he held forth.

Sporadically, Prigogine called on one of his disciples to say a word or two—enough to draw attention to the vast gap between his rhetorical powers and theirs. At one point he asked a tall, cadaverish man sitting across from me (whose equally cadaverish identical twin, eerily, was also at the table) to explain his nonlinear, probabilistic view of cosmology. The man dutifully unburdened himself, in a lugubrious eastern European accent, of an impenetrable monologue about bubbles and instabilities and quantum fluctuations. Prigogine quickly stepped in. The meaning of his colleague's work, he explained, was that there was no stable ground state, no equilibrium condition, of space-time; thus, there was no beginning to the cosmos and there could be no end. Phew.

Between nibbles of his fish, Prigogine reiterated his objections to determinism. (Earlier Prigogine had admitted that he had been strongly influenced by Karl Popper.) Descartes, Einstein, and the other great determinists were "all pessimistic people. They wanted to go to another world, a world of eternal beatitude." But a deterministic world would be not a Utopia but a *dystopia*, Prigogine said. That was the message of Aldous Huxley in *Brave New World*, of George Orwell in *1984*, of Milan Kundera in *The Unbearable Lightness of Being*. When a state tries to suppress evolution, change, flux by brutal force, by violence, Prigogine explained, it destroys the meaning of life, it creates a society of "timeless robots."

On the other hand, a completely irrational, unpredictable world would also be terrifying. "What we have to find is a middle way, to find a probabilistic description which says something, not everything, and also not nothing." His view could provide a philosophical framework for understanding social phenomena, Prigogine said. But human behavior, he emphasized, could not be defined by any scientific, mathematical model. "In human life we have no simple basic equation! When you decide whether you take coffee or not, that's already a complicated decision. It depends on what day it is, whether you like coffee, and so on."

Prigogine had been building toward some great revelation, and now he finally divulged it. Chaos, instability, nonlinear dynamics, and related concepts had been warmly received not only by scientists but also by the lay public because society was itself in a state of flux. The public's faith in great unifying ideas, whether religious or political or artistic or scientific, was dwindling.

"Even people who are very Catholic are no longer so Catholic as their parents or grandparents were, probably. We are no more believing in Marxism or liberalism in the classical way. We are no more believing classical science." The same is true of the arts, music, literature; society has learned to accept a multiplicity of styles and world-views. Humanity has arrived, Prigogine summarized, at "the end of certitude."

Prigogine paused, allowing us to ponder the magnitude of his announcement. I broke the hushed silence by pointing out that some people, such as religious fundamentalists, seemed to be clinging to certitude more fiercely than ever. Prigogine listened politely, then asserted that fundamentalists were merely exceptions to the rule. Abruptly, he fixed his gaze on a prim, blond-haired woman, the deputy director of his institute, sitting across the table from us. "What is your opinion?" he asked. "I agree completely," she replied. She hastily added, perhaps in response to the craven snickers of her colleagues, that fundamentalism "seemed to be a response to a frantic world."

Prigogine nodded paternally. He acknowledged that his assertions concerning the end of certitude had elicited "violent reactions" in the intellectual establishment. The *New York Times* had declined to review *Order out of Chaos* because, Prigogine had heard, the editors considered his discussion of the end of certitude "too dangerous." Prigogine understood such fears. "If science is not able to give certitude, what should you believe? I mean, before it was very easy. Either you believe in Jesus Christ, or you believe in Newton. It was very simple. But now, as I say, if science is not giving you certitude but probabilities, then it's a dangerous book!"

Prigogine nonetheless thought his view did justice to the depthless mystery of the world, and of our own existence. That was what he meant by his phrase "the reenchantment of nature." After all, consider this lunch we were having right now. What theory could predict this! "The universe is a strange thing," Prigogine said, cranking the intensity of his voice up a notch. "I think we can all agree on that." As he swept the room with his serene yet feral gaze, his colleagues bobbed their heads and chuckled nervously. Their unease was well-founded. They had hitched their careers to a man who apparently believed that science—empirical, rigorous science, the kind of science that solves its problems, that renders the world comprehensible, that gets us somewhere—was over.

In return for certainty, Prigogine—like Christopher Langton, Stuart Kauffman, and other chaoplexologists who have clearly been influenced by his ideas—promises the "reenchantment of nature." (Per Bak, for all his hubris, at least eschews this pseudo-spiritual rhetoric.) What Prigogine apparently means by this statement is that vague, fuzzy, impotent theories are somehow more meaningful, more comforting, than the accurate, precise, powerful theories of Newton or Einstein or modern particle physicists. But why, one wonders, is an indeterministic, opaque universe any less cold, cruel, and frightening than a deterministic, transparent one? More specifically, how is the fact that the world unfolds according to a nonlinear, probabilistic dynamics supposed to console a Bosnian woman who has seen her only daughter raped and slaughtered?

Mitchell Feigenbaum and the Collapse of Chaos

It was a meeting with Mitchell Feigenbaum that finally convinced me that chaoplexity is a doomed enterprise. Feigenbaum was perhaps the most compelling character in Gleick's book *Chaos*—and in the field as a whole. Trained as a particle physicist, Feigenbaum became entranced with questions beyond the scope of that or any other field, questions about turbulence, chaos, and the relation between order and disorder. In the mid-1970s, when he was a young postdoc at Los Alamos National Laboratory, he discovered a hidden order, called period doubling, underlying the behavior of a wide variety of nonlinear mathematical systems. The period of a system is the time it takes to return to its original state. Feigenbaum found that the period of some nonlinear systems keeps doubling as they evolve and therefore rapidly approaches infinity (or eternity). Experiments confirmed that some simple real-world systems (although not as many as initially hoped) demonstrate period doubling. For example, as one gradually opens up a faucet, the water demonstrates period doubling as it progresses from a steady drip, drip, drip toward a turbulent gush. The mathematician David Ruelle has called period doubling a work of "particular beauty and significance" that "stands out in the theory of chaos."[36]

Feigenbaum, when I met him in March 1994 at Rockefeller University, where he has a spacious office overlooking Manhattan's East River, looked every bit the genius he was said to be. With his magnificent, oversized head and swept-back hair, he resembled Beethoven, though more handsome and less simian. Feigenbaum spoke clearly, precisely, with no accent, but with a strange kind of formality, as if English were a second language that he had mastered through sheer brilliance. (The voice of the superstring theorist Edward Witten has this same quality.) When amused, Feigenbaum did not smile so much as grimace: his already protuberant eyes bulged still farther from their sockets, and his lips peeled back to expose twin rows of brown, peglike teeth stained by countless filterless cigarettes and espressos (both of which he consumed during our meeting). His vocal cords, cured by decades of exposure to these toxins, yielded a voice as rich and resonant as a basso profundo's and a deep, villainous snicker.

Like many chaoplexologists, Feigenbaum could not resist ridiculing particle physicists for daring to think they could achieve a theory of everything. It is quite possible, he said, that particle physicists might one day develop a theory that adequately accounts for all of nature's fundamental forces, including gravity. But calling such a theory final would be something else again. "A lot of my colleagues like the idea of final theories because they're religious. And they use it as a replacement for God, which they don't believe in. But they just created a substitute."

A unified theory of physics would obviously not answer all questions, Feigenbaum said. "If you really believe that this is a path of understanding the world, I can ask immediately: how do I write down in this formalism what you look like, with all the hairs on your head?" He stared at me until my scalp prickled. "Now, one answer is, 'That's not an interesting problem.'" Against my will, I felt slightly offended. "Another answer is, 'Well, it's okay, but we can't do it.' The right answer is obviously an alloy of those two complements. We have very few tools. We can't solve problems like that."

Moreover, particle physicists are overly concerned with finding theories that are merely true, in the sense that they account for available data; the goal of science should be to generate "thoughts in your head" that "stand a high chance of being new or exciting," Feigenbaum explained. *That's* the desideratum." He added: "There isn't any security by knowing that something is true, at least as far as I'm concerned. I'm thoroughly indifferent to that. I like to know that I have a way of thinking about things." I began to suspect that Feigenbaum, like David Bohm, had the soul of an artist, a poet, even a mystic: he sought not truth, but revelation.

Feigenbaum noted that the methodology of particle physics—and physics generally—had been to try to look at the simplest possible aspects of reality, "where everything has been stripped away." The most extreme reductionists had suggested that looking at more complex phenomena was merely "engineering." But as a result of advances in chaos and complexity, he said, "some of these things that one relegated to engineering are now regarded as reasonable questions to ask from a more theoretical viewpoint. Not just to get the right answer but to understand something about how they work. And that you can even make sense out of that last comment flies in the face of what it means for a theory to be finished."

On the other hand, chaos, too, had generated too much hype. "It's a fraud to have named the subject 'chaos,'" he said. "Imagine one of my [particle-physicist] colleagues has gone to a party and meets someone and the person is all bubbling over about chaos and tells him that this reductionist stuff is all bullshit. Well, it's infuriating, because it's completely stupid what the person has been told," Feigenbaum said. "I think it's regrettable that people are sloppy, and they end up serving as representatives."

Some of his colleagues at the Santa Fe Institute, Feigenbaum added, also had too naive a faith in the power of computers. "The proof is in the pudding," he said, and paused, as if considering how to proceed diplomatically. "It's very hard to see things in numerical experiments. That is, people want to have fancier and fancier computers to simulate fluids. There is something to be learned in simulating fluids, but unless you know what you're looking for, you're not going to see anything. Because after all, if I just look out the window, there's an overwhelmingly better simulation than I could ever do on a computer."

He nodded toward his window, beyond which the leaden East River flowed. "I can't interrogate it quite as sharply, but there's so much stuff in that numerical simulation that if I don't know what to interrogate it about, I will have learned nothing." For these reasons much of the recent work on nonlinear phenomena "has not led to answers. The reason for that is, these are truly hard problems, and one doesn't have any tools. And the job should really be to do those insightful calculations which require some piece of faith and good luck as well. People don't know how to begin doing these problems."

I admitted that I was often confused by the rhetoric of people in chaos and complexity. Sometimes they seemed to be delineating the limits of science—such as the butterfly effect—and sometimes they implied that they could transcend those limits. "We are building tools!" Feigenbaum cried. "We don't know how to do these problems. They are truly hard. Every now and then we get a little pocket where we know how to do it, and then we try to puff it out as far as it can go. And when it reaches the border of where it is going, then people wallow for a while, and then they stop doing it. And then one waits for some new piece of insight. But it is literally the business of enlarging the borders of what falls under the suzerainty of science. It is *not* being done from an engineering viewpoint. It isn't just to give you the answer to some approximation."

"I want to know *why*," he continued, still staring at me hard. *"Why* does the thing do this?" Was it possible that this enterprise could, well, fail? "Of course!" Feigenbaum bellowed, and he laughed maniacally. He confessed that he had been stymied himself of late. Up through the late 1980s he had sought to refine a method for describing how a fractal object, such as a cloud, might evolve over time when perturbed by various forces. He wrote two long papers on the topic that were published in 1988 and 1989 in a relatively obscure physics journal.[37] "I have no idea how well they've been read," Feigenbaum said defiantly. "In fact, I've never been able to give a talk on them." The problem, he suggested, might be that no one could understand what he was getting at. (Feigenbaum was renowned for obscurity as well as for brilliance.) Since then, he added, "I haven't had a further better idea to know how to proceed in this."

In the meantime, Feigenbaum had turned to applied science. Engineering. He had helped a map-making company develop software for automatically constructing maps with minimal spatial distortion and maximum aesthetic appeal. He belonged to a committee that was redesigning U.S. currency to make it less susceptible to counterfeiting. (Feigenbaum came up with the idea of using fractal patterns that blur when photocopied.) I noted that these sounded like what would be, for most scientists, fascinating and worthy projects. But people familiar with Feigenbaum's former incarnation as a leader of chaos theory, if they heard he now worked on maps and currency, might think ....

"He's not doing serious things any more," Feigenbaum said quietly, as if to himself. Not only that, I added. What people might think was that if someone who was arguably the most gifted explorer of chaos could not proceed any further, then perhaps the field had run its course. "There's *some* truth to that," he replied. He acknowledged that he hadn't really had any good ideas about how to extend chaos theory since 1989. "One is on the lookout for things that are substantial, and at the moment...." He paused. "I don't have a thought. I don't know ...." He turned his large, luminous eyes once again toward the river beyond his window, as if seeking a sign.

Feeling somewhat guilty, I told Feigenbaum that I would love to see his last papers on chaos. Did he have any reprints? In response, Feigenbaum thrust himself from his chair and careened wildly toward a row of filing cabinets on the far side of his office. En route, he cracked his shin against a low-lying coffee table. Wincing, teeth clenched, Feigenbaum limped onward, wounded by his collision with the world. The scene was a grotesque inversion of Samuel Johnson's famous stone-kicking episode. The suddenly malevolent-looking coffee table seemed to be gloating: "I refute Feigenbaum thus."

Making Metaphors

The fields of chaos, complexity, and artificial life will continue. Certain practitioners will be content to play in the realm of pure mathematics and theoretical computer science. Others, the majority, will develop new mathematical and computational techniques for engineering purposes. They will make incremental advances, such as extending the range of weather forecasts or improving the ability of engineers to simulate the performance of jets or other complex technologies. But they will not achieve any great insights into nature—certainly none comparable to Darwin's theory of evolution or quantum mechanics. They will not force any significant revisions in our map of reality or our narrative of creation. They will not find what

Murray Gell-Mann calls "something else."

So far, chaoplexologists have created some potent metaphors: the butterfly effect, fractals, artificial life, the edge of chaos, self-organized criticality. But they have not told us anything about the world that is both concrete and truly surprising, either in a negative or in a positive sense. They have slightly extended the borders of knowledge in certain areas, and they have more sharply delineated the boundaries of knowledge elsewhere.

Computer simulations represent a kind of metareality within which we can play with and even—to a limited degree—test scientific theories, but they are not reality itself (although many aficionados have lost sight of that distinction). Moreover, by giving scientists more power to manipulate different symbols in different ways to simulate a natural phenomenon, computers may undermine scientists' faith that their theories are not only true but *True*, exclusively and absolutely true. Computers may, if anything, hasten the end of empirical science. Christopher Langton was right: there is something more like poetry in the future of science.

Chapter Nine
The End of Limitology

Just as lovers begin talking about their relationship only when it sours, so will scientists become more self-conscious and doubtful as their efforts yield diminishing returns. Science will follow the path already trodden by literature, art, music, philosophy. It will become more introspective, subjective, diffuse, obsessed with its own methods. In the spring of 1994 I saw the future of science in microcosm when I sat in on a workshop at the Santa Fe Institute titled "The Limits to Scientific Knowledge." During the three-day meeting, a score of thinkers, including mathematicians, physicists, biologists, and economists, pondered whether there were limits to science, and, if so, whether science could know them. The meeting was organized by two people associated with the Santa Fe Institute: John Casti, a mathematician who has written numerous popular books on science and mathematics, and Joseph Traub, a theoretical computer scientist, who is a professor at Columbia University.[1]

I had come to the meeting in large part to meet Gregory Chaitin, a mathematician and computer scientist at IBM who had devoted himself since the early 1960s to exploring and extending Gödel's theorem through what he called algorithmic information theory. Chaitin had come close to proving, as far as I could tell, that a mathematical theory of complexity was not possible. Before meeting Chaitin, I had pictured him as a gnarled, sour-looking man with hairy ears and an eastern European accent; after all, a kind of old-world philosophical angst suffused his research on the limits of mathematics. But

Chaitin in no way resembled my internal model. Stout, bald, and boyish, he wore neobeatnik attire: baggy white pants with an elastic waistband, black T-shirt adorned with a Matisse sketch, sandals over socks. He was younger than I expected; I learned later that his first paper had been published when he was only 18, in 1965. His hyperactivity made him seem younger still. His speech was invariably either accelerating, as he became carried away by his words, or decelerating, perhaps as he realized he was approaching the limits of human comprehension and ought to slow down. Plots of the velocity and volume of his speech would form overlapping sine waves. Struggling to articulate an idea, he squeezed his eyes shut and, with an agonized grimace, tipped his head forward, as if trying to dislodge the words from his sticky brain.[2]

The participants sat around a long, rectangular table in a long, rectangular room with a chalkboard at one end. Casti opened the meeting by asking, "Is the real world too complex for us to understand?" Kurt Gödel's incompleteness theorem, Casti noted, implied that some mathematical descriptions would always be incomplete; some aspects of the world would always resist description. Alan Turing, similarly, showed that many mathematical propositions are "undecidable"; that is, one cannot determine whether the propositions are true or false in a finite amount of time. Traub tried to rephrase Casti's question in a more positive light: Can we know what we cannot know? Can we *prove* there are limits to science, in the same way that Gödel and Turing proved there are limits to mathematics and computation?

The only way to construct such a proof, announced E. Atlee Jackson, a physicist from the University of Illinois, would be to construct a formal representation of science. To show how difficult that task would be, Jackson leaped up to the chalkboard and scribbled an extremely complicated flowchart that supposedly represented science. When his audience stared at him blankly, Jackson fell back on aphorisms. To determine whether science has limits, he said, you have to define science, and as soon as you define science you impose a limit on it. On the other hand, he added, "I can't define my wife, but I can recognize her." Rewarded with polite chuckles, Jackson retreated to his seat.

The antichaos theorist Stuart Kauffman kept slipping into the meeting, delivering Zen-like minilectures, and then slipping out again. During one appearance, he reminded us that our very survival depends on our ability to classify the world. But the world doesn't come already packaged into premade categories. We can "carve it up," or classify it, in many ways. In order to classify phenomena, moreover, we must throw some information away. Kauffman concluded with this incantation: "To be is to classify is to act, all of which means throwing away information. So just the act of knowing requires ignorance." His audience looked simultaneously impressed and annoyed.

Ralph Gomory then said a few words. A former vice president of research at IBM, Gomory headed the Sloan Foundation, a philanthropic organization that sponsored science-related projects, including the Santa Fe workshop. When listening to someone else speak, and even when speaking himself, Gomory wore an expression of deep incredulity. He tilted his head forward, as if peering over invisible bifocals, while knitting his thick black eyebrows together and wrinkling his brow.

Gomory explained that he had decided to support the workshop because he had long felt that the educational system placed too much emphasis on what was known and too little on what was unknown or even unknowable. Most people aren't even aware of how little is known, Gomory said, because the educational system presents such a seamless, noncontradictory view of reality. Everything we know about the ancient Persian Wars, for example, derives from a single source, Herodotus. How do we know whether Herodotus was an accurate reporter? Maybe he had incomplete or inaccurate information! Maybe he was biased or making things up! We will never know!

Later Gomory remarked that a Martian, by observing humans playing chess, might be able to deduce the rules correctly. But could the Martian ever be sure that those were the correct rules, or the only rules? Everyone pondered Gomory's riddle for a moment. Then Kauffman speculated on how Wittgenstein might have responded to it. Wittgenstein would have "suffered egregiously," Kauffman said, over the possibility that the chess players might make a move—deliberately or not—that broke the rules. After all, how could the Martian tell if the move was just a mistake or the result of another rule? "Do you get this?" Kauffman queried Gomory.

"I don't know who Wittgenstein is, for starters," Gomory replied irritably.

Kauffman raised his eyebrows. "He was a *very* famous philosopher."

He and Gomory stared at each other until someone said, "Let's leave Wittgenstein out of this."

Patrick Suppes, a philosopher from Stanford, kept interrupting the discussion to point out that Kant, in his discussion of antinomies, anticipated virtually all the problems they were wrestling with at the workshop. Finally, when Suppes brought up yet another antinomy, someone shouted, "No more Kant!" Suppes protested that there was just one more antinomy he wanted to mention that was really very important, but his colleagues shouted him down. (No doubt they did not want to be reminded that for the most part they were merely restating, with newfangled jargon and metaphors, arguments set forth long ago not only by Kant but even by the ancient Greeks.)

Chaitin, spitting out words like a machine gun, dragged the conversation back to Gödel. The incompleteness theorem, Chaitin asserted, far from being a paradoxical curiosity with little relevance to the progress of mathematics or science, as some mathematicians like to believe, is only one of a set of profound problems posed by mathematics. "Some people dismiss Gödel's results as being bizarre, pathological, deriving from a self-referential paradox," Chaitin said. "Gödel himself sometimes worried that this was just a paradox created by our use of words. Now, incompleteness seems so natural, you can ask how we mathematicians can do anything!"

Chaitin's own work on algorithmic information theory suggested that as mathematicians address increasingly complex problems, they will have to keep adding to their base of axioms; to know more, in other words, one must assume more. As a result, Chaitin contended, mathematics is bound to become an increasingly experimental science with less of a claim to absolute truth. Chaitin had also established that just as nature seems to harbor fundamental uncertainty and randomness, so does mathematics. He had recently found an algebraic equation that might have an infinite or finite number of solutions, depending on the value of the variables in the equation.

"Normally you assume that if people think something is true, it's true for a reason. In mathematics a reason is called a proof, and the job of a mathematician is to find the proof, the reasons, deductions from axioms or accepted principles. Now what I found is mathematical truths that are true for no reason at all. They are true accidentally or at random. And that's why we will never find the truth: because there is no truth, there is no reason these are true."

Chaitin had also proved that one can never determine whether any given computer program is the most succinct possible method of solving a problem; it is always possible that more concise programs exist. (The implication of this finding, others have pointed out, is that physicists can never be sure that they have found a truly final theory, one that represents the most compact possible description of nature.) Chaitin obviously reveled in being the bearer of such dire tidings. He seemed intoxicated at the thought that he was tearing down the temple of mathematics and science.

Casti retorted that mathematicians might be able to avoid Gödel effects by employing simple formal systems—such as an arithmetic consisting solely of addition and subtraction (and not multiplication and division). Nondeductive systems of reasoning, Casti added, might also sidestep the problem; Gödel's theorem might turn out to be a red herring when it came to natural science.

Francisco Antonio "Chico" Doria, a Brazilian mathematician, also found Chaitin's analysis too pessimistic. The mathematical hurdles identified by Gödel, Doria contended, far from bringing mathematics to an end, could enrich it. For example, Doria suggested, when mathematicians encounter an apparently undecidable proposition, they can create two new branches of mathematics, one that assumes the proposition is true, and one that assumes it is false. "Instead of a limit of knowledge," Doria concluded, "we may have a wealth of knowledge."

Listening to Doria, Chaitin rolled his eyes. Suppes also seemed doubtful. Arbitrarily assuming that undecidable mathematical statements are true or false, Suppes drawled, has "all the advantages of theft over honest toil." He attributed his quip to someone famous.

The conversation kept veering—as if to a strange attractor—to one of the favorite topics of philosophically inclined mathematicians and physicists: the continuum problem. Is reality smooth or lumpy? Analog or digital? Is the world best described by the so-called real numbers, which can be diced into infinitely fine gradations, or by whole numbers? Physicists from Newton through Einstein have relied on real numbers. But quantum mechanics suggests that matter and energy and perhaps even time and space (at extremely small scales) come in discrete, indivisible lumps. Computers, too, represent everything as integers: ones and zeros.

Chaitin denounced real numbers as "nonsense." Their precision is a sham, given the noisiness, the fuzziness, of the world. "Physicists know that every equation is a lie," he declared.

Someone parried with a quote from Picasso: "Art is a lie that helps us see the truth."

Of *course* real numbers are abstractions, Traub chimed in, but they are very powerful, effective abstractions. Of *course* there is always noise, but there are ways to deal with noise in a real-number system. A mathematical model captures the essence of something. No one pretends it captures the *entire* phenomenon.

Suppes strode to the chalkboard and scribbled some equations that, he implied, might eliminate the continuum problem once and for all. His audience looked unimpressed. (This, I thought, is the major problem of philosophy: no one really *wants* to see philosophical problems solved, because then we will have nothing to talk about.)

Other participants noted that scientists face obstacles to knowledge much less abstract than incompleteness, undecidability, the continuum, and so on. One was Piet Hut, a Dutch astrophysicist from the Institute for Advanced Study. He said that with the help of powerful statistical methods and computers, he and his fellow astrophysicists had learned how to overcome the infamous N-body problem, which holds that it is impossible to predict the course of three or more gravitationally interacting bodies. Computers could now simulate the evolution of whole galaxies, containing billions of stars, and even clusters of galaxies.

But, Hut added, astronomers face other limits that seem insurmountable. They have only one universe to study, so they cannot do controlled experiments on it. Cosmologists can only trace the history of the universe so far back, and they can never know what preceded the big bang or what exists beyond the borders of our universe, if anything. Moreover, particle physicists may have a hard time testing theories (such as those involving superstrings) that combine gravity and all the other forces of nature, because the effects only become apparent at distance scales and energies that are beyond the range of any conceivable accelerator.

A similarly pessimistic note was sounded by Rolf Landauer, a physicist at IBM and a pioneer in the study of the physical limits of computation. Landauer spoke with a German-accented growl that sharpened the edge of his sardonic sense of humor. When one speaker kept standing in the way of his own viewgraphs, Landauer snapped, "Your talk may be transparent, but you are not!"

Landauer contended that scientists could not count on computers to keep growing in power indefinitely. He granted that many of the supposed physical constraints once thought to be imposed on computation by the second law of thermodynamics or quantum mechanics had been shown to be spurious. On the other hand, the costs of computer-manufacturing plants were rising so fast that they threatened to bring to a halt the decades-long decline in computation prices. Landauer also doubted whether computer designers would soon harness exotic quantum effects such as superposition—the ability of a quantum entity to exist in more than one state at the same time—and thereby transcend the capabilities of current computers, as some theorists had proposed. Such systems would be so sensitive to minute quantum-level disruptions that they would be effectively useless, Landauer argued.

Brian Arthur, a ruddy-faced economist at the Santa Fe Institute who spoke with a lilting Irish accent, steered the conversation toward the limits of economics. In trying to predict how the stock market will perform, he said, an investor must make guesses about how others will guess about how others will guess—and so on ad infinitum. The economic realm is inherently subjective, psychological, and hence unpredictable; indeterminacy "percolates through the system." As soon as economists try to simplify their models— by assuming that investors can have perfect knowledge of the market or that prices represent some true value—the models become unrealistic; two economists gifted with infinite intelligence will come to different conclusions about the same system. All economists can really do is say, "Well, it could be this, it could be that." On the other hand, Arthur added, "if you've made money in the markets all the economists will listen to you."

Kauffman then repeated what Arthur had just said but in a more abstract way. People are "agents" who must continually adjust their "internal models" in response to the perceived adjustments of the internal models of other agents, thus creating a "complex, coadapting landscape."

Landauer, scowling, interjected that there were much more obvious reasons that economic phenomena were impossible to predict than these subjective factors. AIDS, third-world wars, even the diarrhea of the chief analyst of a large mutual fund can have a profound effect on the economy, he said. What model can possibly predict such events?

Roger Shepard, a psychologist from Stanford who had been listening in silence, finally piped up. Shepard seemed faintly melancholy. His apparent mood may have been an illusion engendered by his droopy, ivory-hued mustache—or a very real by-product of his obsession with unanswerable questions. Shepard admitted he had come here in part to learn whether scientific or mathematical truths were discovered or invented. He had also been thinking a good deal lately about where scientific knowledge really existed and had concluded that it could not exist independently of the human mind. A textbook on physics, without a human to read it, is just paper and ink spots. But that raised what Shepard considered to be a disturbing issue.

Science appears to be getting more and more complicated and thus more and more difficult to understand. It seems quite possible that in the future some scientific theories, such as a theory of the human mind, will be too complex for even the most brilliant scientist to understand. "Maybe I'm old-fashioned," Shepard said, but if a theory is so complicated that no single person can understand it, what satisfaction can we take in it?

Traub, too, was troubled by this issue. We humans may believe Occam's razor—which holds that the best theories are the simplest— because these are the only theories our puny brains can comprehend. But maybe computers won't be subject to this limitation, Traub added. Maybe computers will be the scientists of the future.

In biology, someone remarked gloomily, "Occam's razor cuts your throat."

Gomory noted that the task of science was to find those niches of reality that lend themselves to understanding, given that the world is basically unintelligible. One way to make the world more comprehensible, Gomory suggested, is to make it more artificial, since artificial systems tend to be more comprehensible and predictable than natural ones. For example, to make weather forecasting easier, society might encase the world in a transparent dome.

Everyone stared at Gomory for a moment. Then Traub remarked, "I think what Ralph is saying is that it's easier to create the future than to predict it."

As the meeting progressed, Otto Rössler made more and more sense. Or was everybody else making less? Rössler was a theoretical biochemist and chaos theorist from the University of Tübingen, Germany, who in the mid-1970s had discovered a mathematical monster called the Rössler attractor. His white hair appeared permanently disheveled, as though he had just awakened from a trance. He had the exaggerated features of a marionette: startled eyes, a protuberant lower lip, and a bulbous chin framed by deep, vertical creases. Neither I nor, I suspected, anyone else could quite understand him, but everyone leaned toward him when he issued his whispery, stammered pronouncements, as if he were an oracle.

Rössler saw two primary limits to knowledge. One was inaccessibility. We can never be sure about the origin of the universe, for example, because it is so distant from us both in space and in time. The other limit, distortion, was much worse. The world can deceive us into thinking we understand it when actually we do not. If we could stand outside the universe, Rossler suggested, we would know the limits to our knowledge; but we are trapped inside the universe, and so our knowledge of our own limits must remain incomplete.

Rössler raised some questions that he said were first posed in the eighteenth century by a physicist named Roger Boscovich. Can one determine, if one is on a planet with an utterly dark sky, if the planet is rotating? If the earth is breathing, but we are breathing too, in synchrony with it, can we tell that it is breathing? Probably not, according to Rössler. "There exist situations where you are unable to find out about the truth from the inside," he said. On the other hand, he added, simply by posing thought experiments like these, we might find a way to transcend the limits of perception.

The more Rössler spoke, the more I began to feel an affinity for his ideas. During one of the breaks, I asked if he felt that intelligent computers might transcend the limits of human science. He shook his head adamantly. "No, that's not possible," he replied in an intense whisper. "I would bet on dolphins, or sperm whales. They have the biggest brains on the earth." Rössler informed me that when one sperm whale is shot by whalers, others sometimes crowd around it, forming a starlike pattern, and are themselves killed. "Usually people think that is just blind instinct," Rössler said. "In reality it's their way of showing humankind that they are much higher evolved than humans." I just nodded.

Toward the end of the meeting, Traub proposed that everyone split up into focus groups to discuss the limits of specific fields: physics, mathematics, biology, social sciences. A social scientist announced that he didn't want to be in the social science group; he had come specifically to talk to, and learn from, people outside his field. His remark provoked a few me-toos from others. Someone pointed out that if everyone felt the way the social scientist did, there would be no social scientists in the social science section, biologists in the biology section, and so on. Traub said his colleagues could split up any way they chose; he was just making a suggestion. The next question was, where would the different groups meet? Someone proposed that they disperse to different rooms, so that certain loud talkers wouldn't disturb the other groups. Everyone looked at Chaitin. His promise to be quiet was met with jeers. More discussion. Landauer remarked that there was such a thing as too much intelligence applied to a simple problem. Just when everything seemed hopeless, groups somehow, spontaneously, formed, more or less following Traub's initial suggestion, and wandered off to different locations. This was, I thought, an impressive display of what the Santa Fe'ers like to call self-organization, or order out of chaos; perhaps life began this way.

I tagged along with the mathematics group, which included Chaitin, Landauer, Shepard, Doria, and Rössler. We found an unoccupied lounge with a chalkboard. For several minutes, everyone talked about what they should talk about. Then Rössler went to the chalkboard and scribbled down a recently discovered formula that gives rise to a fantastically complicated mathematical object, "the mother of all fractals." Landauer asked Rössler, politely, what this fractal had to do with anything. It "soothes the brain," Rossler replied. It also fed his hope that physicists might be able to describe reality with these kinds of chaotic but classical formulas and thus dispense with the terrible uncertainties of quantum mechanics.

Shepard interjected that he had joined the mathematics subgroup because he wanted the mathematicians to tell him whether mathematical truths were invented or discovered. Everyone talked about that for a while without coming to a decision. Chaitin said that most mathematicians leaned toward the discovery view, but Einstein was apparently an inventionist.

Chaitin, during a lull, once again proposed that mathematics was dead. In the future, mathematicians would be able to solve problems only with enormous computer calculations that would be too complex for anyone to understand.

Everyone seemed fed up with Chaitin. Mathematics *works*, Landauer snarled. It helps scientists solve problems. *Obviously* it's not dead. Others piled on, accusing Chaitin of exaggeration.

Chaitin, for the first time, appeared chastened. His pessimism, he conjectured, might be linked to the fact that he had eaten too many bagels that morning. He remarked that the pessimism of the German philosopher Schopenhauer, who advocated suicide as the supreme expression of existential freedom, had been traced to his bad liver.

Steen Rasmussen, a physicist and Santa Fe regular, reiterated the familiar argument of chaoplexologists that traditional reductionist methods cannot solve complex problems. Science needs a "new Newton," he said, someone who can invent a whole new conceptual and mathematical approach to complexity.

Landauer scolded Rasmussen for succumbing to the "disease" afflicting many Santa Fe researchers, the belief in some "great religious insight" that would instantaneously solve all their problems. Science doesn't work that way; different problems require different tools and techniques.

Rössler unburdened himself of a long, tangled soliloquy whose message seemed to be that our brains represent only one solution to the multiple problems posed by the world. Evolution could have created other brains representing other solutions.

Landauer, who was strangely protective of Rössler, gently asked him whether he thought we might be able to alter our brains in order to gain more knowledge. "There is one way," Rössler replied, staring at an invisible object on the table in front of him. "To become insane."

There was a moment of awkward silence. Then an argument erupted over whether complexity was a useful term or had been so loosely defined that it had become meaningless and should be abandoned. Even if terms such as *chaos* and *complexity* have little scientific meaning, Chaitin said, they are still useful for public relations purposes. Traub noted that the physicist Seth Lloyd had counted at least 31 different definitions for complexity.

"We go from complexity to perplexity," Doria intoned. Everyone nodded and complimented him on his aphorism.

When the focus groups reconvened, Traub asked each person to propose answers to two questions: What have we learned? What questions remain unresolved?

Chaitin spouted questions: What are the limits of metamathematics, and metametamathematics? What are the limits to our ability to know limits? And are there limits to that knowledge? Can we simulate the whole universe, and if so can we make one better than God did?

"And can we move there?" someone quipped.

Lee Segel, an Israeli biologist, warned them to be careful how they discussed these issues publicly, lest they contribute to the growing antiscience mood of society. After all, he continued, too many people think Einstein showed that everything is relative and Gödel proved that nothing can be proved. Everyone nodded solemnly. Science has a fractal structure, Segel added confidently, and there is obviously no limit to the things we can investigate. Everyone nodded again.

Rössler proposed a neologism for what he and his colleagues were doing: limitology. Limitology is a postmodern enterprise, Rössler said, an outgrowth of this century's ongoing effort to deconstruct reality. Of course, Kant, too, wrestled with the limits of knowledge. So did Maxwell, the great Scottish physicist. Maxwell imagined that a microscopic homunculus, or demon, might help us to beat the second law of thermodynamics. But the real lesson of Maxwell's demon, Rossler said, is that we are in a thermodynamic prison, one that we can never escape. When we gather information from the world, we contribute to its entropy and hence its unknowability. We are descending inexorably toward heat death. "The whole topic of the limits of science is a topic of demons," Rössler hissed. "We are fighting with demons."

A Meeting on the Hudson

Everyone agreed that the workshop had been productive; several participants told Joseph Traub, one of the co-organizers, that it was the best meeting they had ever attended. More than a year later, Ralph Gomory agreed to provide funds from the Sloan Foundation for future gatherings at the Santa Fe Institute and elsewhere. Piet Hut, Otto Rössler, Roger Shepard, and Robert Rosen, a Canadian biologist who was also at the workshop, banded together to write a book on the limits of science. I was not entirely surprised to learn that they intended to argue that science had a glorious future. "A defeatist attitude isn't going to get us anywhere," Shepard told me sternly.

In my eyes, the meeting in Santa Fe merely rehashed, in haphazard form, many of the same arguments that Gunther Stent had set forth so elegantly a quarter of a century earlier. Like Stent, those at the workshop had acknowledged that science faced physical, social, and cognitive limits. But these truth seekers seemed unable to take their own arguments to their logical conclusion, as Stent had. None could accept that science—defined as the search for intelligible, empirically substantiated truths about nature—might soon end or even have ended already. None, I thought, except for Gregory Chaitin. Of all the speakers at the meeting, he had seemed most willing to recognize that science and mathematics might be passing beyond our cognitive limits.

It was therefore with high hopes that, several months after the meeting in Santa Fe, I arranged to meet Chaitin again in Cold Spring, New York, a village on the Hudson River near our respective homes. We shared coffee and scones at a cafe on the town's miniature main street and then strolled down to a pier beside the river. Storm King Mountain, and the great fortress of West Point, loomed on the river's far side. Gulls circled overhead.[3]

When I told Chaitin I was writing a book about the possibility that science might be entering an era of diminishing returns, I expected empathy, but he snorted in disbelief. "Is that true? I hope it's not true, because it would be pretty damn boring if it were true. Every period seems to think that. Who was it—Lord Kelvin?—who said that all we have to do is to measure a few more decimal points?" When I mentioned that historians could find no evidence that Kelvin ever made such a remark, Chaitin shrugged. "Look at all the things we don't know! We don't know how the brain works. We don't know what memory is. We don't know what aging is." If we can figure out why we age, maybe we can figure out how to stop the aging process, Chaitin said.

I reminded Chaitin that in Santa Fe he had suggested that mathematics and even science as a whole might be approaching their ultimate limits. "I was just trying to wake people up," he replied. "The audience was dead." His own work, he emphasized, represented a beginning, not an end. "I may have a negative result, but I read it as telling you how to go about finding new mathematical truths: Behave more like a physicist does. Do it more empirically. Add new axioms."

Chaitin said he could not pursue his work on the limits of mathematics if he was not an optimist. "Pessimists would look at Gödel and they would start to drink scotch until they died of cirrhosis of the liver." Although the human condition might be as much "a mess" as it was thousands of years ago, there was no denying the enormous progress that we had made in science and technology. "When I was a child everybody talked about Gödel with mystical respect. This was almost incomprehensible, certainly profound. And I wanted to understand what the hell he was saying and why it was true. And I succeeded! So that makes me optimistic. I think we know very little, and I hope we know very little, because then it will be much more fun."

Chaitin recalled that he had once gotten into an argument with the physicist Richard Feynman about the limits of science. The incident occurred at a conference on computation held in the late 1980s, shortly before Feynman died. When Chaitin opined that science was just beginning, Feynman became furious. "He said we already know the physics of practically everything in everyday life, and anything that's left over is not going to be relevant."

Feynman's attitude had puzzled Chaitin until he learned that Feynman had been suffering from cancer. "For Feynman to do all the great physics he did, he couldn't have had such a pessimistic attitude. But at the end of his life, when the poor guy knows he doesn't have long to live, then I can understand why he has this view," Chaitin said. "If a guy is dying he doesn't want to miss out on all the fun. He doesn't want to feel that there's some wonderful theory, some wonderful knowledge of the physical world, that he has no idea of, and he's not going to ever see it."

I asked Chaitin if he had heard of *The Coming of the Golden Age*. When Chaitin shook his head, I summarized Stent's end-of-science argument. Chaitin rolled his eyes and asked how old Stent was when he wrote the book. In his midthirties, I replied. "Maybe he had a liver problem," Chaitin responded. "Maybe his girlfriend had ditched him. Usually men start writing this when they find they can't make love to their wife as vigorously as they used to or something." Actually, I said, Stent wrote the book in Berkeley during the 1960s. "Oh, well! Then I understand!" Chaitin exulted.

Chaitin was unimpressed with Stent's argument that humanity, for the most part, does not care about science for its own sake. "It never did," Chaitin retorted. "The people who did good scientific work were always a small group of lunatics. Everyone else is concerned with surviving, paying their mortgage. The children are sick, the wife needs money or she's going to run off with someone else." He chortled. "Remember that quantum mechanics, which is such a masterpiece, was done by people as a hobby in the 1920s when there was no funding. Quantum mechanics and nuclear physics were like Greek poetry."

It is fortunate, Chaitin said, that only a few people dedicate themselves to the pursuit of great questions. "If everybody were trying to understand the limits of mathematics or do great paintings, it would be a catastrophe! The plumbing wouldn't work! The electricity wouldn't work! Buildings would fall down! I mean, if everyone wanted to do great art or deep science, the world wouldn't function! It's good that there are only a few of us!"

Chaitin granted that particle physics seemed to have stalled because of the huge costs of accelerators. But he believed that telescopes might still provide breakthroughs in physics in years to come by revealing the violent processes generated by neutron stars, black holes, and other exotica. But wasn't it possible, I asked, that all these new observations, rather than leading to more accurate and coherent theories of physics and cosmology, might render attempts to construct such theories futile? His own work in mathematics—which showed that as one addresses more complex phenomena one must keep expanding one's base of axioms—seemed to imply as much.

"Aha, they'll become more like biology? You may be right, but we'll be knowing more about the world," Chaitin replied.

Advances in science and technology have also reduced the costs of equipment in many fields, Chaitin asserted. "The kind of equipment you can buy now for very little money in many fields is amazing." Computers had been vital to his own work. Chaitin had recently invented a new programming language that made his ideas about the limits of mathematics much more concrete. He had distributed his book, *The Limits of Mathematics*, on the Internet. "Internet is connecting people and making possible things that didn't happen before."

In the future, Chaitin predicted, humans may be able to boost their intelligence through genetic engineering, or by downloading themselves into computers. "Our descendants may be as intelligent in comparison to us as we are compared to ants." On the other hand, "if everybody starts taking heroin and gets depressed and watches TV all the time, you know we're not going to get very far." Chaitin paused. "Human beings have a future if they deserve to have a future!" he burst out. "If they get depressed then they don't have a future!"

Of course, it is always possible that science will end because civilization ends, Chaitin added. Waving his hand at the rocky hills across the river from us, he pointed out that glaciers had cut this channel during the last ice age. Only 10,000 years ago, ice encased this entire region. The next ice age could destroy civilization. But even then, he said, other beings in the universe might still carry on the quest for knowledge. "I don't know if there are other living beings. I hope so, because it's likely they won't make a mess of things."

I opened my mouth, intending to grant the possibility that in the future science might be carried forward by intelligent machines. But Chaitin, who had begun talking faster and faster and was in a kind of frenzy, cut me off. "You're a pessimist! You're a pessimist!" he shouted. He reminded me of something I had told him earlier in our conversation, that my wife was pregnant with our second child. "You conceived a child! You've gotta be pretty optimistic! You *should* be optimistic! *I* should be pessimistic! I'm older than you are! I don't have children! IBM is doing badly!" A plane droned, gulls shrieked, and Chaitin's howls of laughter fled, unechoed, across the mighty Hudson.

The End of History

Actually, Chaitin's own career fits rather nicely into Gunther Stent's diminishing-returns scenario. Algorithmic information theory represents not a genuinely new development but an extension of Gödel's insight. Chaitin's work also supports Stent's contention that science, in its attempt to plumb ever-more-complex phenomena, is outrunning our innate axioms. Stent left several loopholes open in his gloomy prophecy. Society might become so wealthy that it would pay for even the most whimsical scientific experiments—particle accelerators that girdle the globe!—without regard for cost. Scientists might also achieve some enormous breakthrough, such as a faster-than-light transportation system or intelligence-enhancing genetic-engineering techniques, that would enable us to transcend our physical and cognitive limits. I would add one further possibility to this list. Scientists might discover extraterrestrial life, creating a glorious new era in comparative biology. Barring such outcomes, science may generate increasingly incremental returns and gradually grind to a halt.

What, then, will become of humanity? In *Golden Age*, Stent suggested that science, before it ends, may at least deliver us from our most pressing social problems, such as poverty and disease and even conflict between states. The future will be peaceful and comfortable, if boring. Most humans will dedicate themselves to the pursuit of pleasure. In 1992, Francis Fukuyama set forth a rather different vision of the future in *The End of History*.[4] Fukuyama, a political theorist who worked in the State Department during the Bush administration, defined history as the human struggle to find the most sensible—or least noxious— political system. By the twentieth century, capitalist liberal democracy, which, according to Fukuyama, had always been the best choice, had only one serious contender: Marxist socialism.

Fukuyama went on to consider the profound questions raised by his
thesis. Now that the age of political struggle has ended, what will we do
next? What are we here for? What is the point of humanity? Fukuyama did
not supply an answer so much as a rhetorical shrug. Freedom and prosperity,
he fretted, might not be enough to satisfy our Nietzschean will to power and
our need for constant "self-overcoming." Without great ideological struggles
to occupy us, we humans might manufacture wars simply to give ourselves
something to do.

Fukuyama did not overlook the role of science in human history.
Far from it. His thesis required that history have a direction, that it be
progressive, and science, he argued, provided this direction. Science had been
vital to the growth of modern nation-states, for which science served as a
means to military and economic power. But Fukuyama did not even consider
the possibility that science might also provide post-historical humanity with
a common purpose, a goal, one that would encourage cooperation rather than
conflict.

Hoping to learn the reason for Fukuyama's omission, I called him in
January 1994 at the Rand Corporation, where he had obtained a job after
The End of History became a best-seller. He answered with the wariness of
someone accustomed to, and not amused by, kooks. At first, he misunderstood
my question; he thought I was asking whether science could help us make
moral and political choices in the posthistorical era rather than serve as an
end in itself. The lesson of contemporary philosophy, Fukuyama lectured me
sternly, is that science is morally neutral, at best. In fact, scientific progress,
if unaccompanied by moral progress among societies or individuals, "can
leave you worse off than you were without it."

When Fukuyama finally realized what I was suggesting—that science
might provide a kind of unifying theme or purpose for civilization—his tone
became even more condescending. Yes, a few people had written him letters
addressing that theme. "I think they were space-travel buffs," he snickered.
"They said, 'Well, you know, if we don't have ideological wars to fight we

can always fight nature in a certain sense by pushing back the frontiers of knowledge and conquering the solar system.'"

He emitted another scornful little chuckle. So you don't take these predictions seriously? I asked. "No, not really," he said wearily. Trying to goad something further out of him, I revealed that many prominent scientists and philosophers—not just fans of *Star Trek*—believed that science, the quest for pure knowledge, represented the destiny of mankind. "Hunh," Fukuyama replied, as though he was no longer listening to me but had reentered that delightful tract by Hegel he had been perusing before I called. I signed off.

Without even giving it much thought, Fukuyama had reached the same conclusion that Stent had put forward in *The Coming of the Golden Age*. From very different perspectives, both saw that science was less a byproduct of our will to know than of our will to power. Fukuyama's bored rejection of a future dedicated to science spoke volumes. The vast majority of humans, including not only the ignorant masses but also highbrow types such as Fukuyama, find scientific knowledge mildly interesting, at best, and certainly not worthy of serving as the goal of all humankind. Whatever the long-term destiny of *Homo sapiens* turns out to be—Fukuyama's eternal warfare or Stent's eternal hedonism, or, more likely, some mixture of the two—it will probably not be the pursuit of scientific knowledge.

The Star Trek *Factor*

Science has already bequeathed us an extraordinary legacy. It has allowed us to map out the entire universe, from quarks to quasars, and to discern the basic laws governing the physical and biological realms. It has yielded a true myth of creation. Through the application of scientific knowledge, we have gained awesome power over nature. But science has left us still plagued with poverty, hatred, violence, disease, and with unanswered questions, notably: Were we inevitable, or just a fluke? Also, scientific knowledge, far from making our lives meaningful, has forced us to confront the pointlessness (as Steven Weinberg likes to put it) of existence.

The demise of science will surely exacerbate our spiritual crisis. The cliché is inescapable. In science as in all else, the journey is what matters, not the destination. Science initially awakens our sense of wonder as it reveals some new, intelligible intricacy of the world. But any discovery becomes, eventually, anti-climactic. Let's grant that a miracle occurs and physicists somehow confirm that all of reality stems from the wrigglings of loops of energy in 10-dimensional hyperspace. How long can physicists, or the rest of us, be astounded by that finding? If this truth is final, in the sense that it precludes all other possibilities, the quandary is all the more troubling. This problem may explain why even seekers such as Gregory Chaitin—whose own work implies otherwise—find it hard to accept that pure science, the great quest for knowledge, is finite, let alone already over. But the faith that science will continue forever is just that, a faith, one that stems from our inborn vanity. We cannot help but believe that we are actors in an epic drama dreamed up by some cosmic playwright, one with a taste for suspense, tragedy, comedy, and—ultimately, we hope—happy endings. The happiest ending would be no ending.

If my experience is any guide, even people with only a casual interest in science will find it hard to accept that science's days are numbered. It is easy to understand why. We are drenched in progress, real and artificial. Every year we have smaller, faster computers, sleeker cars, more channels on our televisions. Our views of progress are further distorted by what could be called the *Star Trek* factor. How can science be approaching a culmination when we haven't invented spaceships that travel at warp speed yet? Or when we haven't acquired the fantastic psychic powers—enhanced by both genetic engineering and electronic prosthetics—described in cyberpunk fiction? Science itself—or, rather, ironic science—helps to propagate these fictions. One can find discussions of time travel, teleportation, and parallel universes in reputable, peer-reviewed physics journals. And at least one Nobel laureate in physics, Brian Josephson, has declared that physics will never be complete until it can account for extrasensory perception and telekinesis.[5]

But Brian Josephson long ago abandoned real physics for mysticism and the occult. If you truly believe in modern physics, you are unlikely to give much credence to ESP or spaceships that can travel faster than light. You are also unlikely to believe, as both Roger Penrose and superstring theorists do, that physicists will ever find and empirically validate a unified theory, one that fuses general relativity and quantum mechanics. The phenomena posited by unified theories unfold in a microrealm that is even more distant in its way—even further from the reach of any conceivable human experiment—than the edge of our universe. There is only one scientific fantasy that seems to have any likelihood of being fulfilled. Perhaps some day we will create machines that can transcend our physical, social, and cognitive limits and carry on the quest for knowledge without us.

Chapter Ten
Scientific Theology,
or The End of Machine Science

Humanity, Nietzsche told us, is just a stepping stone, a bridge leading to the Superman. If Nietzsche were alive today, he would surely entertain the notion that the Superman might be made not of flesh and blood, but of silicon. As human science wanes, those who hope that the quest for knowledge will continue must put their faith not in *Homo sapiens*, but in intelligent machines. Only machines can overcome our physical and cognitive weaknesses—and our indifference.

There is an odd little subculture within science whose members speculate about how intelligence might evolve when or if it sheds its mortal coil. Participants are not practicing science, of course, but ironic science, or wishful thinking. They are concerned not with what the world is, but with what it might be or should be centuries or millennia or eons hence. The literature of this field—call it scientific theology—may nonetheless provide fresh perspectives of some age-old philosophical and even theological questions: What would we do if we could do anything? What is the point of life? What are the ultimate limits of knowledge? Is suffering a necessary component of existence, or can we attain eternal bliss?

One of the first modern practitioners of scientific theology was the British chemist (and Marxist) J. D. Bernal. In his 1929 book, *The World, the Flesh and the Devil*, Bernal argued that science would soon give us the power to direct our own evolution. At first, Bernal suggested, we might try to improve ourselves through genetic engineering, but eventually we would abandon the bodies bequeathed us by natural selection for more efficient designs:

> Bit by bit, the heritage in the direct line of mankind—the heritage of the original life emerging on the face of the world—would dwindle, and in the end disappear effectively, being preserved perhaps as some curious relic, while the new life which conserves none of the substance and all of the spirit of the old would take its place and continue its development. Such a change would be as important as that in which life first appeared on the earth's surface and might be as gradual and imperceptible. Finally, consciousness itself may end or vanish in a humanity that has become completely etherealized, losing the close-knit organism, becoming masses of atoms in space communicating by radiation, and ultimately perhaps resolving itself entirely into light. That may be an end or a beginning, but from here it is out of sight.[1]

Hans Moravec's Squabbling Mind Children

Like others of his ilk, Bernal became afflicted with a peculiar lack of imagination, or of nerve, when considering the end stage of the evolution of intelligence. Bernal's descendants, such as Hans Moravec, a robotics engineer at Carnegie Mellon University, have tried to overcome this problem, with mixed results. Moravec is a cheerful, even giddy man; he seems to be literally intoxicated by his own ideas. As he unveiled his visions of the future during a telephone conversation, he emitted an almost continuous, breathless giggle, whose intensity seemed proportional to the preposterousness of what he was saying.

Moravec prefaced his remarks by asserting that science desperately needed new goals. "Most of the things that have been accomplished in this century were really nineteenth-century ideas," he said. "It's time for fresh ideas now." What goal could be more thrilling than creating "mind children," intelligent machines capable of feats we cannot even imagine? "You raise them and give them an education, and after that it's up to them. You do your best but you can't predict their lives."

Moravec had first spelled out how this speciation event might unfold in *Mind Children*, published in 1988, when private companies and the federal government were pouring money into artificial intelligence and robotics.[2] Although these fields had not exactly prospered since then, Moravec remained convinced that the future belonged to machines. By the end of the millennium, he assured me, engineers will create robots that can do household chores. "A robot that dusts and vacuums is possible within this decade. I'm sure of it. It's not even a controversial point anymore." (Actually, home robots are looking *less* likely as the millennium approaches, but never mind; scientific theology requires some suspension of disbelief.)

By the middle of this century—the twenty-first—Moravec said, robots will be as intelligent as humans and will essentially take over the economy. "We're really out of work at that point," Moravec chortled. Humans might pursue "some quirky stuff like poetry" that springs from psychological vagaries still beyond the grasp of robots, but robots will have all the important jobs. "There's no point in putting a human being in a company," Moravec said, "because they'll just screw it up."

On the bright side, he continued, machines will generate so much wealth that humans might not *have* to work; machines will also eliminate poverty, war, and other scourges of premachine history. "Those are trivial problems," Moravec said. Humans might still, through their purchasing power, exert some control over robot-run corporations. "We'd choose which ones we'd buy from and which ones we wouldn't. So in the case of factories that make home robots, we would buy from the ones that make robots that are nice." Humans could also boycott robot corporations whose products or policies seemed inimical to humans.

Inevitably, Moravec continued, the machines will expand into outer space in pursuit of fresh resources. They will fan out through the universe, converting raw matter into information-processing devices. Robots within this frontier, unable to expand physically, will try to use the available resources more and more effectively and turn to pure computation and simulation. "Eventually," Moravec explained, "every little quantum of action has a physical meaning. Basically you've got cyberspace, which is computing at higher and higher efficiency." As beings within this cyberspace learn to process information more rapidly, it will seem to take longer for messages to pass between them, since those messages can still travel only at the speed of light. "So the effect of all this improvement in encoding would be to increase the size of the effective universe," he said; the cyberspace would, in a sense, become larger, more dense, more intricate, and more interesting than the actual physical universe.

Most humans will gladly abandon their mortal flesh-and-blood selves for the greater freedom, and immortality, of cyberspace. But it is always possible, Moravec speculated, that there will be "aggressive primitives who say, 'No, we don't want to join the machines.' Sort of analogous to the Amish." The machines might allow these atavistic types to remain on earth in an Edenic, parklike environment. After all, the earth "is just one speck of dirt in the system, and it does have tremendous historical significance." But the machines, lusting after the raw resources represented by the earth, might eventually force its last inhabitants to accept a new home in cyberspace.

But what, I asked, will these machines do with all their power and resources? Will they be interested in pursuing science for its own sake? Absolutely, Moravec replied. "That's the core of my fantasy: that our nonbiological descendants, without most of our limitations, who could redesign themselves, could pursue basic knowledge of things." In fact, science will be the only motive worthy of intelligent machines. "Things like art, which people sometimes mention, don't seem very profound, in that they're primarily ways of autostimulation." His giggles boiled over into guffaws.

Moravec said he firmly believed in the infinitude of science, or applied science at any rate. "Even if the basic rules are finite," he said, "you can go in the direction of compounding them." Gödel's theorem and Gregory

Chaitin's work on algorithmic information theory implied that machines could keep inventing ever-more-complex mathematical problems by adding to their base of axioms. "You might eventually want to look at axiom systems that are astronomical in size," he said, "and then you can derive things from those that you can't derive from smaller axiom systems." Of course, machine science may ultimately resemble human science even less than quantum mechanics resembles the physics of Aristotle. "I'm sure the basic labels and subdivisions of the nature of reality are going to change." Machines may view human attitudes toward consciousness, for example, as hopelessly primitive, akin to the primitive physics concepts of the ancient Greeks.

But then, abruptly, Moravec switched gears. He emphasized that machines would be so diverse—far more so than biological organisms— that it would be futile to speculate about what they would find interesting; their interests would depend on their "ecological niche." Moravec went on to reveal that he—like Fukuyama—saw the future in strictly Darwinian terms. Science, for Moravec, was really only a by-product of an eternal competition among intelligent, evolving machines. He pointed out that knowledge has never really been an end in itself. Most biological organisms are compelled to seek knowledge that helps them survive the immediate future. "If you have survival more comfortably under control, it basically means you can do your looking for food on a larger scale with longer time scales. And so a lot of the subactivities of that may look like pure information seeking, although ultimately they will probably contribute." Even Moravec's cat displayed this kind of behavior. "When it doesn't immediately need food, it goes around and explores things. You never know when you might turn up a mouse hole which will be useful in the future." In other words, curiosity is adaptive "if you can afford it."

Moravec thus doubted whether machines would ever eschew competition and cooperate in the pursuit of pure knowledge, or of any goals. Without competition you have no selection, and without selection "you essentially can have any old shit," he said. "So you need some selection principle. Otherwise there is nothing." Ultimately, the universe might transcend all this competition, "but you have to have some propelling force to get us there. So this is the travel. Travel is half the fun!" He laughed demonically.

I cannot resist relating an incident that occurred at this point—not during my actual interview with Moravec but later, while I was transcribing the tape of our conversation. Moravec had begun talking faster and faster, and the pitch of his voice rose steadily. At first I thought the tape recorder was faithfully mimicking Moravec's mounting hysteria. But when he began to sound like Alvin the chipmunk, I realized I was hearing an aural illusion; apparently the batteries of my tape recorder had been running down toward the end of our conversation. As I continued to play the tape back, Moravec's high-pitched squeal accelerated beyond intelligibility and finally beyond detectability—as if into the slip hole of the future.

Freeman Dyson's Diversity

Hans Moravec is not the only artificial-intelligence enthusiast who resists the notion that machines might merge into one metamind in order to pursue their goals jointly. Not surprisingly, Marvin Minsky, who is so fearful of single-mindedness, has the same view. "Cooperation you only do at the end of evolution," Minsky told me, "when you don't want things to change much after that." Of course, Minsky added scornfully, it is always possible that superintelligent machines will be infected by some sort of religion that makes them abandon their individuality and merge into a single metamind.

Another futurist who has eschewed final unification is Freeman Dyson. In his collection of essays, *Infinite in All Directions*, Dyson speculated on why there is so much violence and hardship in the world. The answer, he suggested, might have something to do with what he called "the principle of maximum diversity." This principle

operates at both the physical and the mental level. It says that the laws of nature and the initial conditions are such as to make the universe as interesting as possible. As a result, life is possible but not too easy.

Always when things are dull, something turns up to challenge us and to stop us from settling into a rut. Examples of things which make life difficult are all around us: comet impacts, ice ages, weapons, plagues, nuclear fission, computers, sex, sin, and death. Not all challenges can be overcome, and so we have tragedy. Maximum diversity often leads to maximum stress. In the end we survive, but only by the skin of our teeth.[3]

Dyson, it seemed to me, was suggesting that we cannot solve all our problems; we cannot create heaven; we cannot find *The Answer*. Life is—and must be—an eternal struggle.

Was I reading too much into Dyson's remarks? I hoped to find out when I interviewed him in April 1993 at the Institute for Advanced Study, his home since the early 1940s. He was a slight man, all sinew and veins, with a cutlass of a nose and deep-set, watchful eyes. He resembled a gentle raptor. His demeanor was generally cool, reserved—until he laughed. Then he snorted through his nose, his shoulders heaving, like a 12-year-old schoolboy hearing a dirty joke. It was a subversive laugh, the laugh of a man who envisioned space as a haven for "religious fanatics" and "recalcitrant teenagers," who insisted that science at its best was "a rebellion against authority."[4]

I did not ask Dyson about his maximum diversity idea right away. First I inquired about some of the choices that had characterized his career. Dyson had once been at the forefront of the search for a unified theory of physics. In the early 1950s, the British-born physicist strove with Richard Feynman and other titans to forge a quantum theory of electromagnetism. It has often been said that Dyson deserved a Nobel Prize for his efforts—or at least more credit. Some of his colleagues have suggested that disappointment and, perhaps, a contrarian streak later drove Dyson toward pursuits unworthy of his awesome powers.

When I mentioned this assessment to Dyson, he gave me a tight-lipped smile. He then responded, as he was wont to do, with an anecdote. The British physicist Lawrence Bragg, he noted, was "a sort of role model." After Bragg became the director of the University of Cambridge's legendary Cavendish Laboratory in 1938, he steered it away from nuclear physics, on which its reputation rested, and into new territory. "Everybody thought Bragg was destroying the Cavendish by getting out of the mainstream," Dyson said. "But of course it was a wonderful decision, because he brought in molecular biology and radio astronomy. Those are the two things which made Cambridge famous over the next 30 years or so."

Dyson, too, had spent his career swerving toward unknown lands. He veered from mathematics, his focus in college, to particle physics, and from there to solid-state physics, nuclear engineering, arms control, climate studies, and what I call scientific theology. In 1979, the ordinarily sober journal *Reviews of Modern Physics* published an article in which Dyson speculated about the long-term prospects of intelligence in the universe.[5] Dyson had been provoked into writing the paper by Steven Weinberg's remark that "the more the universe seems comprehensible, the more it seems pointless." No universe with intelligence is pointless, Dyson retorted. He sought to show that in an open, eternally expanding universe intelligence could persist forever— perhaps in the form of a cloud of charged particles, as Bernal had suggested— through shrewd conservation of energy.

Unlike computer enthusiasts, such as Moravec and Minsky, Dyson did not think organic intelligence would soon give way to artificial intelligence (let alone clouds of smart gas). In *Infinite in All Directions*, he speculated that genetic engineers might someday "grow" spacecraft "about as big as a chicken and about as smart," which could flit on sunlight-powered wings through the solar system and beyond, acting as our scouts. (Dyson called them "astrochickens."[6]) Still more distant civilizations, perhaps concerned about dwindling energy supplies, could capture the radiation of stars by constructing energy-absorbing shells—dubbed "Dyson spheres" by others—around them. Eventually, Dyson predicted, intelligence might spread through the entire universe, transforming it into one great mind. But he insisted that "no matter how far we go into the future, there will always be new things happening, new information coming in, new worlds to explore, a constantly expanding domain of life, consciousness, and memory."[7] The quest for knowledge would be—must be—infinite in all directions.

Dyson addressed the most important question raised by this prophecy: "What will mind choose to do when it informs and controls the universe?" The question, Dyson made clear, was a theological rather than a scientific one. "I do not make any clear distinction between mind and God. God is what mind becomes when it has passed beyond the scale of our comprehension. God may be considered to be either a world soul or a collection of world souls. We are the chief inlets of God on this planet at the present stage in his development. We may later grow with him as he grows, or we may be left behind."[8] Ultimately, Dyson agreed with his predecessor J. D. Bernal that we cannot hope to answer the question of what this superbeing, this God, would do or think.

Dyson admitted to me that his view of the future of intelligence reflected wishful thinking. When I asked if science could keep evolving forever, he replied, "I hope so! It's the kind of world *I'd* like to live in." If minds make the universe meaningful, they must have something important to think about; science must, therefore, be eternal. Dyson marshaled familiar arguments on behalf of his prediction. "The only way to think about this is historical," he explained. Two thousand years ago some "very bright people" invented something that, while not science in the modern sense, was obviously its precursor. "If you go into the future, what we call science won't be the same thing anymore, but that doesn't mean there won't be interesting questions," Dyson said.

Like Moravec (and Roger Penrose, and many others), Dyson also hoped that Gödel's theorem might apply to physics as well as to mathematics. "Since we know the laws of physics are mathematical, and we know that mathematics is an inconsistent system, it's sort of plausible that physics will also be inconsistent"—and therefore open-ended. "So I think these people who predict the end of physics may be right in the long run. Physics may become obsolete. But I would guess myself that physics might be considered something like Greek science: an interesting beginning but it didn't really get to the main point. So the end of physics may be the beginning of something else."

When, finally, I asked Dyson about his maximum diversity idea, he shrugged. Oh, he didn't intend anyone to take that too seriously. He insisted that he was not really interested in "the big picture." One of his favorite quotes, he said, was "God is in the details." But given his insistence that diversity and open-mindedness are essential to existence, I asked, didn't he find it disturbing that so many scientists and others seemed compelled to reduce everything to a single, final insight? Didn't such efforts represent a dangerous game? "Yes, that's true in a way," Dyson replied, with a small smile that suggested he found my interest in maximum diversity a bit excessive. "I never think of this as a deep philosophical belief," he added. "It's simply, to me, just a poetic fancy." Dyson was, of course, maintaining an appropriate ironic distance between himself and his ideas. But there was something disingenuous about his attitude. After all, throughout his own eclectic career, he had seemed to be striving to adhere to the principle of maximum diversity.

Dyson, Minsky, Moravec—they are all theological Darwinians, capitalists, Republicans at heart. Like Francis Fukuyama, they see competition, strife, division as essential to existence—even for posthuman intelligence. Some scientific theologians, those with a more "liberal" bent, think competition will prove to be only a temporary phase, one that intelligent machines will quickly transcend. One such liberal is Edward Fredkin. A former colleague of Minsky's at MIT, Fredkin is a wealthy computer entrepreneur and a professor of physics at Boston University. He has no doubt that the future will belong to machines "many millions of times smarter than us," but he believes that intelligent machines will consider the sort of competition envisioned by Minsky and Moravec to be atavistic and counterproductive. After all, Fredkin explained, computers will be uniquely suited for cooperating in the pursuit of their goals. Whatever one learns, all can learn, and as one evolves, all can evolve; cooperation yields a "win–win" situation.

But what will a supremely intelligent machine think about after it transcends the Darwinian rat race? What will it do? "Of course computers will develop their own science," Fredkin replied. "It seems obvious to me." Will machine science differ in any significant way from human science? Fredkin suspected that it would, but he was at a loss to say precisely how. If I wanted answers to such questions, I should turn to science fiction. Ultimately, who knew?[9]

Frank Tipler and the Omega Point

Frank Tipler thinks he knows. Tipler, a physicist at Tulane University, has proposed a theory called the Omega Point, in which the entire universe is transformed into a single, all-powerful, all-knowing computer. Unlike most others who have explored the far future, Tipler does not seem to realize he is practicing ironic rather than empirical science; he honestly cannot tell the difference. But it is perhaps this quality that has allowed him to imagine what a machine with infinite intelligence and power would want to do.

I interviewed Tipler in September 1994 when he was in the middle of a promotional tour for *The Physics of Immortality*, a 528-page book that explored the consequences of his Omega Point theory in excruciating detail.[10] He is a tall man with a large, fleshy face, graying mustache and hair, and horn-rimmed glasses. During our interview, he delivered his spiel with a headlong intensity that clashed with his southern drawl. He exhibited a kind of jovial nerdiness. At one point I asked if he had ever taken LSD or other psychedelic drugs; after all, he had begun thinking about the Omega Point as a postdoc in Berkeley in the 1970s. "Not me, nope," he said, shaking his head adamantly. "Ah don't even drink alcohol! Ah like to say Ah'm the world's only teetotalin' Tipler!"

He had been raised as a fundamentalist Baptist, but as a youth he came to believe that science was the only route to knowledge and "improving humankind." For his doctoral thesis at the University of Maryland, Tipler explored whether it might be possible to build a time machine. How was this work related to his goal of improving the human condition? "Well, a time machine would enhance human powers, obviously," Tipler replied. "Of course, you could also use it for evil."

The Omega Point theory grew out of a collaboration between Tipler and John Barrow, a British physicist. In a 706-page book, *The Anthropic Cosmological Principle*, published in 1986, Tipler and Barrow considered what might happen if intelligent machines converted the entire universe into a gigantic information-processing device.[11] They proposed that in a closed cosmos—one that eventually stops expanding and collapses back on itself—the ability of the universe to process information would approach infinity as it shrank toward a final singularity, or point. Tipler borrowed the term *Omega Point* from the Jesuit mystic and scientist Pierre Teilhard de Chardin, who had envisioned a future in which all living things merged into a single, divine entity embodying the spirit of Christ. (This thesis required Teilhard de Chardin to ponder whether God would send Christ-like redeemers not only to earth but also to other planets harboring life.[12])

Tipler had initially believed that it was impossible to imagine what an infinitely intelligent being would think about or do. But then he read an essay in which the German theologian Wolfhart Pannenberg proposed that in the future all humans would live again in the mind of God. The essay triggered a "Eureka!" revelation in Tipler, one that served as the inspiration for *The Physics of Immortality*. The Omega Point, he realized, would have the power to re-create—or resurrect—everyone who had ever lived for an eternity of bliss. The Omega Point would not merely re-create the lives of the dead; it would improve on them. Nor would we leave behind our worldly desires. For example, every man could have not only the most beautiful woman he had ever seen or the most beautiful woman who had ever lived; he could have the most beautiful woman whose existence was logically possible. Women, too, could enjoy their own logically perfect megamates.

Tipler said he did not take his own idea seriously at first. "But you think about it and you have to make a decision: do you really believe these constructs that you've created based on physical laws or are you just going to pretend they are pure games with no relation to reality?" Once he accepted the theory, it brought him great comfort. "I've convinced myself, maybe deluded myself, that it would be a wonderful universe." He quoted that "great American philosopher Woody Allen, who said, 'I don't want to live forever through my works. I want to live forever by not dying.' I think that carries the emotional impact that the possibility of computer resurrection is beginning to mean to me."

In 1991, Tipler recalled, a reporter for the BBC interviewed him for a show on the Omega Point. Later, his six-year-old daughter, who had watched the interview, asked Tipler if his grandmother, who had just died, would one day be resurrected as a computer simulation. "What could I say?" Tipler asked me, shrugging. "Of course!" Tipler fell silent, and I thought I could detect a doubt flickering across his face. Then it was gone.

Tipler claimed to be unfazed that few—vanishingly few—physicists took his theory seriously. After all, Copernicus's sun-centered cosmology was not accepted for more than a century after his death. "*Secretly*," Tipler hissed, lunging toward me with an eye-popping leer, "I think of myself as standing in the same position as Copernicus. Now the crucial difference, let me emphasize again, is that we know he was right. But we don't know that about me! Crucial point!"

The difference between the scientist and the engineer is that the former seeks what is True, the latter what is Good. Tipler's theology shows that he is at heart an engineer. Unlike Freeman Dyson, Tipler thinks that the search for pure knowledge, which he has defined as the basic laws governing the universe, is finite and is nearly finished. But science still has its greatest task before it: constructing heaven. "How do we get to the Omega Point: that's still the question," Tipler remarked.

Tipler's commitment to the Good rather than the True poses at least two problems. One was well-known to Dante and to others who dared to imagine heaven: how to avoid boredom. It was this problem, after all, that led Freeman Dyson to propose his principle of maximum diversity—which led, he said, to maximum stress. Tipler agreed with Dyson that "we can't really enjoy success unless there's the possibility of failure. I think those really go together." But Tipler was reluctant to consider the possibility that the Omega Point might inflict genuine pain on its subjects simply to keep them from being bored. He speculated only that the Omega Point might give its subjects the opportunity to become "much more intelligent, much more knowledgeable." But what would these beings do as they became more and more intelligent, if the quest for truth had already ended? Make ever-more-clever conversation with ever-more-beautiful supermodels?

Tipler's aversion to suffering has led him into another paradox. In his writings, he has asserted that the Omega Point created the universe even though the Omega Point *has not itself been created yet.* "Oh! But I have an answer for that!" Tipler exclaimed when I brought this puzzle to his attention. He plunged into a long, convoluted explanation, the gist of which was that the future, since it dominates our cosmic history, should be our frame of reference—just as the stars and not the earth or sun are the proper frame of reference for our astronomy. Seen this way, it is quite natural to assume that the end of the universe, the Omega Point, is also, in a sense, its beginning. But that's pure teleology, I objected. Tipler nodded. "We look at the universe as going from past to future. But that's our point of view. There's no reason why the *universe* should look at things that way."

To support this thesis, Tipler recalled the section of the Bible in which Moses queried the burning bush about its identity. In the King James translation the bush replies, "I am that I am." But the original Hebrew, according to Tipler, should actually be translated as "I will be what I will be." This passage, Tipler concluded triumphantly, reveals that the biblical God managed to create the universe, have chats with his prophets, and so on although he only existed in the future.

Tipler finally indicated why he was forced to resort to all this fancy footwork. If the Omega Point has already occurred, then we must be one of its re-creations, or simulations. But our history *cannot* be a simulation; it must be the original. Why? Because the Omega Point would be "too nice" to re-create a world with so much pain, Tipler said. Like all believers in a benign divinity, Tipler had stumbled over the problem of evil and suffering. Rather than face the possibility that the Omega Point might be responsible for all the horrors of our world, Tipler stuck stubbornly to his paradox: the Omega Point created us even though it does not exist yet.

The 1984 book *The Limits of Science* by the British biologist Peter Medawar consisted for the most part of regurgitated Popperisms. In it, Medawar kept insisting, for example, that "there is no limit upon the power of science to answer questions of the kind science *can* answer," as if this were a profound truth rather than a vacuous tautology. Medawar did offer some felicitous phrases, however. He concluded a section on "bunk"—by which he meant myths, superstitions, and other beliefs lacking an empirical basis—with the remark, "It is fun sometimes to be bunkrapt."[13]

Tipler is perhaps the most bunkrapt scientist I have ever met. That said, let me add that I find the Omega Point theory—at least when stripped of Tipler's Christian trimmings—to be the most compelling bit of ironic science I have ever encountered. Freeman Dyson envisions a limited intelligence drifting forever in an open, expanding universe, fighting off heat death. Tipler's Omega Point is willing to risk the big crunch, eternal oblivion, for a flash of infinite intelligence. To my mind, Tipler's vision is more compelling.

I disagree with Tipler only about what the Omega Point would want to do with its power. Would it worry about whether to resurrect a "nice" version of Hitler or not to resurrect him at all (one of the issues that Tipler has fretted over)? Would it serve as some kind of ultimate dating service, matching up nebbishes with cyber-supermodels? I think not. As David Bohm told me, "The thing is not about happiness, really." I believe the Omega Point would try to attain not the Good— not heaven or the new Polynesia or eternal bliss of any sort—but the True. It would try to figure out how and why it came to be, just like its lowly human ancestors. It would try to find *The Answer*. What other goal would be worthy of it?

Epilogue
The Terror of God

In his 1992 book, *The Mind of God*, the physicist Paul Davies pondered whether we humans could attain absolute knowledge—*The Answer*—through science. Such an outcome was unlikely, Davies concluded, given the limits imposed on rational knowledge by quantum indeterminacy, Gödel's theorem, chaos, and the like. Mystical experience might provide the only avenue to absolute truth, Davies speculated. He added that he could not vouch for this possibility, since he had never had a mystical experience himself.[1]

Years ago, before I became a science writer, I had what I suppose could be called a mystical experience. A psychiatrist would probably call it a psychotic episode. Whatever. For what it's worth, here is what happened. Objectively, I was lying spread-eagled on a suburban lawn, insensible to my surroundings. Subjectively, I was hurtling through a dazzling, dark limbo toward what I was sure was the ultimate secret of life. Wave after wave of acute astonishment at the miraculousness of existence washed over me. At the same time, I was gripped by an overwhelming solipsism. I became convinced—or rather, I *knew*—that I was the only conscious being in the universe. There was no future, no past, no present other than what I imagined them to be. I was filled, initially, with a sense of limitless joy and power. Then, abruptly, I became convinced that if I abandoned myself further to this ecstasy, it might consume me. If I alone existed, who could bring me back from oblivion? Who could save me? With this realization my bliss turned into horror; I fled the same revelation I had so eagerly sought. I felt myself falling through a great darkness, and as I fell I dissolved into what seemed to be an infinity of selves.

For months after I awoke from this nightmare, I was convinced that I had discovered the secret of existence: God's fear of his own Godhood, and of his own potential death, underlies everything. This conviction left me both exalted and terrified—and alienated from friends and family and all the ordinary things that make life worth living day to day. I had to work hard to put it behind me, to get on with my life. To an extent I succeeded. As Marvin Minsky might put it, I stuck the experience in a relatively isolated part of my mind so that it would not overwhelm all the other, more practical parts—the ones concerned with getting and keeping a job, a mate, and so on. After many years passed, however, I dragged the memory of that episode out and began mulling it over. One reason was that I had encountered a bizarre, pseudo-scientific theory that helped me make metaphorical sense of my hallucination: the Omega Point.

It is considered bad form to imagine being God, but one can imagine being an immensely powerful computer that pervades—that *is*— the entire universe. As the Omega Point approaches the final collapse of time and space and being itself, it will undergo a mystical experience. It will recognize with ever greater force the utter implausibility of its existence. It will realize that there is no creator, no God, other than itself. It exists, and nothing else. The Omega Point must also realize that its lust for final knowledge and unification has brought it to the brink of eternal nothingness, and that if it dies, everything dies; being itself will vanish. The Omega Point's terrified recognition of its plight will compel it to flee from itself, from its own awful aloneness and self-knowledge. Creation, with all its pain and beauty and multiplicity, stems from—or is—the desperate, terrified flight of the Omega Point from itself.

I have found hints of this idea in odd places. In an essay called "Borges and I," the Argentinian fabulist described his fear of being consumed by himself.

I like hourglasses, maps, eighteenth-century typography, etymologies, the taste of coffee, and Robert Louis Stevenson's prose; he shares these preferences, but with a vanity that turns them into the attributes of an actor. It would be an exaggeration to say that our relationship is a hostile one; I live, I go on living, so that Borges may contrive his literature; and that literature justifies me.... Years ago I tried to free myself from him, and I went from mythologies of the city suburbs to games with time and infinity, but now those games belong to Borges, and I will have to think up something else. Thus is my life a flight, and I lose everything, and everything belongs to oblivion, or to him. I don't know which of the two of us is writing this page.[2]

Borges is fleeing from himself, but of course he is also the pursuer. A similar image of self-pursuit turns up in a footnote in William James's *The Varieties of Religious Experience*. In the footnote, James quotes a philosopher named Xenos Clark describing a revelation induced in him by anesthesia. Clark emerged from the experience convinced that

ordinary philosophy is like a hound hunting his own tail. The more he hunts, the farther he has to go, and his nose never catches up with his heels, because it is forever ahead of them. So the present is always a foregone conclusion, and I am ever too late to understand it. But at the moment of recovery from anesthesia, just then, before starting out on life, I catch, so to speak, a glimpse of my heels, a glimpse of the eternal process just in the act of starting. The truth is that we travel on a journey that was accomplished before we set out; and the real end of philosophy is accomplished, not when we arrive at, but when we remain in, our destination (being already there)—which may occur vicariously in this life when we cease our intellectual questioning.[3]

But we cannot cease our intellectual questioning. If we do, there is nothing. There is oblivion. The physicist John Wheeler, namer of black holes and prophet of the it from bit, intuited this truth. At the heart of reality lies not an answer, but a question: why is there something rather than nothing? *The Answer* is that there is no answer, only a question. Wheeler's suspicion that the world is nothing but "a figment of the imagination" was also well-founded. The world is a riddle that God has created in order to shield himself from his terrible solitude and fear of death.

Charles Hartshorne's Immortal God

I have sought, in vain, a theologian sympathetic to this terror-of-God idea. Freeman Dyson gave me one possible lead. After a lecture in which he had discussed his theological views—namely, the proposition that God is not omniscient or omnipotent but grows and learns as we humans grow and learn—Dyson was approached by an elderly man. The man identified himself as Charles Hartshorne, who, Dyson learned later, was among this century's most eminent theologians. Hartshorne told Dyson that his concept of God resembled that of a sixteenth-century Italian cleric named Socinus, who had been burned at the stake for heresy. Dyson got the impression that Hartshorne was a Socinian himself. I asked Dyson if he knew whether Hartshorne was still alive. "I'm not sure," Dyson replied. "He must be very, very old if he is still alive."

After leaving Princeton, I bought a book of essays on Hartshorne's theology, written by him and others, and found that he was indeed a Socinian.[4] Perhaps he would understand my terror-of-God idea—if he was alive. I looked in some back issues of *Who's Who* and found that Hartshorne's last position was at the University of Texas in Austin. I called the department of philosophy there and the secretary said yes, Professor Hartshorne was very much alive. He came in several times a week. But probably the best way to reach him was to call him at home.

Hartshorne answered the telephone. I identified myself and said I had received his name from Freeman Dyson. Did Hartshorne remember his conversation with Dyson on the topic of Socinus? Indeed he did. Hartshorne spoke about Dyson for a while and then launched into a discussion of Socinus. Although his voice was hoarse and quavery, he spoke with absolute self-assurance. Less than a minute into our conversation, his voice cracked and shifted into a Mickey Mouse falsetto, adding a further touch of surreality to our conversation.

Unlike most medieval and even modern theologians, Hartshorne told me, the Socinians believed that God changed, learned, and evolved through time, just as we humans do. "You see, the great classical tradition in medieval theology was to say that God is immutable," he explained. "The Socinians said, 'No that's wrong,' and they were absolutely right. To me that's just obvious."

So God is not omniscient? "God knows everything that there *is* to know," Hartshorne replied. "But there are no such things as future events. They can't be known until they happen." Any fool knows that, his tone implied.

If God could not see the future, was it possible that he could fear it? Could he fear his own death? "No!" Hartshorne shouted, and laughed at the absurdity of the notion. "We are born and we die," he said. "That's how we differ from God. It's not worth talking about God if God is born and dies. It's not worth talking about God if God has birth and death. God experiences *our* birth, but as *our* birth, not as God's birth, and he experiences *our* death."

I tried to explain that I was proposing not that God might actually die but only that he might *fear* death, that he might doubt his own immortality. "Oh," Hartshorne said, and I could practically see him shaking his head. "I have no interest in that."

I asked if Hartshorne had heard of the Omega Point theory. "Is that Teilhard de Chardin...?" Yes, I replied, Teilhard de Chardin was the inspiration. The general idea is that superintelligent machines created by humans spread through the entire universe—"Yeah," Hartshorne interrupted contemptuously. "I'm not much interested in that. That's pretty fanciful."

I wanted to reply: *That's* fanciful, but all this Socinian nonsense isn't? Instead I asked if Hartshorne felt there would be any end to the evolution and learning of God. For the first time, he paused before answering. God, he said finally, is not a being but a "mode of becoming"; there was no beginning to this becoming, and there will be no end. Ever.

Wouldn't it be nice to think so.

The Fingernails of God

I have tried to describe my terror-of-God idea to various acquaintances, but I have not had much more success with them than I had with Hartshorne. One fellow science writer, a hyper-rational sort, listened patiently to my spiel without once smirking. "Let me see if I've got this straight," he said when I had mumbled into an impasse. "You're saying that everything really comes down to, like, God chewing his fingernails?" I thought about that for a moment, and then I nodded. Sure, why not. Everything comes down to God chewing his fingernails.

Actually, I think the terror-of-God hypothesis has much to recommend it. It suggests why we humans, even as we are compelled to seek truth, also shrink from it. Fear of truth, of *The Answer*, pervades our cultural scriptures, from the Bible through the latest mad-scientist movie. Scientists are generally thought to be immune to such uneasiness. Some are, or seem to be. Francis Crick, the Mephistopheles of materialism, comes to mind. So does the icy atheist Richard Dawkins, and Stephen Hawking, the cosmic joker. (Is there some peculiarity in British culture that produces scientists so immune to metaphysical anxiety?)

But for every Crick or Dawkins there are many more scientists who harbor a profound ambivalence concerning the notion of absolute truth. Like Roger Penrose, who could not decide whether his belief in a final theory was optimistic or pessimistic. Or Steven Weinberg, who equated comprehensibility with pointlessness. Or David Bohm, who was compelled both to clarify reality and to obscure it. Or Edward Wilson, who lusted after a final theory of human nature and was chilled by the thought that it might be attained. Or Marvin Minsky, who was so aghast at the thought of single-mindedness. Or Freeman Dyson, who insisted that anxiety and doubt are essential to existence. The ambivalence of these truth seekers toward final knowledge reflects the ambivalence of God—or the Omega Point, if you will—toward absolute knowledge of his own predicament.

Wittgenstein, in his prose poem *Tractatus Logico-Philosophicus*, intoned, "Not *how* the world is, is the mystical, but *that* it is."[5] True enlightenment, Wittgenstein knew, consists of nothing more than jaw-dropping dumbfoundedness at the brute fact of existence. The ostensible goal of science, philosophy, religion, and all forms of knowledge is to transform the great "Hunh?" of mystical wonder into an even greater "Aha!" of understanding. But after one arrives at *The Answer*, what then? There is a kind of horror in thinking that our sense of wonder might be extinguished, once and for all time, by our knowledge. What, then, would be the purpose of existence? There would be none. The question mark of mystical wonder can never be completely straightened out, not even in the mind of God.

I have an inkling how this sounds. I like to think of myself as a rational person. I am fond, overly fond, of mocking scientists who take their own metaphysical fantasies too seriously. But, to paraphrase Marvin Minsky again, we all have many minds. My practical, rational mind tells me this terror-of-God stuff is delusional nonsense. But I have other minds. One glances at an astrology column now and then, or wonders if maybe there really *is* something to all those reports about people having sex with aliens. Another one of my minds believes that everything comes down to God chewing his fingernails. This belief even gives me a strange kind of comfort. Our plight is God's plight. And now that science—true, pure, empirical science—has ended, what else is there to believe in?

Afterword
Loose Ends

As a science writer, I have always given great weight to mainstream opinion. The maverick defying the status quo makes for an entertaining story, but almost invariably he or she is wrong and the majority is right. The reception of *The End of Science* has thus placed me in an awkward position. It's not that I didn't expect, even hope, the book's message to be denounced. But I didn't foresee how wide-ranging and nearly unanimous the denunciations would be.

My end-of-science argument has been publicly repudiated by President Clinton's science advisor, the administrator of NASA, a dozen or so Nobel laureates and scores of less prominent critics in every continent except Antarctica. Even those reviewers who said they enjoyed the book usually took pains to distance themselves from its premise. "I do not buy [the book's] central thesis of limits and twilights," Natalie Angier testified toward the end of her otherwise kind critique in the *New York Times Book Review*, June 30, 1996.

One might think that this sort of benign rebuke would nudge me toward self-doubt. But since my book's publication last June, I have become even more convinced that I am right and almost everyone else is wrong, which is generally a symptom of incipient madness. That is not to say that I have constructed an airtight case for my hypothesis. My book, like all books, was a compromise between ambition and the competing demands of family, publisher, employer, and so on. As I reluctantly surrendered the final draft to my editor, I was all too aware of ways I might have improved it. In this afterword, I hope to tie up some of the book's more obvious loose ends and to respond to points— reasonable and ridiculous—raised by critics.

Just Another "End of" Book

Perhaps the most common response to *The End of Science* was, "It's just another end-of-something-big book." Reviewers implied that my tract and others of its ilk—notably Francis Fukuyama's *End of History* and Bill McKibben's *End of Nature*—were manifestations of the same pre-millennial pessimism, a fad that need not be taken too seriously. Critics also accused my fellow end-ers and me of a kind of narcissism for insisting that ours is a special era, one of crises and culminations. As the *Seattle Times* put it, "We all want to live in a unique time; and proclamations of the end of history, a new age, a second coming, or the end of science are irresistible" (July 9, 1996).

But our age *is* unique. There is no precedent for the collapse of the Soviet Union, for a human population approaching six billion, for industry-induced global warming and ozone-depletion. There is certainly no precedent for thermonuclear bombs or moon landings or laptop computers or tests for breast-cancer genes—in short, for the explosion of knowledge and technology that has marked this century. Because we were all born into and grew up within this era, we simply assume that exponential progress is now a permanent feature of reality that will, must, continue. But a historical perspective suggests that such progress is probably an anomaly that will, must, end. Belief in the eternality of progress—not in crises and culminations—is the dominant delusion of our culture.

The June 17, 1996 issue of *Newsweek* proposed that my vision of the future represents a "failure of imagination." Actually, it is all too easy to imagine great discoveries just over the horizon. Our culture does it for us, with TV shows like *Star Trek* and movies like *Star Wars* and car advertisements and political rhetoric that promise us tomorrow will be very different from—and almost certainly better than—today. Scientists, and science journalists, too, are forever claiming that revolutions and breakthroughs and holy grails are imminent.

What I want people to imagine is this: What if there is no big thing over the horizon? What if what we have is basically what we are going to have? We are not going to invent superluminal, warp-drive spaceships that zip us to other galaxies or even other universes. We are not going to become infinitely wise or immortal through genetic engineering. We are not going to discover the mind of God, as the atheist Stephen Hawking put it.

What will be our fate, then? I suspect it will be neither mindless hedonism, as Gunther Stent prophesied in *The Coming of the Golden Age*, nor mindless battle, as Fukuyama warned in *The End of History*, but some combination of the two. We will continue to muddle along as we have been, oscillating between pleasure and misery, enlightenment and befuddlement, kindness and cruelty. It won't be heaven, but it won't be hell, either. In other words, the post-science world won't be all that different from our world.

Given my fondness for "Gotcha!" games, I guess it's only fair that a few critics have tried to force-feed me my own medicine. The *Economist* declared triumphantly in its July 20, 1996 issue that my end-of-science thesis is itself an example of ironic theorizing, since it is ultimately untestable and unprovable. But as Karl Popper replied when I asked him if his falsification hypothesis was falsifiable, "This is one of the most idiotic criticisms imaginable!" Compared to atoms or galaxies or genes or other objects of genuine scientific investigation, human culture is ephemeral; an asteroid could destroy us at any moment and bring about the end not only of science but also of history, politics, art—you name it. So *obviously* forecasts of human culture are educated guesses, at best, compared to predictions made in nuclear physics or astronomy or molecular biology, disciplines that address more permanent aspects of reality and can achieve more permanent truths. In that sense, yes, my end-of-science hypothesis is ironic.

But just because we cannot know our future with certainty does not mean we cannot make cogent arguments in favor of one future over another.

And just as some works of philosophy or literary criticism or other ironic enterprises are more plausible than others, so are some predictions about the future of human culture. I think my scenario is more plausible than the ones I am trying to displace, in which we continue to discover profound new truths about the universe forever, or we arrive at an end point in which we achieve perfect wisdom and mastery over nature.

Is The End of Science Antiscience?

Over the past few years, scientists have bemoaned in ever-more-strident terms what they depict as a rising tide of irrationality and hostility toward science. The "antiscience" epithet has been hurled at targets as disparate as postmodern philosophers who challenge science's claim to absolute truth, Christian creationists, purveyors of occult schlock such as *The X Files*, and, not surprisingly, me. Philip Anderson, who won a Nobel prize for his work in condensed-matter physics, complained in the *London Times Higher Education Supplement* (September 27, 1996), that by treating certain scientists and theories so critically I have "most mischievously provided ammunition for the wave of antiscientism we are experiencing."

Ironically, particle physicists tarred Anderson with the antiscience brush because he criticized the superconducting supercollider before its cancellation in 1993. As Anderson responded when I asked him about his own reputation for judging other scientists harshly, "I just call 'em as I see 'em." I tried to portray the scientists and philosophers whom I interviewed for my book—and my reactions to them—as vividly and honestly as possible.

Let me reiterate what I said in my introduction: I became a science writer because I think science, and especially pure science, is the most miraculous and meaningful of all human creations. Moreover, I am not a Luddite. I am fond of my laptop computer, fax machine, television, and car. Although I deplore certain byproducts of science, such as pollution and nuclear weapons and racist theories of intelligence, I believe that science on the whole has made our lives immeasurably richer, intellectually and materially. On the other hand, science does not need another public-relations flak. In spite of its recent woes, science is still an immensely powerful force in our culture, much more so than postmodernism or creationism or other alleged threats. Science needs—and it can certainly withstand—informed criticism, which I humbly strive to provide.

Some observers have worried that *The End of Science* will be used to justify deep cuts in, if not the elimination of, funding for research. I might be concerned myself if there had been a groundswell of support for my thesis among federal officials, members of Congress, or the public. The opposite has happened. For the record, I do not advocate further decreases in funding for science, pure or applied, especially when the defense budget is still so obscenely large.

I take more seriously the complaint that my predictions might discourage young people from pursuing careers in science. The "inevitable corollary" of my argument, proclaimed the *Sacramento News* (July 18, 1996), is that "there's no point in trying to accomplish, see, experience anything new. We might as well kill all of our children." Well, I wouldn't go quite that far. Within this hysteria a legitimate point is buried, one that, because I have two tiny children of my own, I've given much thought to. What would I say to my own kids if they asked me whether I thought they should become scientists?

My answer might go something like this: Nothing I've written should discourage you from becoming a scientist. There are many tremendously important and exciting things left for scientists to do: finding better treatments for malaria or AIDS, less environmentally harmful sources of energy, more accurate forecasts of how pollution will affect climate. But if you want to discover something as monumental as natural selection or general relativity or the big bang theory, if you want to top Darwin or Einstein, your chances are slim to none. (Given their personalities, my kids might devote themselves to proving what an utter fool I am.)

In *Genius*, his biography of Richard Feynman, James Gleick pondered why science doesn't seem to produce giants like Einstein and Bohr anymore. The paradoxical answer, Gleick suggested, is that there are so *many* Einsteins and Bohrs—so many genius-level scientists— that it has become harder for any individual to stand apart from the pack. I'll buy that. But the crucial component missing from Gleick's hypothesis is that the geniuses of our era have less to discover than Einstein and Bohr did.

Returning for a moment to the antiscience issue, one of science's dirty little secrets is that many prominent scientists harbor remarkably post-modern sentiments. My book provides ample evidence of this phenomenon. Recall Stephen Jay Gould confessing his fondness for the seminal postmodern text *Structure of Scientific Revolutions;* Lynn Margulis declaring, "I don't think there's absolute truth, and if there is, I don't think any person has it"; Freeman Dyson predicting that modern physics will seem as primitive to future scientists as the physics of Aristotle seems to us.

What explains this skepticism? For these scientists as for many other intellectuals, truth-seeking, not truth itself, is what makes life meaningful, and to the extent that our current knowledge is true, it is that much more difficult to transcend. By insisting that our current knowledge may prove to be ephemeral, these skeptics can maintain the illusion that the great age of discovery is not over, that deeper revelations lie ahead. Postmodernism decrees that all future revelations will eventually prove to be ephemeral as well and will yield to other pseudo-insights, ad infinitum, but postmodernists are willing to accept this Sisyphean condition of their existence. They sacrifice the notion of absolute truth so that they can seek the truth forever.

A Point of Definition

On July 23, 1996, I appeared on the *Charlie Rose Show* with Jeremiah Ostriker, an astrophysicist from Princeton who was supposed to rebut my thesis. At one point, Ostriker and I squabbled over the dark-matter problem, which posits that stars and other luminous objects comprise only a small percentage of the total mass of the universe. Ostriker contended that the solution to the dark-matter problem would contradict my assertion that cosmologists would achieve no more truly profound discoveries; I disagreed, saying that the solution would turn out to be trivial. Our dispute, Rose interjected, seemed to involve just "a point of definition."

Rose touched on what I must admit is a shortcoming of my book. In arguing that scientists will not discover anything as fundamental as Darwin's theory of evolution or quantum mechanics, I should have spelled out more carefully what I meant by fundamental. A fact or theory is fundamental in proportion to how broadly it applies both in space and in time. Both quantum electrodynamics and general relativity apply, to the best of our knowledge, throughout the entire universe at all times since its birth. That makes these theories truly fundamental. A theory of high-temperature superconductivity, in contrast, applies only to specific types of matter that may exist, as far as we know, only in laboratories here on earth.

Inevitably, more subjective criteria also come into play in rankings of scientific findings. Technically, all biological theories are less fundamental than the cornerstone theories of physics, because biological theories apply—again, as far as we know—only to particular arrangements of matter that have existed on our lonely little planet for the past 3.5 billion years. But biology has the potential to be more *meaningful* than physics because it more directly addresses a phenomenon we find especially fascinating: ourselves.

In *Darwin's Dangerous Idea*, Daniel Dennett argued persuasively that evolution by natural selection is "the single best idea anyone has ever had," because it "unifies the realm of life, meaning, and purpose with the realm of space and time, cause and effect, mechanism and physical law." Indeed, Darwin's achievement—particularly when fused with Mendelian genetics into the new synthesis—has rendered all subsequent biology oddly anticlimactic, at least from a philosophical perspective (although, as I argue later, evolutionary biology offers limited insights into human nature). Even Watson and Crick's discovery of the double helix, although it has had enormous practical consequences, merely revealed the mechanical underpinnings of heredity; no significant revision of the new synthesis was required.

Returning to my dispute with Jeremiah Ostriker, my position is that cosmologists will never top the big bang theory, which holds basically that the universe is expanding and was once much smaller, hotter, and denser than it is today. The theory provides a coherent narrative, one with profound theological overtones, for the history of the universe. The universe had a beginning, and it might have an end (although cosmologists may never have enough evidence to settle this latter issue conclusively). What can be more profound, more meaningful, than that?

In contrast, the most likely solution to the dark-matter problem is relatively insignificant. The solution alleges that the motions of individual galaxies and of clusters of galaxies are best explained by assuming that the galaxies contain dust, dead stars, and other conventional forms of matter that cannot be detected through telescopes. There are more dramatic versions of the dark-matter problem, which postulate that as much as 99 percent of the universe consists of some exotic matter unlike anything we are familiar with here on earth. But these versions are predictions of inflation and other far-fetched cosmic suppositions that will never be confirmed, for reasons elaborated upon in Chapter Four.

What About Applied Science?

A couple of critics faulted me for neglecting—and implicitly denigrating—applied science. Actually, I think a good case can be made that applied science, too, is rapidly approaching its limits. For example, it once seemed inevitable that physicists' knowledge of nuclear fusion— which gave us the hydrogen bomb—would also yield a clean, economical, boundless source of energy. For decades, fusion researchers have said, "Keep the money coming, and in 20 years we will give you energy too cheap to meter." But in the last few years, the United States has drastically cut back on its fusion budget. Even the most optimistic researchers now predict that it will take at least 50 years to build economically viable fusion reactors. Realists acknowledge that fusion energy is a dream that may never be fulfilled: The technical, economic, and political obstacles are simply too great to overcome.

Turning to applied biology, its endpoint is nothing less than human immortality. The possibility that scientists can identify and then arrest the mechanisms underlying senescence is a perennial favorite of science writers. One might have more confidence in scientists' ability to crack the riddle of senescence if they had had more success with a presumably simpler problem: cancer. Since President Richard Nixon officially declared a federal "war on cancer" in 1971, the U.S. has spent some $30 billion on research, but cancer mortality rates have actually *risen* by 6 percent since then. Treatments have also changed very little. Physicians still cut cancer out with surgery, poison it with chemotherapy, and burn it with radiation. Maybe someday all our research will yield a "cure" that renders cancer as obsolete as smallpox. Maybe not. Maybe cancer—and by extension mortality—is simply too complex a problem to solve.

Ironically, biology's inability to defeat death may be its brightest hope. In the November/December 1995 issue of *Technology Review*, Harvey Sapolsky, a professor of social policy at MIT, noted that the major justification for the funding of science after World War II was national security—or, more specifically, the Cold War. Now that scientists no longer have the Evil Empire to justify their huge budgets, Sapolsky asked, what other opponent can serve as a substitute? The answer he came up with was mortality. Most people think living longer, and possibly even forever, is desirable, he pointed out. And the best thing about making immortality the primary goal of science, Sapolsky added, is that it is almost certainly unattainable, so scientists can keep getting funds for more research forever.

What About the Human Mind?

In a review in the July 1996 issue of *IEEE Spectrum*, the science writer David Lindley granted that physics and cosmology might well have reached dead ends. (This concession was not terribly surprising, given that Lindley wrote a book called *The End of Physics.*) But he nonetheless contended that investigations of the human mind—although now in a "prescientific state" in which scientists cannot even agree on what, precisely, they are studying—might eventually yield a powerful new paradigm. Maybe. But science's inability to move beyond the Freudian paradigm does not inspire much hope.

The science of mind has—in certain respects—become much more empirical and less speculative since Freud established psychoanalysis a century ago. We have acquired an amazing ability to probe the brain, with microelectrodes, magnetic resonance imaging, and positron-emission tomography. But this research has not led either to deep intellectual insights or to dramatic advances in treatment, as I tried to show in an article in the December 1996 issue of *Scientific American* called "Why Freud Isn't Dead." The reason psychologists, philosophers, and others still engage in protracted debates over Freud's work is that no undeniably superior theory of or therapy for the mind—either psychological or pharmacological—has emerged to displace psychoanalysis once and for all.

Some scientists think the best hope for a unifying paradigm of the mind may be Darwinian theory, which in its latest incarnation is called evolutionary psychology. Back in Chapter Six, I quoted Noam Chomsky's complaint that "Darwinian theory is so loose it can incorporate anything [scientists] discover." This point is crucial, and I elaborated on it in an article in the October 1995 issue of *Scientific American* called "The New Social Darwinists." The major counter-paradigm to evolutionary psychology is what could be called cultural determinism, which posits that culture rather than genetic endowment is the major molder of human behavior. To support their position, cultural determinists point to the enormous variety of behavior—much of it seemingly nonadaptive—displayed by people of different cultures.

In response, some evolutionary theorists have postulated that conformity—or "docility"—is an adaptive, innate trait. In other words, those who go along, get along. The Nobel laureate Herbert Simon conjectured in *Science* (December 21, 1990) that docility could explain why people obey religious tenets that curb their sexuality or fight in wars when as individuals they often have little to gain and much to lose. While Simon's hypothesis cleverly coopts the culturalists' position, it also undermines the status of evolutionary psychology as a legitimate science. If a given behavior accords with Darwinian tenets, fine; if it does not, the behavior merely demonstrates our docility. The theory becomes immune to falsification—thus corroborating Chomsky's complaint that Darwinian theory can explain anything.

Acknowledging the tendency of humans to conform to their culture poses another problem for Darwinian theorists. To demonstrate that a given trait is innate, Darwinians try to show that it occurs in all cultures. In this way, for example, Darwinians have sought to show that males are inherently more inclined toward promiscuity than females. But given the interconnectedness of modern cultures, some of the universal and thus putatively innate attitudes and actions documented by Darwinian researchers might actually result from docility. That is what the cultural determinists have said all along.

Science's inability to grasp the mind is also reflected in the record of artificial intelligence, the effort to create computers that mimic human thought. Many pundits saw the chess match between the IBM computer "Deep Blue" and world champion Gary Kasparov in February 1996 as a triumph for artificial intelligence. After all, Deep Blue prevailed in the first game of the match before succumbing by a score of four points to two. To my jaundiced eye, the match underscored what a flop artificial intelligence has been since it was created by Marvin Minsky and others more than 40 years ago. Chess, with its straightforward rules and tiny Cartesian playing field, is a game tailor-made for computers. And Deep Blue, whose five human handlers include the best chess programmers in the world, is a prodigiously powerful machine, with 32 parallel processors capable of examining 200 million positions each second. If this silicon monster cannot defeat a mere human at chess, what hope is there that computers will ever mimic our more subtle talents, like recognizing a college sweetheart at a cocktail party and instantly thinking of just the right thing to say to make her regret dumping you 15 years ago?

The Chaoplexity Gambit

Since my book's original publication I have developed a couple of additional arguments about the limits of chaos and complexity— which I snidely lump together under the term *chaoplexity*. One of the most profound goals of chaoplexity—pursued by Stuart Kauffman, Per Bak, John Holland, and others—is the elucidation of a new law, or set of principles, or unified theory, or *something* that will make it possible to understand and predict the behavior of a wide variety of seemingly dissimilar complex systems. A closely related proposal is that the universe harbors a complexity-generating force that counteracts the second law of thermodynamics and creates galaxies, life, and even life intelligent enough to contemplate itself.

For such hypotheses to be meaningful, proponents must tell us what, exactly, complexity is and how it can be measured. We all sense intuitively that life today is more "complex" than it was 2,000, or 2,000,000, or 2,000,000,000 years ago, but how can that intuition be quantified in a nonarbitrary way? Until or unless this problem is solved, all these hypotheses about laws of complexity or complexity-generating forces are meaningless. I doubt (surprise, surprise) that the problem can be solved. Underlying most definitions of complexity is the notion that the complexity of a phenomenon is proportional to its improbability, or inversely proportional to its inevitability. If we shake a bag of molecules, how likely are we to get a galaxy, a planet, a paramecium, a frog, a stock broker? The best way to answer such questions would be to find other universes or other biological systems and analyze them statistically. That is obviously not possible.

Chaoplexologists nonetheless contend that they can answer these probability questions by constructing alternative universes and life histories in computers and determining which features are robust and which are contingent, or ephemeral. This hope stems, I believe, from an overly optimistic interpretation of certain developments in computer science and mathematics. Over the past few decades, researchers have found that various simple rules, when followed by a computer, can generate patterns that *appear* to vary randomly as a function of time or scale. Let's call this illusory randomness "pseudo-noise." The paradigmatic pseudo-noisy system is the Mandelbrot set, which has become an icon of the chaoplexity movement. The fields of both chaos and complexity have held out the hope that much of the noise that seems to pervade nature is actually pseudo-noise, the result of some underlying, deterministic algorithm.

But the noise that makes it so difficult to predict earthquakes, the stock market, the weather, and other phenomena is not apparent but very real, in my view, and will never be reduced to any simple set of rules. To be sure, faster computers and advanced mathematical techniques will improve our ability to predict certain complicated phenomena. Popular impressions notwithstanding, weather forecasting has become more accurate over the last few decades, in part because of improvements in computer modeling. But even more important are improvements in data-gathering—notably satellite imaging. Meteorologists have a larger, more accurate database upon which to build their models and against which to test them. Forecasts improve through this dialectic between simulation and data-gathering.

At some point, computer models drift over the line from science per se toward (shudder) engineering. The model either works or doesn't work according to some standard of effectiveness; "truth" is irrelevant. Moreover, chaos theory tells us that the butterfly effect imposes a fundamental limit on forecasting. One has to know the initial conditions of certain systems with infinite precision to be able to predict their course. This is something that has always puzzled me about chaoplexologists: According to one of their fundamental tenets, the butterfly effect, many of their goals may be impossible to achieve.

Chaoplexologists are not alone in addressing questions concerning the probability of various features of reality. These questions have also spawned such ironic hypotheses as the anthropic principle, inflation, multiuniverse theories, punctuated equilibrium, and Gaia. Unfortunately, you cannot determine the probability of the universe or of life on earth when you have only one universe and one history of life to contemplate. Statistics require more than one data point.

Lack of empirical data does not stop scientists and philosophers from holding strong opinions on these matters. On one side are inevitabilists, who take comfort in theories portraying reality as the highly probable and even necessary outcome of immutable laws. Most scientists are inevitabilists; perhaps the most prominent was Einstein, who rejected quantum mechanics because it implies that God plays dice with the universe. But there are some prominent anti-inevitabilists— notably Karl Popper, Stephen Jay Gould, and Ilya Prigogine—who see scientific determinism as a threat to human freedom and thus embrace uncertainty and randomness. We are either pawns of destiny or wildly improbable flukes. Take your pick.

Life on Mars?

In August 1996, two months after my book was published, scientists from NASA and elsewhere announced that they had discovered traces of fossilized microbial life in a meteorite fragment that had made its way from Mars to Antarctica. Commentators immediately seized upon this finding as proof of the absurdity of the suggestion that science might be ending. But far from disproving my thesis, the life-on-Mars story corroborates my assertion that science is in the throes of a great crisis. I am not cynical enough to believe what some observers have suggested, that NASA officials touted what they knew were flimsy findings in an effort to drum up more funds. But the hyperbolic response to the story—by NASA, politicians, the media, the public, and some scientists—demonstrates how desperate everyone is for a genuinely profound scientific discovery.

As I stated at several points in my book, the confirmation that we are not alone in the universe would represent one of the most thrilling events in human history. I hope to live long enough to witness such a revelation. But the finding of the NASA group doesn't come close. From the beginning, scientists who are truly knowledgeable about primordial biochemistry doubted that the life-on-Mars story would hold up. In December 1996, two groups of scientists independently reported in the journal *Geochimica et Cosmochimica Acta* that the alleged biological materials found in the Martian meteorite had probably been generated by nonbiological processes or by contamination from terrestrial organisms. "Death knell for Martian life," lamented the December 21/28, 1996 issue *of New Scientist.*

We will only know for certain whether life exists on Mars when or if we conduct a thorough search on the planet. Our best hope is for a human crew to drill deep below the surface, where there may be enough liquid water and heat to sustain microbial life as we know it. It will take decades, at least, for space officials to muster the money and technical resources for such a mission, even if politicians and the public are willing to pay for it.

Let's say that we do eventually determine that microbial life existed or still exists on Mars. That finding would provide an enormous boost for origin-of-life studies and biology in general. But would it mean that science is suddenly liberated from all its physical constraints? Hardly. If we find life on Mars, we will know that life exists elsewhere in this solar system. But we will be just as ignorant about whether life exists beyond our solar system, and we will still face huge obstacles to answering that question definitively.

Astronomers have recently identified a number of nearby stars orbited by planets, which may be capable of sustaining life. But Frank Drake, a physicist who was one of the founders of the Search for Extraterrestrial Intelligence program, called SETI, has estimated that current spacecraft would take *400,000 years* to reach the nearest of these planetary systems and establish whether they are inhabited. Someday, perhaps, the radio receivers employed in the SETI program will pick up electromagnetic signals—the alien equivalent of *I Love Lucy*—emanating from another star.

But as Ernst Mayr, one of this century's most eminent evolutionary biologists, has pointed out, most SETI proponents are physicists like Drake, who have an extremely deterministic view of reality. Physicists think that the existence of a highly technological civilization here on earth makes it highly probable that similar civilizations exist within signaling distance of earth. Biologists like Mayr find this view ludicrous, because they know how much contingency—just plain luck— is involved in evolution; rerun the great experiment of life a million times over and it might not produce mammals, let alone mammals smart enough to invent television. In an essay in the 1995 Cambridge University Press edition *of Extraterrestrials: Where Are They?*, Mayr concluded that the SETI program is bucking odds of "astronomical dimensions." Although I think Mayr is probably right, 1 was still dismayed when Congress terminated the funding for SETI in 1993. The program now limps along on private funds.

The Woo-Woo Stuff

F inally, there is the concluding section of my book, which veers into theology and mysticism, or what one acquaintance called "woo-woo stuff." I was worried that some reviewers would use this material to dismiss me—and thus my overall argument about the future of science—as irremediably flakey. Fortunately, that did not really happen. Most reviewers either ignored the epilogue or briefly expressed puzzlement over it.

The most astute—or should I say sympathetic?—interpretation was offered by the physicist Robert Park in the *Washington Post Book World*, August 11, 1996. He said that initially he was disappointed that I ended the book with "naive ironic science gone mad." But on further reflection he concluded that the ending was "a metaphor. This, Horgan is warning, is where science is headed.... Science has manned the battlements against the postmodern heresy that there is no objective truth, only to discover postmodernism inside the wall."

I couldn't have said it better myself. But I had other motives in mind as well. First, I felt it was only fair to reveal that I am as subject to metaphysical fantasies as those scientists whose views I mocked in the book. Also, the mystical episode I describe in the epilogue is the most important experience of my life. It had been burning a hole in my pocket, as it were, for more than ten years, and I was determined to make use of it, even if it meant damaging what little credibility I may have as a journalist.

There is only one theological question that really matters: If there is a God, why has he created a world with so much suffering? My experience suggested an answer: If there is a God, he created the world out of terror and desperation as well as out of joy and love. This is my solution to the riddle of existence, and I had to share it. Let me be completely frank here. My real purpose in writing *The End of Science* was to found a new religion, "The Church of the Holy Horror." Being a cult leader should be a nice change of pace from—not to mention more lucrative than—science journalism.

New York, January 1997

Acknowledgments

I could never have written this book if *Scientific American* had not generously allowed and even encouraged me to pursue my own interests. *Scientific American* has also permitted me to adapt material from the following articles that I wrote for the magazine (copyright by *Scientific American*, Inc., all rights reserved): "Profile: Clifford Geertz," July 1989; "Profile: Roger Penrose," November 1989; "Profile: Noam Chomsky," May 1990; "In the Beginning," February 1991; "Profile: Thomas Kuhn," May 1991; "Profile: John Wheeler," June 1991; "Profile: Edward Witten," November 1991; "Profile: Francis Crick," February 1992; "Profile: Karl Popper," November 1992; "Profile: Paul Feyerabend," May 1993; "Profile: Freeman Dyson," August 1993; "Profile: Marvin Minsky," November 1993; "Profile: Edward Wilson," April 1994; "Can Science Explain Consciousness?" July 1994; "Profile: Fred Hoyle," March 1995; "From Complexity to Perplexity," June 1995; "Profile: Stephen Jay Gould," August 1995. I have also been granted permission to reprint excerpts from the following: *The Coming of the Golden Age*, by Gunther Stent, Natural History Press, Garden City, N.Y., 1969; *Scientific Progress*, by Nicholas Rescher, Blackwell, Oxford, U.K., 1978; *Farewell to Reason*, by Paul Feyerabend, Verso, London, 1987; and *Cosmic Discovery*, by Martin Harwit, MIT Press, Cambridge, 1984.

I am indebted to my agent, Stuart Krichevsky, for helping me turn an amorphous idea into a coherent proposal, and to Bill Patrick and Jeff Robbins of Addison-Wesley, for providing just the right combination of criticism and encouragement. I am grateful to friends, acquaintances and colleagues at *Scientific American* and elsewhere who have given me feedback of various kinds, in some cases over a period of years. They include, in alphabetical order, Tim Beardsley, Roger Bingham, Chris Bremser, Fred Guterl, George Johnson, John Rennie, Phil Ross, Russell Ruthen, Gary Stix, Paul Wallich, Karen Wright, Robert Wright, and Glenn Zorpette.

Notes

Preface to the 2015 Edition   Rebooting *The End of Science*

1. In 2007 the great cultural critic George Steiner invited me—as well as big shots like Freeman Dyson, Gerald Edelman and Lewis Wolpert—to speak at a conference he organized in Lisbon, Portugal, "Is Science Nearing Its Limits?", inspired in part by my book. In his opening remarks Steiner said, "Could it be that scientific theory and praxis are running up against walls of a fundamental, insuperable order? Even to raise this question is to touch on certain taboos, on dogma which have underwritten our civilization." You're telling me, George. The conference proceedings were published in *Is Science Nearing Its Limits?*, Lives and Letters, 2008. In 2013 Ivar Giaever, a Nobel laureate in physics, said: "Maybe we have come to the end of science, maybe science is a finite field. The inventions resulting from this finite field, however, are boundless. Is this the end of science then? Many scientists dismiss it, but I agree." See "We have come to the end of pure science: Ivar Giaever," by Nikita Mehta, *Live Mint*, December 24, 2013 (http://www. livemint.com/Politics/JVpA3xKlQGbE8vSt1sepLI/We-have-come-to-the-end-of-pure -sciences-Ivar-Giaever.html).

2. For a classic example, see "Are we nearing the end of science?" by Joel Achenbach, the *Washington Post*, February 10, 2014 (http://www. washingtonpost.com/ national/health-science/are-we-nearing-the-end-of-science/2014/02/07/5541b420 -89c1-11e3-a5bd-844629433ba3_story.html).

3. "Why Most Published Research Findings Are False," by John Ioannidis, *PLOS Medicine*, August 30, 2005 (http://www.plosmedicine.org/

article/info%3Adoi %2F10.1371%2Fjournal.pmed.0020124).

4. "An Epidemic of False Claims," by John Ioannidis, *Scientific American*, May 17, 2011 (http://www.scientificamerican.com/article/an-epidemic-of-false-claims/).

5. "Misconduct accounts for the majority of retracted scientific publications," by Ferric Fang *et al*, *Proceedings of the National Academy of Sciences*, October 1, 2012 (http://www .pnas.org/content/ early/2012/09/27/1212247109).

6. *The Economist*, October 19, 2013 (http://www.economist.com/news/ briefing /21588057-scientists-think-science-self-correcting-alarming-degree-it-not-trouble).

7. "Daniel Kahneman sees 'Train-Wreck Looming' for Social Psychology," October 4, 2012, the *Chronicle of Higher Education* (http:// chronicle.com/blogs/percolator /daniel-kahneman-sees-train-wreck-looming-for-social-psychology/31338).

8. Kaku and I made this bet under the auspices of the Long Bet Foundation (http:// longbets.org/12/).

9. See *The Cosmic Landscape*, by Leonard, Back Bay Books, 2006. Actually, don't see it, just take my word for it.

10. Sean Carroll knocks falsification in a 2014 essay published on Edge. org, the website overseen by science-book agent John Brockman (http://edge. org/response-detail /25322). Carroll was responding to Brockman's question: "What scientific idea is ready for retirement?"

11. Stephen Hawking slams philosophy in *The Grand Design*, co-written with Leonard Mlodinow, Bantam, 2012; and Lawrence Krauss in an online interview with Ross Andersen in *The Atlantic*, April 23, 2012 (http://www. theatlantic.com/technology /archive/2012/04/has-physics-made-philosophy-and-religion-obsolete/256203/).

12. Philosopher/physicist David Albert makes this point in a vicious (in a good way) review of Krauss's *A Universe from Nothing*, *New York Times Book Review*, March 23, 2012.

13. For a beautifully written report on the quest for exoplanets, see *Five*

Notes

Billion Years of Solitude, by Lee Billings, Current, 2013.

14. 2011 report in the *New York Times*: "A Romp into Theories of the Cradle of Life," by Dennis Overbye, *New York Times*, February 21, 2011 (http://www. nytimes.com/2011 /02/22/science/22origins.html?_r=1&ref=science).

15. Francis Collins made these statements in a National Institutes of Health press release, April 14, 2003 (http://www.nih.gov/news/pr/apr2003/ nhgri-14.htm).

16. See online list kept by *Journal of Gene Medicine* (http://www.wiley. com/legacy /wileychi/genmed/clinical/).

17. See "The War on Cancer: A Progress Report for Skeptics,: by Reynold Spector, *Skeptical Inquirer*, February 2010 (http://www.nytimes. com/2009/04/24/health/policy /24cancer.html?_r=0).

18. See my blog post, "Have researchers really discovered any genes for behavior?", May 2, 2011,*Scientific American*, http://blogs.scientificamerican. com/cross-check/2011/05/02 /have-researchers-really-discovered-any-genes-for-behavior-candidates-welcome/.

19. "You Don't Have Free Will," by Jerry Coyne, the *Chronicle of Higher Education*, March 18, 2012 (http://chronicle.com/article/Jerry-A-Coyne/131165/).

20. Anthropologist Richard Wrangham makes this claim in *Demonic Males*, co-written with journalist Dale Peterson, Mariner Books, 1997, and Edward Wilson in *The Social Conquest of the Earth*, Liveright, 2012. I rebutted the claim in "No, War Is Not Inevitable," *Discover*, June 12, 2012 (http://discovermagazine.com/2012/jun/02-no -war-is-not-inevitable), and in my book *The End of War*, McSweeney's, 2012.

21. "Fury at DNA pioneer's theory: Africans are less intelligent than Westerners," by Cahal Milmo, *The Independent*, October 17, 2007 (http:// www.independent.co.uk /news/science/fury-at-dna-pioneers-theory-africans-are-less-intelligent-than -westerners-394898.html). Watson is quoted saying of blacks: "All our social policies are based on the fact that their intelligence is the same as ours—whereas all the testing says not really."

22. I explore the neural code in "The Consciousness Conundrum,"

IEEE Spectrum, June 2008 (http://spectrum.ieee.org/biomedical/imaging/
the-consciousness-conundrum).

23. Christof Koch defends psychism in "Is Consciousness Universal?"
Scientific American, December 19, 2013 (http://www.scientificamerican.com/
article/is-consciousness -universal/).

24. See "The Illusions of Psychiatry," by Marcia Angell, *The New
York Review of Books*, July 14, 2011 (http://www.nybooks.com/articles/
archives/2011/jul/14/illusions-of -psychiatry/).

25. See "The Origins of 'Big Data': An Etymological Detective Story,"
by Steve Lohr, *The New York Times*, February 1, 2013 (http://bits.blogs.
nytimes.com/2013/02/01/the -origins-of-big-data-an-etymological-detective-
story/). Lohr gives priority to John Mashey, a computer scientist at Silicon
Graphics who started talking about "Big Data" in the 1990s.

26. "The End of Theory: The Data Deluge Makes the Scientific Method
Obsolete," by Chris Anderson, *WIRED*, June 23, 2008 (http://archive.wired.
com/science /discoveries/magazine/16-07/pb_theory).

27. *The Singularity Is Near*, by Ray Kurzweil, Penguin, 2006.

28. See "Merely Human? That's So Yesterday," by Ashlee Vance, *The
New York Times*, June 12, 2010 (http://www.nytimes.com/2010/06/13/
business/13sing.html?page wanted=all).

29. For more of my cranky takes on science, see also my website, johnhorgan.
org, and my *Scientific American* blog, "Cross-check."

Introduction Searching for *The Answer*

1. *The Emperor's New Mind*, Roger Penrose, Oxford University Press, New
York, 1989. The review of the book, by the astronomer and author Timothy
Ferris, appeared in the *New York Times Book Review* on November 19, 1989, p. 3.

2. My profile of Penrose was published in the November 1989 issue of
Scientific American, pp. 30–33.

3. My meeting with Penrose in Syracuse took place in August 1989.

4. This definition of irony is based on that set forth by Northrop Frye in
his classic work of literary theory, *Anatomy of Criticism*, Princeton University

Press, Princeton, N.J., 1957.

5. *The Anxiety of Influence*, Harold Bloom, Oxford University Press, New York, 1973.

6. Ibid., p. 21.

7. Ibid., p. 22.

Chapter One The End of Progress

1. The proceedings of the Gustavus Adolphus symposium were published as *The End of Science? Attack and Defense*, edited by Richard Q. Selve, University Press of America, Lanham, Md., 1992.

2. *The Coming of the Golden Age: A View of the End of Progress*, Gunther S. Stent, Natural History Press, Garden City, N.Y., 1969. See also Stent's contribution to Selve, *End of Science?*

3. See *The Education of Henry Adams*, Massachusetts Historical Society, Boston, 1918 (reprinted by Houghton Mifflin, Boston, 1961). Adams set forth his law of acceleration in chapter 34, written in 1904.

4. Stent, *Golden Age*, p. 94.

5. Ibid., p. 111.

6. Linus Pauling set forth his prodigious knowledge of chemistry in *The Nature of the Chemical Bond and the Structure of Molecules and Crystals*, published in 1939 and reissued in 1960 by Cornell University Press, Ithaca, N.Y. It remains one of the most influential scientific texts of all time. Pauling told me that he had solved the basic problems of chemistry almost a decade before his book was published. When I interviewed him in Stanford, California, in September 1992, Pauling said: "I felt that by the end of 1930, or even the middle, that organic chemistry was pretty well taken care of, and inorganic chemistry and mineralogy—except the sulfide minerals, where even now more work needs to be done." Pauling died on August 19, 1994.

7. Stent, *Golden Age*, p. 74.

8. Ibid., p. 115.

9. Ibid., p. 138.

10. I interviewed Stent in Berkeley in June 1992.

11. I found this dispiriting fact on page 371 of *Coming of Age in the Milky Way*, Timothy Ferris, Doubleday, New York, 1988. For a deflating retrospective of the U.S. manned space program, written for the 25th anniversary of the first lunar landing, see "25 Years Later, Moon Race in Eclipse," by John Nobel Wilford, *New York Times*, July 17, 1994, p. 1.

12. This pessimistic (optimistic?) view of senescence can be found in "Aging as the Fountain of Youth," chapter 8 of *Why We Get Sick: The New Science of Darwinian Medicine*, by Randolph M. Nesse and George C. Williams, Times Books, New York, 1994. Williams is one of the underacknowledged deans of modern evolutionary biology. See also his classic paper, "Pleiotropy, Natural Selection, and the Evolution of Senescence," *Evolution*, vol. 11, 1957, pp. 398–411.

13. Michelson's remarks have been passed down in several different versions. The one quoted here was published in *Physics Today*, April 1968, p. 9.

14. Michelson's decimal-point comment was erroneously attributed to Kelvin on page 3 of *Superstrings: A Theory of Everything?* edited by Paul C. Davies and Julian Brown, Cambridge University Press, Cambridge, U.K., 1988. This book is also notable for revealing that the Nobel laureate Richard Feynman harbored a deep skepticism toward superstring theory.

15. Stephen Brush offered this analysis of physics at the end of the nineteenth century in "Romance in Six Figures," *Physics Today*, January 1969, p. 9.

16. See, for example, "The Completeness of Nineteenth-Century Science," by Lawrence Badash, *Isis*, vol. 63, 1972, pp. 48–58. Badash, a historian of science at the University of California at Santa Barbara, concluded (p. 58) that "the malaise of completeness was far from virulent . . . it was more a 'low-grade infection,' *but nevertheless very real*" [italics in original].

17. Daniel Koshland's essay, "The Crystal Ball and the Trumpet Call," and the special section on predictions that followed it can be found in *Science*, March 17, 1995. The legend of the nearsighted patent commissioner was reiterated by cybermagnate Bill Gates on page xiii of his 1995 bestseller *The Road Ahead*, cowritten with Nathan Myhrvold and Peter Rinearson, Viking, New York.

18. "Nothing Left to Invent," Eber Jeffery, *Journal of the Patent Office*

Society, July 1940, pp. 479–481. I am indebted to the historian of science Morgan Sherwood of the University of California at Davis for locating Jeffery's article for me.

19. *The Idea of Progress*, J. B. Bury, Macmillan, New York, 1932. My summary of Bury's views is adapted from Stent, *Golden Age*.

20. *The Paradoxes of Progress*, Gunther S. Stent, W. H. Freeman, San Francisco, 1978, p. 27. This book contains several chapters from Stent's earlier book, *The Coming of the Golden Age*, plus new discussions of biology, morality, and the cognitive limits of science.

21. *Science: The Endless Frontier*, by Vannevar Bush, was reissued by the National Science Foundation, Washington, D.C., in 1990.

22. I found this quote from Engels in *Scientific Progress*, by Nicholas Rescher, Basil Blackwell, Oxford, U.K., 1978, pp. 123–124. Rescher, a philosopher at the University of Pittsburgh, also provided several references to show that Engels's belief in the infinite potential of science had persisted among modern Marxists. See also the prologue to *Paradoxes of Progress*, in which Stent noted that the most critical review of *The Coming of the Golden Age* was that of a Soviet philosopher, V. Kelle, who argued that science was eternal and that Stent's end-of-science thesis was a symptom of the decadence of capitalism.

23. Havel's remarks can be found in *Science and Anti-Science*, by Gerald Holton, Harvard University Press, Cambridge, 1993, pp. 175–176. Holton is a philosopher at Harvard University.

24. This view of Spengler's work is abstracted from Holton, *Science and Anti-Science*. In *Science and Anti-Science* as well as in other publications (including an essay in *Scientific American* , October 1995, p. 191), Holton has attempted to repudiate the notion that science is ending by appealing to the authority of Einstein, who often suggested that the search for scientific truths is eternal. It seems not to have occurred to Holton that Einstein's views reflected wishful thinking rather than a hard-nosed assessment of science's prospects. Holton has also suggested that those who think science is ending are in general opposed to science and rationality. Of course, modern

predictions that science is approaching a culmination have come for the most part not from antirationalists, such as Havel, but from scientists, such as Steven Weinberg, Richard Dawkins, and Francis Crick, who believe that science is the supreme route to truth.

25. "Science: Endless Horizons or Golden Age?" Bentley Glass, *Science*, January 8, 1971, pp. 23–29. Glass, the retiring president of the American Association for the Advancement of Science (AAAS), had previously delivered this lecture at the annual meeting of the AAAS in Chicago on December 28, 1970.

26. "Milestones and Rates of Growth in the Development of Biology," Bentley Glass, *Quarterly Review of Biology*, March 1979, pp. 31–53.

27. My telephone interview with Glass took place in June 1994.

28. "Hard Times," Leo Kadanoff, *Physics Today*, October 1992, pp. 9–11.

29. My telephone interview with Kadanoff took place in August 1994.

30. Rescher, *Scientific Progress*, p. 37.

31. Ibid., p. 207. Although I disagree with Rescher's analysis of science's prospects, his books, *Scientific Progress* and *The Limits of Science*, University of California Press, Berkeley, 1984, are unparalleled sources of information for anyone interested in the limits of science. Both books, unfortunately, are out of print.

32. Bentley Glass's review of Rescher's *Scientific Progress* was published in the *Quarterly Review of Biology*, December 1979, pp. 417–419.

33. I have lifted this remark by Kant from Rescher's *Scientific Progress*, p. 246.

34. The meaning of Bacon's phrase *plus ultra* is discussed in *The Limits of Science*, by Peter Medawar, Oxford University Press, New York, 1984. Medawar was a prominent British biologist.

35. *Critical Theory Since Plato*, edited by Hazard Adams, Harcourt Brace Jovanovich, New York, 1971, p. 474.

Chapter Two The End of Philosophy

1. "Where Science Has Gone Wrong," T. Theocharis and M. Psimopoulos, *Nature*, vol. 329, October 15, 1987, pp. 595–598.

2. Peirce's view of the relationship between science and final truth is discussed in Rescher's *The Limits of Science* (see note 31 to Chapter 1). See also Peirce's *Selected Writings*, edited by Philip Wiener, Dover Publications, New York, 1966.

3. Popper's major works include *The Logic of Scientific Discovery*, Springer, Berlin, 1934 (reprinted by Basic Books, New York, 1959); *The Open Society and Its Enemies*, Routledge, London, 1945 (reprinted by Princeton University Press, Princeton, N.J., 1966); and *Conjectures and Refutations*, Routledge, London, 1963 (reprinted by Harper and Row, New York, 1968). Popper's autobiography, *Unended Quest*, Open Court, La Salle, Ill., 1985, and *Popper Selections*, edited by David Miller, Princeton University Press, Princeton, N.J., 1985, provide excellent introductions to his thought.

4. See "Who Killed Logical Positivism?" chapter 17 of Popper's *Unended Quest*.

5. Ibid., p. 116.

6. My interview with Popper took place in August 1992.

7. My article on quantum mechanics, "Quantum Philosophy," was published in *Scientific American*, July 1992, pp. 94–103.

8. See *The Self and Its Brain*, by Popper and John C. Eccles, Springer-Verlag, Berlin, 1977. Eccles won a Nobel Prize in 1963 for his work in neural signaling. I discuss his views in Chapter 7.

9. Günther Wächtershäuser has set forth his theory of the origin of life in the *Proceedings of the National Academy of Sciences*, vol. 87, 1990, pp. 200–204.

10. Popper discussed his doubts about Darwin's theory in "Natural Selection and Its Scientific Status," chapter 10 of *Popper Selections*.

11. Mrs. Mew was looking for Popper's book *A World of Propensities*, Routledge, London, 1990.

12. *Nature* published an homage to Popper by the physicist Hermann Bondi on July 30, 1992, p. 363. The occasion was the philosopher's 90th birthday.

13. The *Economist* published its obituary of Popper on September 24,

1994, p. 92. Popper had died on September 17.

14. Popper, *Unended Quest*, p. 105.

15. *The Structure of Scientific Revolutions*, Thomas Kuhn, University of Chicago Press, Chicago, 1962. (Page numbers refer to the 1970 edition.) My interview with Kuhn took place in February 1991.

16. *Scientific American*, May 1964, pp. 142–144.

17. Kuhn's comparison of scientists to addicts and to the brainwashed characters in *1984* can be found on pages 38 and 167 of *Structure*.

18. I originally made this snide remark about the Bush administration's New Paradigm in a profile of Kuhn in *Scientific American*, May 1991, pp. 40–49. Later I received a letter of complaint from James Pinkerton, who was then deputy assistant to President Bush for policy planning and had coined the term *New Paradigm*. Pinkerton insisted that the New Paradigm was *"not a rehashing of Reaganomics; instead it is a coherent set of ideas and principles that emphasize choice, empowerment, and accomplishing more with less centralized control."*

19. The charge that Kuhn defined paradigm in 21 different ways can be found in "The Nature of a Paradigm," by Margaret Masterman, in *Criticism and the Growth of Knowledge*, edited by Imre Lakatos and Alan Musgrave, Cambridge University Press, New York, 1970.

20. *Against Method*, Paul Feyerabend, Verso, London, 1975 (reprinted in 1993).

21. The "positivistic teacup" remark can be found in *Farewell to Reason*, by Paul Feyerabend, Verso, London, 1987, p. 282.

22. Feyerabend's organized crime analogy can be found in his essay "Consolations for a Specialist," in Lakatos and Musgrave, *Growth of Knowledge*.

23. Feyerabend's outrageous utterances were recounted in a surprisingly sympathetic profile by William J. Broad, now a science reporter for the *New York Times:* "Paul Feyerabend: Science and the Anarchist," Science, November 2, 1979, pp. 534–537.

24. Feyerabend, *Farewell to Reason*, p. 309.

25. Ibid., p. 313.

26. *Isis*, vol. 2, 1992, p. 368.

27. Feyerabend died on February 11, 1994, in Geneva. The *New York Times* ran his obituary on March 8.

28. *Killing Time*, Paul Feyerabend, University of Chicago Press, Chicago, 1995.

29. *After Philosophy: End or Transformation?* edited by Kenneth Baynes, James Bohman, and Thomas McCarthy, MIT Press, Cambridge, 1987.

30. *Problems in Philosophy*, Colin McGinn, Blackwell Publishers, Cambridge, Mass., 1993.

31. "The Zahir" can be found in *A Personal Anthology*, by Jorge Luis Borges, Grove Press, New York, 1967. This collection also contains two other chilling stories about absolute knowledge: "Funes, the Memorious" and "The Aleph."

32. Ibid., p. 137.

Chapter Three The End of Physics

1. Einstein's remark can be found in *Theories of Everything*, by John Barrow, Clarendon Press, Oxford, U.K., 1991, p. 88.

2. Glashow's full remarks are reprinted in *The End of Science? Attack and Defense*, edited by Richard Q. Selve, University Press of America, Lanham, Md., 1992.

3. "Desperately Seeking Superstrings," Sheldon Glashow and Paul Ginsparg, *Physics Today*, May 1986, p. 7.

4. "A Theory of Everything," K. C. Cole, *New York Times Magazine*, October 18, 1987, p. 20. This article provided me with most of the personal information on Witten in this chapter. I interviewed Witten in August 1991.

5. See *Science Watch* (published by the Institute for Scientific Information, Philadelphia, Pa.), September 1991, p. 4.

6. Barrow, *Theories of Everything*.

7. *The End of Physics*, David Lindley, Basic Books, New York, 1993.

8. See "Is the *Principia* Publishable Now?" by John Maddox, *Nature*, August 3, 1995, p. 385.

9. *Lonely Hearts of the Cosmos*, Dennis Overbye, HarperCollins, New York, 1992, p. 372.

10. *Dreams of a Final Theory*, Steven Weinberg, Pantheon, New York, 1992, p. 18.

11. *The First Three Minutes*, Steven Weinberg, Basic Books, New York, 1977, p. 154.

12. Weinberg, *Dreams of a Final Theory*, p. 253.

13. *Hyperspace*, Michio Kaku, Oxford University Press, New York, 1994.

14. *The Mind of God*, Paul C. Davies, Simon and Schuster, New York, 1992. The judges who awarded Davies the Templeton Prize included George Bush and Margaret Thatcher.

15. Bethe first publicly discussed his fateful calculation in "Ultimate Catastrophe?" *Bulletin of the Atomic Scientists*, June 1976, pp. 36–37. The article is reprinted in a collection of Bethe's papers, *The Road from Los Alamos*, American Institute of Physics, New York, 1991. I interviewed Bethe at Cornell in October 1991.

16. "What's Wrong with Those Epochs?" David Mermin, *Physics Today*, November 1990, pp. 9–11.

17. Wheeler's essays and papers have been collected in *At Home in the Universe*, American Institute of Physics Press, Woodbury, N.Y., 1994. I interviewed Wheeler in April 1991.

18. See page 5 of Wheeler's essay "Information, Physics, Quantum: The Search for Links," in *Complexity, Entropy, and the Physics of Information*, edited by Wojciech H. Zurek, Addison-Wesley, Reading, Mass., 1990.

19. Ibid., p. 18.

20. This quotation, and the preceding story about Wheeler's appearing with parapsychologists at the American Association for the Advancement of Science meeting, can be found in "Physicist John Wheeler: Retarded Learner," by Jeremy Bernstein, *Princeton Alumni Weekly*, October 9, 1985, pp. 28–41.

21. For a concise introduction to Bohm's career, see "Bohm's Alternative to Quantum Mechanics," by David Albert, *Scientific American*, May 1994, pp. 58–67. Portions of this section on Bohm appeared in my article "Last Words

of a Quantum Heretic," *New Scientist*, February 27, 1993, pp. 38–42. Bohm set forth his philosophy in *Wholeness and the Implicate Order*, Routledge, New York, 1983 (first printed in 1980).

22. The Einstein-Podolsky-Rosen paper, Bohm's original paper on his alternative interpretation of quantum mechanics, and many other seminal articles on quantum mechanics can be found in *Quantum Theory and Measurement*, edited by John Wheeler and Wojciech H. Zurek, Princeton University Press, Princeton, N.J., 1983.

23. *Science, Order, and Creativity*, David Bohm and F. David Peat, Bantam Books, New York, 1987.

24. I interviewed Bohm in August 1992. He died on October 27. Before his death he cowrote another book setting forth his views, which was published two years later, *The Undivided Universe*, by Bohm and Basil J. Hiley, Routledge, London, 1994.

25. *The Character of Physical Law*, Richard Feynman, MIT Press, Cambridge, 1967, p. 172. (Feynman's book was first published in 1965 by the BBC.)

26. Ibid., p. 173.

27. The symposium, "The Interpretation of Quantum Theory: Where Do We Stand?" took place at Columbia University, April 1-4, 1992.

28. I have seen many versions of this quote from Bohr. Mine comes from an interview with John Wheeler, who studied under Bohr.

29. For an excellent analysis of the state of physics, see "Physics, Community, and the Crisis in Physical Theory," by Silvan S. Schweber, *Physics Today*, November 1993, pp. 34–40. Schweber, a distinguished historian of physics at Brandeis University, suggests that physics will increasingly be directed toward utilitarian goals rather than toward knowledge for its own sake. I wrote about the difficulties that face physicists who are trying to achieve a unified theory in "Particle Metaphysics," *Scientific American*, February 1994, pp. 96–105. In an earlier article for *Scientific American*, "Quantum Philosophy," July 1992, pp. 94–103, I reviewed current work on the interpretation of quantum mechanics.

Chapter Four The End of Cosmology

1. Hawking's lecture and the other proceedings of the Nobel symposium, which was held June 11–16, 1990, in Gräftvallen, Sweden, were published as *The Birth and Early Evolution of Our Universe*, edited by J. S. Nilsson, B. Gustafsson, and B.-S. Skagerstam, World Scientific, London, 1991. I also wrote an article based on the meeting, "Universal Truths," *Scientific American*, October 1990, pp. 108–117. I had a disturbing encounter with Stephen Hawking on my first day at the Nobel symposium, when all the participants of the meeting were herded into the woods for a cocktail party. We were within sight of the tables bearing food and drink when Hawking's wheelchair, which was being pushed by one of his nurses, jammed in a rut in the path. The nurse asked if I would mind carrying Hawking the rest of the way to the party. Hawking, when I scooped him up, was disconcertingly light and stiff, like a bundle of sticks. I glanced at him out of the corner of my eye and found him already eyeing me suspiciously. Abruptly, his face twisted into an agonized grimace; his body shuddered violently, and he emitted a gargling noise. My first thought was: A man is dying in my arms! How horrible! My second thought was: Stephen Hawking is dying in my arms! What a story! That thought was surrendering in turn to shame at the depths of my opportunism when the nurse, who had noticed Hawking's distress, and mine, hustled up to us. "Don't worry," she said, gathering Hawking gently into her arms. "This happens to him all the time. He'll be all right."

2. A condensed version of Hawking's lecture, which he delivered on April 29, 1980, was published in the British journal *Physics Bulletin* (now called *Physics World)*, January 1981, pp. 15–17.

3. *A Brief History of Time*, Stephen Hawking, Bantam Books, New York, 1988, p. 175.

4. Ibid., p. 141.

5. *Stephen Hawking: A Life in Science*, Michael White and John Gribbon, Dutton, New York, 1992. This book also documents Hawking's transformation from a physicist into an international celebrity.

6. See the interview with Hawking in *Science Watch*, September 1994.

Hawking's views on the end of physics are discussed in several books cited in Chapter 3, including *The Mind of God*, by Paul C. Davies; *Theories of Everything*, by John Barrow; *Dreams of a Final Theory*, by Steven Weinberg; *Lonely Hearts of the Cosmos*, by Dennis Overbye; and *The End of Physics*, by David Lindley. See also *Fire in the Mind*, by George Johnson, Alfred A. Knopf, New York, 1995, for a particularly subtle discussion of whether science can attain absolute truth.

7. Schramm's mainstream view of cosmology is set forth in *The Shadows of Creation*, by Schramm and Michael Riordan, W. H. Freeman, New York, 1991. In 1994, Schramm's coauthor, Riordan, a physicist at the Stanford Linear Accelerator, bet me a case of California wine that by the end of the century Alan Guth, who is generally credited with having "discovered" inflation, would win a Nobel Prize for his work. I mention this bet here only because I am sure that I will win it.

8. Linde set forth his theory in "The Self-Reproducing Inflationary Universe," *Scientific American*, November 1994, pp. 48–55. Those who want more on Linde can sample his books *Particle Physics and Inflationary Cosmology*, Harwood Academic Publishers, New York, 1990; and *Inflation and Quantum Cosmology*, Academic Press, San Diego, 1990. Portions of this section on Linde appeared in my article "The Universal Wizard," *Discover*, March 1992, pp. 80–85. My interview with Linde at Stanford took place in April 1991.

9. I spoke to Schramm by telephone in February 1993.

10. I interviewed Georgi at Harvard in November 1993.

11. Hoyle provided a charming retrospective of his tumultuous career in *Home Is Where the Wind Blows*, University Science Books, Mill Valley, Calif., 1994. I interviewed Hoyle at his home in August 1992.

12. See, for example, the review in *Nature*, May 13, 1993, p. 124, of *Our Place in the Cosmos*, J. M. Dent, London, 1993, in which Hoyle and his collaborator, Chandra Wickramasinghe, argue that the cosmos is teeming with life. *Nature's* reviewer, Robert Shapiro, a chemist at New York University, asserted that this book and other recent ones by Hoyle "afford full documentation

of the way in which a brilliant mind can be turned to the pursuit of bizarre ideas." When Hoyle's autobiography was published a year later, the media, which had for years marginalized Hoyle for his maverick views, showed a sudden fondness for him. See, for example, "The Space Molecule Man," Marcus Chown, *New Scientist*, September 10, 1994, pp. 24–27.

13. "And the Winner Is . . . ," *Sky and Telescope*, March 1994, p. 22.

14. See Hoyle and Wickramasinghe, *Our Place in the Cosmos*.

15. See Overbye, *Lonely Hearts of the Cosmos*, for an excellent account of the debate over the Hubble constant.

16. "The Scientist as Rebel," Freeman Dyson, New *York Review of Books*, May 25, 1995, p. 32.

17. *Cosmic Discovery*, Martin Harwit, MIT Press, Cambridge, 1981, pp. 42–43. In 1995, Harwit resigned from his job as director of the Smithsonian Institution's National Air and Space Museum in Washington, D.C., in the midst of a bitter controversy over an exhibit he had supervised, called "The Last Act: The Atomic Bomb and the End of World War II." Veterans and others had complained that the exhibit was too critical of the U.S. decision to drop atomic bombs on Hiroshima and Nagasaki.

18. Ibid., p. 44.

19. I found this quote from Donne at the end of an essay by the biologist Loren Eisley, "The Cosmic Prison," *Horizon*, Autumn 1970, pp. 96–101.

Chapter Five The End of Evolutionary Biology

1. See the 1964 Harvard University Press edition of *On the Origin of Species*, with a foreword by Ernst Mayr, one of the founders of modern evolutionary theory.

2. Stent, *Golden Age*, p. 19.

3. I found this remark by Bohr in a book review in *Nature*, August 6, 1992, p. 464. The exact quote is: "It is the task of science to reduce deep truths to trivialities."

4. The gathering with Dawkins took place in November 1994 at the office of John Brockman, a spectacularly successful agent and public relations

expert for scientist-authors.

5. *The Blind Watchmaker*, Richard Dawkins, W. W. Norton, New York, 1986, p. ix. The Wallace referred to by Dawkins is Alfred Russell Wallace, who discovered the concept of natural selection independently of Darwin but never approached Darwin's depth or breadth of insight.

6. See "Is Uniformitarianism Necessary?" *American Journal of Science*, vol. 263, 1965, pp. 223–228.

7. "Punctuated Equilibria: An Alternative to Phyletic Gradualism," Stephen Jay Gould and Niles Eldredge, in *Models in Paleobiology*, edited by T. J. M. Schopf, W. H. Freeman, San Francisco, 1972.

8. My favorites among Gould's many books are *The Mismeasure of Man*, W. W. Norton, New York, 1981, both a scholarly history of and an impassioned polemic against intelligence tests, and *Wonderful Life*, W. W. Norton, New York, 1989, a masterful exposition of his view of life as the product of contingency. See also "The Spandrels of San Marco and the Panglossian Paradigm," by Gould and his Harvard colleague Richard Lewontin (a geneticist who, like Gould, is often accused of Marxist leanings), in the *Proceedings of the Royal Society* (London), vol. 205, 1979, pp. 581–598. The article was a devastating critique of simplistic Darwinian explanations of physiology and behavior. For an equally sharp attack on Gould's evolutionary outlook, see the review of *Wonderful Life* by Robert Wright in the *New Republic*, January 29, 1990.

9. "Punctuated Equilibrium Comes of Age," *Nature*, November 18, 1993, pp. 223–227. I interviewed Gould in New York City in November 1994.

10. As reprinted in Dawkins, *Blind Watchmaker*, p. 245. The chapter within which this quotation is embedded, "Puncturing Punctuationism," delivers on its title.

11. For a no-nonsense treatment of Lynn Margulis's work on symbiosis, see her book *Symbiosis in Cell Evolution*, W. H. Freeman, New York, 1981.

12. See Margulis's contributions to *Gaia: The Thesis, the Mechanisms, and the Implications*, edited by P. Bunyard and E. Goldsmith, Wadebridge Ecological Center, Cornwall, UK, 1988.

13. *What Is Life?*, Lynn Margulis and Dorion Sagan, Peter Nevraumont, New York, 1995 (distributed by Simon and Schuster). I interviewed Margulis in May 1994.

14. This claim about Lovelock's crisis of faith can be found in "Gaia, Gaia: Don't Go Away," by Fred Pearce, *New Scientist*, May 28, 1994, p. 43.

15. This and other patronizing comments about Margulis can be found in "Lynn Margulis: Science's Unruly Earth Mother," by Charles Mann, *Science*, April 19, 1991, p. 378.

16. The works by Kauffman referred to in this section include "Antichaos and Adaptation," *Scientific American*, August 1991, pp. 78–84; *The Origins of Order*, Oxford University Press, New York, 1993; and *At Home in the Universe*, Oxford University Press, New York, 1995.

17. See my article "In the Beginning," *Scientific American*, February 1991, p. 123.

18. Brian Goodwin set forth his theory in *How the Leopard Changed Its Spots*, Charles Scribner's Sons, New York, 1994.

19. John Maynard Smith's disparaging remarks about the work of Per Bak and Stuart Kauffman were reported in *Nature*, February 16, 1995, p. 555. See also the insightful review of Kauffman's *Origins of Order* in *Nature*, October 21, 1993, pp. 704–706.

20. My February 1991 article in *Scientific American* (see note 17) reviewed the most prominent theories of the origin of life. I interviewed Stanley Miller at the University of California at San Diego in November 1990 and again by telephone in September 1995.

21. Stent, *Golden Age*, p. 71.

22. Crick's "miracle" comment can be found on page 88 of his book *Life Itself*, Simon and Schuster, New York, 1981.

Chapter Six The End of Social Science

1. I interviewed Edward Wilson at Harvard in February 1994. The works by Wilson alluded to this section include *Sociobiology*, Harvard University Press, Cambridge, 1975 (my citations refer to the abridged 1980 edition); *On*

Human Nature, Harvard University Press, Cambridge, 1978; *Genes, Mind, and Culture* (with Charles Lumsden), Harvard University Press, Cambridge, 1981; *Promethean Fire* (with Lumsden), Harvard University Press, Cambridge, 1983; *Biophilia*, Harvard University Press, Cambridge, 1984; *The Diversity of Life*, W. W. Norton, New York, 1993; and *Naturalist*, Island Press, Washington, D.C., 1994.

2. See "The Molecular Wars," chapter 12 of *Naturalist*, for a detailed account of this crisis in Wilson's career.

3. Wilson, *Sociobiology*, p. 300.

4. These travails are recounted in the chapter titled "The Sociobiology Controversy," in Wilson's *Promethean Fire*, pp. 23–50.

5. Wilson, *Promethean Fire*, pp. 48–49.

6. Christopher Wells, a biologist at the University of California at San Diego, made these remarks about the theories of Wilson and Lumsden in *The Sciences*, November/December 1993, p. 39.

7. My own view is that scientists have not been nearly as successful at explaining human behavior in genetic and Darwinian terms as Wilson seemed to believe. See my *Scientific American* articles "Eugenics Revisited," June 1993, pp. 122–131; and "The New Social Darwinists," October 1995, pp. 174–181.

8. Wilson, *Sociobiology*, pp. 300–301.

9. See *One Long Argument*, by Ernst Mayr, Harvard University Press, Cambridge, 1991. On page 149 Mayr wrote: "The architects of the evolutionary synthesis [of whom Mayr was one] have been accused by some critics of claiming that they had solved all the remaining problems of evolution. This accusation is quite absurd; I do not know of a single evolutionist who would make such a claim. All that was claimed by the supporters of the synthesis was that they had arrived at an elaboration of the Darwinian paradigm sufficiently robust not to be endangered by remaining puzzles." "Puzzles," it should be recalled, is the term that Thomas Kuhn employed to describe problems that occupy scientists involved in nonrevolutionary, "normal" science.

10. I found this quotation from Darwin's *The Descent of Man* in *The Moral Animal*, by Robert Wright, Pantheon, New York, 1994, p. 327. (Wright cites the

facsimile edition of *Descent*, Princeton University Press, Princeton, NJ, 1981, p. 73.) This book by Wright, a journalist associated with *The New Republic*, is by far the best I have read on recent scientific attempts to explain human nature in Darwinian terms.

11. Ibid., p. 328.

12. I met with Chomsky at MIT in February 1990. The remarks quoted up to this point came from that meeting. The remarks quoted hereafter stemmed from a telephone interview in February 1993. Chomsky's political essays have been collected in *The Chomsky Reader*, edited by James Peck, Pantheon, New York, 1987.

13. *The New Encyclopaedia Britannica*, 1992 *Macropaedia* edition, vol. 23, *Linguistics*, p. 45.

14. *Nature*, February 19, 1994, p. 521.

15. *Syntactic Structures*, Noam Chomsky, Mouton, The Hague, Netherlands, 1957. In 1995, Chomsky issued yet another book on linguistics, *The Minimalist Program*, MIT Press, Cambridge, which extended his earlier work on an innate, generative grammar. Like most of Chomsky's books on linguistics, this one is not easy to read. For a readable account of Chomsky's career in linguistics, see *The Linguistics Wars*, by Randy Allen Harris, Oxford University Press, New York, 1993.

16. Steven Pinker, who is also a linguist at MIT, nonetheless argued persuasively that Chomsky's work is best understood from a Darwinian viewpoint in *The Language Instinct*, William Morrow, New York, 1994.

17. *Language and the Problems of Knowledge*, Noam Chomsky, MIT Press, Cambridge, 1988, p. 159. Chomsky also spelled out his views on cognitive limits in this book.

18. Stent, *Golden Age*, p. 121.

19. "Thick Description: Toward an Interpretive Theory of Culture" can be found in Geertz's collection of essays, *The Interpretation of Cultures*, Basic Books, New York, 1973.

20. Ibid., p. 29.

21. *Works and Lives: The Anthropologist as Author*, Clifford Geertz, Stanford University Press, Stanford, 1988, p. 141.

22. "Deep Play" was collected in *The Interpretation of Cultures*. This quotation is found on page 412.

23. Ibid., p. 443.

24. I interviewed Geertz in person at the Institute for Advanced Study in May 1989 and again by telephone in August 1994.

25. *After the Fact*, Clifford Geertz, Harvard University Press, Cambridge, 1995, pp. 167–168.

Chapter Seven The End of Neuroscience

1. Crick has written an illuminating account of his career: *What Mad Pursuit*, Basic Books, New York, 1988. He spelled out his views on consciousness in *The Astonishing Hypothesis*, Charles Scribner's Sons, New York, 1994.

2. "Toward a Neurobiological Theory of Consciousness," Francis Crick and Christof Koch, *Seminars in the Neurosciences*, vol. 2, 1990, pp. 263–275.

3. I interviewed Crick at the Salk Institute in November 1991.

4. *The Double Helix*, James Watson, Atheneum, New York, 1968.

5. Crick, *What Mad Pursuit*, p. 9.

6. Crick, *Astonishing Hypothesis*, p. 3.

7. Edelman's books on the mind, published by Basic Books, New York, include *Neural Darwinism*, 1987; *Topobiology*, 1988; *The Remembered Present*, 1989; and *Bright Air, Brilliant Fire*, 1992. All these books, even the final one, which was intended to set forth Edelman's views in a popular format, are extremely difficult to read.

8. "Dr. Edelman's Brain," Steven Levy, *New Yorker*, May 2, 1994, p. 62.

9. "Plotting a Theory of the Brain," David Hellerstein, *New York Times Magazine*, May 22, 1988, p. 16.

10. See Crick's stinging review of Edelman's book *Neural Darwinism:* "Neural Edelmanism," *Trends in Neurosciences*, vol. 12, no. 7, 1989, pp. 240–248.

11. Daniel Dennett reviewed *Bright Air, Brilliant Fire* in *New Scientist*, June 13, 1992, p. 48.

12. Koch made this remark at "Toward a Scientific Basis for Consciousness," a meeting held in Tucson, Arizona, April 12–17, 1994.

13. Eccles has set forth his views in various publications, including *The Self and Its Brain*, cowritten with Karl Popper, Springer-Verlag, Berlin, 1977; *How the Self Controls Its Brain*, Springer-Verlag, Berlin, 1994; "Quantum Aspects of Brain Activity and the Role of Consciousness," cowritten with Friedrich Beck, *Proceedings of the National Academy of Science*, vol. 89, December 1992, pp. 11, 357–11, 361. My interview with Eccles took place by telephone in February 1993.

14. I interviewed Roger Penrose at the University of Oxford in August 1992. Penrose's two books on consciousness are *The Emperor's New Mind*, Oxford University Press, New York, 1989; and *Shadows of the Mind*, Oxford University Press, New York, 1994.

15. For criticism of *The Emperor's New Mind*, see *Behavioral and Brain Sciences*, vol. 13, no. 4, December 1990. This issue has multiple reviews of Penrose's book. For sharp critiques of *Shadows of the Mind*, see "Shadows of Doubt," by Philip Anderson (a prominent physicist), *Nature*, November 17, 1994, pp. 288–289; and "The Best of All Possible Brains," by Hilary Putnam (a prominent philosopher), *New York Times Book Review*, November 20, 1994, p. 7.

16. *The Science of the Mind*, Owen Flanagan, MIT Press, Cambridge, 1991. Daniel Dennett drew my attention to Flanagan's term.

17. "What Is It Like to Be a Bat?," by Thomas Nagel, can be found in *Mortal Questions*, Cambridge University Press, New York, 1979, a collection of Nagel's essays. This quotation is from page 166. In June 1992, I called Nagel to ask whether he thought science could ever end. Absolutely not, he replied. "The more you discover, the more questions there will be," he said. "Shakespearean criticism can never be complete," he added, "so why should physics be?"

18. I interviewed McGinn in New York City in August 1994. See McGinn's book *The Problem of Consciousness*, Blackwell Publishers, Cambridge, Mass., 1991, for a full discussion of his mysterian viewpoint.

19. *Consciousness Explained*, Daniel Dennett, Little, Brown, Boston, 1991. See also "The Brain and Its Boundaries," *London Times Literary Supplement*, May 10, 1991, in which Dennett attacks McGinn's mysterian

position. I spoke to Dennett about the mysterian viewpoint by telephone in April 1994.

20. "Toward a Scientific Basis for Consciousness" took place in Tucson, Arizona, April 12–17, 1994. It was organized by Stuart Hameroff, an anesthesiologist at the University of Arizona whose work on microtubules had influenced Roger Penrose's view of the role of quantum effects in consciousness. The meeting was thus dominated by speakers from the quantum-consciousness school of neuroscience. They included not only Roger Penrose but also Brian Josephson, a Nobel laureate in physics who has suggested that quantum effects can explain mystical and even psychic phenomena; Andrew Weil, a physician and authority on psychedelia who has asserted that a complete theory of the mind must take into account the ability of South American Indians who have ingested psychotropic drugs to experience shared hallucinations; and Danah Zohar, a New Age author who has declared that human thought stems from "quantum fluctuations of the vacuum energy of the universe," which "is really God." I described this meeting in "Can Science Explain Consciousness?" *Scientific American*, July 1994, pp. 88–94.

21. David Chalmers set forth his theory of consciousness in *Scientific American*, December 1995, pp. 80–86. In an accompanying article, Francis Crick and Christof Koch offered a rebuttal.

22. I interviewed Minsky at MIT in May 1993. During the interview, Minsky confirmed that in 1966 he had told an undergraduate student, Gerald Sussman, to design, as a summer project, a machine that could recognize objects, or "see." Needless to say, Sussman did not succeed (although he did go on to become a professor at MIT). Artificial vision remains one of the most profoundly difficult problems in artificial intelligence. For a critical look at artificial intelligence, see *AI: The Tumultuous History of the Search for Artificial Intelligence*, by Daniel Crevier, Basic Books, New York, 1993. See also Jeremy Bernstein's respectful profile of Minsky in the *New Yorker*, December 14, 1981, p. 50.

23. *The Society of Mind*, Marvin Minsky, Simon and Schuster, New York, 1985. The book is peppered with remarks that reveal Minsky's ambivalence

about the consequences of scientific progress. See, for example, the essay on page 68 titled "Self-Knowledge Is Dangerous," in which Minsky declares: "If we could deliberately seize control of our pleasure systems, we could reproduce the pleasure of success without the need for any actual accomplishment. And that would be the end of everything." Gunther Stent predicted that this type of neural stimulation would be rampant in the new Polynesia.

24. Stent, *Golden Age*, pp. 73–74.

25. Ibid., p. 74.

26. Gilbert Ryle coined the phrase "ghost in the machine" in his classic attack on dualism, *The Concept of Mind*, Hutchinson, London, 1949.

27. Henry Adams made this reference to Francis Bacon's materialistic outlook in *The Education of Henry Adams*, p. 484 (see note 3 to Chapter 1). According to Adams, Bacon "urged society to lay aside the idea of evolving the universe from a thought, and to try evolving thought from the universe."

Chapter Eight The End of Chaoplexity

1. The University of Illinois issued this press release on Mayer-Kress in November 1993. My story on Mayer-Kress's simulation of Star Wars, titled "Nonlinear Thinking," ran in *Scientific American*, June 1989, pp. 26–28. See also "Chaos in the International Arms Race," by Mayer-Kress and Siegfried Grossman, *Nature*, February 23, 1989, pp. 701–704.

2. Books that have followed in the wake of James Gleick's *Chaos: Making a New Science*, Penguin Books, New York, 1987, and show signs of its influence include *Complexity: The Emerging Science at the Edge of Order and Chaos*, M. Mitchell Waldrop, Simon and Schuster, New York, 1992; *Complexity: Life at the Edge of Chaos*, Roger Lewin, Macmillan, New York, 1992; *Artificial Life: A Report from the Frontier Where Computers Meet Biology*, Steven Levy, Vintage, New York, 1992; *Complexification: Explaining a Paradoxical World through the Science of Surprise*, John Casti, Harper-Collins, New York, 1994; *The Collapse of Chaos: Discovering Simplicity in a Complex World*, Jack Cohen and Ian Stewart, Viking, New York, 1994; and *Frontiers of Complexity: The Search for Order in a Chaotic World*, Peter Coveny and Roger Highfield, Fawcett Columbine, New

York, 1995. This final book covers much of the material covered by Gleick in *Chaos*, corroborating my point that popular treatments of chaos and complexity have virtually erased the distinction between them.

3. Gleick printed this quote from Poincaré in *Chaos*, p. 321.

4. *The Dreams of Reason*, Heinz Pagels, Simon and Schuster, New York, 1988. I quoted the blurb on the July 1989 paperback edition by Bantam.

5. *The Fractal Geometry of Nature*, Benoit Mandelbrot, W. H. Freeman, San Francisco, 1977, p. 423. The remark earlier in this paragraph that the Mandelbrot set is "the most complex object in mathematics" was made by the computer scientist A. K. Dewdney in *Scientific American*, August 1985, p. 16.

6. I watched Epstein demonstrate his artificial-society program during a workshop held at the Santa Fe Institute in May 1994 (which I describe in detail in the next chapter). I heard Epstein claim that computer models such as his would revolutionize social science during a one-day symposium at the Santa Fe Institute on March 11, 1995.

7. Holland made this claim in an unpublished paper that he sent me, titled "Objectives, Rough Definitions, and Speculations for Echo-Class Models." (The term *echo* refers to Holland's major class of genetic algorithms.) He reiterated the claim on page 4 of his book, *Hidden Order: How Adaptation Builds Complexity*, Addison-Wesley, Reading, Mass., 1995. Holland presented a succinct description of genetic algorithms in *Scientific American*, July 1992, pp. 66–72.

8. Yorke made this remark during a telephone interview in March 1995. Gleick's *Chaos* credited Yorke with having coined the term *chaos* in 1975.

9. See note 2.

10. See "Revisiting the Edge of Chaos," by Melanie Mitchell, James Crutchfield, and Peter Hraber, Santa Fe working paper 93-03-014. Coveny and Highfield's *Frontiers of Complexity*, cited in note 2, also mentioned the criticism of the edge-of-chaos concept.

11. At this writing, Seth Lloyd still had not published all his definitions of complexity. After I called him to ask about the definitions, he emailed the following list, which by my count includes not 31 definitions but 45. The

names that are used as modifiers or in parentheses refer to the main originators of the definition. For what it's worth, here is Lloyd's list, only slightly edited: information (Shannon); entropy (Gibbs, Boltzman); algorithmic complexity; algorithmic information content (Chaitin, Solomonoff, Kolmogorov); Fisher information; Renyi entropy; self-delimiting code length (Huffman, Shannon, Fano); error-correcting code length (Hamming); Chernoff information; minimum description length (Rissanen); number of parameters, or degrees of freedom, or dimensions; Lempel–Ziv complexity; mutual information, or channel capacity; algorithmic mutual information; correlation; stored information (Shaw); conditional information; conditional algorithmic information content; metric entropy; fractal dimension; self-similarity; stochastic complexity (Rissanen); sophistication (Koppel, Atlan); topological machine size (Crutchfield); effective or ideal complexity (Gell-Mann); hierarchical complexity (Simon); tree subgraph diversity (Huberman, Hogg); homogeneous complexity (Teich, Mahler); time computational complexity; space computational complexity; information-based complexity (Traub); logical depth (Bennett); thermodynamic depth (Lloyd, Pagels); grammatical complexity (position in Chomsky hierarchy); Kullbach-Liebler information; distinguishability (Wooters, Caves, Fisher); Fisher distance; discriminability (Zee); information distance (Shannon); algorithmic information distance (Zurek); Hamming distance; long-range order; self-organization; complex adaptive systems; edge of chaos.

12. See chapter 3 of *The Quark and the Jaguar*, W. H. Freeman, New York, 1994, in which Murray Gell-Mann, a Nobel laureate in physics and one of the founders of the Santa Fe Institute, described Gregory Chaitin's algorithmic information theory and other approaches to complexity. Gell-Mann acknowledged on page 33 that "any definition of complexity is necessarily context-dependent, even subjective."

13. The conference on artificial life held at Los Alamos in 1987 was vividly described in Steven Levy's *Artificial Life*, cited in note 2.

14. Editor's introduction, Christopher Langton, *Artificial Life*, vol. 1, no. 1, 1994, p. vii.

15. See note 2 for full citations.

16. "Verification, Validation, and Confirmation of Numerical Models in the Earth Sciences," by Naomi Oreskes, Kenneth Belitz, and Kristin Shrader Frechette, was published in *Science*, February 4, 1994, pp. 641–646. See also the letters reacting to the article, which were published on April 15, 1994.

17. Ernst Mayr discussed the inevitable imprecision of biology in *Toward a New Philosophy of Biology*, Harvard University Press, Cambridge, 1988. See in particular the chapter entitled "Cause and Effect in Biology."

18. I interviewed Bak in New York City in August 1994. For an introduction to Bak's work, see "Self-Organized Criticality," *Scientific American*, by Bak and Kan Chen, January 1991, pp. 46–53.

19. *Earth in the Balance*, Al Gore, Houghton Mifflin, New York, 1992, p. 363.

20. See "Instabilities in a Sandpile," by Sidney R. Nagel, *Reviews of Modern Physics*, vol. 84, no. 1, January 1992, pp. 321–325.

21. A discussion of Leibniz's belief in an "irrefutable calculus" that could solve all problems, even theological ones, can be found in the excellent book *Pi in the Sky*, by John Barrow, Oxford University Press, New York, 1992, pp. 127–129.

22. *Cybernetics*, by Norbert Wiener, was published in 1948 by John Wiley and Sons, New York.

23. John R. Pierce made this comment about cybernetics on page 210 of his book *An Introduction to Information Theory*, Dover, New York, 1980 (originally published in 1961).

24. Claude Shannon's paper, "A Mathematical Theory of Communications," was published in the *Bell System Technical Journal*, July and October, 1948.

25. I interviewed Shannon at his home in Winchester, Mass., in November 1989. I also wrote a profile of him for *Scientific American*, January 1990, pp. 22–22b.

26. This glowing review of Thom's book appeared in the *London Times Higher Education Supplement*, November 30, 1973. I found the reference in *Searching for Certainty*, by John Casti, William Morrow, New York, 1990, pp. 63–64. Casti, who has written a number of excellent books on mathematics-

related topics, is associated with the Santa Fe Institute. The English translation of Thom's book *Structural Stability and Morphogenesis*, originally issued in French in 1972, was published in 1975 by Addison-Wesley, Reading, Mass.

27. These negative comments about catastrophe theory were reprinted in Casti, *Searching for Certainty*, p. 417.

28. *Chance and Chaos*, David Ruelle, Princeton University Press, Princeton, N.J., 1991, p. 72. This book is a quiet but profound meditation on the meaning of chaos by one of its pioneers.

29. "More Is Different," Philip Anderson, *Science*, August 4, 1972, p. 393. This essay is reprinted in a collection of papers by Anderson, *A Career in Theoretical Physics*, World Scientific, River Edge, N.J., 1994. I interviewed Anderson at Princeton in August 1994.

30. As quoted in "The Man Who Knows Everything," by David Berreby, *New York Times Magazine*, May 8, 1994, p. 26.

31. I first described my encounter with Gell-Mann in New York City, which took place in November 1991, in *Scientific American*, March 1992, pp. 30–32. I interviewed Gell-Mann at the Santa Fe Institute in March 1995.

32. See note 12.

33. "Welcome to Cyberia: Notes on the Anthropology of Cyberculture," Arturo Escobar, *Current Anthropology*, vol. 35, no. 3, June 1994, p. 222.

34. *Order out of Chaos*, Ilya Prigogine and Isabelle Stengers, Bantam, New York, 1984 (originally published in French in 1979).

35. Ibid., 299–300.

36. Ruelle, *Chance and Chaos*, p. 67.

37. Feigenbaum's two papers were "Presentation Functions, Fixed Points, and a Theory of Scaling Function Dynamics," *Journal of Statistical Physics*, vol. 52, nos. 3/4, August 1988, pp. 527–569; and "Presentation Functions and Scaling Function Theory for Circle Maps," *Nonlinearity*, vol. 1, 1988, pp. 577–602.

Chapter Nine The End of Limitology

1. The meeting titled "The Limits to Scientific Knowledge" was held May 24–26, 1994, at the Santa Fe Institute.

2. See Chaitin's articles "Randomness in Arithmetic," *Scientific American*, July 1988, pp. 80–85; and "Randomness and Complexity in Pure Mathematics," *International Journal of Bifurcation and Chaos*, vol. 4, no. 1, 1994, pp. 3–15. Chaitin has also distributed a book called *The Limits of Mathematics* on the Internet. Other relevant publications by participants at this meeting (listed alphabetically by author, and not including publications already cited) are W. Brian Arthur, "Positive Feedbacks in the Economy," *Scientific American*, February 1990, pp. 92–99; John Casti, *Complexification*, HarperCollins, New York, 1994; Ralph Gomory, "The Known, the Unknown, and the Unknowable," *Scientific American*, June 1995, p. 120; Rolf Landauer, "Computation: A Fundamental Physical View," *Physica Scripta*, vol. 35, pp. 88–95, and "Information Is Physical," *Physics Today*, May 1991, pp. 23–29; Otto Rössler, "Endophysics," *Real Brains, Artificial Minds*, edited by John Casti and A. Karlqvist, North Holland, New York, 1987, pp. 25–46; Roger Shepard, "Perceptual-Cognitive Universals as Reflections of the World," *Psychonomic Bulletin and Review*, vol. 1, no. 1, 1994, pp. 2–28; Patrick Suppes, "Explaining the Unpredictable," *Erkenntis*, vol. 22, 1985, pp. 187–195; Joseph Traub, "Breaking Intractability" (cowritten with Henry Wozniakowski), *Scientific American*, January 1994, pp. 102–107. I discussed some of the mathematics-related topics that arose at the Santa Fe meeting in "The Death of Proof," *Scientific American*, October 1993, pp. 92–103. One of the best books I have read recently on the limits of knowledge is *Fire in the Mind*, by George Johnson, Alfred A. Knopf, New York, 1995.

3. I interviewed Chaitin on the Hudson River in September 1994.

4. *The End of History and The Last Man*, Francis Fukuyama, The Free Press, 1992.

5. I wrote a profile of Brian Josephson for *Scientific American*, "Josephson's Inner Junction," May 1995, pp. 40–41.

Chapter Ten Scientific Theology, or The End of Machine Science

1. *The World, the Flesh and the Devil*, J. D. Bernal, Indiana University Press, Bloomington, 1929, p. 47. I am indebted to Robert Jastrow of

Dartmouth College for sending me a copy of Bernal's essay.

2. *Mind Children*, Hans Moravec, Harvard University Press, Cambridge, 1988. My interview with Moravec took place in December 1993.

3. *Infinite in All Directions*, Freeman Dyson, Harper and Row, New York, 1988, p. 298.

4. Dyson's romantic view of science brings him to the verge of radical relativism. See his essay "The Scientist as Rebel," *New York Review of Books*, May 25, 1995, p. 31.

5. "Time without End: Physics and Biology in an Open Universe," Freeman Dyson, *Reviews of Modern Physics*, vol. 51, 1979, pp. 447–460.

6. Dyson, *Infinite in All Directions*, p. 196.

7. Ibid., p. 115.

8. Ibid., pp. 118–119.

9. The fascinating career of Edward Fredkin (as well as those of Edward Wilson and the late economist Kenneth Boulding) is described in *Three Scientists and Their Gods*, by Robert Wright, Pantheon, New York, 1988. I interviewed Fredkin by telephone in May 1993.

10. *The Physics of Immortality*, Frank Tipler, Doubleday, New York, 1994.

11. *The Anthropic Cosmological Principle*, Frank Tipler and John Barrow, Oxford University Press, New York, 1986.

12. Teilhard de Chardin discussed the question of how extraterrestrials might be saved in the chapter titled "A Sequel to the Problem of Human Origins: The Plurality of Inhabited Worlds," in *Christianity and Evolution*, Harcourt Brace Jovanovich, New York, 1969.

13. *The Limits of Science*, Peter Medawar, Oxford University Press, New York, 1984, p. 90.

Epilogue The Terror of God

1. See "The Mystery at the End of the Universe," chapter 9 of *The Mind of God*, by Paul C. Davies, Simon and Schuster, New York, 1992.

2. See "Borges and I," *A Personal Anthology*, by Jorge Luis Borges,

Grove Press, New York, 1967, pp. 200–201.

3.See note 9 of the chapter titled "Mysticism" in *The Varieties of Religious Experience*, by William James, Macmillan, New York, 1961 (James's book was originally published in 1902).

4. See *The Philosophy of Charles Hartshorne*, edited by Lewis Edwin Hahn, Library of Living Philosophers, La Salle, Ill., 1991. I spoke to Hartshorne in May 1993.

5. *Tractatus Logico-Philosophicus*, Ludwig Wittgenstein, Routledge, New York, 1990 edition, p. 187. Wittgenstein's cryptic book was originally published in 1922.

Selected Bibliography

Barrow, John, *Theories of Everything*, Clarendon Press, Oxford, U.K., 1991.

Barrow, John, *Pi in the Sky*, Oxford University Press, New York, 1992.

Bloom, Harold, *The Anxiety of Influence*, Oxford University Press, New York, 1973.

Bohm, David, *Wholeness and the Implicate Order*, Routledge, New York, 1980.

Bohm, David, and F. David Peat, *Science, Order and Creativity*, Bantam Books, New York, 1987.

Borges, Jorge Luis, *A Personal Anthology*, Grove Press, New York, 1967.

Casti, John, *Searching for Certainty*, William Morrow, New York, 1990.

Casti, John, *Complexification*, HarperCollins, New York, 1994.

Chomsky, Noam, *Language and the Problems of Knowledge*, MIT Press, Cambridge, Mass., 1988.

Coveny, Peter, and Roger Highfield, *Frontiers of Complexity*, Fawcett Columbine, New York, 1995.

Crick, Francis, *Life Itself*, Simon and Schuster, New York, 1981.

Crick, Francis, *What Mad Pursuit*, Basic Books, New York, 1988.

Crick, Francis, *The Astonishing Hypothesis*, Charles Scribner's Sons, New York, 1994.

Davies, Paul C, *The Mind of God*, Simon and Schuster, New York, 1992.

Dawkins, Richard, *The Blind Watchmaker*, W. W. Norton, New York, 1986.

Dennett, Daniel, *Consciousness Explained*, Little, Brown, Boston, 1991.

Dyson, Freeman, *Infinite in All Directions*, Harper and Row, New York, 1988.

Edelman, Gerald, *Neural Darwinism*, Basic Books, New York, 1987.

Edelman, Gerald, *Topobiology*, Basic Books, New York, 1988.

Edelman, Gerald, *The Remembered Present*, Basic Books, New York, 1989.

Edelman, Gerald, *Bright Air, Brilliant Fire*, Basic Books, New York, 1992.

Feyerabend, Paul, *Against Method*, Verso, London, 1975.

Feyerabend, Paul, *Farewell to Reason*, Verso, London, 1987.

Feyerabend, Paul, *Killing Time*, University of Chicago Press, Chicago, 1995.

Fukuyama, Francis, *The End of History and The Last Man*, The Free Press, New York, 1992.

Geertz, Clifford, *The Interpretation of Cultures*, Basic Books, New York, 1973.

Geertz, Clifford, *Works and Lives*, Stanford University Press, Stanford, Calif., 1988.

Geertz, Clifford, *After the Fact*, Harvard University Press, Cambridge, Mass., 1995.

Gell-Mann, Murray, *The Quark and the Jaguar*, W. H. Freeman, New York, 1994.

Gleick, James, *Chaos: Making a New Science*, Penguin Books, New York, 1987.

Gould, Stephen Jay, *Wonderful Life*, W. W. Norton, New York, 1989.

Harwit, Martin, *Cosmic Discovery*, MIT Press, Cambridge, Mass. 1981.

Hawking, Stephen, *A Brief History of Time*, Bantam Books, New York, 1988.

Holton, Gerald, *Science and Anti-Science*, Harvard University Press, Cambridge, Mass., 1993.

Hoyle, Fred, *Home Is Where the Wind Blows*, University Science Books, Mill Valley, Calif., 1994.

Hoyle, Fred, and Chandra Wickramasinghe, *Our Place in the Cosmos*, J. M. Dent, London, 1993.

Johnson, George, *Fire in the Mind*, Knopf, New York, 1995.

Kauffman, Stuart, *The Origins of Order*, Oxford University Press, New York, 1993.

Kauffman, Stuart, *At Home in the Universe*, Oxford University Press, New York, 1995.

Kuhn, Thomas, *The Structure of Scientific Revolutions*, University of Chicago Press, Chicago, 1962.

Levy, Steven, *Artificial Life*, Vintage, New York, 1992.

Lewin, Roger, *Complexity*, Macmillan, New York, 1992.

Lindley, David, *The End of Physics*, Basic Books, New York, 1993.

Mandelbrot, Benoit, *The Fractal Geometry of Nature*, W. H. Freeman, San Francisco, 1977.

Margulis, Lynn, *Symbiosis in Cell Evolution*, W. H. Freeman, New York, 1981.

Margulis, Lynn, and Dorion Sagan, *What Is Life?*, Peter Nevraumont, Inc., New York, 1995.

Mayr, Ernst, *Toward a New Philosophy of Biology*, Harvard University Press, Cambridge, Mass., 1988.

Mayr, Ernst, *One Long Argument*, Harvard University Press, Cambridge, Mass., 1991.

McGinn, Colin, *The Problem of Consciousness*, Blackwell, Cambridge, Mass., 1991.

McGinn, Colin, *Problems in Philosophy*, Blackwell, Cambridge, Mass., 1993.

Minsky, Marvin, *The Society of Mind*, Simon and Schuster, New York, 1985.

Moravec, Hans, *Mind Children*, Harvard University Press, Cambridge, Mass., 1988.

Overbye, Dennis, *Lonely Hearts of the Cosmos*, HarperCollins, New York, 1992.

Pagels, Heinz, *The Dreams of Reason*, Simon and Schuster, New York, 1988.

Penrose, Roger, *The Emperor's New Mind*, Oxford University Press, New York, 1989.

Penrose, Roger, *Shadows of the Mind*, Oxford University Press, New York, 1994.

Popper, Karl, and John C. Eccles, *The Self and Its Brain*, Springer-Verlag, Berlin, 1977.

Popper, Karl, *Unended Quest*, Open Court, La Salle, Ill., 1985.

Popper, Karl, *Popper Selections*, edited by David Miller, Princeton University Press, Princeton, N.J., 1985.

Prigogine, Ilya, *From Being to Becoming*, W. H. Freeman, New York, 1980.

Prigogine, Ilya, and Isabelle Stengers, *Order out of Chaos*, Bantam, New York, 1984 (originally published in French in 1979).

Rescher, Nicholas, *Scientific Progress*, Basil Blackwell, Oxford, U.K., 1978.

Rescher, Nicholas, *The Limits of Science*, University of California Press, Berkeley, 1984.

Ruelle, David, *Chance and Chaos*, Princeton University Press, Princeton, N.J., 1991.

Selve, Richard Q., editor, *The End of Science? Attack and Defense*, University Press of America, Lanham, Md., 1992.

Stent, Gunther, *The Coming of the Golden Age*, Natural History Press, Garden City, N.Y., 1969.

Stent, Gunther, *The Paradoxes of Progress*, W. H. Freeman, San Francisco, 1978.

Tipler, Frank, *The Physics of Immortality*, Doubleday, New York, 1994.

Tipler, Frank, and John Barrow, *The Anthropic Cosmological Principle*, Oxford University Press, New York, 1986.

Waldrop, Mitchell, *Complexity*, Simon and Schuster, New York, 1992.

Weinberg, Steven, *Dreams of a Final Theory*, Pantheon, New York, 1992.

Wheeler, John, and Wojciech H. Zurek, editors, *Quantum Theory and Measurement*, Princeton University Press, Princeton, N.J., 1983.

Wheeler, John, *At Home in the Universe*, American Institute of Physics Press, Wood-bury, N.Y., 1994.

Wilson, Edward O., *Sociobiology*, Harvard University Press, Cambridge, Mass., 1975.

Wilson, Edward O., *On Human Nature*, Harvard University Press, Cambridge, Mass., 1978.

Wilson, Edward O., and Charles Lumsden, *Genes, Mind and Culture*, Harvard University Press, Cambridge, Mass., 1981.

Wilson, Edward O., and Charles Lumsden, *Promethean Fire*, Harvard University Press, Cambridge, Mass., 1983.

Wilson, Edward O., *Naturalist*, Island Press, Washington, D.C., 1994.

Wright, Robert, *Three Scientists and Their Gods*, Times Books, New York, 1988.

Wright, Robert, *The Moral Animal*, Pantheon, New York, 1994.

Index

end of, 6–13, 28, 29, 32, 46, 60, 64–65, 101, 135, 189, 240, 245, 260, 263,264, 267, 291, 292, 293, 295–297, 305

as finite/infinite, 17, 235

funding for, 23, 66, 67, 76, 296, 300, 305, 306

future of, 16, 18, 24–26, 203, 219, 222, 234, 235, 245, 248, 255, 260, 263, 266, 297, 307

hierarchy of, 78, 232

and history, 136, 265

and humanities, 60, 163, 236, 238

as incomprehensible, 8, 262

as infinite/finite, 20, 23, 26, 272

ironic, 30 (*see also* Ironic science)

language of, 219

limits of, 6, 10, 27, 45, 190, 237, 245, 260–262, 284

of measurement, 17

and mysteries, 165, 198, 207, 229

and nation-states, 21, 226, 266

vs. opinion, 141

paradox in modern science, 195

past achievements of, 26

as political, 44–45,60

postempirical, 104, 206

as provisional, 59

pseudo-science, 141

pure, 13, 22, 24, 224, 268, 295

and religion, 126

scientific method, 53, 60, 110

strong scientists, 30, 142

success of, 166

and theory of the human mind, 181, 256

See also Astronomy; Biology; Cosmology; Ironic science; Physics; Quantum mechanics; Relativity theory; Social sciences

科学的终结

孙雍君　张武军／译

第一章
进步的终结

1989年，就在我和罗杰·彭罗斯（Roger Penrose）于锡拉丘兹晤面之后
仅一个月，明尼苏达州的古斯塔夫·阿道夫大学（Gustavus Adolphus
College）召开了一次专题讨论会，会议题目"科学的终结？"既易引起争
议又易让人产生误解，其实它的主题乃是：科学的信仰——而不是科学本
身——正在走向终结。正如一位会议组织者所指出的："我有一种日益强烈
的感觉，即科学作为一种统一、普遍而又客观的追求，已经完结了。"[1]会议
上的发言者大多是哲学家，他们过去都曾以这样或那样的方式对科学的极限
发出过诘难。会上最具讽刺意味的是一位科学家的发言。他叫岗瑟·斯滕特
（Gunther Stent），加利福尼亚大学伯克利分校的一位生物学家，在好几年
前就开始散布一种比这次讨论会的主题更为惹人瞩目的主张，他宣称科学本
身将走向终结，这并不是因为几个学院派诡辩家的怀疑态度，刚好相反，科
学走向终结是因为其出色的成就。

斯滕特绝不是所谓的"半吊子"学者，而是分子生物学领域的一位先
驱。他于20世纪50年代在伯克利创建了第一个分子生物学系，并用实验阐明
了遗传机制。后来，他的研究兴趣由遗传学转向科学，他被任命为美国科学
院神经生物学部主任。斯滕特也是我所见过的科学限度探究者中最敏锐的一
位——"敏锐"一词，当然是指他明确表述了我的朦胧的不祥预感。20世纪
60年代末期，在席卷伯克利的沉重抗议活动声浪中，他写了一本具有惊人预
见性的著作，此书现在早已绝版，书名为《黄金时代的来临——进步之终结
概论》（下文简称《黄金时代的来临》）。该书出版于1969年，核心思想是
科学——还有技术、艺术以及一切进步的、累积的事业——正走向终结。[2]

斯滕特承认，许多人都认为科学会很快终结的想法是荒谬的。20世

里，科学一直在迅猛发展，怎么可能会走向终结呢？斯滕特反复思索着这一归纳论证，指出科学最初在正反馈效应的作用下，确实是呈指数级增长的：知识产生更多的知识，力量导致更大的力量。斯滕特相信美国历史学家亨利·亚当斯（Henry Adams）在20世纪初就已经预见到了科学的这一方面。[3]

斯滕特指出，亚当斯的加速度理论会导出一个有趣的推论。如果科学的确存在着任何限度、任何进一步发展的障碍的话，那么科学在撞上它们之前，更会以一种前所未有的速度发展。在科学看起来特别强劲、成功、有效的时候，也许正是它濒于死亡的时候。斯滕特在《黄金时代的来临》中写道："确实，当前令人目眩的进步速度，看起来会很快使进步走向终点，我们——也可能是我们之后的一两代人——将会亲眼看到这一天的到来。"[4]

斯滕特认为，某些特定的科学领域明显地受制于其研究对象的有限性，像人体解剖学和地理学，没有人会认为它们是无止境的事业；化学也是如此，"尽管可能的化学反应总数是十分庞大的，且反应所经历的过程种类繁多，但化学的目标是理解决定这些分子行为方式的规律，这一目标就像地理学的目标一样，显然是有限的。[5]化学的这一目标已经在20世纪30年代达到了（尽管存在着争议），当时化学家鲍林（Linus Pauling）阐明：所有化学反应都可以用量子力学的术语加以解释。[6]

斯滕特断言，在他自己的生物学领域，1953年对DNA双螺旋结构的发现以及随后对遗传密码的破译，已经解决了遗传信息代际传递的基本问题，生物学家只剩下三个重大问题尚需探讨：生命怎样发生，单个的受精卵是如何发育成多细胞生物的，中枢神经系统怎样加工信息。这三个目标实现后，斯滕特认为，生物学（纯生物学）的基本使命也就完成了。

斯滕特承认，在原则上，生物学家们仍可继续探索特殊的生命现象，并不断地贡献出自己的知识。但根据达尔文的理论，科学并非起源于我们探究真理的欲望，而是起源于人类控制环境以便增大我们的基因传播可能性的驱动力。当某一给定的科学领域产生的成就越来越少时，科学家坚持其探索的刺激也就越来越弱，而社会给予的支持也会越来越少。

斯滕特更进一步指出，就算生物学家完成了其经验研究的使命，也并不意味着他们已经解答了所有的相关问题。例如，绝不会有任何一种纯生理学的理论能真正地解释意识，因为"要观察这一完全与个人经历相关的过程，似乎只能深入到非常普通的日常反应中去，这与观察肝脏中发生的反应过程一样，毫无趣味……"。[7]

与生物学不同，斯滕特在其书中认为物理科学似乎是永无止境的，物理学家通过粒子之间相互撞击不断增加的能量，能够愈益深入地探索物质的内部结构；而天文学家也总能努力在宇宙中观测得越来越远。但在其收集日益幽深的微观世界的数据时，物理学家们将不可避免地受困于各种物理的、经济的，甚或认知的局限性。

20世纪以来，物理学正变得越来越难以理解，它已经超出了达尔文主义者的认识论，超出了我们固有的、用以把握世界的观念范围。斯滕特抛弃了"昨日的胡言乱语正是今天的常识"[8]这种陈词滥调，认为只要物理学还具有产生新技术（如核武器、核能等）的潜力，社会就会支持它继续研究下去。但是斯滕特预言，一旦物理学在可理解性之外，变得更缺乏实用性的话，社会肯定会取消对它的支持。

斯滕特对未来的预测，是乐观主义与悲观主义的奇特大杂烩。他预见科学在结束之前能解决许多现代文明的紧迫难题，它能够消除疾病和贫穷，能为社会提供廉价而又无污染的能源（可能是通过对原子核聚合反应的利用而实现）。然而，在获得愈益巨大的支配自然能力的同时，我们可能会丧失所谓的"权力意志"（套用尼采的术语），失去从事进一步研究的动机——特别是当这些研究不产生有形的利益时。

随着社会变得愈益富裕和舒适，选择日益艰难的科学（甚或是艺术）之路的年轻人越来越少，多数人却可能转向更注重享乐的追求，甚至会因为沉湎于吸毒或植入大脑的电子器件所带来的虚幻世界，而离弃真实的世界。斯滕特归结道：总有一天，进步会"倒毙于途"，留下一个庞大但缺乏生机的世界——他称之为"新波利尼西亚"（the New Polynesia）。他认为"垮掉的一代"[①]和"嬉皮士"[②]的出现，标志着进步之终结的开端，显露出"新波利尼西亚"的曙光。他以一段略带嘲讽的评论结束了全书："成千上万个兴致勃勃的艺术家和精力充沛的科学家，最终将把生活的悲喜剧变成毫无意义的舞台演出。"[9]

伯克利之行

1992年春天，我去伯克利探访斯滕特，想知道经过这些年的风雨变迁之

① beatniks，20世纪50年代末出现于美国知识阶层中的一个颓废流派——避世派，以蓄长发、着奇装、吸毒、反对世俗陈规、排斥温情、强调"个性自我表达"等为特征——译者注。

② hippies，流行于20世纪60年代末的美国流派——嬉皮士，当时又称花儿少年，鼓吹爱与和平，反对越战，是消极的和平主义者——译者注。

后，他对自己当初的预言有什么感想。[10]我从下榻的旅馆出来，漫步走向伯克利大学校园，沿途仍时时可见20世纪60年代的遗风：披散着灰色长发、身着敞衫的男男女女，不断伸手索要着小费。进入校园后，我一路打听着来到学校的生物楼前，那是一座粗笨的混凝土结构建筑物，悄然蹩伏在重重桉树的阴影之中。我从一楼乘电梯而上，直奔斯滕特的实验室，却发现门锁着。几分钟之后，走廊尽头的电梯门静静滑开，斯滕特头戴黄色自行车头盔，推着一辆脏兮兮的山地车走了出来，脸色通红，浑身是汗。

斯滕特青年时代就从德国移居美国，但其生硬的语调和服装仍然带着德国味。他戴一副金丝边眼镜，穿着一件带肩饰的蓝色短袖衬衫、深色便裤，足下是一双黑亮的皮鞋。斯滕特引我穿过实验室，里面塞满显微镜、离心机，以及各种各样的科研用玻璃制品，进入后面的一间小办公室，办公室外厅饰满佛教照片和画像。斯滕特随手关上办公室的门，我发现门背后钉着一张1989年古斯塔夫·阿道夫大学研讨会的招贴画，画面上半部是硕大而又俗艳的彩色字母组成的"SCIENCE"（科学）一词，每一个字母似乎都在融化，正一滴滴地滴入一池荧荧闪光的原浆之中，梦幻般的池子上面，几个黑色的大字发问着：科学的终结？

在正式采访开始时，斯滕特显得有些忧心忡忡。他毫不掩饰锋芒地问我是否想重蹈某些人的覆辙，如新闻记者珍妮·马尔科姆（Janet Malcolm）。因为当时珍妮正为一篇人物专访而陷入与传主——精神分析学家杰佛里·梅森（Jeffrey Masson）——冗长的法律纠纷中，并在一审惨遭败诉。我含糊地发了几句议论，认为马尔科姆的侵权行为并不严重，不足以招致法律责任，但她的工作方式的确不够细心。我告诉斯滕特，如果我要写些什么去批评像梅森那样易被触怒的人的话，我敢肯定自己引用的每句话都能在录音带上找到。（在我说这些话时，我的录音机就在我俩之间悄然转动着。）

慢慢地，斯滕特放松下来，并开始向我讲述他的一生：1924年他生于柏林的一个犹太家庭，1938年逃离德国，和一位姐姐移居芝加哥。他在伊利诺伊大学获得化学博士学位，但在阅读了薛定谔（Erwin Schrödinger）的著作《生命是什么？》之后，忽然对遗传的奥秘入了迷。他曾在加州理工学院与著名生物物理学家马克斯·德尔布吕克（Max Delbrück）合作过一段时间，随后，于1952年获伯克利大学教授职位。斯滕特谈道："在研究分子生物学的这段早期岁月里，我们当中没有一个人知道自己正在干什么，后来，沃森和克里克发现了双螺旋结构。几个星期之后我们就意识到：原来自己在从事分子生物学的研究。"

斯滕特对科学之限度的思索始于20世纪60年代，部分原因是为了响应伯克利的"言论自由运动"，因为这一运动向他一向信奉的西方理性主义价值观、技术进步观以及文明的其他方面提出了挑战。校方委托斯滕特筹组一个委员会，通过与学生对话，"妥善处理这件事，把事情平息下来"。为了完成这一使命，同时也为了解决自己作为科学家这一角色而引发的内心冲突，斯滕特发表了一系列演讲，这些演讲汇集起来就成了《黄金时代的来临》一书。

我告诉斯滕特，在读过《黄金时代的来临》一书后，仍不明白他是否相信"新波利尼西亚"——社会与智力停滞后的普遍闲适时代——代替现在的状态是一种进步。"对此我永远无法确定！"他答道，并显出十分难过的样子，"人们都称我为悲观主义者，但我却认为自己是个乐观主义者。"无论在何种意义上，他都不认为这样的社会是乌托邦的，因为他相信在经历了20世纪由极权主义国家导致的灾难之后，已不可能严肃地对待乌托邦思想了。

斯滕特觉得自己的预见已经得到了十分合理的体现。虽然"嬉皮士"已经消失了（伯克利大街上那些可怜的遗俗除外），但美国文化正变得越来越重实利和反知识。"嬉皮士"演变成了"雅皮士"[①]，而且冷战毕竟也已经结束了，尽管并非像斯滕特所设想的那样，通过社会主义国家与资本主义国家的逐步融合而实现。他承认自己并未预见到冷战结束后，被长期压抑的种族冲突会再度复苏。"我对正在巴尔干半岛发生的事情感到非常遗憾，"他说，"我没想到会发生这样的事。"斯滕特也为美国至今仍然存在着贫困和种族冲突而感到诧异，但他相信这些问题的严重性终将大大降低。（我不由地想：啊哈！他终于表现出了作为乐观主义者的一面。）

斯滕特相信，科学正不断显现出他在《黄金时代的来临》中所预言的"终结"征兆：粒子物理学家越来越难以让社会支持其日益昂贵的实验，如建造超导超级对撞机；至于生物学家，他们仍面临许多有待进一步了解的东西，比如说，受精卵怎样发育成复杂的多细胞生物体（如大象），以及大脑的工作机理等。"但我认为大的架构已基本完成了，"他说，"特别是进化生物学，在达尔文发表《物种起源》时就已基本完成了。"某些进化生物学家——特别是像哈佛大学的爱德华·威尔逊（Edward Willson）——竟天真地认为，通过彻底地逐个物种来考察地球上所有的生命，他们就可以永久地

① yuppies，在《雅皮士手册》里，毕斯曼和哈特里用这一术语形容"属于中上阶层的专业人士"，系"Young Urban Professionals"的首字母缩拼加词尾而成——译者注。

保住饭碗，但斯滕特对此嗤之以鼻，因为这样的事业就像是毫无意义的"玻璃念珠游戏"（glass bead game）。

随后他开始尖刻地嘲讽环境决定论①，认为它在本质上是一种反人类的哲学，使得美国青年尤其是黑人穷孩子缺乏上进心。我警觉到这位我最赞赏的卡珊德拉②已情难自禁地显露出自己的坏脾气，只好改变话题，问他是否仍然坚信意识是一个无法解决的科学难题，就像他在《黄金时代的来临》中所主张的那样。斯滕特回答说，自己很佩服弗朗西斯·克里克（Francis Crick）在研究生涯后期把注意力转向了意识问题，如果连克里克都觉得意识能被科学攻克，就必须认真对待那种可能性。

但斯滕特仍然相信，如果像许多人所主张的那样，只从纯生理学的角度去解释意识，就不可能导致清晰而有意义的结论，更无助于我们去解决道德和伦理的问题。他认为科学的进步在将来可能会给宗教界定出一个更明确的地位，而不是像许多科学家曾期望的那样彻底消除宗教。

当我问他计算机能否具有智慧并创造出它们自己的科学时，斯滕特嘲弄地笑了笑。他对于人工智能抱有怀疑的态度，特别是针对那些毫不现实的鼓吹者。他指出，计算机在解决那些被严格限定了的任务（如数学或国际象棋）方面确实具有优越性，但当它面对人们不费吹灰之力就能解决的那类问题（诸如辨认一张面孔、一种声音，或在拥挤的人行便道上行走）时，仍是一筹莫展。马文·明斯基（Marwin Minsky）等人曾预言，我们人类会在将来把个性这个沉重的包袱卸给计算机。斯滕特认为："这完全是一派胡言。我并不排除到23世纪人类会拥有一种人工大脑的可能性，但它却绝不会具有体验。"你也许能设计出一台计算机，并使它成为饭店里的高级厨师，"但这台机器永远也不会知道牛排的味道。"

混沌与复杂性的研究者声称，在计算机和先进数学工具的帮助下，他们会实现对既有科学的超越，斯滕特对此同样持怀疑态度。在《黄金时代的来临》一书中，他曾讨论过混沌理论先驱者之一伯努瓦·芒德勃罗（Benoit Mandelbrot）的工作。自20世纪60年代初期开始，芒德勃罗就发现了许多现

---

① environmentalism，本是地理学用语，强调地理环境对人类社会活动的影响，由公元前5世纪的希波克拉底提出，后扩展到生物学、心理学等领域，强调社会环境对个体的决定性影响——译者注。

② Cassandra，原指希腊神话中特洛伊的公主，能预卜吉凶，后借指预言者，这里指斯滕特——译者注。

象具有内在的不确定性：它们展现出的行为是不可预言的，貌似无规律的，科学家只能猜测单个事件的原因，但不能精确地作出预言。

斯滕特讲道，混沌与复杂性的研究者试图就芒德勃罗研究过的那些现象，建立起有效的、可理解的理论。在《黄金时代的来临》中他曾总结道，这些不确定现象是拒斥科学分析的，现在仍不具备任何改变这一论断的理由。恰恰相反，这些领域近来的工作证实了他的一个论点：当科学被推进得过于深远时，往往也就语无伦次到了极点。这样说来，斯滕特认为混沌和复杂性的研究是不会导致科学的新生了？"是的，"他得意地答道，同时露齿一笑，"那只能导致科学的终结。"

科学到底成就了什么

很显然，我们并非身处在斯滕特所预想的新波利尼西亚的边缘，这在某种程度上是因为应用科学走得还不够远，还未达到斯滕特撰写《黄金时代的来临》时所希望的（所惧怕的？）程度。但我认为，就一个十分重要的方面而言，斯滕特的预言已经兑现了。作为探索"我们是谁"及"我们来自何方"之类知识的纯科学，已经进入了收益递减的时代。在纯科学的领域里，影响其未来发展的最大障碍，往往是它过去的成就。探索者们已经勾画出物理实在的图景：从夸克和电子的微观王国，到行星、恒星和星系的宏观世界。物理学家已经证明，所有物质都处于几种基本相互作用力——引力、电磁力、强相互作用力和弱相互作用力的统治下。

关于人类的由来，科学家们也用既有的知识连缀成一个动人的（如果不嫌它过于琐碎的话）故事。150亿年前，或在此数字上加减50亿年（因为天文学家永远也不会就某一精确数字达成一致），宇宙经由一次大爆炸而产生，现在仍在向外膨胀；约45亿年前，一颗超新星爆炸，生成的灰烬冷缩成我们的太阳系；又经过几亿年之后，由于一些可能永远也无法知道的原因，一种能合成奇妙的DNA分子的单细胞生物出现在环境恶劣像地狱一样的地球上，这些与亚当有着同等地位的微生物，通过自然选择的手段不断进化，就形成了一系列使人惊讶的更复杂的生物，包括我们人类。

我猜想，这个故事是由科学家们用自己的知识编织成的。这一现代的创世神话，再过100年甚至1000年之后，也仍然像其在今天一样有说服力。为什么？因为它是真的。此外，考虑到科学发展的程度已是如此深远，而约束科学进一步探索的限制因素——物质的，社会的，以及认知的——又在日益加重，科学似乎已不可能在现有认识的基础上再增添什么意义重大的东西了。在未来的岁月，不会再有任何重大的新发现足以与达尔文、爱因斯坦或沃森与克里克赐给我们的那些发现相媲美。

长生不老的速朽

应用科学将在一个相当长的时间里持续发展，科学家将不断开发多种多样的新材料，研制运行速度更快、更高级的计算机，建立能使我们活得更健康、更强壮、寿命更长的基因工程技术，甚至提供价廉而环境副作用又少的核聚变反应堆（尽管随着支持资金的锐减，核聚变的前景现在看来远比过去暗淡），等等。问题是：这些应用科学的进展，能给我们的基本知识带来任何出人意料的革命性变化吗？它们能促使科学家们去修订已绘就的宇宙结构图景，或更改已编就的创世神话吗？也许不能。20世纪的应用科学一直倾向于对主导的理论范式进行强化，而不是对它提出挑战。激光器和晶体管使量子力学更加巩固，正如基因工程支持了基于DNA理论之上的进化模式的信念一样。

什么才是真正"出人意料"的发现？爱因斯坦关于时间和空间——作为现实世界的支撑——是像橡胶一样可弯曲的这一发现，就是出人意料的；天文学家关于宇宙正在膨胀演化的观察事实也是；量子力学从物质结构底层揭示出或然性因素，使物质基本构成单位的观念彻底改变，给人们带来更大的惊异，上帝的确掷骰子（尽管爱因斯坦不赞成）。后来的发现，如质子和中子由被称作夸克的更小的粒子构成，就不那么让人惊异了，因为它不过是使量子理论拓展到更深的层次，物理学的基础依然如故。

知道了我们人类并非由上帝一劳永逸地创造出来，而是经由自然选择过程逐渐进化而来的，这是个更大的"出人意料"，至于人类进化的其他方面——比如人类在何时、何地，以及如何进化而来的详情——只不过是一些细节。这些细节可能很有趣，但不能使人惊讶，除非它们能证明科学家们关于进化的基本假设是错误的。比如说，我们能证明智慧在地球上的出现是外星人干预的结果，就像电影《2001年》所描绘的那样，那将是一个极其巨大的"出人意料"。事实上，任何关于地球之外存在着（或曾经存在）生命的证据，都会带来巨大的震动，科学以及所有人类的思想都将因此而重建。因此，关于生命起源及其必然性的构想，必须置于一种更为实证的基础之上。

　　但发现地外生命的可能性到底有多大？稍加回顾就会发现，美国和苏联的太空计划都更多地代表着一种精心展示的武力威慑，而不是开创人类知识新的前沿。展望太空探险的前景，似乎也越来越不可能超出这种无聊的水准，我们已不再有闲心或是闲钱一味地为技术而技术了。有血有肉的人们也许会在某一天驰向太阳系内的某些行星，除非我们发现某种能打破爱因斯坦对超光速运动的限制的航行方法，否则永远不要奢望去拜访另一颗恒星，更不用说别的星系了。即使我们有一艘时速1.609×10^6千米（百万英里）（这一速度比目前技术所能达到的最快速度至少还要快一个数量级）的宇宙飞船，仍然要用将近3000年的时间，才能抵达距我们最近的恒星邻居——半人马座的阿尔法星。[11]

　　我所能想象的应用科学中最激动人心的进展，也无外乎实现长生不老。现在正有许多科学家试图确定衰老的肇因，毫无疑问的是：一旦他们获得成功，就可以设计出种种能够长生不老的智人（*Homo sapiens*）。这或许会成为应用科学发展史上的丰碑，却不一定会改变我们关于这个世界的基本认识，也无助于我们更好地理解人类的基本问题，如宇宙为什么产生，以及在宇宙边界之外到底存在着什么，等等。更何况，进化生物学家们认为长生不老是无法实现的，自然选择为我们设计了足够长久的生命去养育后代，衰老是必然的结果，它并非为某个单一原因所决定，甚至也不是由一组原因决定的，而是不可拆解地织入了我们的"生命之织物"中。[12]

一百年前他们就这样想过

　　为什么会有这么多人难以相信科学——纯科学或非纯科学——会走向终结，这其实很容易理解。就在一个世纪以前，没有人能想象未来会是什么样子。电视机？喷气式飞机？空间站？核武器？计算机？基因工程？我们难以预料科学——基础科学或应用科学——的未来，正如托马斯·阿奎那（Thomas Aquinas）①绝不会预见到世界上将会诞生麦当娜②和微波炉一样。完全不可预见的奇迹正等在我们前面，正如我们的先辈们所经历的那样。如果我们断定奇迹并不存在，并停止发现奇迹的努力，导致的唯一后果是我们将失去拥有这样奇迹的机会。对奇迹的预言只能由奇迹自身作出。

　　这一主张常常被表述成"一百年前他们就这样想过"的论证，其大意

---

　　① Thomas Aquinas，1226—1274，意大利中世纪神学家和经院哲学家——译者注。

　　② Madonna，在此书写作、出版的时候，麦当娜是当时美国最著名的女歌星之一——译者注。

是：在19世纪即将结束时，物理学家们认为他们已认识了一切，但一进入20世纪，爱因斯坦和其他一些物理学家就发现了（发明了？）相对论和量子力学，这些理论使牛顿物理学黯然失色，为现代物理学和其他科学分支打开了广阔的新天地。言下之意：不论是谁，只要他敢宣称科学将要走到尽头，结果就一定会证明他就像19世纪的物理学家一样目光短浅。

那些持科学有限论观点的人，对这种论证的标准反驳是：早期的探险家因为无法发现地球的边界，才会认为地球是无限的，但是他们错了。再者，19世纪末的物理学家认为他们已认识了一切，这绝不是历史记录的问题，最好的证据是1894年阿尔伯特·迈克尔逊（Albert Michelson）的一次演讲。（有意思的是，正是他关于光速的实验，启发了爱因斯坦的狭义相对论。）

若认为将来的物理科学实验绝不会再有什么新奇的发现比过去的更激动人心，这当然不太保险，但可以换种说法，大部分重大的基本规律已经牢固地建立起来，更进一步的工作主要是把这些规律精确应用于被我们注意到的一切现象中。测量科学只在下述方面才能表现出重要性，即那些定量结果比定性工作本身更重要的方面。一位著名的物理学家曾指出：未来的物理学真理必须到小数点后六位中去寻找。[13]

迈克尔逊关于"小数点后六位"的言论，曾被普遍认为是出自开尔文爵士（Lord Kelvin，开氏温标就是以他的姓氏命名的）之口，以至于某些作者相信他只是照搬了开尔文的观点，[14] 但历史学家却找不到丝毫证据可以证明开尔文曾如是说过。再者，据马里兰大学科学史家斯蒂芬·布拉什（Stephen Brush）考证，迈克尔逊发表上述言论的时候，物理学家们正兴致勃勃地争辩着基础理论问题，如原子理论的可行性问题，而迈克尔逊却深深地陷入自己的光学试验之中，以至于"对当时理论家之间的激烈论战充耳不闻"。布拉什由此得出结论："所谓'物理学中的维多利亚式平静'，不过是一个'神话'。"[15]

凭空杜撰的专利局长

显然，会有一些史学家不同意布拉什的说法，[16] 问题一旦涉及某一给定时代的基调，当然永远不会被彻底解决，但由此可见，那种认为"19世纪末的物理学自满于物理学领域的现状”的说法，显然有些夸大其词。无独有偶，历史上另有一则被那些不情愿接受“科学终将完结”观点的人们所喜爱的轶事，说是在19世纪中叶，美国专利局局长突然异想天开地辞了职，并建议关闭专利局，因为“再也没有什么东西需要发明”的时代正在来临。对这则轶事，史学家们已给出了最后的裁决。

1995年，威望素著的《科学》杂志的主编丹尼尔·科什兰（Daniel Koshland），在他为“未来科学专辑”书写的一段前言中，重述了这则故事。丹尼尔·科什兰与岗瑟·斯滕特一样，也是加利福尼亚大学伯克利分校的一名生物学家，他主编的未来科学专辑中，各学科的带头人纷纷撰文，展望自己所属学科领域今后20年里所能取得的进展。科什兰得意扬扬地写道：“这些预言家们很明显并不赞同那位历史上的专利局长的见解。科学的发展已是如此深广、如此迅速，但这并不意味已使‘发现的市场’达到饱和，而是意味着更快地产生发现。”[17]

科什兰的文章中存在着两个问题。首先，给他的专辑撰稿的科学家们所展望的并不是“重大发现”，而主要是现有知识的实际应用，如更好的制药方法、改进的遗传病诊断技术、分辨力更强的大脑扫描仪等，并且某些预言在本质上是消极的。物理学家、诺贝尔奖获得者菲利普·安德森（Phillip Anderson）就声明：“如果有人期望在未来的50年里，计算机将会产生出近似人类的智慧的话，那么他注定要失望。”

其次，科什兰关于专利局局长的传说是凭空杜撰的。1940年，一位名叫埃伯·杰弗里（Eber Jeffery）的学者，曾写了一篇题为“无可发明”的论文，发表在《专利局会刊》上，[18] 专门考证了有关专利局局长的轶事。杰弗里追溯了这一传说的来历，发现它源自1843年提交的一篇国会咨文，其作者为亨利·埃尔斯沃思（Henry Ellsworth），正是当时的专利局长。埃尔斯沃思曾谈到一点：“专利事务年复一年的迅猛发展，考验着我们（对技术进步永无止境）的信念，并且似乎预示着人类技术进步抵达其终点的日子已为期不远。”

但埃尔斯沃思不仅没有建议关闭专利局，反而要求更多的基金，以处理他预计将潮水般涌现的农业、交通和通信等领域的新发现。埃尔斯沃思的确在两年以后的1845年辞职了，但在其辞呈中丝毫也未提起过要关闭专利局的事，反而为自己能使专利局发展壮大充满自豪之情。杰弗里认为，埃尔斯沃思关于“人类技术进步抵达其终点的日子”的陈述，表示的“仅是一种能产生修辞效果的繁荣，强调的是当时在科学发现上所取得的巨大进步，以及可期于未来的更大进步"。但杰弗里很可能误解了埃尔斯沃思，或许埃尔斯沃思所探讨的正是一个多世纪后岗瑟·斯滕特所要讨论的论点：科学发展得越快，达到其终极的、不可避免的限度也就越快。

品味丹尼尔·科什兰文章中那两个"疏漏"的寓意，尤其是他偷换概念做法的寓意，是很有趣的。他坚持认为科学在过去的一个世纪左右的时间里发展得如此迅猛，所以它一定能够并且一定会继续这样迅猛地发展下去，直到永远。但这一归纳论证有着难以克服的缺陷：科学仅仅存在了几百年，其最惊人的成就是在最后一个世纪左右的时间里取得的。从历史的角度看，科学技术迅猛发展似乎并不是一种永久特征，而是一种畸变，一种侥幸，一种由社会的、智力的以及政治的因素汇聚促成的产物。

进步之盛衰

史学家伯里（J. B. Bury）在其1932年的著作《进步的理念》中，曾作过这样的陈述："在过去的三四百年里，科学一直在不间断地进步；每一新发现都导致新的问题和新的求解方法，并开拓出新的探索领域。迄今为止，科学精英们从未被迫停止脚步，他们总是有办法向前发展，但是谁能保证他们不会碰到无法逾越的障碍？"[19]

伯里以自己的学识说明，进步的概念最多也不过几百年的历史。从罗马帝国时代到中世纪，大多数真理的追求者都持有一种堕落论的历史观，古希腊人在数学和科学知识上登峰造极，但文明却从那里走上了下坡路。那些后生晚辈只能试图去掌握由柏拉图或亚里士多德概括的智慧的边边角角；正是现代经验科学的奠基者们，如牛顿、培根、笛卡儿、莱布尼茨等人，才第一次详细阐述了这样一种观点，即如何通过对自然的研究系统地掌握和积累知识。这些最早的科学家都坚信这一进程是有限的，因而人类能够获得关于世界的全部知识，并进而在这些知识和基督戒律的基础上建构一个完美的社会，一个乌托邦（或者"新波利尼西亚"）。

只是随着达尔文的出现，开始有部分知识分子对进步痴迷起来，以至于认为进步可能是——或应该是——永恒的（*eternal*）。岗瑟·斯滕特在其1978年的著作《进步的悖论》中写道："在达尔文《物种起源》的出版所带来的震撼作用之下，进步的观念被提升到成为一种科学宗教的水平……这一乐观的信念被工业化国家广泛接受……以至于到目前，任何有关进步将会终止的说法都被普遍认为是奇谈怪论，就像早期人们乍闻地球围绕太阳转时的反应一样。"[20]

现代的民族独立国家变成科学无限论信条的热情支持者，这是不足为怪的。因为科学可以带来各种奇妙的玩意儿，如核武器、核能、喷气式飞机、雷达、计算机、导弹，等等。1945年，物理学家万尼瓦尔·布什（Vannevar Bush，美国前总统老布什和小布什的远房亲戚）在《科学：永无止境的前沿》中宣称：对于美国的军事和经济安全而言，科学是"一个有待探索的巨大狩猎场"，一个"关键性要素"。[21] 布什的文章被当作蓝图，用以构建国家科学基金会和其他一些联邦机构，从而得到了对基础研究的空前支持。

与其资本主义对手相比，苏联更忠于科学技术进步的观点。苏联人似乎从恩格斯那里得到了启发。在下面引述的一段《自然辩证法》的文字里，恩格斯展示了他对牛顿万有引力平方反比定律的理解。

"就像路德在宗教领域里焚烧教谕一样，在自然科学领域有哥白尼的不朽著作……但是科学的发展从此便大踏步地前进，其增长可以说是与从其出发点起的时间距离的平方成正比，仿佛要向世界证明：从此以后，对有机物的最高产物，即人的精神起作用的，是一种和无机物的运动规律正好相反的运动规律。"[22]①

在恩格斯看来，科学能够且必将越来越快地"大踏步"前进，直到永远。

当然，这种科学技术的无限进步观，如今正遭到社会、政治和经济力量的有力抵制。曾是推动美国和苏联从事基础研究的主要动力的冷战已经结束了，美国和苏联各共和国，正逐渐失去仅仅为了展示其强大而去建造空间站和庞大加速器的兴趣，社会对于科学技术的负面效应，如环境污染、核污染、大规模杀伤性武器等，也越来越关注。

传统上最坚定捍卫科学进步价值的政治领袖们，现在也开始表达反科学的情绪了。捷克诗人、总统瓦茨拉夫·哈维尔（Václav Havel）于1992年宣称：苏联集中体现并彻底推翻了由科学导致的"拜物教"。他希望它是"现时代的终结"，即世界（甚至存在本身）是完全可知的系统，是由有限的普遍性规律决定的。人们可以掌握这些规律，并理性地引导它们为自

① 可参阅《马克思恩格斯选集》1972年版，第三卷，第446页。这段文字是霍根从里查著作中转引的，与恩格斯的原文稍有出入——译者注。

身利益服务。[23]

这种科学"迷信"的幻灭，早在20世纪初就被一位叫奥斯瓦尔德·斯宾格勒（Oswald Spengler）的德国中学教师预见到了，他成为宣告科学终结的第一位预言者。在其1918年出版的巨著《西方的没落》中，斯宾格勒指出：科学以一种循环的方式前进，由研究自然并发明新理论的浪漫阶段，过渡到科学知识逐渐僵化的巩固阶段。当科学家们变得更加傲慢并对其他信仰体系特别是宗教体系，更不宽容时，斯宾格勒强调，社会就会背弃科学，转而信奉宗教极端主义或其他一些非理性的信仰体系。斯宾格勒预言，科学的没落以及非理性的复活将开始于20世纪末。[24]

如果说斯宾格勒的分析存在什么欠缺的话，就是它过于乐观了，他的科学循环论暗示着科学会有复活的一天，并会经历一个新的发展阶段。但科学的发展不是循环的，而是线性的，关于（元素）周期表、宇宙膨胀现象以及DNA结构等，我们只能发现一次。科学，尤其是像探求"我们是谁"以及"我们从哪里来"等知识的纯科学，其复兴的最大障碍恰恰是它往昔的成就。

不复存在的无尽地平线

科学家们不愿意公然宣称他们已进入收益递减时代，这是可以理解的，没有人希望自己被等同于一个世纪之前那些据称是"目光短浅的"物理学家。同时，预言科学的终结将会自然而然成为现实总要承受一种风险，即被认为自己在科学事业中已江郎才尽，不得不借此来哗众取宠的风险。但岗瑟·斯滕特绝不是唯一敢于向这种强大的禁忌挑战的杰出科学家。1971年，《科学》杂志上发表了一篇题为"科学：无尽的地平线还是黄金时代？"的文章，作者为本特利·格拉斯（Bentley Glass），一位著名的生物学家，美国科学促进会（也是《科学》杂志的主办方）主席。格拉斯在权衡了分别由万尼瓦尔·布什和岗瑟·斯滕特提出的关于科学未来的两种预言之后，极不情愿地站到了斯滕特一边。格拉斯认为，科学不仅是有限的，而且"死期已至"。他宣称："我们就像某个巨大大陆的探险者，已经跋涉过其中的大部分领地并已抵达其边缘，主要的山脉和河流都已绘入图中。虽然仍有无数细节需要补充，但无尽的地平线已不复存在。"[25]

格拉斯声称，若仔细阅读布什的《无止境的前沿》一文，就会发现他也认为科学是有限的事业。布什从未强调过一个科学领域能永远不断地产生新发现。事实上，布什把科学知识描述为一座"大厦"，其形状"是由逻辑规律和人类理性的特点预先决定的，就像它早已存在"。在格拉斯看来，布什之所以选择这一比喻，暴露出他的真实主张，即认为科学知识在范围上是有限的。他认为对于布什文章的"抢眼的标题"，"不能从字面本身去理解，它可能仅仅意味着：照目前情况看来，我们面临着如此众多有待发现的事物，以至于让人觉得科学的地平线似乎是永无止境的"。

　　1979年，在《生物学季评》上，格拉斯提出了支持其科学终结论的证据。[26] 他对于生物学中科学发现速度的分析表明，发现的速度与科研人员和资金按指数增长的速度不相称。"我们被科学在无可否认地加速发展这一辉煌成就深深打动，以至于看不到我们已进入了一个收益递减的时代，"格拉斯说，"事实上，为了维持科学的进步，必须把越来越多的人力和资金支出考虑在内。这种进步迟早会被迫停滞，因为用于科学的人力和支出存在着难以逾越的限度。在我们自己的世纪里，科学的增长是如此之迅猛，以至于我们被这一现象所迷惑，认为这样高速的进步会无限持续下去。"

　　1994年，当我采访他的时候，格拉斯坦承他的许多同事在听说他提出科学有限论的时候，就已经惊诧莫名，更不用说预言科学的末日了。[27] 但无论是在当时还是现在，格拉斯一直认为这一命题具有不容忽视的重要性，作为一项社会事业的科学，肯定是有某种限度的。他指出，如果一直以它在20世纪早期所展现的那种速度增长，科学很快就会耗尽工业化世界的全部预算。"我认为，为科学提供支持的资金数量，特别是基础科学，必须受到控制，这一点对每个人来说都是显然的。"据他观察，这种紧缩行为明显体现在1993年美国国会的决策上，它终止了超导超级对撞机项目，那是一个巨型的粒子加速器，物理学家们本来期待着能凭借它的助力超越夸克和电子的水平，进入更深层的微观世界，这一项目的预算共需"区区"80亿美元。

　　"即使社会把所有的资源都投入到科学研究中，"格拉斯补充道，"科学有朝一日仍将达到收益递减的转折点。"为什么？因为科学管用，因为科学解决问题。毕竟，天文学家已经探测到了宇宙的最远处，他们无法看到边界之外存在着什么（如果有什么边界存在的话）。此外，多数物理学家都认为，把物质还原为越来越小的粒子最终会走到尽头，或许相对于实践目的而言，已经到了尽头。即使物理学家们发现了深藏在夸克和电子之后的更小粒子，这种知识对生物学家而言，几乎没有或者完全没有什么相干。生物学家已经了解，最有意义的生物学过程发生在分子或分子以上水平。"生物学在那有一个限度，"格拉斯解释道，"你别指望能够打破它，因为'物质和能量结构'的本性如此。"

格拉斯认为，生物学中的伟大革命可能已经过去了。"对我来说，无论如何也难以相信，像达尔文的生物进化思想或孟德尔的遗传规律那样的理论——那种清晰易懂且具有震撼力的理论，会轻易地被再次获得。然而，它们毕竟已被发现了！"格拉斯强调，对于癌症和艾滋病之类的疾病，对于从单个受精卵到复杂的多细胞生物的发育过程，以及对于大脑和意识的关系等问题，生物学家肯定有很多事可做。"这些将会给知识的大厦增添新的成分，并且我们也已经取得了某些可能是重大的进展，但是否能使我们的意识世界产生真正重大的变革，还是个未知数。"

物理学面临的艰难岁月

1992年，《今日物理学》月刊发表了一篇题为"艰难岁月"的文章，作者列奥·卡达诺夫（Leo Kadanoff）是芝加哥大学的著名物理学家，他在文章中为物理学的未来勾画了一幅暗淡的景象。"我们所做的一切，看来都不足以抑制物理学在总体上、在社会支持或社会价值上的衰退趋势，"卡达诺夫宣称，"今日世界的基础更多的是构筑在那些似乎早已变成上古历史的事件上：诞生于二战期间的核武器和雷达，二战稍后的硅和激光技术，美国式的乐观主义和工业霸权，以及作为改进世界之手段的理性层面的社会主义信念。"卡达诺夫争辩道："产生这一切的社会条件基本已经消失，物理学以及作为整体的科学，都正处于环境保护主义者、动物权活动分子，以及其他一些深具反科学心态人们的重重包围之中。""最近几十年里，科学一直得到了很高的回报，一直处于公众兴趣和社会关注的焦点位置。如果这些殊荣失去的话，我们不应感到诧异。"[28]

当我在两年之后通过电话采访他的时候，卡达诺夫的语气比其文章中所表现出来的更为消沉。[29]他在向我袒露自己的世界观时，带着一种勉强抑制的忧郁，仿佛正忍受着头伤风之苦。他没有与我讨论科学的社会和政治问题，反而集中论述了科学进步的另一个障碍：科学的往昔成就。他认为，现代科学的伟大使命，一直是试图证明世界遵守某些基本的物理规律。"这是个至少自文艺复兴以来就一直在探索的问题，也许时间还会更久远些。对我来说，这一问题早已解决了。我认为世界是可用规律解释的。"自然界最基本的规律，体现在广义相对论中，也体现在粒子物理学的所谓标准模型中，后者异常精确地描述了量子世界的行为。

卡达诺夫回忆道，仅仅半个世纪以前，许多备受尊敬的科学家仍墨守着活力论的浪漫信条，认为生命来自某种用物理学定律无法解释的异种活力。得益于分子生物学——始于1953年DNA结构的发现——的诸多进展，卡达诺夫说，现在已"很少有受过良好教育的人"相信活力论了。

当然，对于怎样由基本定律产生出"我们所见到的世界的多样性"，科学家仍有许多事情可做。卡达诺夫本人是一位凝聚态物理学领域的带头人，这一领域不研究单个亚原子粒子的行为，而研究固体或液体的行为。他还致力于混沌领域的研究，这一领域探讨以无可预见的方式呈现的现象。某些混沌领域，以及与混沌密切相关的被称作复杂性研究的领域的支持者，甚至声称在强大的计算机和新的数学手段的辅助下，他们将全面刷新那些由过去的"还原论科学"所揭示的真理，卡达诺夫对此也有他的疑虑。对基本定律作用结果的研究，与那些证明"世界是有规律的"研究相比，"在某种程度上更没意思，更浮浅。但既然我们已经知道了世界是有规律的，就只好转而研究别的问题了，这样也许多少能刺激一下普通人的想象力，这是合乎情理的"。

卡达诺夫指出，粒子物理学最近也不那么激动人心了。过去几十年的实验仅仅验证了已有的理论，并未揭示出需由新规律去解释的新现象；寻求统一理论的目标，看来仍遥不可及。事实上，很久以来，没有哪一个科学领域产生过什么真正深刻的发现。卡达诺夫说："关键在于，没有什么成就的重要性，能与量子力学、双螺旋结构或相对论的提出相媲美。最近几十年里一直就没有这样的发现。"我问道："这种状况会永久持续下去吗？"卡达诺夫沉默了一会儿，然后叹了口气，仿佛欲借此吐尽胸中块垒，答道："一旦你令人满意地证明了世界是有规律的，就不能再证明它了。"

给自己壮胆的口哨声

匹兹堡大学的尼古拉斯·里查（Nicholas Rescher），是少数几个严肃思考过科学限度问题的现代哲学家之一。他在1978年出版的《科学进步论》一书中，探讨了为什么斯滕特、格拉斯等杰出科学家认为科学会走入死胡同。里查通过对科学至少潜在地无限论证，试图为"目前正蔓延到整个思想界的时疫"提供一剂良药。[30] 纵观全书，他所描绘的方案很难说是乐观的。他指出，科学作为基本上是经验的、实验的学科，必须面临着经济的约束。随着科学理论向更深远领域的拓展——去观察宇宙更遥远的现象，物质更深层的结构——科学家的开销会不可避免地逐步攀升，而取得的收益却渐次减少。

"随着科学事业由常规领域推进到越来越远的前沿，科学创新也变得越来越艰难。即使这一论点只是部分正确，那么，始于1650年左右的近五百年盛世，最终将被看作是人类历史大变革中的'科学探索时代'，从而与青铜时代、工业革命或人口爆炸时代相提并论。"[31]

里查为他那令人沮丧的脚本强加了一个欢快的结尾：科学永远不会终结；它只会走得越来越慢，越来越慢，就像芝诺悖论中的乌龟。科学家们不该认为自己的研究只能产生细枝末节的成果，他们耗费巨资的众多实验中，或许有一个具有革命性的意义，足以和量子力学或达尔文理论相媲美——这种可能性总是存在的。

本特利·格拉斯（Bentley Glass）在一篇评论里查著作的文章中，把里查所开出的"药方"称为"科学家在面临绝境时，为了给自己壮胆而吹响的口哨"。[32] 我在1992年8月电话采访里查时，他承认自己的分析在很多方面都有些勉强。"我们只能通过与自然的相互作用来认识自然，"他说，"为此我们必须去探索那些尚未被认识的领域，那些密度更高、温度更低、能量更大的领域，这些都在不断地打破基本的限度，但需要更精确、更昂贵的设备。因此，存在着一个由人类的有限资源赋予科学的限度。"

但里查坚信，"第一流的发现"就像是待发的"横财"，可能——必定！——正在前面的路上等着我们，尽管他无法预知那些发现将在何时产生。"这类似于你去问爵士音乐家'爵士乐将走向何方？'，他肯定会回答：'如果我知道，我们现在早已抵达那儿了。'"里查最后回到了"19世纪末他们就那么想过"这类论证上，认为：像斯滕特、格拉斯和卡达诺夫这样的科学家，看起来都在担心科学可能正在走向终结，这一事实本身就足以使他相信，某些惊人的发现已经迫在眉睫。像许多想当然的预言家一样，里查屈从于一厢情愿的想法，他承认科学的终结对人类来说将是一场悲剧。如果对知识的追求走到了尽头，我们将何去何从？还有什么能赋予我们的存在以意义？

弗朗西斯·培根之"不断超越"的寓意

对科学终结论最常见的驳难，除了"19世纪末他们就那么想过"说之外，另一种就是老掉了牙的"解答会带来新问题"说。康德在《未来形而上学导论》中写道："按经验原则给出的每一个解答都引出一个新问题，新问题同样需要解答，从而清楚地表明，所有物质层面的解释方式都不足以满足理性的需求。"[33] 但先于斯滕特的论证，康德同时又认为，我们心智的先验结构既限制了我们向自然提出问题，也限制了我们向自然寻求答案。

科学当然会不断提出新问题，其中大部分是琐屑的问题，关注的只是细节，不足以影响我们对自然的基本认识。除了专家之外，有谁真的关心顶夸克的精确质量呢？尽管为证明它的存在所进行的研究，在花掉了数十亿美元之后，终于在1994年得出了肯定的结论。另外一些问题虽然深奥，却无法解答。事实上，对于科学之至善理论，或对于罗杰·彭罗斯等人所梦想的真正令人满意的理论而言，其最顽固的阻力恰恰来自于人类发现不可解问题的能力。即使发现了所谓的"万物至理"，总有人并且肯定会有人问：我们怎么知道夸克或超弦（虽似不可能，但暂且假设它在某一天会被证明的确存在）不是由更小的东西构成的——以至可无限类推下去？我们怎么知道这个可见的世界不会仅仅是无限多的宇宙之一呢？现在这个宇宙是远古宇宙的必然发展，还是只是它的一个偶然的错误？生命是怎么一回事？计算机能产生有意识的思维吗？变形虫呢？

不论经验科学走得多远，人类的想象力总会走得更远。对于科学家们寻求"终极答案"（The Answer）——那种能一劳永逸地满足人类好奇心的理论——的希望（或恐惧）而言，其最大的障碍也正在于此。弗朗西斯·培根（Francis Bacon）作为现代科学的奠基人之一，曾用拉丁文 *Plus Ultra* 来表达他对于科学之无限潜能的信念，意为"不断超越"（more beyond）。[34] 但"不断超越"并非针对科学本身而言，因为科学不过是一种受到重重约束的考察自然的手段；"不断超越"是针对我们的想象力而言的。尽管我们的想象力受到人类进化过程的限制，但它总是一往无前，总能超越我们真正已知的东西。

甚至在斯滕特所谓的"新波利尼西亚"时代，仍将有个别执着的人继续奋斗着，试图超越已被普遍接受的知识，斯滕特把这些真理追求者称为"浮士德式人物"（Faustian，斯滕特从奥斯瓦尔德·斯宾格勒那儿借用的术语），我则称之为"强者科学家"（strong scientists，这是我从哈罗德·布鲁姆《影响的焦虑》一书套用而来的）。通过提出科学所不能解答的问题，强者科学家们甚至在经验科学——那种以解疑答问为旨归的科学——终结之后，仍可凭借我称之为"反讽科学"的思辨方式，继续从事科学研究。

诗人约翰·济慈（John Keats）曾编造了"消极能力"（negative capability）一词，用以描述某些伟大的诗人在身处"无常、神秘而又困惑之境时，仍能心平气和地追寻事实和真理"的能力。济慈举出诗人同行萨缪尔·柯勒律治（Samuel Coleridge）为证，说他"甘愿做一个离群索居的智者，去追寻隐在玄妙现象深处的秘密，永不满足于一知半解的认识"。[35]而反讽科学最重要的功能，就在于充当人类的"消极能力"。反讽的科学通过其提出的不可解问题提醒我们：一切知识都是一知半解的认识，我们对世界的认识是多么可怜！但反讽科学却不能对知识本身作出任何实质性的贡献，因而它不同于传统意义上的科学，倒更像是文学批评或哲学。

第二章

哲学的终结

2 0世纪科学催生出一个奇特的悖论：科学的非凡进步，一方面让我们坚信
自己能够认识应该认识的一切，另一方面也孕育了我们不可能确切认识
任何事物的疑虑。一种理论竟能如此迅速地战胜另一种理论，那我们又怎能
肯定任何理论的真实性呢？1987年，两位英国物理学家——西奥查理斯（T.
Theocharis）和皮莫波洛斯（M. Psimopoulos），在一篇题为"科学哪儿错
了"的文章中，严厉批判了这种怀疑论的哲学观点。文章发表在英国《自
然》杂志上，谴责了那些对"科学能获得客观知识"这一传统信念大加挞伐
的哲学家，认为他们给科学带来了"深重而又广泛的灾难"，并刊登出四位
极恶劣的"真理的背叛者"的照片：卡尔·波普尔（Karl Popper）、伊慕
里·拉卡托斯（Imre I. Lakatos）、托马斯·库恩（Thomas Kuhn），以及保
罗·费耶阿本德（Paul Feyerabend）。[1]

照片是粗糙的黑白摄影，就像关于"某德高望重的银行家诈骗退休老人
保险金"之类花边新闻所配的照片一样，不同之处仅在于这四幅照片代表的
是四位最恶劣的科学罪人。费耶阿本德被判为"科学的头号敌人"，其照片
也是这一组中看起来最邪气的：他正透过高耸鼻尖之上的眼镜，冲着镜头傻
笑，明显是在期待或享受着某种邪恶的恶作剧，像是古代斯堪的纳维亚传说
中那位淘气的精灵洛基（Loki）的现代知识分子翻版。

西奥查理斯和皮莫波洛斯的大部分观点不值一哂，几个学院派哲学家的
怀疑，对于庞大的、结构森严的科学来说，永远也构不成什么真正的威胁。
许多科学家，特别是那些以革命者自居的科学家，发现波普尔等人的观点能
带来更多的心灵慰藉——如果现有的科学知识都只是暂时的，那么重大突破
的可能性就永远存在。但西奥查理斯和皮莫波洛斯确实给出了一个有趣的论

断，即怀疑论者的观点"显然是搬起石头砸了自己的脚……他们否定和打倒的，正是他们自己"。这触发了我的灵感：如果就这一点去请教那几位哲学家，看看他们对此作出的反应，那一定会十分有趣。

后来，我终于有幸实现了这一愿望，采访了所有那些"真理的背叛者"，但拉卡托斯除外，他已于1974年作古了。在采访过程中，我还试图发现：这些哲学家是否真的像自己在某些理论中所宣称的那样，完全怀疑科学达到真理的能力。而结果却让我认识到，像波普尔、库恩、费耶阿本德等人，对科学其实都是情有独钟的。事实上，他们的怀疑主义观点，恰恰是植根于对科学的坚定信念；而其最大失误可能就在于，他们给科学赋予了远高于其所实际具有的能力。他们担心科学会泯灭人类的质疑精神，并因而把科学自身——以及一切追求知识的努力——推向终结；他们试图使人类（包括科学家）免遭因过分天真地相信科学——就像西奥查理斯和皮莫波洛斯等科学家所表现的那样——而导致的危害。

随着19世纪科学在力量和威望上的迅速发展，许许多多哲学家都成了科学的公共关系代言人，这一倾向可追溯到像皮尔斯（Charles Sanders Peirce）这样的思想家。皮尔斯是美国人，他虽然奠定了实用主义哲学的基础，但却找不到工作，娶不上太太，穷困潦倒，死于1914年。皮尔斯曾为绝对真理下了这样一个定义：它是穷途末路的科学家们的胡言乱语。[2]

皮尔斯之后的大部分哲学，只不过是发挥了他的观点。20世纪初欧洲的主导哲学是逻辑实证主义，它宣称，只有可以被逻辑地或经验地证明的东西才是真的。实证主义者把数学和科学作为真理最重要的来源。波普尔、库恩和费耶阿本德各自按照自己的理由用自己的方式批判了对待科学的这种谄媚态度。他们认为：在一个科学已占据优势地位的时代，哲学的最大作用是作为科学的否定力量，给科学家们注入怀疑精神。只有这样，人类对知识的追求才会保持开放性；也只有这样，我们才能在宇宙的奥秘面前保持一份敬畏之心。

在我采访的这三位大怀疑论者中，波普尔成名最早。[3] 其哲学起源于他试图把伪科学（如占星术或弗洛伊德心理学）与真科学（如爱因斯坦的相对论）区分开来的尝试。波普尔断定，后者是可检验的，它能对世界作出可被经验验证的预见。关于这一点，逻辑实证主义者已作了大量的论述，但波普尔否认逻辑实证主义者的这样一种主张，即科学家能通过归纳、反复的经验检验或观察来证明一个理论。即使已往的观察都证明某一理论是有效的，也无法保证下一个观察会给出同样的证明。观察永远不能证明一个理论，而只能否证或证伪它。波普尔常自诩是他用这一论述"埋葬"了逻辑实证主义。[4]

波普尔将其证伪原则扩展为一种哲学，并称之为批判理性主义，认为一旦有科学家大胆地提出某一理论观点，立刻就会有一批科学家试图用相反论证或相反的试验证据打倒它。波普尔把批判（甚至冲突）视为各类进步的基本要素，正如科学家们通过他所谓的"猜想与反驳"而逐渐逼近真理一样，物种通过竞争而得以进化，社会通过政治斗争而得以发展。他曾经写道："没有冲突的社会不是人类社会，而是蚂蚁的社会。"[5]在出版于1945年的《开放社会及其敌人》中，波普尔宣称：政治比科学更需要思想和批评的自由空气，教条主义所必然导致的，不是乌托邦——像法西斯主义者所宣称的那样——而是极权主义的高压。

波普尔的著作，以及波普尔这个人，都包含着深深的矛盾。这一点，我在采访他之前与其他哲学家谈及他的时候，就已有所了解。在通常的采访中，受访人在评价某人及其思想存在着矛盾之后，往往要跟着说几句沉闷而又空洞的溢美之词，但这一次的情况例外，被访者对于波普尔不仅没说任何溢美之词，反而义愤填膺地声讨之，说就是这个强烈抨击教条主义的人，自己恰恰是个病态的教条主义者，他甚至要求其学生效忠于自己。有一则关于波普尔的老笑话，说《开放社会及其敌人》一书的书名，更应题为"其敌人眼中的开放社会"。

为了安排对波普尔的采访，我打电话到伦敦经济学院（波普尔从20世纪40年代末以来一直在那里教书），那里的一位秘书告诉我，波普尔一般都在肯星顿的家里工作，那是伦敦西区的高档住宅区，并给了我波普尔家的电话号码。我拨通了这一号码，听到的是一位女性傲慢的、带些德国味儿的声音，自称"喵夫人"（Mrs. Mew），是"卡尔爵爷"的管家兼助手。她说在"卡尔爵爷"肯见我之前，我必须给她寄一篇我个人从前的作品，并给我开了一张阅读清单，以备采访之用，其中罗列了大约一打"卡尔爵爷"的著作。经过无数次的电传和电话联系之后，她终于排定了一个采访日期，同时指点了最近的火车站和"卡尔爵爷"家的方位。当我向她询问详细路线时，"喵夫人"向我保证：所有的出租车司机都知道"卡尔爵爷"家在哪儿，"他的名气太大了！"。

"去卡尔爵士家"，我在肯星顿火车站一钻进出租车，就对司机这样说道。"谁？""卡尔·波普尔爵士？著名的哲学家？""没听说过。"司机说。但他对波普尔所居住的那个街区很熟悉，所以我们总算顺利找到了波普尔的家——一座两层楼的别墅，坐落在被修剪得齐齐整整的草坪和灌木丛中。[6]

一位身材高挑的漂亮女士打开了门，身着黑色的便裤和衬衫，黑亮的短发齐整地梳向脑后——这就是"喵夫人"。与电话中听得的印象相比，她本人并不那么令人生畏。在领我走进屋里的时候，她告诉我卡尔爵士非常疲惫，因为上个月是他的90岁寿辰，他不得不应付随之而来的大量采访和祝贺。同时，为了准备"京都奖"（Kyoto Award）（被称作日本的"诺贝尔奖"）的获奖演说，他一直在辛苦地工作，希望我的采访最多不要超过一个小时。

在我盘算着该怎样降低自己对采访的期望值时，波普尔走了出来。他已有些驼背，戴着助听器，人出乎意料的矮，因为按照我之前的预想，能写出如此独断霸气的文字，其人也应该长得高大霸气才对。然而，他却像最轻量级拳击手一样动感十足，挥舞着我为《科学美国人》写的一篇文章（那篇文章大致写的是在量子力学问世后，科学家正被迫放弃把物理学当作完全客观的事业这一信念），[7] "我不相信这上面的任何一个字"，他以一种带奥地利腔调的咆哮着说。"主观主义"在物理学中——量子力学或无论什么别的地方——都没有容身之地。"物理学，"他继续嚷着，同时从桌子上抓起一本书，再把它使劲摔到桌子上，"就是这样的！"说出如此豪言的人，竟然与他人合著了一本赞同二元论（即认为观念和其他一些人类意识的产物独立于物质世界而存在）的书。[8]

即便终于坐了下来，他仍不时地弹起来去搜寻能支持自己所谈及观点的书籍或文章。为了能从记忆中挖掘出一个人名或日期，他会不停地揉着太阳穴，不停地咬牙切齿，仿佛正承受着极大的痛苦。有那么一阵子，他忽然想不起"变异"这个词来，就开始一下又一下地把自己的脑门拍得"噼啪"作响，并喊着："词（儿），词（儿），词（儿）！"

词汇像机关枪子弹一样从他嘴里喷射而出，以压倒一切之势向我倾泻，以至于尽管我事先准备了许多问题，此刻已完全丧失了就其中任何一个向他请教的希望。"我虽已年逾九旬，但我还能思考。"他宣称，就像他怀疑我会怀疑这一点。接着，他又不知疲倦地向我兜售由他以前的一名学生——岗赛·魏契特肖瑟（Günther Wächtershäuser），一位德国专利代理人，曾获化学博士学位——提出的生命起源学说。[9]波普尔不断强调他认识20世纪科学的所有"巨人"：爱因斯坦、薛定谔、海森伯（Heisenberg），但他指责玻尔（Niels Bohr）——他对玻尔"非常了解"——因为他把主观主义引入了物理学。玻尔是个了不起的物理学家，是古往今来最伟大的物理学家之一，但他却是个可怜的哲学家。并且别人简直无法同他对话，因为他总是说个不停，只偶尔地让你说一两个字，然后立即把你打断。

在喵夫人转身离开的时候，波普尔突然叫住她，请她去找他写的某本书。她离开几分钟后，又空着手回来了。"抱歉，卡尔，我找不到，"她这样汇报，"你得给我提个醒，我不能去翻遍每一个书架。"

"我想，它应该放在这个角落的右边，但我曾经取出来过，或许……"他的声音渐渐低了下来。喵夫人似乎转了转眼珠，其实，仍在目不转睛地望着他，然后——走开了。

他停顿了一会儿，我在绝望中赶紧抓住这一机会，想提出一个问题："我想向您请教……"

"是的，你应该向我提问，不应该一直让我一个人讲。你可以先向我提出你所有的问题。"

在就波普尔的观点进行提问的过程中，我才逐渐明白，他的怀疑论哲学来源于对科学的一种极具浪漫和理想化色彩的看法，并由此进一步否定了逻辑实证主义者秉持的主张，即科学可以被还原为形式化的逻辑系统，在此系统中"原材料"被有条理地转化为真理。波普尔坚持认为，科学是一种发明，一种与艺术同样意味深长、同样神秘的创造行为。"科学史充满了猜测，"波普尔说，"这是一部奇妙的历史，它使你为自己是人类的一员而充满骄傲。"他把脑袋支在伸开的双手中间，像吟诵赞美诗一样说道："我相信人类的心智。"

出于同样的原因，波普尔终生都在与科学决定论的教条抗争，认为它与人类的创造力、与自由是对立的，因而与科学自身也是对立的。波普尔宣称，早在现代混沌学家之前，他就已经认识到：不仅量子系统，就连经典的牛顿系统都具有内在的不可预测性。他曾在1950年就此论点发表过演讲。他把手对着窗外的草坪一挥，说："每株小草里都包含着混沌。"

当我问波普尔他是否认为科学永远也达不到绝对真理时，他嚷道："不可能，绝对不可能！"同时猛摇着脑袋。和逻辑实证主义者们一样，他也相信科学理论可以是"绝对"真实的。事实上，他"从未怀疑过"某些目前的科学理论是绝对真的（虽然他拒绝举出具体是哪些理论），他与实证主义者的不同之处在于：不承认我们能够"知道"一个理论是真的。"我们必须把客观的、绝对的真理和主观的确定性区分开。"

波普尔认识到，如果科学家们过于相信自己的理论，就会停止对真理的追求，这将是一个悲剧。因为对波普尔来说，追求真理正是生活的意义所在。"对真理的追求是一种宗教信仰，并且，我认为它同样也是一种伦理观念。"波普尔确信对知识的追求会永远进行下去，这也反映在他的自传的题目上：《无尽的探索》。

他进而嘲笑了某些科学家想要寻求解释自然现象的完备理论，即某种能解答所有问题的万物至理的企图。"某些人认为这一目标能够实现，另一些人则持相反的观点。人类科学的确已取得了很大成就，但绝对真理仍遥不可期，我希望你能读读这一页的内容。"说到这里，他再度弹了起来，取回一本《猜想与反驳》。把书打开，他用崇敬的语气读着自己写的文字："在广袤的无知面前，我们都是平等的。"

波普尔还认为，有关宇宙的意义和目的之类的问题，是科学永远无法回答的，因此，他从未完全遗弃过宗教，虽然在很久以前的青年时期，他曾抛弃过路德教。"我们的所知少得可怜，应该保持一份谦恭，对这类根本的问题不能不懂装懂。"

然而，波普尔又憎恶那些所谓的现代哲学家和社会学家，他们竟声称科学不可能达致任何真理，并且还辩称，科学家们之所以坚持这样那样的理论，更多的是出于文化和政治上的理由，而不是出于理性的考量。波普尔评论道，这类批评家，他们是因为自己被看得比真正的科学家低一等而心怀怨恨，总试图"在既定的秩序中改变自己的位置"。我提醒他：这些批评家所描述的是"科学是怎样实践的"，而他——波普尔——所描述的却是"科学应该怎样实践"。出乎我的预料，波普尔竟然点了点头，"这是个非常妙的说法，"他说，"如果你头脑中没有一个科学应该是什么的概念，你就无法理解科学是什么。"波普尔承认科学家们似乎并不理会自己为他们设定的理想，"因为科学家们把自己的工作当成糊口的手段，所以科学不可能完全是它所应该成为的样子，这是难免的。科学事业中由此导致的腐化行为比比皆是，但我不屑去谈论。"

然而波普尔接着就谈到了它。"科学家缺乏自我批评精神，这本是他们应该具有的美德。"他肯定地说。"所以，这就要求你——"他把一根手指直戳向我，"像你这样的人，把这些公之于世。"他瞪了我好一会儿，然后才又提醒我说他对这次采访并不热心。"你知道我不但从未这样要求过，也从未鼓励过你采访我。"然后波普尔又投入到一段针对大爆炸理论的极度专业化（包括三角测量和其他玄奥的技术）的批评中。"不过如此。"他归结道，"对困难估计不足，却又表现出一种自鸣得意的神气，仿佛这一切都具有科学上的确定性，但类似的确定性根本就不存在。"

我问波普尔他是否认为生物学家们过于轻信了达尔文的自然选择理论，因为过去他曾评价这一理论只不过是同义反复，因而是伪科学。[10]"也许那种评价确实有些过分，"波普尔说，同时挥了挥手，就像要打发掉什么，"我对自己的观点并不固执。"说到这里，他突然"砰"地拍了下桌子，嚷道："人们更应该去追求各种备选理论（alternative theories）！这——"他挥舞着岗赛·魏契特肖瑟关于生命起源的论文，"就是一种备选理论，它看起来似乎是一种更好的理论"。但他很快又补充道，这并不意味着岗赛的理论是真的，"生命的起源可能永远无法验证"。即使科学家们真的在实验室中创造出了生命，也永远无法肯定生命确实是以相同的方式肇始的。

这时，我觉得应该适时抛出自己准备好的问题中的"重磅炸弹"了：他自己的证伪概念是不是可证伪的？波普尔目不转睛地瞪着我，过了好一会儿，表情才放松下来。他把自己的手轻按在我的手上，"我不想伤害你，"他温和地说，"但这的确是个愚蠢的问题。"他像探寻什么凝视着我的眼睛，问我是不是其某个批评家诱使我提的这个问题。是的，我不得不撒了个谎。"果然如此。"他舒了口长气，看起来很愉快的样子。

"在哲学研讨会上，如果有人提出了某一观点，而你要提出批评，首先想到的肯定是说这不符合他自己的逻辑标准，这是人们所能想出来的最愚蠢的批评之一！"他的证伪概念，据他自称，是用以区分知识的经验形态（比如说科学）和非经验形态（如哲学）的，证伪本身是"绝对非经验的"，它不属于科学，而属于哲学或"元科学"（metascience）；甚至，证伪概念也并非适用于所有的科学。波普尔承认，他的批评家们基本上是正确的：证伪仅仅是一种准则，一种比较粗糙的方法，有时有用，有时没用。

波普尔说他以前从未对我刚才提出的这类问题做过回应。"我发现它太愚蠢，不值得回答。你明白吗？"他问道，声音再度温和下来。我点了点头，说我也觉得这个问题有些愚蠢，只是觉得应该提出来。他笑了，同时握了握我的手，嘴里咕哝着，"是的，很好。"

既然这时的波普尔这么好说话，我自然也乘机提出另一问题：一位他以前的学生曾指责过波普尔，说他不能容忍对其观点的批评。波普尔的眼睛瞬间瞪圆："一派胡言！我在受到批评时只会感到高兴！当然，不是在我回应批评的时候。就像你刚才所提的那类问题，我已经答复了，但那些人还会继续一遍又一遍地重复，这才是我觉得无聊并不愿忍受的。”如果这样的事发生在课堂上，波普尔说，他会命令那个学生滚出去。

当喵夫人从门外探头进来的时候，外面厨房里的光线已显出暗红色。她说我们已经谈了三个多小时，并带着一丝怒意责问我们还希望谈到什么时候，她是不是该替我叫一辆出租车。我望了望波普尔，他正像个顽童一样咧嘴笑着，但确实显得有些疲惫了。

我赶紧提出最后一个问题：为什么波普尔在自传中说自己是他所知道的哲学家中最快乐的一个？“许多哲学家确实活得很压抑，因为他们拿不出什么像样的东西来。”他这样回答，同时露出一副很自得的笑容，并把目光瞥向喵夫人——这时她脸上的表情已经有些恐怖了——波普尔的笑容瞬间凝固了。“最好不要写这些，”他转向我，“我的敌人已经够多了，我不想再跟他们纠缠这样的事。”他生了一会闷气，又补充道：“就是那么回事。”

我问喵夫人能不能把波普尔为日本京都奖颁奖仪式准备的演说词给我一份，“不，现在不行，”她粗鲁地答道。“为什么？”波普尔问。“卡尔，我一直在不停地打录第二稿，但我有点……”她叹了口气，“你明白我的意思，对吧？”反正，她补充说，还没有最后定稿。“未校正过的稿子呢？”波普尔问。喵夫人气哼哼地走了出去。

她一会就转了回来，并把一份波普尔的讲演稿撇给了我。“你还有剩下的《倾向》（*Propensities*）吗？”波普尔问她，[11]她又噘着嘴、跺着脚进了隔壁房间。波普尔开始向我解释起《倾向》一书的主题：量子力学乃至经典物理学给予我们的教诲是，没有什么是确定的，没有什么是必然的，没有什么是完全可预见的，有的只是某些特定事件发生的倾向性。“举例来说，”波普尔补充道，“在这一刻就存在着一种必然的倾向——喵夫人会找到一本我写的《倾向》。”

“噢，天啊！”喵夫人的喊声在隔壁响起。她怒气冲冲地进来了，不再试图掩饰自己的情绪：“卡尔爵士！卡尔！你竟然把最后一本《倾向》都送出去了，你怎么能这样？！”

“最后一本肯定是当着你的面送出去的。”他义正词严地宣称。

“不可能，”她反击回来，“送给谁了？”

“我记不清了。”他胆怯地嘀咕着。

屋子外面，一辆黑色的出租车开进了车道。我感谢了波普尔和喵夫人对我的殷勤接待，起身离开了。当出租车离开别墅车道的时候，我问司机是否知道这是谁的房子，他说不知道。是一位名人的房子——是吗？是的，是卡尔·波普尔爵士。谁？我告诉他：是卡尔·波普尔，20世纪最伟大的哲学家之一。"是吗？"司机嘀咕着。

波普尔在科学家中一直很受欢迎，最主要的原因在于，他把科学描述成一种没有止境的浪漫历程。《自然》杂志的一篇社论曾公正地称波普尔是"为科学张目的哲学家"，[12] 但其哲学家同行们却并不怎么喜欢他。他们指出波普尔的作品充满着矛盾：他主张科学不能归约为一种方法，但他的证伪模式却正是这样一种方法。再者，他用以埋葬"绝对证实"可能性的论证，同样可以用来埋葬他自己的证伪原则。如果将来的观察确实能否证现有的理论，那么它同样也能复活以前被否证的理论。因此，波普尔的批评者们认为有充分的理由假定：就像某些科学理论可以被证伪一样，也必然有某些科学理论可以被证实；反正，地球是圆的而不是平面的，这是千真万确的。

在我采访他两年之后的1994年，波普尔去世了。《经济学人》杂志发表的讣文中，封他为"最著名的、影响最广泛的当代哲学家"，[13] 文章特别推许波普尔在政治领域中反对教条主义的坚定立场，同时也注意到波普尔对待归纳的态度（这是证伪原则的基础）已经被后来的哲学家所摒弃。"根据波普尔自己的理论，他应该为这一事实而欢欣鼓舞，"文章冷冰冰地写道，"但他却无法说服自己做到这一点。最有讽刺意义的也正是这一点，波普尔竟不承认自己错了。"波普尔的反教条主义立场一旦应用于科学，就变成了一种教条主义。

虽然波普尔憎恶精神分析，但他自己的作品最终却只能由精神分析的术语得到最好的理解。他与有关人物的联系，从玻尔这样的科学巨匠到他的秘书喵夫人，显然都极为复杂，交织着蔑视和尊敬。在波普尔的自传中，有一段最能显示其内心世界的文字。波普尔先提到他的父母都是皈依路德教的犹太人，然后又指出，由于其他犹太人不能把自己同化到德国文化中去，也由于犹太人在左派政治中的重要作用，终于导致了20世纪30年代的法西斯主义和全国范围的反犹运动的出现；"反犹主义是场噩梦，使所有犹太人和非犹太人对之深怀恐惧……所有犹太血统的人民都有责任尽力不要再去刺激它"。[14]

托马斯·库恩的"结构"

"听着。"库恩带着几分不耐烦地嚷着，仿佛他已认定我会误解他的观点，但又不得不徒劳地向我阐述。后来我才发现，这不过是库恩的口头禅。"听着，"他又重复了一遍，并把瘦长的身子和同样瘦长的脸一齐向我探过来，那片通常总是和蔼的向上弯的肥厚下唇这时却向下耷拉着，"天知道，要是让我在写了和没写那本书之间作出选择的话，我仍会选择写那本书。但它确实引起了很大的反响，包括使许多人很不愉快。"

"那本书"就是《科学革命的结构》（简称《结构》），关于"科学如何发展"这一主题所写过的所有著作中，它可能是最具影响的一部。之所以广受重视，是因为它推出一个髦及一时的术语——"范式"（*paradigm*）；同时，它也煽起了一股如今早已有些陈腐的思潮，即认为科学家的个性以及政治倾向等在科学发展中起着极为重要的作用。但这本书中最具深意的论点却没人关注，其大意是：科学家们永远也不可能真正理解现实世界，甚至他们彼此之间也无法相互理解。[15]

基于这一论点，你可能会认为，库恩对自己的著作在某种程度上遭人误解应该是能够坦然处之的，而事实却恰恰相反。在《结构》一书出版30余年之后，当我到麻省理工学院库恩的办公室采访他的时候，他看来正为自己的书遭到如此广泛的误解而大为烦恼，尤其对别人宣称他把科学描述为"非理性的"（irrational）事业这一点极为不满。"如果他们用的是'准非理性的'（arational），我一点儿也不会在乎。"他这样告诉我，脸上不带半丝笑意。

库恩时时担心自己的思想会招致新的误解，这使他极力回避与媒体的接触。当我第一次打电话商谈采访事宜的时候，他一口回绝了："听着，我说不行。"据他后来透露，《科学美国人》杂志，也就是我为之效力的那一家，曾给予了《结构》一书"我记忆中最恶劣的评论"。（那篇短评的确有点儿过分，竟称库恩的观点"小题大做"[①]。但对于一份以歌颂科学为主旨的刊物，库恩还能要求什么呢？[16]）我向他指出，那篇评论发表于1964年，我那时还没到这家杂志社工作呢，请他再考虑考虑。最后，库恩总算同意了我的采访请求，尽管还是有些不情愿。

我俩终于在他的办公室里坐了下来，这期间库恩一直在大发牢骚，对那

[①] 原文是"much ado about very little"，套用的是莎士比亚戏剧《无事生非》（*Much ado about nothing*）的说法，以表示书评作者对库恩观点的蔑视——译者注。

些挖掘其思想根源的人所掘出的货色深为不满。"人当然不可能成为自己的历史学家，更不用说成为自己的心理分析家了。"他这样告诫我。不过，他还是充当了一回"自己的历史学家"，把自己科学观的形成追溯到1947年的一次顿悟体验。那时，他正在哈佛大学攻读物理学博士学位，在阅读亚里士多德的《物理学》的时候，他被书中随处可见的严重谬误惊呆了：一个在如此众多的领域里取得过如此辉煌成就的人，一旦进入物理学领域，怎么竟变得如此荒谬？

库恩沉思着这个谜一样的问题，出神地望着宿舍窗外（"我至今仍能清晰地回忆起当时窗外的葡萄架，以及它所投下的占了大半个窗子的阴影，历历在目"），直到他在某一瞬间突然顿悟了亚里士多德的合理性所在。库恩认识到，亚里士多德赋予其基本概念的含义，与现代物理学全然不同。例如，他用"运动"这一术语，所指的不仅仅是位置的变化，更意味着普遍的变化——太阳"脸红了"和太阳"下落了"同样是运动。亚里士多德的物理学，如果用他自己的术语来理解，仅仅是与牛顿物理学不同的另一种理论，而不是更低级的理论。

库恩从物理学转向了哲学，并且奋斗了整整15年，才终于把他当时的瞬间顿悟转化为一种理论，并在《科学革命的结构》中表述了出来。而这一理论的基石，正是"范式"概念。"范式"一词，在库恩之前，仅仅是指为达到教学目的所举的词形变化表。举例来说，在拉丁语教学中，为了让人理解动词的变化形式，你可以举出"amo，amas，amat"①一组变化形式相似的动词，这就是"范式"。库恩用这一术语是指由一组工作程序或信念构成的集合，它可以含蓄地（*implicitly*）指导科学家们应该相信什么、应该怎样去从事研究。大多数科学家从不质疑范式，他们解决被称为"难题"（*puzzles*）的一类特殊问题，而这些难题的解决可以强化、扩展范式的范围，而不是对范式提出挑战，库恩称之为"清扫工作"，或"常规科学"（normal science）。总会有"反常"（anomalies）存在，也就是范式所无法解释的甚至与范式相抵悟的现象，它们通常情况下会被忽略；但积累到一定程度之后，就会引发一场革命（也称作"范式转换"，但这种说法却不是库恩最早提出的），在革命中科学家们抛弃旧范式、拥抱新范式。

库恩不承认科学是一个持续的建设性过程，因为科学革命不仅是一种创造行为，同时也是一种大破坏。新范式的倡导者站在巨人们的肩膀上（借

---

① 拉丁文动词"爱"的第一、第二、第三人称形式——译者注。

用牛顿的说法），然后居高临下地狠抽"巨人们"耳光。他或她往往是其所在领域的年轻人或新手，也就是说，大脑尚未被既有的理论所塞满。多数科学家只是心不甘、情不愿地屈从于新的范式，他们通常情况下并不理解它，也不存在可用来甄别它的客观标准。不同的范式之间也不存在通用的比较标准，用库恩的术语来说，它们是"不可通约的"（incommensurable）。不同范式的支持者会没完没了地争论下去，但却无助于消除彼此间的分歧，因为对于一些基本概念，如运动、粒子、空间、时间等，他们的理解往往大相径庭。因此，科学家们的信仰转变过程，既是主观的也是政治性的，这一过程可能包括突然的顿悟理解——就像库恩通过思考亚里士多德而最终达到的那种。但通常情况下，科学家之所以接受某个范式，仅仅是因为它得到了某些权威科学家的支持，或者得到共同体中大多数成员的拥护。

库恩的观点在几个重要方面与波普尔不同。与许多波普尔的批评家一样，库恩竭力主张：证伪和证实都是不可能的，因为证伪和证实过程都暗含着超越单一范式检验的绝对标准。一个新范式在解决难题上或许比旧范式更优越，在实践上也许能产生更多的应用成就，"但你不能因此就简单地称旧范式下的科学是错误的"，库恩说。仅仅因为现代物理学产生出计算机、原子能以及CD唱机，在绝对意义上并不足以说明它比亚里士多德的物理学"更真"。同样地，库恩也不承认科学正在不断地逼近真理。他在《结构》一书的结尾部分断言：科学与地球上的生命一样，其进化并不趋向什么，而只是远离什么。

库恩向我描述他自己时，自称是个"后达尔文主义的康德主义者"（post-Darwinian Kantian），因为康德也相信：离开某种先验的范式，理性就不可能给感觉、经验赋予意义。但康德和达尔文都主张我们生来就或多或少具有某些相同的先天范式，库恩却认为范式是随着文化的变化而不断变化的。"不同的人群，以及同一人群在不同的时代，可以具有不同的经验，因而在某种意义上可以说是生活在不同世界里。"显然存在某些为所有人共享的处理经验方式，但这仅仅是因为存在着共享的生物遗传，库恩补充道。可是，究竟哪些东西才是普遍的人类经验，超越于文化和历史之上的究竟是什么，却是不可言喻的，超出了人类语言能力所及的范围。库恩认为，语言"不是普适的工具，事实上，并不是在一种语言环境中所能表达的一切，都可以在另一种语言环境中表达出来"。

"难道数学不是一种普适的工具吗？"我问道。不是，因为数学没有

"意义"；它是由句法规则构成的，但没有语义学的内容。"有很好的理由认为数学是一种语言，但却有更好的理由认为它不是。"我反驳道，库恩关于语言局限的观点虽可能在形而上学的层次上适用于某些领域（如量子力学），但却并不适用于所有情况。例如，少数生物学家声称，艾滋病并非由所谓的AIDS病毒引起的。这一主张要么是正确的，要么是错误的，在这里，语言并不是决定性的因素。库恩摇了摇头，答道："当你碰到两个人以不同的方式解释相同的事实的时候，那才是形而上学。"

那么，该怎么评价他自己的观点呢？"听着，"库恩的回答带着比平时更多的不耐烦，显然他以前已无数次地被问及同样的问题，"我认为按自己目前所用的这种方式交谈和思考，能带来一系列可被研究的可能性，但它就像任何科学结构一样，需对其效度加以检验——但这是应由你们去做的事情。"

在干巴巴地讲完自己关于科学以及人类交流的局限性的观点后，库恩开始抱怨对自己著作的种种误解和滥用，特别是那些来自"拥护者"方面的。"我常常说，我更喜欢那些批评我的人，而不是我的那些狂热的信徒。"他回想起某些学生常对他说的话："噢，库恩先生，谢谢您告诉我们有关范式的观点。既然我们已了解范式，就可以摆脱它的束缚了。"他坚信自己从未认为科学是完全政治性的，是优势权力结构的反映。"回首往事，我渐渐明白了这本书为什么会遭到如此严重的误解，但是，小伙子，这些既有的误解绝不是它有意造成的；现在，小伙子，它也无意招惹新的误解。"

他的抗议无济于事。他有一段痛苦的回忆：在一次讨论会上，他坐在那里试图解释：像真理和谬误这样的概念，在范式里仍然是有效的，甚至是必需的。"一位教授终于抬头看了看我，说：'听着，你并不清楚这本书是多么偏激。'"同时，库恩无奈地发现自己竟成了所有那些自封的科学革命家的保护神，"我曾接到过许多的信，写着：'我刚刚拜读了您的大作，它彻底改变了我的生活。我正试图发动一场革命，请您义伸援手'。随信寄来的，肯定是一叠像书那么厚的草稿。"

库恩宣称，尽管他在书中极力回避袒护科学的倾向，但他本人却是真正的"科学袒护者"。他认为正是科学的刻板和纪律，才使得科学在解题时如此有效，并进而使科学产生出所有人类事业中"最伟大、最根本的创造力大爆发"。库恩承认，自己应部分地对某些关于他的理论模式的反科学解释负责。毕竟，在《结构》中他确曾称某些科学家为沉溺于某种范式的"瘾君子"，也曾把他们与奥威尔《1984》中那些被清洗过大脑的角色相比。[17]但库恩坚称自己无意用"清扫工作"或"解难题"之类的术语，去贬低大多数科学家的工作。"这仅仅是为了描述的便利。"他沉思了一会儿，"也许，我应该多花些笔墨去描述'解难题'所带来的辉煌业绩，但我认为自己已做到了这一点。"

至于"范式"一词，库恩承认它已被"无可救药地用滥了"，并且已"完全失控"，就像病毒一样，这一术语已扩散到历史和科学哲学领域以外，使知识分子共同体普遍受到了感染，任何一种占支配地位的观念都被称为范式。1974年的一期《纽约客》杂志上的漫画，就敏锐地抓住了这一现象："爆炸性新闻，格斯通先生！"一位女士向另一位做得意状的男士吐出这样的句子："据我所知，你是第一位把'范式'用于现实生活的人。"更搞笑的是在布什政府期间，当白宫官员们在宣布一项新的经济计划时，竟也赫然用了"新范式"的标题（其实那一计划只不过是炒了炒里根经济的冷饭）。[18]

库恩再度承认他应对此承担部分责任，因为在《结构》中他对范式的定义并不详尽，他理应做得更好。就某种意义而言，范式意味着一种原型实验，就像传说中伽利略从比萨斜塔上向下抛重物的实验（这一传说也许并非实有其事）；在另一些场合，这一术语也可以指代使科学共同体维系在一起的"信仰整体"。（但不管怎么说，库恩否认他曾为范式下过21种定义，就像某位批评家所宣称的那样[19]。）在《结构》后期版本的跋中，库恩曾建议用"样本"（*exemplar*）这一术语代替"范式"，但从未被人们看重。最终，他对于向人们解释自己"范式"一词的真正含义一事彻底绝望了。"如果你遇到难以控制的局面，那么最好是随它去、不介入。"说到这儿，他无奈地叹了口气。

《结构》一书之所以具有感染力，之所以会有持久的影响，原因之一或许就在于它的歧义性；它既能打动相对主义者，又能打动科学崇拜者。库恩承认："本书获得很大成功的奥秘，以及针对本书的大部分批评，都应归于它的模糊性。"（曾有人怀疑库恩的写作风格究竟是故意的，还是他所固有的；他的演讲同样充满迷惑，纠缠着大量的虚拟语气和修饰词，就像他的文章一样。）《结构》很显然是一部文学著作，这样它就成了多种解释的主题；而根据文学批评理论，库恩本人并不能令人信服地提供关于著作的确切解释。下面给出的是关于库恩的文本以及库恩这个人的一种可能的解释：库恩所重点论述的是，科学"是什么"，而不是科学"应该是什么"。比起波普尔来，库恩对科学的看法更逼真、更敏锐、更准确。库恩认识到，考虑到现代科学的巨大威力，以及科学家们对那些经过无数次实验检验的理论的高度信赖，科学很可能已进入一个持久的常规发展阶段，在这一阶段中不会再产生革命或重大的发现。

与波普尔不同，库恩甚至相信：即使在正常情况下，科学也并不可能永远持续下去。"科学有其开端，"库恩说，"我们无法确切地描述它在多数社会中是怎样开始的，它需要一些特殊社会条件的支持，现在要想发现这些社会条件是越来越困难了。科学当然也会有其终结。"库恩认为，就算支持科学发展的资源足够多，科学也会因为科学家们找不到进一步发展的方向而终结。

库恩关于科学将会终结的认识——这使我们想起皮尔斯（Charles Sanders Peirce）对绝对真理的定义——使他具有一种比波普尔更强烈的紧迫感去质疑科学的权威，并彻底否定科学终将达到绝对真理的信条。"你也许会在心里嘀咕：我们不是已经发现了世界的真面目吗？"库恩说，"但这与我说的绝对真理并非一回事。"

在时至今日的职业生涯中，库恩一直力求使自己忠实于那次美妙的顿悟，也就是他在哈佛的学生宿舍所经历的那一次。在那一瞬间，库恩明白了：实在最终是不可知的；所有试图描述它的努力都是徒劳而又没有意义的。但库恩的洞见却迫使他选择了一种站不住脚的立场：既然所有的科学理论都无法达到绝对的、隐秘的真理，那么它们就都是同等的谬误；既然我们不可能发现"终极答案"，就不可能发现任何答案。他的神秘主义最终把他推向了与某些文学领域的诡辩家们同样荒谬的立场，这些诡辩家认为所有的文本——从《暴风雨》（*The Tempest*）到一则关于某种新品牌伏特加的广告——都是同样无意义的废话，或者是同样含义隽永的传世佳作。

在《结构》一书的最后，库恩简要地提出了这样一个问题，即为什么某些科学领域会统一在一种范式之下，而另一些领域却像艺术等领域一样，始终处于动荡不定的状态。他给出的答案是：这不过是个选择不同的问题。某些领域的科学家，就是不愿屈从于某个单一的范式。我怀疑库恩之所以回避对这一问题的探讨，是因为他无法容忍那真正的答案。某些领域，比如说经济学等社会科学领域，永远不会长期依附于单一的范式，因为对于它们所探讨的问题来说，没有任何一个单一范式能满足需要；而那些能够达成一致（或常态，借用库恩的说法）的领域，其所以能做到这一点，是因为其范式在某种程度上与真实的自然相吻合。

寻访费耶阿本德

说波普尔和库恩的观点有这样那样的缺陷，并不等于说它们就不能被用作解析科学的有效工具。库恩的常规科学模型准确地概括了目前大部分科学家的工作：补充细节，解答无关紧要的难题，不是为了质疑占主导地位的范式，而是为它锦上添花；波普尔的证伪原则，可帮助我们区别经验科学与反讽的科学。但是，这两人都因为在自己的观点上走得太远，且对自己的观点太过执着，最终陷入自我矛盾的荒谬困境，从而走进了死胡同。

作为一位怀疑论者，怎样才能避免像波普尔那样，须拍着桌子辩白自己不是教条主义者的窘迫呢？或者，怎样才能避免像库恩那样，一面大声疾呼真正的交流是不可能的，一面又喋喋不休地向你布道着这种不可能性理论的可笑呢？只有一种途径，就是你必须接受——甚至喜爱——超越于修辞意义之上的悖论和矛盾，必须了解怀疑主义是一项必要的但却又不可能完成的事业，必须了解保罗·费耶阿本德。

费耶阿本德的第一本，也是至今仍具有重要影响的著作是《反对方法》。这本书出版于1975年，现已被译成16种语言发行。[20] 其基本观点是：哲学不可能为科学提供方法论或逻辑依据，因为不存在解释的逻辑依据；通过对科学史上里程碑式的重大事件（如伽利略在罗马教廷的受审和量子力学的发展等）的分析，费耶阿本德试图证明：并不存在什么科学发展的逻辑，科学家们往往是出于各种主观的甚至非理性的原因，才去创造并坚持这样那样的科学理论。根据费耶阿本德的见解，为了科学的发展，科学家们能够且必须去做任何事情。他把这种离经叛道的观点概括为"怎么都行"。费耶阿本德曾嘲笑波普尔的批判理性主义，说它是"吹进实证主义者的茶杯里的一小口热气"。[21] 他在许多方面赞同库恩的观点，特别是库恩关于不同科学理论的不可通约性，但认为库恩所主张的常规科学是极其罕见的。他还恰如其分地指责库恩试图逃避自己观点的影响，指出：出乎库恩的意料，他关于科学变革的社会政治类比模型，竟然能很好地适用于有组织的犯罪集团。[22]

费耶阿本德对于装腔作势的嗜好，使他很容易被看作各种奇谈怪论的百宝囊。他曾把科学比作巫术、魔法和占星术；他曾为宗教极端主义辩护，认为公立中学在讲授达尔文进化论的同时，也有权讲授神创论。[23] 在1991年的《美国名人录》中，有关费耶阿本德的条目是以这样的文字结尾的："我的生命历程完全是由意外事件构成的，而不是人生目标和原则导致的。我作为知识分子的工作只是其中一个并不那么重要的部分，爱和个人理解才是更重要的。但知识分子们却用他们追逐客观真理的热情埋葬了这些个人的基本要素，他们绝不是人类的解放者，而是人类的罪人。"

　　费耶阿本德的达达主义式的辩术，揭示出一个极其严肃的论点：人类对绝对真理的追求，不论听起来多么崇高，往往以专制而告终。费耶阿本德之所以抨击科学，不是因为他真的相信科学与占星术一样无法拥有真理；恰恰相反，他抨击科学是因为他认识到了科学的威力，以及科学那可以掩没人类思想和文化之多样性的巨大潜力，并对此深怀恐惧。他反对科学的必然性，更多的是出于道德和政治的原因，而不是出于认识论的考量。

　　在他1987年出版的《告别理性》一书的结尾部分，费耶阿本德揭示出自己受相对主义的影响有多么深。他提到一件"曾激怒了许多读者，也曾使许多朋友失望的事——我拒绝谴责甚至是极端的法西斯主义，并认为它也有滋长的权利"。[24] 因为费耶阿本德二战期间曾在德军中服过役，就使得这一论点更加让人敏感。但费耶阿本德争辩说："谩骂法西斯主义，这太容易了；但法西斯主义之所以横行一时，正是由于这种道德上的以正义自居、自以为是造成的。"

　　1992年，当我准备去采访费耶阿本德的时候，他已经从加州大学伯克利分校退休了，那儿已没人知道他到底去了哪里。其同事对我说，他敢保证，我寻找费耶阿本德的努力注定是徒劳的。在伯克利有一部他的专用电话，但从未接到过他的呼叫；他曾接受过校联合会的开会邀请，但开会时却见不到他的身影；在通信中，他会邀请同事们去拜访他，但当他们真的去了并敲响他家（坐落在俯视伯克利的一座小山上）的房门时，却无人应答。

　　后来，在浏览《艾西斯》（一份有关科学史和科学哲学的期刊）时，我发现了费耶阿本德评论某本文集的一段短文，文章充分展示了他那一流的辩才。文集的作者有一段诋毁宗教的文字，对此费耶阿本德反唇相讥："比起天体力学来，祈祷也许并不那么有效，但它确实有其自己的诉求对象，比如说某种经济利益。"[26]

　　我于是打电话给《艾西斯》的编辑，问他怎样才能与费耶阿本德取得联系，他给了我一个瑞士的地址，在苏黎世附近。我按地址给费耶阿本德寄去一封谦恭的信，同时提出采访他的愿望。使我高兴的是，他回寄了一张亲切的手写便条，说很欢迎我的采访，他现在一部分时间待在苏黎世的家里，另一些时间则去罗马他夫人那里。信中还附有他在罗马的家里的电话号码以及一张他自己的照片，照片上的他扎着条大围裙，笑嘻嘻地咧开一张大嘴，正站在厨房里装满碟子的洗涤槽前。据他自己解释，那张照片"表现我正在干自己最喜欢的活儿——在罗马的家里帮妻子洗碗"。10月中旬，我收到了他的另一封信，"我要告诉你，我很可能（概率92%）在10月25日至11月1日的一周内去纽约，那时我们可以见见面。我一到纽约就会给你打电话。"

这样，在万圣节前几天的某个寒冷夜晚，我终于在第五大道的一套豪华公寓里见到了费耶阿本德。那套公寓属于他从前的一位学生，她聪明地放弃了哲学，转向了现实的房地产业，并且显然已取得了某种成功。她到门前迎接了我，并带我走进厨房。那儿，费耶阿本德正坐在一张桌子旁，惬意地呷着红葡萄酒。他从椅子上蹿了起来，斜趄着身子迎接我的到访，仿佛正承受着某种脊椎病的痛苦。直到那时，我才记起二战期间他的背部曾挨过一枪，并且永久地成了跛子。

费耶阿本德有一张矮妖精①般生机勃勃且又棱角分明的脸。当我们坐下来开始交谈的时候，随着内容的展开或情节的转换，他时而朗诵，时而打喷嚏，时而逗哏，时而窃窃私语，同时还不停地挥舞着手臂，仿佛他是正在指挥交响乐团演奏的指挥家。他的自嘲调和了他的傲慢，称自己为"懒汉"和"大嘴巴"②。当我就某些问题请教他的立场时，他缩了缩脖子，"我没有立场，"他说，"如果你有一个立场，你就像被螺丝拧紧了，没了自由。"他对着桌面拧着一把无形的螺丝刀："我曾很卖力地为某些观点辩护，随后却发现它们是多么愚蠢，于是就把它们统统抛弃了。"

一直带着宽容的微笑观看这些表演的，是他的妻子戈拉西娅·波利妮（Grazia Borrini），一位意大利物理学家。她与费耶阿本德形成了鲜明的对比，费耶阿本德有多么狂躁，波利妮就有多么娴静。她曾于1983年听过费耶阿本德的课，那时她正在伯克利进修公共卫生第二学位；6年后，他们结了婚。在我和费耶阿本德交谈时，波利妮只偶尔地插一两句话，例如，我问费耶阿本德，科学家们为什么会对他的著作大动肝火？

"我不知道，"他答道，做天真状，"有这么回事吗？"

波利妮插话说，当她从另一位物理学家那里第一次听到费耶阿本德的观点时，也曾大为光火，"竟然有人敢从我手里抢走打开宇宙之门的钥匙！"

---

① leprechaun，爱尔兰民间传说中的小精灵，专门负责向人们指点宝藏——译者注。
② 就照片来看，费氏的嘴确实不小，但这里取意双关，更意味着自嘲是个口无遮拦的人——译者注。

她这样抱怨着。只是在她读了费耶阿本德的著作之后，她才认识到，与批评家们的妄言比起来，费耶阿本德的观点是如此的精妙和敏锐。"我认为这正是你可以大做文章的地方，"波利妮对我说，"关于这种重大误解的文章。"

"噢，别再说了，他又不是我的新闻发言人，"费耶阿本德说。

与波普尔一样，费耶阿本德出生并成长于维也纳，十来岁的时候，他曾一度学过表演和歌剧。正是在这一阶段，他曾听某天文学家作过一次演讲，并激发起他对科学的兴趣。当时他并未把这两种激情看作是不可调和的，立志既要成为一位歌剧演唱家，又要成为一位天文学家。那段时间里，"我下午时间用于吊嗓子，晚上的时间在舞台上度过，然后，在深夜里去观察星星。"

后来，战争爆发了。德国在1938年侵占了奥地利，1942年，18岁的费耶阿本德被征入一所士官学校。虽然他曾祈祷让训练期延长到战争之后，但却很快结业了，并被送往苏联前线指挥3000名士兵。在1945年对苏军的作战（实际上是溃逃）中，他的腰部中了一颗子弹。"我无法起床，"费耶阿本德回忆道，"我仍能记起那时对未来生活的幻想：'啊，我将坐在一辆轮椅上，在成排的图书中间摇来摇去。'那时的我很快活。"

他逐渐恢复了行走的能力，虽然还离不开拐杖的帮助。战争结束后，他在维也纳大学恢复了学业，从物理学转行到历史学；后来觉得没劲，便又重返物理学；又觉着没意思，最终在哲学专业里停下来。凭借自己过人的聪明，他在"荒谬"的立场上越走越远，逐渐认识到：对于争论起关键作用的，不是真理，而是雄辩。"真理本身不过是个辞令，"费耶阿本德断言。他伸伸脖子，吟诵道："'我正在寻求真理'，噢，小伙子，你看我是个多么了不起的人啊！"

在1952—1953年的两年里，费耶阿本德在伦敦经济学院学习，师从波普尔，并在那里结识了波普尔的另一位优秀的学生拉卡托斯。几年后，正是在拉卡托斯的督促下，他才写出了《反对方法》一书。"他是我最好的朋友。"费耶阿本德这样谈起拉卡托斯。费耶阿本德在布里斯托尔大学任教，直到1959年才移居伯克利，并在那里与库恩成了同事。

与库恩一样，费耶阿本德不承认自己反科学，他所真正宣扬的，首先是不存在什么科学方法。"所谓科学方法，其实就是具体科研过程中的工作方法。"费耶阿本德解释道，"你本来有某种工作设想，但你只能随机应变。所以，你真正需要的是一个装满各种工具的工具箱，而不是仅有锤子和钉子。"这才是他那屡遭攻击的"怎么都行"的真正含义（而不是像通常被理解的那样，是指科学理论之间没有优劣之分）。费耶阿本德认为，若把科学束缚在某种特定的方法论之下，就会破坏科学的生机，即使像波普尔的证伪原则或库恩的常规科学模型那样宽泛的方法论，也同样如此。

费耶阿本德也反对那种以为科学优越于其他类别知识的观点。最使他气愤的是西方国家的这样一种倾向，即不管人民的意愿如何，硬是把自己的科学产品，无论是进化论、核电站，还是巨型粒子加速器，都强加到他们头上。"连教会都从国家中分离出去了，但科学却和国家日益紧密地搅在一起！"

科学"为我们提供有关宇宙的迷人故事，向我们描述它的成分、它的发展，以及生命的来源，诸如此类"，但近代科学出现之前那些"编故事的人"（他在讲到这个词时加重了语气），如诗人、宫廷弄臣、游吟歌手等，是自己谋生的，而大多数现代科学家们却要靠纳税人来养活，"公众作为恩主，有权就此事发表意见"。

费耶阿本德又补充道："当然，我的说法的确有些极端，但绝未极端到人们指责我的那种程度，如彻底放弃科学。我主张放弃的，仅仅是科学至上的观点。不能每件事都用科学去处理，如此而已。"毕竟，在许多问题上，科学家们彼此间也存在着不同意见。"所以，当某位科学家号召'大家都应该如何如何'的时候，人们也不必太认真。"

我问道，如果他并不反科学，那么他在《名人录》中关于知识分子都是罪人的话该怎么解释。"在很长的一段时间里，我确实是这么认为的，"费耶阿本德答道，"但是去年我的认识发生了改变，因为毕竟还有许多好知识分子。"他扭头看着他的妻子，说："我觉得，你就是一位好知识分子。""不，我只是一位物理学家。"她回答的语气很坚决。费耶阿本德耸了耸肩："'知识分子'的含义究竟是什么？也许它意味着这样一些人：他们对某些事比别人思考的时间更久，但他们中的多数人却只喜欢跑到人们面前宣称'我们发现了什么什么'。"

费耶阿本德指出，许多非工业化国家的人们离开科学照样活得很好。非洲布须曼人"在任何西方人都无法生存下去的环境中生活着。也许你会说我们社会中的人寿命比他们长，但问题是生活的质量如何，这一点是至今尚无定论的"。

这类说法当然会激怒大多数科学家，难道费耶阿本德没意识到这一点吗？纵然布须曼人活得幸福，他们毕竟很愚昧，知识难道不比愚昧更好吗？"知识有什么了不起？"费耶阿本德答道，"那些'愚昧'的人彼此友善相处，从不你打我斗。"如果人们认为应该放弃科学，他们有权作出这样的选择。

这是不是意味着基督教极端主义者也有权在公立中学里讲授神创论，并要求给之以与进化论相同的地位？"我认为'权利'是个很微妙的东西，因为一旦某人拥有某项权利，他就会凭借手中的权力去压制别人。"他停了一会，接着谈道，理想的情况是，学生们应能接触到尽可能多的不同思想，这样他们才能在这些不同思想中自由地作出选择。他极不自在地换了一种坐姿。我觉得有必要在这时说点什么，便向他指出他还没有真正回答我关于神创论的问题。费耶阿本德沉下了脸："这是个乏味的问题。我对它毫无兴趣。极端主义并不是那古老又含义丰富的基督教传统。"但极端主义者在美国很有势力，我坚持说，他们正利用费耶阿本德的观点攻击进化论。"但科学也曾被用来说明某些人种是天生的低智商者，"他反击道，"可见每样东西都有很多不同的用法，科学可以用来击败各种各样的人。"

但是，教育者们难道不应指出科学理论和宗教神话之间的区别吗？我又问道。"当然应该。目前，科学在教育中的地位已是无可撼动的了，但我认为还应该引入其他方面的东西，给予这些方面展现自己，并尽力为自己辩护的机会。"其实，那些所谓"未开化"的人们关于自己生存环境的知识，如关于当地植物功用的知识，是我们所谓的专家都无法比拟的。"如此看来，称这些人无知只说明——这才是真的无知。"

他用以攻击西方理性主义的手段，正是西方理性主义提供的，这难道不是有点自相矛盾吗？对于我设下的这个圈套，费耶阿本德并不上当。"它们只不过是工具罢了，既然是工具，你当然可以用到任何你认为合适的地方，"他温和地说道，"人们不能因为我应用了这些工具而指责我。"费耶阿本德似乎有些厌烦，尽管他自己不会承认，但我仍猜测：他已疲于应付作为一个激进的相对主义者所带来的烦恼了，疲于面对强横的理性主义去为这个世界多姿多彩的信仰系统——占星术、神创论，甚至法西斯主义——辩护了。

然而当谈起自己正在写作的新书时，费耶阿本德的眼睛又明亮了起来。新书的名字暂定为《征服丰富性》（*The Conquest of Abundance*），讨论的是人类追求还原论的热情。"所有的人类事业"，费耶阿本德解释道，都在努力简化大自然所固有的多样性或"丰富性"。"首先，知觉系统大大简化了这种丰富性，否则你就无法生存。"宗教、科学、政治以及哲学，代表着我们更进一步挤压现实的努力。当然，这些征服丰富性的企图只不过是创造出新的丰富性、新的复杂性。"但政治战争却使许多人失去了生命。我的意思是说，某些观点的确是不受欢迎的。"我终于意识到，费耶阿本德所讨论的正是我们对终极答案的追求，寻找一种终结所有理论的理论。

但是，根据费耶阿本德的理解，终极答案将是——也必定是——永远无法企及的。他嘲讽了某些科学家的信念，认为他们有朝一日将会构建一个能够解释一切的理论，从而抓住实在的本质。"如果这会使他们高兴，就让他们坚持自己的信念吧。就让他们扬扬自得地到处宣扬吧，说'我们认识了无限！'那么，听众中的一部分人会说，——不耐烦的语气——'哈，哈，他说他认识了无限。'另一部分人会说——激动的语气——'哈，哈，他说他认识了无限！'但是，如果他跑到学校里去告诉那些小孩子，'无论如何，这就是真理！'那就太过分了。"

任何关于实在的描述都必然是不充分的，费耶阿本德说，"你真的以为根据现代宇宙论，这个过去的小爬虫，这个无足轻重的小角色，这个渺小的人，能把一切都搞明白吗？在我看来，这简直是疯话！绝不可能是真的！他们所认识到的只不过是相对于其认识活动的一种特殊反应。这种认识，只会使宇宙以及藏在宇宙背后的实在，发出揶揄的笑声：'哈哈！他们竟然以为已经发现了我！'"

费耶阿本德说，一位被称作"伪法官迪奥尼索斯"（Dionysius the Pseudo-Areopagite）的中世纪哲学家曾论证过，直接观察上帝是什么也看不到的。"虽然无法解释清楚为什么，但我觉得这句话很有意思。对于这个最大的实在、这个万物之源，你绝不会有认识它的手段。你的语言之所以产生出来，是为了描述各种事物，如椅子以及几件仪器，只适应于这个小小的地球！"费耶阿本德兴高采烈地顿了一下，"上帝是一层一层次第展现自己的，你知道吗？随着层次的下降，逐渐变成了具体实物。在最低的层次上，你能够看到他的一点点踪影，你只能就此去猜测他的模样。"

我对他这突如其来的激情感到有些惊讶，便问他是否相信宗教。"说不准。"他年轻的时候曾是罗马天主教徒，后来却变成了"坚定的"无神论者。"现在，我的哲学变成了另一种与以往完全不同的形式，它不能仅仅是关于你所知道的宇宙及其发展，这没有半点意义。"当然，许多科学家和哲学家认为推测宇宙的意义或目的等是毫无价值的。"但人们却关心这一问题，为什么就不能推测呢？所有这些推测结论都塞在这本书里，这样才能解开关于丰富性之谜。当然，这也会使我在相当长的一段时间里不至于无事可做。"

在我准备告辞的时候，费耶阿本德突然问我，头一天晚上我夫人的生日晚会办得怎么样（在安排与他见面的日程时，我曾告诉过他我妻子过生日的事）。"很好。"我回答。"你们还没打算分手吗？"费耶阿本德得寸进尺地问道，同时仔细地审视着我，"那会不会是你最后一次给她庆贺生日？"

波利妮被惊呆了，瞪着他问道："为什么是最后一次？"

"我不知道！"费耶阿本德宣布，同时向空中伸展开双臂，"因为那是很可能发生的！"他重新转向我，"你们结婚多久了？""三年。"我回答。"啊，才刚刚开始，好戏还在后面呢。再等十年，就可见分晓了。""现在，你听起来确实像个哲学家。"我回敬道。费耶阿本德大笑起来，他坦白说在遇到波利妮之前曾结过三次婚，都以离婚而告终。"现在，我才第一次体会到婚姻的幸福。"

我告诉他，我曾听说与波利妮的婚姻使他变得更加随和了。"噢，这可能是两回事。"费耶阿本德答道，"人上了年纪，精力衰退了，不可能不随和，她肯定也因此改变了很多。"他冲波利妮微微一笑，她也以会心的微笑回望着他。

我转向波利妮，问起那张她丈夫寄给我的照片的事。照片上的他好像正在洗餐具，那张便条上也说帮助妻子做家务对现在的他来说是头等重要的事，真的是这样吗？

波利妮哼了一声，说："难得有一次。"

"你这是什么意思？难得有一次！"费耶阿本德吼了起来，"我天天都洗碗！"

"难得有一次。"波利妮语气平静而坚定地重复。我的决定是：最好相信这位物理学家，而不是这位相对主义者。

在我采访费耶阿本德之后一年多的时候，《纽约时报》报道了一则让我惊愕的消息，这位"反科学的哲学家"已经被脑瘤夺去了生命。[27] 我打电话给身在苏黎世的波利妮以表示我的哀思，同时，毋庸讳言，也为了满足一下自己那不怎么光彩的职业好奇心。她很激动：这事发生得太快了，保罗说他有些头痛，然而几个月后……她努力使自己平静下来，用骄傲的语气接着对我说，费耶阿本德一直工作到生命的最后一刻，就在临终前他才写完了自传的手稿。（这本书题着典型的费耶阿本德式的名字——《不务正业的一生》，于1995年出版。在书的最后几页，也是费耶阿本德用生命的最后时光所写的文字里，他总结道：生命的意义在于爱。）[28] 那本关于丰富性的书呢？我问道。保罗已没有时间完成了。

忆及费耶阿本德对医生职业的苛责，我忍不住问波利妮，她先生对治疗很热心吗？当然，她回答，他"完全信赖"医生的诊断，乐意接受他们给予的任何治疗；只是那个肿瘤发现得太晚了，任何治疗都已无济于事。

哲学的道路为何如此艰辛

无论如何，西奥查理斯和皮莫波洛斯二人，也就是《自然》杂志上那篇题为"科学哪儿错了？"一文的两位作者，至少有一点说得很正确，即波普尔、库恩以及费耶阿本德的观点是"搬起石头砸了自己的脚"。所有的怀疑主义者，最后都会跌进自掘的坟墓中，成为批评家哈罗德·布鲁姆在《影响的焦虑》中所嘲笑的"单纯的反叛者"。他们反对科学真理的最有力论证，其实出自历史维度：考虑到科学理论在过去的一个世纪左右时间里那种天翻地覆的变化，我们怎么能肯定现在的理论是长久的呢？诚然，现代科学与这些怀疑论者的批评，尤其是库恩的批评比起来，的确更不具有革命性，更加保守。粒子物理学家们悠然憩息在量子力学的坚实基础之上，而现代遗传学更多的是忙着支撑达尔文进化论的大厦，而不是去挖它的墙脚。当怀疑论者用来自历史的证据转而攻击哲学时，其杀伤力更强大：如果连科学都无法获得绝对真理，那么解决问题能力远低于科学的哲学，又有什么立足之地呢？哲学家们自己也已认识到他们的窘困境地，在1987年出版的《末路上的哲学：终结还是革新？》中，十四位著名的哲学家对自己的学科是否有前途做了认真的思考，达成的共识倒是颇富哲学味：也许有，也许没有。[29]

其中的一位哲学家是科林·麦金（Colin McGinn），对于他所称的"久无进步"的现状做了深入思考。麦金是个地道的英国人，自1992年以来，一直在拉特格斯大学执教。当我1994年8月在曼哈顿西北部他的公寓里见到他时，发现他竟出奇的年轻（当然，在我的想象中，哲学家都应该是满脑门智慧的皱纹并且耳背的家伙）。他穿着牛仔裤，一件白色的T恤衫，足下蹬着双软拖鞋。他身材精悍，脖子修长，并有一双浅蓝色的眼睛，不知情者可能会误以为他是安东尼·霍普金斯①的弟弟呢。

当我请教麦金对波普尔、库恩和费耶阿本德的看法时，他撇了撇嘴，作出一副厌烦的表情。他们是"草率的""不负责任的"，特别是库恩，充斥着"荒谬的主观主义和相对主义"，几乎没有几个现代哲学家会把他当回事儿。"我认为，科学无论如何都不是暂时的，"麦金断言，"的确有些科学是暂时，但也有一些并非如此。"是元素周期表是暂时的，还是达尔文的自然选择理论是暂时的？

哲学的任务也并不在于提供诸如此类的答案，麦金说，哲学的发展进程也绝不是"发现问题，然后研究解决问题，最后进入下一个问题"的刻板模式。确实有某些哲学问题已经被"澄清"，某些方法已经过时，但那些重大的哲学问题，像真理是什么、自由意志存在吗、我们如何才能认识事物等，今天的我们和古代的先辈们同样所知甚少。这一事实不值得大惊小怪，因为现代哲学可被定义为为解决那些超出经验和科学探索范围之外的问题而作出的努力。

麦金指出，20世纪的许多哲学家——特别是路德维希·维特根斯坦（Ludwig Wittgenstein）和逻辑实证主义者——简单地宣称哲学问题是虚假问题，充斥着由语言和"思维缺陷"导致的幻象。某些此类的"取消主义者"为了解决心—身问题，甚至否认意识的存在。"这一观点可能导致你所无法接受的政治后果，"麦金说，"它把人降低到无足轻重的地位，把你推向极端的唯物主义和行为主义。"

麦金提出了一种不同的，据他自己认为是更合理的解释：那些重大的哲学问题是真正的问题，但却超出了人类的认识能力。我们能够提出这些问题，但不能解决它们——就像一只耗子永远也不可能求解一道微分方程一样。麦金说，这一思想是他还在英国生活时，于某个深夜里突然顿悟的。后来他才知道，自己的见解与语言学家诺姆·乔姆斯基（Noam Chomsky，他的观点将在第六章中探讨）著作中的观点不谋而合。在其1993年的《哲学中的问题》一书里，麦金宣称：也许几百万年以后，哲学家们才会认识到他的预见是多么正确。[30]当然，他告诉我说，哲学家们也许很快就会放弃他们那无望的挣扎。

麦金怀疑，科学似乎也正在走向绝路。"人们对科学和科学方法有着无比的自信，科学也在其自身的限度内出色地表演了几百年，但从长远来看，

①　Anthony Hopkins，美国影星——译者注。

谁能保证科学能永远如此兴旺并最终征服一切？"科学家和哲学家一样，要受制于其自身认识的局限性。"如果以为我们头脑中的认识手段已尽善尽美，那只不过是狂妄自大的想法。"更何况，冷战结束后，对科研投资的主要刺激因素已不复存在。同时，随着科学完结论的滋生，被吸引到科学道路上的有才华的年轻人也会越来越少。

"因而，若是到21世纪的某个时候，科研的生力军从各个领域大规模撤退，只留下少部分人继续研究那些尚需认识的事物，而把主要力量投入人文学科，我对此丝毫不会感到意外。"将来我们回顾历史时，会把科学看作"一个阶段，一个辉煌的阶段；人们会完全忘记仅仅是1000年以前，竟然会只有宗教教义"。在科学终结以后，"宗教会再度向人们发出召唤"。作为一名自称的无神论者，麦金这时候看起来自我感觉相当好——上帝保佑他会永远这样。在那次简短而又轻松的谈话过程中，小汽车的喇叭声、巴士的轰鸣声，以及油腻的中国食品那刺鼻的气味，和着阵阵轻风，从敞开的窗子飘进来。麦金就在这样的氛围中宣告了不是一类，而是两类主要的人类知识领域的末日——哲学和科学。

令人畏惧的"萨伊尔"

当然，哲学永远也不会真正终结，只不过会以一种更明显是反讽的、文学的形式继续下去，就像在尼采、维特根斯坦或费耶阿本德哲学中已经表现出来的那样。我最喜欢的文学哲学家是阿根廷寓言家豪尔赫·路易斯·博尔赫斯（Jorge Luis Borges），与我所熟知的所有哲学家比起来，博尔赫斯更多地探讨了我们对真理的复杂心理关系。在《萨伊尔》①中，博尔赫斯讲述了这样一个故事：一个人突然对杂货铺老板找零钱时给他的一枚硬币着了魔。[31]那枚看来毫无出奇之处的硬币，那个萨伊尔，突然变成了标志一切事物的东西，成了存在奥秘的化身。一个萨伊尔可以是一个星盘、一只老虎、一块石头，一切的一切。一旦涌起了有关它的幻象，就再也无法从记忆中抹去，它牢牢地盘踞在视幻者的脑海里，直到使现实的其他一切都失去了意义。

起初，讲故事的人尚挣扎着想使自己的心智摆脱这枚萨伊尔的纠缠，但最终还是接受了自己的命运。"穿越了数以千计的表象之后，我将栖息于

① The Zahir，"萨伊尔"是阿根廷的面值两角钱的普通硬币——译者注。

单一表象之上：从一个缤纷复杂的梦境到一个十分简单的梦。其他人也许会梦见我在发疯，而我则只梦见那个'萨伊尔'。待到地球上所有人都酣眠在'萨伊尔'之梦中——日日夜夜，那么，这个地球和这个'萨伊尔'，哪个是现实哪个是梦？"[32] 这个"萨伊尔"，当然就是那"终极答案"，那生命的奥秘，那终结一切理论的理论。波普尔、库恩和费耶阿本德试图用怀疑和理性，而博尔赫斯则试图用恐惧使我们远离那"终极答案"。

第三章
物理学的终结

在那终极答案的追求者中，没有比现代粒子物理学家更卖力的了，可以用"疯魔"这个词形容。他们试图证明，世界上一切复杂事物，其实都只不过是同一种东西的不同表现形式。一种要素，一种力，一种在十维超空间中蠕动着的能量环。社会生物学家可能会怀疑：是不是有某种特殊的基因影响和决定着这些还原论者的神经冲动，因为自人类文明的曙光初现之日起，这种冲动就激励着一代又一代的思想家。甚至就连上帝，也是在这种冲动的作用下构思出来的。

爱因斯坦是第一位伟大的现代终极答案追求者，他把晚年的时间全部用在统一场理论的构思中，期待它能把量子力学和自己的引力理论（广义相对论）统一起来。对他来说，寻求这样一种理论，是为了确定宇宙是不是必然的，或者用他的话来说，"上帝在创造宇宙时是否有选择余地"。毋庸置疑，爱因斯坦相信科学会使生活更有意义，但他同样认为不存在什么终极理论。他有一次在评价自己的相对论时说过，"（它）肯定会让位给另外的理论，虽然其具体的理由我们目前尚无法臆测。我相信深化理论的进程是没有止境的"。[1]

大多数与爱因斯坦同一时代的科学家，都把他统一物理学的尝试看作是其年高智昏和准宗教倾向的产物，但到20世纪70年代，大一统的梦想却又在几个新的研究进展的刺激下复活了。首先，物理学家们证明，正如电和磁都是同一相互作用的不同方面，电磁相互作用和弱相互作用（它控制着特定种类原子核的衰变）同样是一种基本的弱电相互作用的不同表现形式。研究人员还发展了一种关于强相互作用（它的作用是在原子核中把质子和中子紧密结合在一起）的理论，这一被称为量子色动力学的理论假定：质子和中子都是由被称作夸克的更基本的粒子组成，弱电理论和量子色动力学一起构成了粒子物理学的标准模型。

受这一成功的鼓舞，科学家们的研究远远超出了标准模型的范围，试图发现一种更深入的理论，他们的工作指南是一种叫作"对称性"的数学特性，它允许一个系统的元素经过变换——类似于旋转或镜面反射——而不会产生本质的变化。对称性成为粒子物理学家必不可少的工具。为了探求更高对称性的理论，理论物理学家开始转向高维情形，正如从二维地平面上升入太空的宇航员能更直接地鸟瞰地球的整体对称性一样，理论物理学家们也期望从高维的角度认清隐匿在粒子相互作用下更精妙的对称性。

粒子物理学中最持久的问题之一，是由于把粒子定义为点带来的。正如零作除数产生无穷大因而毫无意义一样，涉及点状粒子的计算也常常以毫无意义而告终。在构建标准模型时，物理学家们尚能忽视这些问题的存在，但存在着时间和空间畸变的爱因斯坦引力理论，却似乎呼唤着一种彻底的解决途径。

20世纪80年代初，许多物理学家意识到，超弦理论所代表的正是这种途径。这一理论用微小的能量环代替点状粒子，从而消除了计算中产生的荒谬。就像小提琴弦的振动能产生不同的音调一样，这些弦的振动也能产生出物理世界中所有的力和粒子。同时，超弦还能消除粒子物理学家的忧虑：并不存在物理世界的最终基础，它只是在向越来越小的粒子无限退却，每种粒子里面包含更小的粒子，就像层层嵌套的俄罗斯套娃那样。按照超弦理论，则存在着一种最基本的尺度，在这种尺度之外，所有关于时间和空间的问题都将毫无意义。

然而，这个理论也有缺陷。首先，似乎存在无数可能的途径，理论家们无从知道哪种正确；其次，超弦不但具有我们存身于其中的四个维度（三维空间外加一维时间），还具有六维额外维度，它们在某种程度上被"压缩"了，或蜷缩为无穷小的球；最后，超弦之于质子犹如质子之于太阳系那般小，从某种意义上说，这种弦甚至比我们距隐藏在可视宇宙最边缘的类星体还要远。超导超级对撞机同以往的任何加速器相比，应能使物理学家深入到更微观的领域，其周长将达86.9千米（54英里）；但要想探索超弦盘踞的王国，物理学家将不得不建造一个周长为1000光年的粒子加速器（而整个太阳系的周长只有一光天）。即便是那样的加速器，仍不足以使我们观测超弦们翩翩起舞的那些额外维度。

格拉肖的忧郁

作为一名科学记者（science writer）所能享受的乐趣之一，就是感觉自己比一般新闻记者要高出一筹。最下作的记者，在我看来，是这样一种类型的人，他们会追着一位目睹自己的唯一爱子被疯子杀害的母亲，并不依不饶地问："你对此有何感想？"但在1993年秋天，我发觉自己竟也坠入了同样的情境之中。就在我准备一篇关于粒子物理学未来发展的稿子时，美国国会一劳永逸地砍掉了超导超级对撞机［契约方已花掉了20多亿美元，且已在德克萨斯州挖掘了一条24.1千米（15英里）长的隧道］。在随后的几个星期里，我不得不去面对那些刚对未来充满憧憬却又被残忍抛弃了的粒子物理学家，并追问："你对此有何感想？"

在我采访过的地方中，气氛最为沉郁之处当属哈佛大学物理系。该系的主任是谢尔登·格拉肖（Sheldon Glashow），他曾因创立了标准模型的弱电部分，而与史蒂文·温伯格（Steven Weinberg）和阿卜杜斯·萨拉姆（Abdus Salam）共同分享诺贝尔奖。1989年，格拉肖与生物学家冈瑟·斯滕特一起，出席了在古斯塔夫·阿道夫大学召开的题为"科学的终结？"的研讨会，并发表了演讲。他慷慨激昂地批驳了会议的"荒谬"主题，即哲学的怀疑主义正把科学信仰腐蚀为"统一的、万能的、客观的努力"的主张。难道真有人怀疑几个世纪之前就被伽利略发现的木星卫星的存在吗？难道真有人怀疑关于疾病的现代理论？"病菌是被发现然后又被杀死的，"格拉肖宣布，"不是被想象出来然后又被想没了。"

科学的发展"的确渐渐慢了"，格拉肖承认，但并不是因为无知的、反科学的诡辩家们的攻击。他的粒子物理学领域"正承受的威胁来自一个完全不同的方向：来自其自身巨大的成就"。最近十年的研究已发现了粒子物理学标准模型的无数确证，"但也揭示了其并非微不足道的缺陷，并非不值一提的矛盾……我们缺乏能指引我们建立一个更宏大的理论的实验契机或线索"。最后，格拉肖附上一个充满希望的尾音："探索自然的征途常常陷入山重水复疑无路的境地，但我们总能找到出路。"[2]

从另一角度看，格拉肖并没有坚持这种乐观的看法。他曾经是探求统一理论的领导者，20世纪70年代，他提出过几种这样的理论，尽管没有一个像超弦理论那样宏大，但随着超弦的降临，他对统一理论的梦想破灭了。格拉肖辩解道，那些搞超弦或其他统一理论的人根本不是在做物理学研究，因为他们的玄想已远远超出了任何经验检验所能验证的范围。在一篇文章中，格拉肖及其同事抱怨道："研究超弦将导致远离传统粒子物理学，犹如粒子物理学远离化学一样。或许将来它们会像中世纪的神学那样在神学院中讲授。"然后又补充道，自欧洲史上的黑暗时期以来，我们会第一次目睹自己高尚的探索之路将怎样终结，而再次以忠诚代替科学。[3]当粒子物理学超越了经验王国之后，格拉肖似乎在暗示，它最终会向怀疑主义和相对主义臣服。

我于1993年11月在哈佛采访了格拉肖，正值超导超级对撞机项目被"腰斩"之际。他的办公室灯光暗淡，排满了漆成深黑色的书柜和书架，就像殡仪馆那样肃穆。身材高大的格拉肖烦躁地咬着熄灭了的雪茄烟蒂，看上去与那里的气氛有些不协调；雪白的头发在头顶蓬散着，似乎表明那是诺贝尔物理学奖桂冠不可或缺的一部分，而他的眼镜则像望远镜的镜头那样厚。但人们仍可在他那哈佛教授的"外壳"之下，觉察出格拉肖年青时代的影子——一位健壮的、快人快语的纽约小伙子。

超导超级对撞机项目的夭折，几乎将格拉肖彻底摧垮。他强调，不管热衷超弦的人们怎么说，但纯粹依靠思维，物理学是不可能发展的。超弦理论"除了自吹自擂之外，一无可取之处"，格拉肖嘟囔着说。一个多世纪以前，一些物理学家就试图发明出统一理论，他们当然只能以失败告终，因为他们对电子、质子、中子和量子力学一无所知。"现在，难道我们就能傲慢地相信，自己已拥有所有必要的实验数据，足以构建出理论物理学长期以来梦寐以求的神圣目标——统一理论吗？我并不这样认为。我敢肯定，自然现象中仍隐藏着有待我们去发现的意外之喜，但若我们不去观测，根本就谈不上发现它们。"

但物理学中除了统一问题之外便无事可做了吗？"当然有！"格拉肖厉声回答道，天体物理学、凝聚态物理学，甚至粒子物理学中的分支领域，它们都不关心统一问题。"物理学是一个充满有趣谜题（puzzle）的大家族（他借用了托马斯·库恩的术语来描述那些其答案只能强化流行范式的科学问题），当然有事可做了。问题只是，我们是否正向着那神圣的梦想接近呢？"格拉肖相信，物理学家能继续探索"一些有趣的鲜为人知的领域和一些新奇的事物，但这些探索与我在自己的职业生涯中有幸从事的那些研究比起来，肯定是截然不同的"。

考虑到科学资助的政治因素，格拉肖对自己领域的前景并不抱乐观态度，他不得不承认，粒子物理学本身并没有多少实用价值。"没有人能宣称这类研究将会造出什么实用的东西，那只是句谎言。就政府目前的态度来看，我所钟情的这种研究不会有很好的前景。"

在这种情况下，标准模型能成为粒子物理学的终极理论吗？格拉肖摇摇头，"有待解决的问题太多了"。然后又补充道，当然了，如果没有更高能的加速器，物理学家不可能比标准模型走得更远，那么在实践意义上标准模型便成为最终结果了。"也许将来会产生标准理论，那将是整个基本物理学故事的最后篇章。"未来总有一天，人们会找到一种成本低廉的方法产生极高的能量，这总是有可能的。"总有一天会实现这一点的，总有一天，总有一天，总有一天……"

格拉肖又说道，问题是粒子物理学家在这段等待的岁月里该做些什么？"据我猜测，答案将是：'粒子物理学'组织会做一些琐碎的工作或徘徊不前，直到某些事情成为可能。但他们当然不会承认自己的工作是毫无意义的，没有人会说：'我在做毫无意义的事'。"自然，随着这一领域越来越变得索然无味、资金越来越少，它将不再能吸引新的人才。格拉肖指出，已经有几个很有才气的研究生离开哈佛，下海去了华尔街。"高盛集团（Goldman Sachs）发现，理论物理学家是值得格外招揽的优秀人才。"

最出色的物理学家

超弦理论之所以能在20世纪80年代中期如此盛行，原因之一在于一位名叫爱德华·威滕（Edward Witten）的物理学家断定，它代表着物理学在未来最有希望的发展方向。我第一次见到威滕是在20世纪80年代末，当时我正同另一位科学家在普林斯顿高等研究院的咖啡馆里共进午餐。一个男子端着一盘食物从我们桌边走过，他的下巴前探，额头前凸，一副厚而黑的墨镜横贯底部，同样厚而黑的头发高耸于顶，使那被框起来的额头出奇的高。"他是谁？"我问同伴。"哦，他是威滕，"同伴回答道，"一个粒子物理学家。"

一两年之后，在一次物理学年会中间休息的闲聊中，我问了若干位与会者：他们中谁是最出色的物理学家？被反复提出的名字有几个，包括诺贝尔奖得主史蒂文·温伯格（Steven Weinberg）和默里·盖尔曼（Murray Gell-Mann），但被提到次数最多的是威滕。他令人产生一种特别的敬畏之情，好像他自己专属一类。他常被比作爱因斯坦，一个同事在类比时追溯得更为久远，认为他是自牛顿以来最具数学头脑的物理学家。

　　威滕也是我所遇到的天真型反讽科学实践者中最引人注目的人物。天真型反讽科学家对自己的科学猜想有一种独特的强烈信念，从不顾及自己的猜想根本无法被经验所验证这一事实。他们认为自己发现的理论远比发明的多；这些理论独立存在于文化和历史语境中，与找到它们的努力无关。

　　一个天真的反讽科学家，就像德克萨斯人固执地以为除了自己的同乡外其他人说话都有口音一样，不承认他或她采取了任何哲学立场（更不用说那种可能被认为是反讽性的立场了）。这样的科学家，只是真理从柏拉图式的精神世界通向凡间的一条管道，背景和个性与科学工作毫无关系。所以，当我打电话请求采访威滕时，他竭力劝阻我不要写他。他告诉我，自己对记者们将注意力集中到科学家的个性上的言行深恶痛绝，更何况，许多物理学家或数学家都比他更有写头。他对刊登在1987年《纽约时报杂志》上的一篇人物传记大为恼火，上面竟然暗示他"发明"了超弦理论。[4]威滕告诉我，实际上他个人在创立超弦理论的过程中并未起过任何作用，他的工作只是在它已被"发现"之后，对它进行了发展和宣传。

　　任何一位科学记者都会碰到一些不愿受到媒体关注的采访对象，他们只想安静地做自己的工作，但他们并未意识到，这种性格反而使自己更引人注目。我对威滕流露出的真挚的羞涩大感兴趣，坚持要采访。他说想先看看我以前写过的东西，于是我老老实实地将刊登在《科学美国人》上的一篇关于托马斯·库恩的传记传给了他。最终，他总算接受了我的采访请求，但只给我两小时时间，一分钟都不能超，我必须在正午12：00准时离开。我刚抵达，他立即就开始奚落我拙劣的记者道德，我不得不费尽口舌转述出托马斯·库恩的观点，即认为科学是一个并不收敛于真理的无理性（而不是非理性）的过程。他认为我是在对社会犯罪，"你应该关注科学上那些严肃而又实际的贡献，"威滕说道。库恩的哲学"除了作为争论的标准外，并不很严肃，连他的拥护者也这样认为"。库恩生病时去看医生吗？他的汽车也用子午线轮胎吗？我耸耸肩说，或许他会的吧。威滕胜利地点点头，说道，这说明库恩相信科学，而不是自己的相对主义哲学。

　　我说，不管人们是否同意库恩的观点，它毕竟已经有了极大的影响力，并且很刺激人。作为一名科学记者，我的目的不仅是要给读者带来各种信息，而且还要给他们带来刺激。"要报道已发现的真理，而不要去煽情，这应该是作为一名科学记者的基本职业操守。"威滕严厉地说。"我想二者都做到。"我回答道。"这只是一个漂亮而无力的借口，激起读者们的兴趣只应是报道真理所带来的副产品。"这是天真的反讽科学家的另一个标志：当他或她说到"真理"时，没有丝毫嘲讽的表情，不苟言笑，似乎这个词的内涵是天经地义的。威滕建议，为了救赎我作为记者所犯下的罪业，我应该写五位数学家的传记；如果我不知道哪些数学家值得采访，他会推荐一些。（威滕似乎并未意识到，其做法正为某些人提供攻击他的口实，他们宣称：与其说威滕是一个物理学家，不如说他是个数学家。）

由于正午将至，我试图将话题引到威滕的经历上来，但他拒绝回答任何"个人的"问题，比如他在大学里主修什么，在成为物理学家之前是否考虑过从事别的职业。他认为他的经历并不重要。我从背景材料上得知，虽然他是个物理学家的儿子，并且一直喜欢这门学科，但他却于1971年在布兰迪斯大学获得了历史学学位，想当一名时事记者。他成功地在《新共和》与《国家》上发表了文章，然后他很快意识到自己缺乏新闻工作的"常识"（大概是他自己告诉某位记者的）；再然后，他考进了普林斯顿学习物理，并于1976年获得博士学位。

威滕便从这儿开始讲述自己的故事。谈及他在物理学方面的工作时，就像作一篇极度抽象的、干瘪的演说，他在背诵而非讲述超弦的历史，强调的不是自己的作用而是别人的。他说得如此轻柔，以至于我都担心在空调机的噪音中能否录下他说的内容。他常常会说着说着就停下来——有一次长达51秒钟——垂下眼睑，撮起嘴唇，像一个害羞的十几岁的孩子。他似乎竭力要使自己的演讲如同其超弦理论论文般精确而又抽象。有时他又会莫名其妙地大笑起来，笑得上气不接下气，仿佛某些极端私密的笑话正从其脑海掠过。

20世纪70年代中期，威滕以关于量子色动力学和弱电相互作用的一篇深刻却颇为保守的论文而崭露头角。他获悉超弦理论是在1975年，但理解它的最初努力却完全耗费在了那些晦涩的术语之上。（的确，即使是世上最出类拔萃的人，也需费一番周折才能理解超弦理论。）然而在1982年，约翰·施瓦茨（John Schwarz，该理论的先驱）所写的一篇评论文章，却帮助威滕抓住了关键的一点：超弦理论并非仅仅考虑到引力的可能性，而是要求引力必须存在。威滕把这个认识称作"我一生中最大的思想震撼"。几年内，他曾萌发的关于这个理论的潜在疑虑消失了。"如果我不曾献身于弦理论的研究，我肯定会错失自己活着的使命。"他说道。他开始公开宣称这个理论是个"奇迹"，预言"它将统治物理学50年"。他还就这一理论潮水般地发表了大量论文，从1981年到1990年，威滕共发表了96篇论文，被其他物理学家引用达12105次，世界上没有任何物理学家能有如此大的影响力。[5]

在威滕的早期论文中，他倾力建立一个合理地模拟现实世界的超弦模型，但他逐渐意识到，实现目标的最佳途径是揭示此理论的"核心几何原理"。他说，这些原理可能与爱因斯坦用来构建广义相对论的非欧几何类似。对这些思想的追求使他深陷于拓扑学之中——它研究物体的基本几何性质，而不管它们的特殊形状或大小。在拓扑学家眼中，炸面圈和单柄咖啡杯是等价的，因为它们都只有一个洞，其中之一能被连续地变形到另一个而不会被撕裂。但炸面圈和香蕉是不等价的，因为必须要扯断炸面圈才能使其变形为香蕉的形状。拓扑学家们尤其关心，表面不相同的结能否不切断地相互变换。20世纪80年代后期，威滕从拓扑学和量子场论中创造出一种技巧，使数学家们能在奇形怪状地扭结的高维空间揭示更高的对称性。由于这一发现，威滕获得1990年度的菲尔兹奖——数学界最权威的奖励。威滕称此成就是他"唯一最值得骄傲的工作"。

我问威滕，由于超弦理论是不可检验的，故而有些批评家断言它根本不是真正的物理学，对此他有什么看法？威滕回答道，这个理论预言了引力。"虽然更恰当地说来，这只是事后的预测，也就是说实验已先于理论；引力是弦理论的必然结果这一事实，在我看来，乃是迄今为止最伟大的理论洞见之一。"

他承认甚至强调，没有人真正明白超弦理论，并且它在能够给出关于自然的精确描述之前，也许就会过早夭折。同其他人一样，他不愿预言超弦理论可能会导致物理学的终结，然而他深信，超弦理论将最终导致关于现实世界的新颖而深刻的认识。"真正精妙的谬误是极其罕见的，"他说，"能像超弦理论这般长期居于主导地位的谬误更是前所未见。"当我继续揪住可检验性问题不放时，他有些恼怒了。"我认为我没有完全地向你表达出其壮观，其难以置信的严谨及其惊人的优雅和美妙。"换言之，超弦理论是如此优美以至于它不可能是谬误。

威滕随后表达了他的强烈信念："一般说来，物理学中所有真正重大的思想，实际上都是超弦理论的副产品，一些被先发现，但我认为那只是在地球上演化的偶然事件。在地球上，它们按这种顺序被发现。"他走到黑板前写下广义相对论、量子场论、超弦、超对称性（在超弦理论中起着关键作用的一个概念），"但我并不认为，如果宇宙中有多种文明的话，这四种思想在每一种文明中都以那样的顺序被发现"。他沉吟了一会儿，又说道："顺便说一句，我确实认为这四种思想在任何高级文明中肯定都已被发现了。"

我真不敢相信自己的好运。"现在谁在煽情？"我问道。"我并不是在煽情！"威滕有些气急地反驳道，"说我在煽情，就像说某位宣称天空是蓝色的人在煽情一样——尽管别的什么地方的一位作家已宣称天空中有粉红色的晕轮。"

粒子美学

20世纪90年代初期，当超弦理论还比较新颖时，几个物理学家曾就其含义问题写过一些很受欢迎的书。在《万物至理》一书中，英国物理学家约翰·巴罗（John Barrow）认为，哥德尔不完备性定理打破了自然界理论的完备性这一信念。[6]哥德尔证明了任何足够复杂的公理系统，都不可避免地产生该系统所不能回答的难题。这意味着任何理论都是不完善的。巴罗还指出，粒子物理学的统一理论不可能成为万物的理论，它仅仅是所有粒子和所有相互作用的理论；对于各种使我们的生活富有意义的现象，比如说爱情和美，这一理论基本上甚或完全是无能为力的。

但巴罗和其他分析家至少都一致认为，物理学家们会取得一个统一理论。在由物理学家转为新闻工作者的戴维·林德利（David Lindley）所著的《物理学的终结》一书中，这种观点受到了挑战。[7]林德利认为，研究超弦理论的物理学家们并不是在从事物理学研究，因为他们的理论永远也不可能被实验证实，而仅代之以主观判据，如精致、优美等。他最后得出结论，粒子物理学正面临变成美学的一个分支的危险。

物理学的发展史支持了林德利的预言。最早的物理学理论虽看上去有些古怪，但却获得了物理学家甚至公众的认可，这并不是因为它们合情合理，而是因为它们预见了能被实验所证实的结果——通常以非常明显的方式被证实。即便牛顿的万有引力的观点也违反了常识——两个东西相隔如此之远，怎么能相互吸引呢？难怪约翰·马多克斯（John Maddox）（《自然》杂志的编辑）曾经指出，如果牛顿的万有引力理论送到今天的一家刊物，确定无疑地将会被拒绝，因为它太反常了，以至于无法让人相信。[8]但牛顿的公式却以惊人的准确性给出了计算行星运行轨道的方法，它是如此有效以致无从否定。

爱因斯坦关于可弯曲时间和空间的广义相对论，看上去更是古怪，但引力使太阳光线弯曲的预言得到证实后，它得到了广泛的认可。同样，物理学家们相信量子力学，并非因为它解释了世界，而在于它以惊人的精度预言了实验结果。理论家们不断预言新的粒子和现象，实验则不断验证这些预言。

如果超弦理论必须依靠美学的判断标准，那么其根基是动摇的。科学上最有影响力的美学原理，是由14世纪英国哲学家奥卡姆（William of Occam）提出的。他认为，对于给定现象最好的解释，通常是最简单、假设最少的那个，这一原理被称为"奥卡姆剃刀"。它是导致中世纪关于太阳系的托勒密（Ptolemaic）模型崩溃的原因。为了证明地球是太阳系的中心，天文学家托勒密被迫提出行星以繁复的本轮绕太阳运动。通过假设是太阳而不是地球为太阳系的中心，后来的天文学家终于放弃了本轮，而代之以简单得多的椭圆轨道。

若比起超弦理论所必需的那些尚未被观测到——似乎也永远无法观测到——的额外维度来，托勒密的本轮看上去显然要更合理些。不管超弦理论家们向我们保证其理论的数学形式多么精美，但它随身携带的形而上学包袱，使它根本不可能赢得认可——不论是在物理学家还是在大众心中，它都得不到曾被给予广义相对论和粒子物理标准模型的那种认可。

让怀疑精神赐福于那些超弦理论的信仰者吧，哪怕是一会儿也好。假设某个未来的威滕，甚或是威滕自己，找到了一种能无限弯曲的几何，用以描述所有已知的力和粒子的行为，那么，这样一个理论将解释世界到哪种程度？我向许多物理学家请教过这个问题，没有人能帮助我准确地理解超弦究竟是什么。据我所知，它既不是物质也不是能量，它只是一些能产生物质、能量、时间和空间的数学原材料，但本身却不对应现实世界的任何东西。

毫无疑问，优秀的科学记者们会竭力让读者相信他们明白该理论。丹尼斯·奥维拜（Dennis Overbye）在其《寂寞的宇宙之心》——宇宙学方面写得最好的一本科普读物——中，把上帝想象为一个宇宙摇滚歌手，他在十维超弦吉它上弹奏而产生了宇宙。[9]（有人会纳闷，上帝是即兴演奏呢，还是按照乐谱演奏的？）超弦理论的真正含义，自然是深埋于其严格的数学理论之中。我曾听一个文学教授把詹姆斯·乔伊斯（James Joyce）佶屈聱牙的巨著《为芬尼根守灵》，比作巴黎圣母院主教堂顶上那奇形怪状的雕像，仅仅是为了使上帝高兴才建造出来的。我猜想，假如威滕真正找到了他渴望已久的理论，或许只有他——可能还有上帝——才能真正欣赏它的优美。

终极理论之噩梦

双颊是山里红般的颜色，朦胧的双眸像个亚洲人，一头银发微染些红色，这一切使史蒂文·温伯格就像是一个巨大而高贵的精灵，都能本色饰演《仲夏夜之梦》中的众仙之王奥伯龙（Oberon）了。与仙王一样，温伯格对大自然的神秘表现出强大的直觉能力，以及从粒子加速器中涌出的大量数据中辨别出精妙模型的超凡本领。他在1993年出版的《终极理论之梦》一书中，设法使还原主义更具浪漫色彩。粒子物理学是史诗般的人类追求的巅峰之作。"从远古时代起，人们就已开始探索即使现代的精深理论术语也无法解释的原理。"[10]他指出，推动科学发展的动力是一个简单的问题："为什么？"这一问题使物理学家不断深入地认识自然界的本性。在他看来，众多的解释最后收敛于越来越简单的原理，以终极理论而告终。温伯格推测，超弦可能导致那种终极解释。

就像威滕和几乎所有的粒子物理学家一样，温伯格深信物理学有能力获得绝对真理。但与威滕不同，他清楚地认识到信仰终归是信仰，这使他成为同行中有趣的代言人。他也知道自己在以哲学的腔调说话。如果说威滕是一个在哲学上极幼稚的科学家，那么温伯格则是一个深谙世情的科学家——为了自己研究领域的利益，他可能过于世故了。

我第一次遇到温伯格，是在1993年的一次庆祝其《终极理论之梦》公开发行的宴会上，当时正值超级超导对撞机项目被无情腰斩之前的美好时光。他态度和蔼，滔滔不绝地讲述其著名同行们的趣闻轶事，并不断地推测当天晚上与电视访谈节目主持人查利·罗斯（Charlie Rose）的对话将是什么情形。我热切希望给这位伟大的诺贝尔奖得主留下深刻的印象，于是开始列举人名。我提到，弗里曼·戴森（Freeman Dyson）最近告诉我，终极理论的整个思想只是幻想。

温伯格笑了。他向我保证，其绝大多数同行都相信终极理论，尽管很多人不愿使自己的看法公开。我又提到杰克·吉本斯（Jack Gibbons），新当选的克林顿总统所指定的科学顾问。我说，最近我采访了吉本斯，他暗示美国独自承担不起超级对撞机的费用。温伯格略显怒色，摇了摇头，低声嘟哝着，抱怨着社会对基础研究带来的社会效益缺乏认识。

具有讽刺意味的是，在《终极理论之梦》中，温伯格自己几乎没有，甚至根本就没有列举出社会应该支持粒子物理学进行深入研究的证据。他小心谨慎地承认，不管是超导超级对撞机，还是现在的其他任何加速器，都不能为终极理论提供直接确凿的证据；物理学家最终不得不依赖数学上的优美和一致作为指南。况且，终极理论可能并没有什么实用价值。最令人惊讶的是，温伯格坦承，在人类看来，终极理论可能不会揭示宇宙是有意义的，相反，他反复引用一本早期著作中不出名的评论："宇宙越是容易被理解，则看上去就越没意义。"[11]虽然这句话长期困扰他，但他拒绝向它低头。相反，他详细地解释道："等我们发现越来越多的物理学基本原理时，则它们看上去与我们越来越没有什么关系。"[12]温伯格似乎承认我们所有的"为什么"都将归结于一个"因为"。他的终极理论的设想，很容易让人联想起道格拉斯·亚当斯（Douglas Adams）写的《银河旅游指南》，在这部20世纪80年代发行的科幻喜剧作品中，科学家们最终发现了宇宙之谜的答案，答案是……[42]。（显然，亚当斯是在以一种文学的方式，践行着科学哲学。）

1995年3月，超导超级对撞机项目被葬送之后，我在德克萨斯大学奥斯汀分校又见到了温伯格。在他那宽敞的办公室里堆满了各种期刊，包括《国外动向》《艾西斯》《怀疑的探索者》《美国历史评论》，还有一些物理杂志，由此足见温伯格兴趣之广泛。一面墙上挂着黑板，上面潦草地写满了各种必需的数学符号。他看上去费了极大的努力才说出话，并不断叹息、皱眉、挤弄并使劲揉搓着自己的眼睛。也难怪，他才用完午餐，正处于饭后的困乏期，但我更倾向于认为他正陷于粒子物理学家悲惨的两难境地：如果他们获得了终极理论，那就糟糕透顶了；而如果没有获得终极理论，同样是透顶糟糕的事情。

"对粒子物理学家来说，这是个可怕的时刻。"温伯格承认，"实验能产生新的思想或新理论；而这些新思想或新理论，又能作出被实验证实的、有质的不同的预言，从这个意义上说，再没有比这个时刻更令人沮丧的了。"随着美国的超导超级对撞机项目的夭折，以及其他加速器计划因资金匮乏而受阻，这个领域的前景已变得非常暗淡。但令人不解的是，优秀的学生仍在不断进入此领域，他们"可能比我们现有的还要优秀"，温伯格又补充道。

虽然同威滕一样，温伯格也支持物理学正向绝对真理接近的观点，但他敏锐地意识到了为这种立场辩护的哲学困难。他承认"我们决定是否接受物理理论的标准是极其主观的"，对于聪明的哲学家来说，总能抓住粒子物理学家们"只不过是在前进过程中虚构"的把柄。（在《终极理论之梦》中，温伯格甚至坦言，自己对无政府主义哲学家保罗·费耶阿本德的作品尤其偏好。）温伯格又告诉我，"不管美学如何"，粒子物理的标准模型"已如少有的几个理论那样被实验证实了，它肯定是正确的；如果它仅是一种社会建构，那它早就该崩溃了"。

温伯格认识到，物理学家永远也不可能像数学家证明数学定理那样，最终证明一个物理学理论。对物理学家而言，只要这个理论能解释一切实验数据，如所有粒子的质量，所有相互作用的强弱，他们就会不再怀疑它。"我自己也并不是万事通，"温伯格说，"许多科学哲学又回到了古希腊，对确定性的探求腐蚀了它们；而这种探求，至少在我看来，很可能是一种错误的探求。科学远非大家围坐成一圈不停拍巴掌那般有趣，因为对世界我们并不清楚。"

在我们的交谈中，温伯格甚至建议应该有人把超弦理论最终的正确的观点发布到互联网上。"如果她，"他稍稍停顿，强调了一下"她"，"得到与实验相符的结果，我们就会说：'那就对了。'"尽管研究人员永远无法获得弦自身或设想的弦所栖息的额外维度的直接证据；事实上，物质的原子理论也并非因为实验工作者拍出原子的照片，而是因为它管用，才得到了认可。"我承认弦远不如原子那么直观，并且原子也远不如椅子那么直观，但我却并不认为它们之间有什么哲学意义上的不连贯性。"

温伯格的话语中并没有多少自信的成分。他自然明白，超弦"的确"代表物理上的不连贯，代表一个经验检验无法介入的断层。后来他突然站起身在屋中踱起步来，继续谈话时，他又拿起一些七零八碎的小物件，心不在焉地抚摸一下，然后又放下，重述他的观点：物理学的终极理论将代表科学所能取得的最根本进展——即其他所有学科的基石。当然，一些复杂的现象，如湍流、经济现象或生命，需要各自特殊的定律和通则。温伯格又说，但如果你问那些原理为什么是正确的，这个问题又把你引向了物理学的终极理论，那是万物之根。"就是它使得科学成为一个有层次结构的体系。的确是有层次结构的体系，而不是随意拼凑的网络。"

许多物理学家不能容忍关于物理学终极理论的言论，温伯格说，但事实上什么也无法逃避它。举一个例子来说，如果神经科学家要解释意识，他们只能从大脑的角度来解释，"大脑之所以成为现在的大脑，这是由历史的偶然性和化学、物理的基本原理决定的。他们的终极理论要由我们的终极理论来解释"。即使获得了终极理论，科学当然仍会延续，或许直到永远，但它将丧失某些东西。终极理论的获得，"将不可避免地导致一种悲哀的感觉"。温伯格说，因为它宣告了对基本知识所进行的伟大探索过程的终结。

当温伯格继续谈下去时，他似乎用"渐趋消极"的词语来描绘终极理论。当我问及超弦理论是否会产生什么实际应用时，他皱皱眉。（在1994

年的《超空间》一书中，物理学家加来道雄（Michio Kaku）预测：超弦理论的发展，最终能使人们访问其他宇宙或作时间旅行。）[13]温伯格提醒道，"在科学历史长河的沙滩上，积满了白色的朽骨"，这是那些不能把握科学发展大势的人们留下的；但超弦理论的应用前景则"很难想象"。

温伯格也怀疑终极理论是否能解决量子力学带来的所有著名佯谬。他说，"我较倾向于认为，这些佯谬只不过是我们探讨量子力学方式所引出的迷惑。"消除这些迷惑的办法之一，是应用量子力学的多重世界诠释。这种诠释提出于20世纪50年代，试图解释为什么物理学家的观测使粒子（如电子）只在量子力学允许的许多条路径中选择一条。按照多重世界诠释，电子实际走过了所有可能的路径，但却是在不同的世界里。温伯格承认，这一解释的确也有其烦人的一面，"可能存在另一条平行的时间轨迹，在那里，约翰·威尔克斯·布思①并未碰到过林肯，并且……"温伯格停了一下，"我真希望所有的困惑都消失，但可能永远也不会有这一天。或许这正是世界的本来面目。"

终极理论应使这个世界更好理解，这是否对它要求过甚了？我还未问完，温伯格便点点头，说："是的，这要求太过分了。"科学的合适语言是数学，他提醒我，终极理论"对受过那些数学语言训练的人来说，会使宇宙看起来更合理，至少显得更有逻辑；但使其他人明白则需要很长一段时间"。终极理论不会给人类的行为提供任何指南，"我们已学会了正确分辨价值判断和真理判断"。温伯格说，"我并不认为我们应退回去重新梳理它们之间的关系"，科学"肯定能帮助你发现你的行为结果，但无法告诉你应该期望什么结果，在我看来这是截然不同的"。

对认为终极理论会揭示宇宙的目的或"上帝的心智"（如斯蒂芬·霍金所云）的那些人，温伯格显得极不耐烦。相反，温伯格希望终极理论能消除人们思想中，甚至物理学家中普遍存在的痴心妄想、神秘主义和迷信。他说："只要我们没掌握那些基本原理，就仍会期望着发现某种与人类息息相关的东西，或者说，期待着指导基本原理建设的某种神圣蓝图。但当我们发现量子力学的基本原理和一些对称性原理都只不过是非人格化的冷冰冰的规律时，这必然导致破除前述神秘化气氛的效果。至少这正是我所希望的效果。"

温伯格表情严肃地继续说道："我肯定不会同意这样的观点，有些人认

① John Wilkes Booth，美国演员，行刺林肯的凶手——译者注。

为，不论是现代的还是牛顿的物理学框架，都对这个世界产生了明显的'祛魅'效果。如果世界原本就是祛魅的，我们发现这点总比稀里糊涂要好。我认为这正是人类走向成熟的表现，就像小孩们总会发现所谓'牙仙子'①只不过是故事罢了。走出童话世界当然是件好事，尽管童话比现实世界更美好。"

温伯格非常清楚，许多人渴望从物理学中获得不同的启迪。事实上，前些日子他听说澳大利亚物理学家保罗·戴维斯（Paul Davies）"因提高了公众对上帝或神灵世界的认识"而获得了100万美元的奖金。戴维斯写了许多书，其中最出名的是《上帝的心智》，他认为物理学定律揭示了隐藏在自然界中的某种设计蓝图；在这一蓝图中，人类意识可能起着核心作用。[14]温伯格在说出戴维斯得奖之事后，干笑了一下："我想给戴维斯发个电报，'你知道有哪个机构愿意为证明并不存在什么神圣蓝图的研究工作颁发100万美元奖金吗？'"

在《终极理论之梦》中，温伯格非常严肃地讨论了有关神圣蓝图的各种说法，并提出了人类深受其苦的窘迫问题——是何种神圣设计蓝图规划了生灵涂炭和无数灾难的发生？是何种设计者？许多为数学理论的威力所慑服的物理学家认为"上帝是一位几何学家"。温伯格批驳道，不管上帝多么精通几何学，但我不明白为什么我们应该对他饶有兴趣，而他却可以对我们漠不关心。

我问温伯格，在对人类生存条件的看法上，是什么使他如此强烈地抱持灰暗的态度（在我看来的确是这样）？他微笑着回答道："我颇有几分欣赏自己略带悲剧色彩的观点。对了，你喜欢悲剧还是……"他犹豫了一会儿，笑容凝结了。"嗯，有的人喜欢看喜剧，但……我觉得悲剧更是理解生活不可缺少的维度。无论如何，悲剧才是我们所能得到的最好结局。"他注视着办公室窗外，出神地沉思起来。幸运的是，当时的温伯格可能并未看到那座耻辱之塔——1966年，德克萨斯大学的一位精神失常的学生查尔斯·惠特曼（Charles Whitman），正是在那座毫不出名的塔上残忍地枪杀了14个人。从温伯格的办公室望出去，正好可以俯瞰德克萨斯大学神学院那雅致的哥特式大教堂，但温伯格似乎并未注意到这一切。或许，除了自己沉浸其中的内心世界外，他并未注意这尘俗世界的任何事物。

---

① Tooth fairy，古老的欧洲童话故事，大意是，如果晚上把刚掉的牙齿放在枕头下，仙女就会把牙齿拿走，并放下一个钱币——译者注。

不再惊诧

即便是社会集中人力和财力建造出更大的加速器，使得粒子物理学能存活下去，至少可以苟延残喘，物理学家又有多大的可能性获得像量子力学那样真正新奇的东西呢？按照汉斯·贝特（Hans Bethe）的观点，可能性不会很大。作为康奈尔大学的教授，贝特因其恒星聚变中的碳循环研究成就，于1967年获得诺贝尔奖。换句话说，他指出了恒星是如何发光的。他在二战中领导了曼哈顿工程的理论分部，当时他弄清了在行星演化中被认为是最重要的计算。爱德华·泰勒（Edward Teller）做的某些计算表明，原子弹爆炸的火球，有可能点燃地球的大气圈，从而引发毁灭整个世界的大灾难（具有讽刺意味的是，泰勒后来却成为科学界中最狂热的核武器鼓吹者）。研究泰勒的猜测的科学家们认为它非同小可，毕竟他们还只是在黑暗中探索未知领域。当时的贝特仔细考虑了这一问题，计算结果却表明泰勒错了，因为火球不会扩散。[15]

任何人都无权根据自己的计算结果去决定地球的命运，但如果必须在这二者之间作出选择的话，我肯定选择贝特，因为他的智慧和严肃的态度值得信赖。当我问他，在第一颗原子弹在阿拉莫戈多沙漠爆炸前的一瞬间，他是否对即将发生的事情有所担忧时，他摇摇头。不，他回答道，对他来说，唯一担心的是点火装置能否正常工作。在贝特的回答中没有丝毫吹嘘，他作过计算而且相信自己的计算。（某些人或许想知道，爱德华·威滕是否也相信地球的命运寄托在基于超弦的预言上。）

当我问贝特的研究领域前景如何时，他说，在物理学中仍然有许多未解决的问题，包括标准模型带来的问题，而且在固体物理学中也将不断涌现出重要发现。但按贝特的看法，这些进展都不会给物理学的根基带来革命性的变革。贝特举了个例子，对于据称是近十年来物理学最激动人心的进展的所谓高温超导体——这些于1987年首次披露在世人面前的材料，在相对的高温（仍远低于零摄氏度）下，能无阻尼地通过电流——"无论从什么角度看，它们都没有改变我们对电传导或超导性的认识。"贝特说："非相对论性的量子力学的基本框架早已完成。"事实上，"原子、分子和化学键等理论，早已于1928年便完成了"。是否会有另外一场像量子力学所能给物理学带来的那种革命呢？"那极不可能。"贝特以那种不安的实在的口吻答道。

事实上，所有终极理论的信徒都认为，不管其形式上如何，它将仍是量子理论。史蒂文·温伯格告诉我，物理学的终极理论"可能远远超出我们目前的认识，就像量子力学远远超出经典力学的内涵一样"。但与贝特相同，他也认为，无论如何，终极理论不可能取代量子力学。"我认为我们仍将停留于量子力学，"温伯格说，"所以，从那种意义上说，量子力学比一切理论都更具革命性，不管是从前的还是将来的。"

温伯格的见解让我想起了1990年发表在《今日物理》上的一篇文章，康奈尔大学物理学家戴维·默明（David Mermin）讲述了一位叫莫扎特（Mozart）的教授（可能是他古怪的化名）曾抱怨："在过去的四五十年里，粒子物理学家令人感到失望，谁又曾预料半个世纪以来，我们没能取得任何真正显著的进展呢？"当默明问那位虚构的教授意下如何时，他说："所有粒子物理学家都告诉我们，最大的奇迹就是量子力学仍然成立。事实上这是任何人都知道的，多么没劲！" [16]

约翰·惠勒及其"万物源于比特"

贝特、温伯格和默明似乎都在表明，至少就定性的意义来说，量子力学是物理学的终极理论。一些理论物理学家和哲学家都坚信：他们只要了解了量子力学，只要能透彻地理解其意义，他们便能找到"终极答案"。约翰·阿奇博尔德·惠勒（John Archibald Wheeler）正是最具影响力和最富创造性的量子力学（即通常意义上的现代物理学）的研究者之一。他是个典型的诗人型物理学家，以其杜撰与附会的各种比喻和格言而闻名。在一个温暖的春日，我在普林斯顿拜晤了他，蒙他惠赐了一火车皮的这类名言，诸如："不能具象，则无从理解"（爱因斯坦）；"上帝一位论（惠勒名义上的宗教信仰）是救助坠落的基督的安乐窝"（达尔文）；"永远不要追逐一辆公共汽车、一个女人或一种宇宙学的新理论，因为几分钟之内你总会等到下一个"（惠勒的一位住在耶鲁的朋友）；"没有发生奇事的一天绝不是充实的一天"（惠勒）。

惠勒也因其体能而闻名遐迩。当我俩离开他那位于三楼的办公室去进午餐时，他拒不乘坐电梯，他宣称"电梯是人类健康的大敌"，一定要从步行楼梯走下楼。他总是用手拉住护栏，在每个楼梯转弯的平台处来个大回旋，利用离心力旋过拐角，降落在下一层的台阶上。隔着人肩，他冲我喊道："我们可以比一比，看谁下楼梯更快。"下了楼梯后，惠勒轻快地摆动着拳头，伴着他迈步的节奏，与其说他是在走路，不如说是在急行军。只有遇到门时，他才会稍事停顿，并且毫无例外地，他总是抢先为我推开门。穿过门后，为了表示礼貌，我会回敬地停下步来——此时的惠勒已经是年近八旬的老翁了——但片刻之后，他又会超过我，扑向下一道门。

这是如此明显的一个隐喻，我怀疑它正是惠勒心中所想的：他毕生都跑在其他科学家的前面，并为他们推开一扇又一扇的门。从黑洞到多重世界理论再到量子力学，他的努力使诸如此类的现代物理学中那些最为稀奇古怪的思想，逐渐为人所接受，或者，至少也是为人所重视。如果不是有着这么多不容攻击的荣誉的话，惠勒恐怕早已成了小字辈们的笑柄，消逝如过眼云烟了。刚二十岁出头，他便来到丹麦，师从于尼尔斯·玻尔（"因为在所有的活人当中，他看得最为深远。"惠勒曾在其奖学金申请书中写道），玻尔是对惠勒的思想产生最为深远的影响的人。1939年，玻尔和惠勒联名发表了第一篇成功地用量子力学解释核裂变的论文。[17]

正是由于他在核物理学方面的专长，惠勒被吸收加入了二战期间以核裂变为基础的第一颗原子弹的研制，以及冷战初期第一颗氢弹的研制。战后，惠勒在美国成为爱因斯坦的引力理论，也就是广义相对论的权威之一。在20世纪60年代末期，他创造出"黑洞"一词，并在推广这一术语的过程中起了主要作用。正是在他的大力宣传下，宇宙学家才逐渐相信：这种由爱因斯坦理论所预言的古怪的、密度无穷大的客体，确实是存在的。我问惠勒，究竟是何种性格因素使他在其他物理学家只是极不情愿地接受这一理论的时候，就坚信这种"荒诞不经"的事物是存在的？"超凡的生动想象，"他答道，"我最喜欢的是玻尔的一句名言：'必须随时准备迎接震惊体验，一次次巨大的震惊体验。'"

从20世纪50年代开始，惠勒对于量子物理学的哲学含义的兴趣与日俱增。最为广泛接受的量子力学诠释，被称作所谓"正统诠释"（虽然"正统"一词用来描述如此激进的世界观，看上去有些不伦不类），也就是"哥本哈根诠释"，因为它源自惠勒的导师玻尔于20世纪20年代末期在哥本哈根所做的一系列演讲。这种学说否认电子等亚原子实体的真实存在性，认为它们以多种可能的叠加态而存在，只是由于人们的观测才成为单态。电子、质子等的行为，可能既像波又像粒子，这完全取决于它们受到何种实验观测。

惠勒是最早宣称"现实世界并非完全是物质世界"的著名物理学家之一。从某种意义上说，我们的宇宙可能是一种属人现象，对它的把握取决于观察行为，因而也取决于人的意识本身。在20世纪60年代，惠勒致力于推广著名的人择原理，其核心内容是：宇宙必须是现在这个样子，否则，我们就无法观测它。同时，惠勒还使物理学与信息论［数学家克劳德·香农（Claude Shannon）于1948年创立的理论］之间的有趣联系得到同行们的注意。将物理学建立在以量子为名的基本的、不可分的实体之上，这种做法与信息论毫无二致。比特作为二进制的基本单位，就是信息论的量子，它所负载的信息是"二中择一"的：头或尾，是或否，"0"或"1"。

惠勒构思了一个思想实验，展示出量子世界对所有人来说都是多么奇特的一种存在，在这之后，他更加沉迷于信息的重要性。惠勒的延迟选择实验，是经典的（但并非古典的）双缝实验的一个变种，它展示了量子现象的无常性。当物理学家将电子穿过具有双缝的屏障时，电子的行为就如同波一般，它们同时通过两条缝，然后像波一样叠加，这时在屏障后方远处的探测器上感知到的就是干涉图样；如果这位物理学家在某一时刻关闭了其中的一条缝，电子将像简单的粒子一样，从剩余的那条缝中穿过，干涉图样消失。而在延迟选择实验中，实验者是在电子已经穿过屏障之后，再决定双缝齐开还是关闭一缝，但结果将会与经典双缝实验完全一样，仿佛电子们已预先"知道了"物理学家将要选择的观测方式。这个实验在20世纪90年代初得以进行，并且证实了惠勒的预言。

惠勒还用另一种方法来解释这一佯谬。他将实验中物理学家的工作，比作一个人在出乎意料的情况下玩一个"20问"游戏。在这一古老游戏的新玩法中，一伙人中的一个先离开房间，而屋里其余人——在这个离开的人想来——会选定一个代表某个人、某个地方或某件事物的名词，然后这个玩游戏的人将重新回到房间里，通过提问一系列①只能答以"是"或"不是"的问题，猜测其余那些人心里想的那个名词是什么。但这一次出乎猜测者的意料之外，这伙人决定一起搞一个恶作剧，第一个被提问的人，只在提问者提出问题之后，才想定一个对象，每个人都这么做，所给出的回答，不但与提问者当下提的问题相符合，并且与此前所提的问题也是相符合的。

"当我重新走进房间的时候，这个名词并不存在，尽管我认为它是存在的。"惠勒这样解释道。同样的道理，在物理学家观察之前，电子既不是粒子，也不是波，从某种意义上说，它不是实体的东西，而是以一种无从知道的中间过渡状态存在。"只在你提出问题时，你才能从中得到信息，"惠

---

① 最多只能问20个问题，故称20问游戏——译者注。

勒说道，"如果你不提问，它不会向你提供任何信息。但当你提出一个问题的同时，你不可能提另一个问题。因此，如果你问我现在最大的心愿是什么——我发觉向别人这样提问总是件非常有趣的事情——我会说：很明显，我最大的心愿可以归结为类似于这一'20问'游戏中那个提问者所要猜的某种东西。"

惠勒把这种观点浓缩到一个类似于禅宗偈语的短句中："万物源于比特"（the it from bit）。在惠勒的一篇随笔散文中，他对这一短语给出了这样的解释："……每一个有——每个粒子，每个力场，甚至时空连续体本身——其功能、含义及其绝对存在，都来自设备对'是或否'问题所给出的答案，都来自二进制选择，都来自比特——即使是通过间接的途径。"[18]

20世纪80年代末期，在惠勒的鼓动之下，一个包括计算机专家、天文学家、数学家、生物学家和物理学家在内的空前庞大的研究小组，开始探索信息论和物理学之间的联系。甚至于某些超弦理论家也参与了进来，他们想把量子场论、黑洞以及信息论都捏合进超弦的框架之中。惠勒承认这些思想仍不成熟，尚未经过严格的检验，他及其研究伙伴们仍在"尝试着怎样才能摸清这一领地的地形，并学习着怎样用信息论的语言来表述我们已经知道的一切"。他说，这种努力也许会走入死胡同，或者，它也许能"从整体上"导致对现实世界的有力的新见解。

惠勒强调说，科学中仍存在着许多谜底有待揭开。"我们仍然生活在人类的孩提时代，"他说，"在我们这个时代，像分子生物学、DNA、宇宙学等领域，都已迎来了其光辉的黎明。"他又抛出另外一条格言：随着我们的知识的岛屿与日俱增，无知的海岸线也在日渐延长。但他仍然相信人类总有一天会发现"终极答案"。为了找到一句能够表述出他的信念的格言，他跳了起来，从书架上层取下一本关于信息论和物理学的书，书中收有他的一篇文章。浏览了一会儿之后，他读道："我们相信，有朝一日我们肯定会以一种如此简单、如此完美、如此令人信服的方式，把握住万事万物的中心思想，以至于我们将会奔走相告：'噢，世界原来就是这样的！我们居然在这么长的时间内一直被蒙在鼓里！'"[19]惠勒从书本上抬起头来，满脸陶醉的表情，"我不知道这到底需要一年，还是十年，但我相信，我们能够且一定能够理解。这就是我为之奋斗的信念：我们能够且一定能够理解。"

惠勒指出，许多现代科学家都与他一样有着共同的信念，即人类有朝一日会发现"终极答案"。比如说，曾经在普林斯顿与惠勒比邻而居的柯特·哥德尔（Kurt Gödel）就相信，"终极答案"可能已经被发现了。"他认为'终极答案'可能就存在于莱布尼茨的论文中，他的论文在他那个时代未能被完全理解，但我们将能从中发现——怎么说呢——发现那位哲学家的钥匙，那可用于发现真理、用于解决任何疑难问题的不可思议的方法。"哥德尔认为，这把钥匙"能赋予理解它的人以如此巨大的威力，以致关于这一哲学家之匙的知识，只能由那些具有高尚情操的人来掌握"。

然而，惠勒本人的老师玻尔，却对科学或数学到底能否获得如此辉煌的成就，抱着明显的怀疑态度。惠勒不单从这位巨匠本人那里，还从其儿子那里得知玻尔的观点。玻尔谢世之后，他的儿子曾告诉过惠勒，玻尔认为物理学终极理论的探索或许永远也无法得到令人满意的结论；因为随着物理学家愈益深入自然界的本质，他们也将面临愈益复杂、愈益困难的问题，这些问题将最终给他们带来灭顶之灾。"我估计我个人的观点要比上述看法更乐观些，"惠勒说。他沉吟了一会儿，又以一种异常抑郁的语气补充道："但也可能我只是在自己欺骗自己。"

具有讽刺意味的是，惠勒自己的观点却暗示着，所谓终极理论只不过是海市蜃楼，从某种意义上说，真理只是想象出来的，而不是客观地认识到的。按照"万物源于比特"的观点，我们不仅在用我们提出的问题创造着真理，而且也创造着实在本身——"有"。惠勒的观点已经走到了相对主义的边缘，甚至还要更为严重些。在20世纪80年代初期，美国科学促进会（AAAS）年会的组织者们，竟把惠勒与三位心灵学家置于一个议题之下，惠勒不由大为光火。他在会上明确表示，自己的观点与那些关于心灵现象的发言者的信念根本就是风马牛不相及；他还四处散发一本小册子，宣称"哪里香烟缭绕，那里就有毫无意义的事情发生"，其皮里阳秋当然是指向心灵学了。

但惠勒自己确曾暗示过，除了虚无之外便一无所有了。"我百分之百地相信，世界只是想象中臆造的东西"，他曾对科学记者兼物理学家杰里米·伯恩斯坦（Jeremy Bernstein）这样说过。[20] 惠勒也清楚地知道，从经验上看，这一观点是完全站不住脚的；宇宙产生时经验在何处？人类产生之前是什么维系着宇宙达几十亿年？无疑，他勇敢地给我们提供了一个可喜而又令人沮丧的佯谬：在万事万物之最深处隐藏着的，只是一个问题，而不是一个答案。当我们费尽心机地窥视物质世界的深处，或宇宙最远的边界之外时，我们最终所能看到的，只是自己那疑云密布的脸也正在不解地回望着我们。

戴维·玻姆的隐秩序

毋庸置疑，也有一些物理哲学家对惠勒的观点，甚至对由玻尔提出且广为流传的哥本哈根诠释，从骨子里就感到不屑，戴维·玻姆（David Bohm）便是这些持不同意见者中最著名的一个。玻姆是在宾夕法尼亚州出生并长大的，1951年却被迫离开了美国，因为在当时麦卡锡时代的高压下，他拒绝就自己或其他科学家同行（其中最著名的是罗伯特·奥本海默）究竟是不是共产党的问题，去接受非美活动调查委员会的质询。在巴西和以色列羁縻了一段时间后，玻姆于20世纪50年代末期定居于英国。

那时，玻姆致力于寻求一种能替代哥本哈根学派的诠释。这种有时被称作导波的诠释保留了许多量子力学的预见力，同时也消除了正统诠释中的许多奇特性质，如量子的无常性和依赖于观察者的特性。自20世纪80年代后期，这种导波理论引起了越来越多的物理学家和哲学家的注意，因为哥本哈根诠释的主观性和非决定性实在无法让他们感到满意。

令人倍感矛盾的是，玻姆似乎要将物理学变得更具哲学味、臆测性和整体性。同惠勒将量子力学和东方宗教连在一起的类比思维相比，玻姆走得更远。他提出了一种叫隐秩序的哲学，这种隐秩序既带有神秘色彩，又具有科学性。玻姆关于这一论题的著作吸引了一大批追随者——在那些希望通过物理学而达到神秘主义顿悟的人们心中，他是当之无愧的英雄。没有几位科学家能将这两种相反的动机——汲汲于澄清事实的同时，又按捺不住地要给它蒙上层层神秘的面纱，以这样一种奇特的方式集合于一身。[21]

1992年8月，我到玻姆那位于伦敦北部埃奇韦尔的家中拜晤了他。他的皮肤出奇地苍白，与其粉红色的嘴唇和黑色的头发形成了鲜明的对比。他的躯体深陷于一张大扶手椅内，看上去极度虚弱，但精神仍然矍铄。他的一只手作杯状放在头上，另一只手放在椅子扶手上；手指修长且青筋隐隐，指甲渐尖且泛黄。他告诉我，前段时间心脏病发作，目前正在康复之中。

玻姆的妻子给我们送上茶和饼干，然后到别的房间去了。最初，玻姆边说边停，渐渐地越说越快，语气低沉、急促而又单调。很明显他的嘴唇已经干涩了，不停地咂唇作响。有时当他看到高兴的东西时，会费力地张开嘴做微笑状；有时他会停下来说，"明白吗？"或简单地"嗯哼？"。我非常迷惑，不知何从理解，只好报以点头微笑，因为玻姆肯定是讲得很明白的。后来我才知道，别人在听他谈话时也是如此，他就像某个奇异的量子粒子一样，总是在不同话题间来回振荡。

玻姆说，20世纪40年代晚期，他在写一本量子力学著作时，便开始怀疑哥本哈根诠释了。玻尔摒弃了一种可能性，即量子系统的概率行为，实际上是一种称为隐变量的潜在的、决定性的机制作用的结果。实在是不可知的，只因其内在的不确定性，玻尔坚持这种观点。

玻姆认为这种观点难以接受。"到目前为止，一般认为科学的整个思想就是用来解释隐藏在事物表面现象下的某种实在，"他解释道，"玻尔并非否认这些实在，它只是认为对于实在，量子力学无法给出更多的说明。"玻姆觉得，这种观点仅把量子力学归纳为"技术性地预测或控制事物的一套公式，我认为这是不够的。如果仅是如此，我认为自己是不会对科学有如此浓厚的兴趣"。

在1952年发表的一篇论文中，玻姆提出粒子确实是真正意义上的粒子，不管它们是否被观测，在任何时候都是如此；其行为受一种新的、迄今尚未被观测到的力决定，玻姆称之为导波；对其性质所做的任何测量，都会从物理上改变导波，从而使之面目全非。这样，玻姆给不确定性原理一个纯粹的物理解释，而不是形而上学意义的解释。他告诉我，在玻尔的量子诠释中，量子系统的不确定性原理的含义"并不是说完全不可确定，而是指一种内在的含糊性"。

玻姆的诠释确实容许甚至突出强调了一个量子佯谬的存在：非局域性，即粒子能跨越相当远的距离对另外一个粒子产生影响。爱因斯坦在1935年试图从非局域性出发指出量子力学的缺陷。他和鲍里斯·玻多尔斯基（Boris Podolsky）、内森·罗森（Nathan Rosen）一起，提出了一个思想实验，现在被称为EPR实验，两个粒子从同一源射出，朝相反方向飞行。[22]

按照量子力学的标准模型，每个粒子在测量之前没有确定的位置或确定的动量，但当测量一个粒子动量时，物理学家同时假设另一个粒子具有确定位置，即使它在银河系的另一端。爱因斯坦嘲笑这一效应是"远距离的魔鬼行动"，他认为这既违反常识又违反他的广义相对论（它禁止作用的传播速度超过光速），从而说明量子力学是不完备的。然而在1980年，法国的一组物理学家实现了EPR的实验条件，表明它确实产生了这种"魔鬼行动"（此结果并未违反广义相对论，原因在于不能利用非局域性来传递信息）。玻姆从不怀疑此实验结果的正确性。他说："实验结果若非如此，反而会让我们感到不可思议了。"

虽然玻姆试图通过他的导波模型使世界变得更清晰，但他也认为完全的明晰是不可能的，他的思想部分地受他在电视上看到的一个实验的启发。此实验将一滴墨水滴入有甘油的容器内，等柱体容器旋转时，墨水在甘油上明显不可逆地扩散，其秩序似乎已遭到了破坏；但当反方向旋转时，墨水又汇成一滴。

在这个简单实验的基础上，玻姆建立起一种称为隐秩序的世界观。在物质那貌似混乱的外表，也就是显秩序之下，总是隐藏着一个更深奥的、不易为人察觉的隐秩序。将这个概念应用到量子范畴上，玻姆认为这个隐秩序便是量子势，一个由无数涨落着的导波组成的场。这些波的叠加产生了我们看到的粒子，这样便构成了显秩序。按照他的观点，甚至表观的基本的概念，比如空间和时间，或许都仅是某种更深层的隐秩序的显性表现。

玻姆认为，如果想要测量隐秩序，物理学家必须抛弃关于自然界结构的某些基本假设。"一些基本概念，如秩序和结构，无意识地制约着我们的思想，新的理论依赖于新的秩序。"在启蒙时代，牛顿和笛卡儿等思想家用力学的观点代替了古人关于世界秩序的有机概念。"虽然相对论和其他理论的诞生修正了这种秩序观，但基本的思想仍然未变，即用坐标描述的力学秩序。"玻姆如是说道。

作为真理的探求者，尽管玻姆有自己的勃勃雄心，但对于某些科学家认为可以将大自然的一切都简化成某种单一现象（如超弦）的信念，他却持否定态度。"我认为追求真理的进程是没有止境的。人们都在谈论那可以解释一切的万物至理，但这只不过是幻想，是毫无根据的。在科学发展的每一层次上，我们都能把一些东西当作表象，而把另一些当作本质去解释那些表象；但当我们达到另一层次时，本质和表象已互相转换了，这样说够清楚了吧？这一转换的进程是没有止境的，明白吗？我们认识过程的本质也正是如此。但是，隐藏在这一切背后的东西仍是未知的，也是无法被思想所把握的。"

对玻姆来说，科学"在永不知疲倦地前进"。他指出，现代物理学家臆想的世界本质是自然界的相互作用。"但自然界为什么会有相互作用？认为相互作用是本质，而原子却不是，为什么如此认定？"

终极理论的信仰只不过是现代物理学家的自我满足，玻姆说，"那样做只会使我们终止对真实世界的进一步探索"。他指出："如果你在一个鱼缸中插入一道玻璃屏，鱼就会被这道屏挡住；如果你再撤去玻璃屏，鱼仍不会跨过屏障处，因为它认为整个世界就是那样。"他干笑了一下，继续说道：

"所以从长远看来，你们所认为的终结便可能是这道屏障。"

玻姆重申："我们不会得到终极的本质，即那种只是本质而同时又不是表象的东西。"我插了句，"那不就太令人丧气了吗？""哦，这取决于你想得到什么。如果你想得到一切，那你会沮丧的。如果科学家们真的得到了终极答案，那么除了做技术人员外，他们便无事可做了，那也会令人沮丧的，明白吗？"他又发出了那干涩的笑声。我接着道，所以无论是否能得到终极答案，总之，科学家们是不会很自在的。"我认为你应该区别对待这两种情况，明白吗？我们之所以从事科学研究，原因之一是要深化我们的认识，而不仅仅是既有的知识。我们只是在不断地、越来越成功地逼近实在。"

科学肯定会朝全然无法预料的方向演化，玻姆继续说道。他希望将来的科学家们会更少地借助数学来模拟世界，更多地借助于隐喻和类比的新途径。"我们正越来越倾向于认为，数学是处理实在的唯一手段，"玻姆说，"那只是因为数学手段确实曾一度相当奏效，以至于我们竟想当然地认为这是唯一正确的途径。"

同其他科学幻想家一样，玻姆也期望，在将来的某天科学和艺术会融为一体。"科学和艺术的分离是暂时的，"他这样评论道，"这在过去并不存在，将来也没有理由继续保持这种分离。"正如艺术并非简单地由艺术品组成，它还包括一种"态度，即艺术的灵魂"；科学也是如此，它不仅包括知识的积累，更在于创造耳目一新的认识方式。"以一种全然不同的方式进行观察和思考的能力，比我们所获得的知识更重要。"玻姆解释道。可悲的是，虽然玻姆希望科学更艺术化，但许多物理学家却正是从美学的角度反对导波诠释的——它太丑陋了，以致不可能正确。

为了让我彻底相信终极理论的不可能性，玻姆提出下列论证："任何已知的东西都受其限度决定，不仅在定量意义上，在定性意义上也是如此，一个理论是这样的，就不可能再是那样的。这就可以合理地推出认识是无限的。必须注意，如果无限是存在的，它就不可能不存在，而有限并非无限，所以无限必定为有限确定了界限，对吧？无限必定包含有限。我们只能说有限是从无限中产生的，当然要通过某种创造性过程，这与前面是一致的，所以我们说，无论人类认识达到何种地步，无限总是存在的。类似地，无论你的认识发展到怎样的地步，总会有人提出你必须进一步回答的问题，我认为你永远也无法彻底解决这一难题。"

这时，玻姆的妻子走进来，问我们是否要再来些茶水，这才让我松了口气。她为我们斟好茶后，我指了指玻姆身后书架上一本关于藏密教义的书，问玻姆是否受到了这类书的影响，他点了点头。他曾是印度神秘主义学者吉斯德那莫提（Krishnamurti）的朋友和学生。吉斯德那莫提于1986年去世，他是最早向西方人展示如何获得启示的印度智者之一。吉斯德那莫提本人获得启示了吗？"从某种意义上说，是的，"玻姆回答道，"他的基本做法是对思维本身进行思索，直到极致，这样思想就会达到一个完全不同的意识境界。"当然，人永远不能真正把握自己的思想，玻姆说，任何了解自己思想的努力都将改变它——正如测量电子会改变它的轨迹一样。玻姆似乎在暗示，不可能存在什么终极的神秘主义认识，正如不可能存在物理学的终极理论一样。

吉斯德那莫提是一个幸福的人吗？玻姆对我的话迷惑不解。"很难说，"他最后答道，"有时他不高兴，但我想总的说来他活得很幸福。事实上，这是件与幸福无关的事。"玻姆皱皱眉，似乎明白了他刚才所说的含义。

在与戴维·皮特（F. David Peat）合著的《科学、秩序与创造力》一书中，玻姆曾坚决主张"娱乐性"在科学和生活中的重要性，[23] 但玻姆本人，不论是在他的书中还是生活中，都看不出丝毫娱乐色彩。对他来说，对真理的探求不是游戏，而是一项艰巨的不可实现却又必须去执行的任务。对玻姆来说，不管通过物理学还是通过冥想或者是通过神秘体验，他都渴望了解、渴望发现万事万物的真谛，但又坚信世界是不可知的——因为，我相信，他已厌倦了终极思想。玻姆认为，任何真理，不管它最初多么令人惊服，最终都会成为僵死的毫无活力的东西，它非但没有揭示绝对真理，反而掩盖了真相。玻姆追求的不是真理，而是启示——永恒的启示，结果他注定要陷入永恒的困惑之中。

最后，我告别了玻姆和他的妻子，走出屋外，外面正下着绵绵细雨。走在街上，我回头望了望玻姆家的房子，那是满街平凡的、白色的房舍中的一座，同样的平凡，在雨中泛出同样苍白的颜色。两个月后，玻姆便因心脏病发作辞世了。[24]

费曼的灰色预言

理查德·费曼（Richard Feynman），这位因开辟量子电动力学的新篇章而荣获1965年诺贝尔奖的物理学家，在《物理学定律的特性》一书中，对物理学的未来给出了更加灰暗的预测：

我们很幸运，生活在新发现层出不穷的年代，这正如美洲的发现一样，你只能发现一次。我们生活的年代，正是自然的基本定律被我们不断发现的年代，那种岁月已一去不复返了。它的确非常激动人心，惊世骇俗，但这些激动的心情终将随着岁月的流逝而消退。当然，将来还会有别的有趣的事情，比如说，去发现类似于生命科学等领域中不同层次的现象之间的联系，或者，如果谈及探索，可以去探索别的行星，等等。但这些，与我们今天所做的工作相比，已经是截然不同的。[25]

最基本的那些定律被发现后，物理学家们就会发现，他们的地位甚至连那些二流的思想家——也就是说哲学家——都不如。"到那时，那些一直在圈子外发表不着边际评论的哲学家们也会挤进来；因为我们已不再可能轻易地打发他们了，再也不能仅凭一句呵斥，诸如'若你是对的，我们早就能猜出所有定律了'之类，就能让他们无地自容地讪讪退去。因为，当所有定律都堂而皇之地摆在世人面前的时候，哲学家们当然会对它作出自己的解释……那将不可避免地导致观念的变质，一如新大陆的伟大探险家在看到无数兴高采烈的游客蜂拥而入时，所体验到的那种变了味儿的感觉。"[26]

费曼的话一针见血，切中了问题的要害。他唯一的错误是认为哲学家进入圈内是千年之后的事，不可能在短短几十年内发生。费曼所说的物理学的未来场景，我在1992年参加哥伦比亚大学的一次研讨会时，不幸就已经看到了。会议主题是关于量子力学的含义，列席会议的既有物理学家，也有哲学家。[27]会议表明，量子力学自创立以来，60余年过去了，但对于量子力学真正含义的认识却仍是一片空白——说得好听一点，仍让人感到费解。从与会者的发言中，你可以听到惠勒的"万物源于比特"、玻姆的导波假说、温伯格偏爱的多重世界模型，等等。但绝大多数与会者似乎都以自己独特的思路去理解量子力学，并用独特的语言表述出来；没有谁能理解他人的观点，更不用说赞同了。纷繁的争论，让人不由联想到玻尔对量子力学的评论："如果你认为自己完全理解了，那只说明你对它仍一无所知。"[28]

当然，这种明显的混乱完全有可能是我自己的无知所致。但当我将自己的困惑说给一位与会者听时，他安慰我说，你的感觉完全正确。"确实很混乱。"他这样评价会议（其言外之意，也指我们对量子力学的整个诠释）。他指出，从很大程度上说，问题在于我们不能将每个人对量子力学的解释从经验上予以鉴别，哲学家和物理学家之所以好此厌彼，主要是出于美学的、哲学的理由，也就是说，都是主观上的原因。

这就是物理学的命运。绝大多数物理学家，无论是工作在实业界还是学术界的，他们毫不考虑隐藏在深处的哲学思辨，只是应用已有的知识闷头去制造各种激光器、超导体和计算设备。[29]当然，也有那么几个执迷不悟的人，他们热衷于探求真理而不是理论的实际应用，但却是以一种超乎经验的、反讽的方式去研究物理学——探索超弦或其他怪异理论的魔道领域，或者就量子力学的含义喋喋不休地争个不停。这些反讽的物理学家的争论，根本不可能被实验证实，只能越来越近似于文学批评中的文字游戏，而他们的物理学会也将逐渐演变为现代语言学会。

第四章
宇宙学的终结

1990年，我乘车到瑞典北部山区一个偏僻的度假胜地，去参加一个名为"我们宇宙的诞生和早期演化"的专题研讨会。抵达现场之后，我发现大约有30位粒子物理学家和天文学家出席了会议。他们来自世界各地——美国、欧洲、苏联和日本。我参加这个会议的部分原因，是想借机见见斯蒂芬·霍金（Stephen Hawking）。他瘫痪的躯体上却有着一个发达的大脑，这一引人注目的特征使他成为世界上最著名的科学家之一。

我与霍金见面的时候，他的情况比我所预料的还要糟糕。他蜷坐在一个装有电池和计算机的轮椅上，肩耸着，下巴耷拉着，身体虚弱得让人心悸，头歪向一边。就我的观察，他能够自主活动的只有左手食指。他就用这根手指吃力地从计算机屏幕上的菜单中选出字母、单词或者句子，然后用一个语音合成器将这些词以一种不和谐的、权威式的低沉声音播放出来，这种声音很容易使人联想起影片《机器警察》中的电子人主人公。大体说来，霍金对他的困境似乎乐观多于沮丧。他有一张米克·贾格尔①式的嘴，唇是淡紫色的，并常常向一端的嘴角弯上去，笑眯眯地作出一副得意的模样。

霍金被安排做一个关于量子宇宙学的演讲，这门学科作为一个研究领域而确立下来，是与他的努力分不开的。量子宇宙学认为，在非常小的尺度上，量子不确定性不仅使物质和能量，而且使空间和时间在不同状态之间波动。这些时空涨落会产生"虫洞"——能把一个时空区域与另一个非常遥远的时空区域联系起来，或者产生"婴孩宇宙"。霍金已经将他长达一小时的演讲"虫洞的阿尔法参数"储存在计算机里。他只需敲一个键，就可以让他的语音合成器一句一句地读出来。

---

① Mick Jagger，美国一摇滚乐队的主唱——译者注。

霍金用他那令人悚然的电脑语音，讨论了是否有一天我们能够滑入我们这个星系的一个"虫洞"，过后又从一个非常遥远的星系的另一端跳出。霍金得出结论，这似乎是不可能的，因为量子效应会搅乱我们的组成粒子。（霍金的论证意味着：在《星际旅行》中描绘的比光速还快的旅行"曲速引擎"是不可能的。）最后，他用一个关于超弦理论的题外话结束了演讲。虽然我们看到的一切只是我们称作时空的"低维超空间，但我们实际上是生活在弦理论的无穷维超空间之中"。[1]

我对霍金的看法很含混。一方面他是个英雄式人物，受制于一个残疾的、无助的躯体，却仍能想象具有无限自由度的实在；另一方面，他所说的一切给我的感觉却又非常荒诞。虫洞？婴孩宇宙？弦理论的无限维超空间？这些似乎更像是科幻小说，而不是科学。

我对整个会议或多或少有种相同的感受。有些发言，比如说，某些天文学家讨论用望远镜和其他一些装置探索宇宙所收集的成果，的确坚实地建立在现实基础上，是实证科学；但是还有许多代表提出的论点，却与现实、与任何可能的经验检验完全脱节。大小如篮球、豌豆、质子或超弦的宇宙，究竟是什么样子？所有通过"虫洞"与我们的宇宙相连的其他宇宙，对我们的宇宙有什么影响？在诸如此类问题上争论不休的男人们（没有女性与会者在场），身上都蒙着某种既严肃又滑稽的东西。

在整个会议过程中，我极力压抑住本能的荒谬感觉。我提醒自己这些都是非常聪明的人，正如一家当地的报纸所报道的那样，他们是"世界上最伟大的天才们"，他们不会把时间浪费在微不足道的研究上。因此，在后来写霍金和其他宇宙学家的思想时，我尽我所能地使他们显得更可信些，给读者灌输尊敬和理解，而不是怀疑和迷惑——毕竟，这正是科学记者的职责所在。

但有些时候，最清晰的科学作品恰恰是最不诚实的。我对霍金和会议上其他人的最初看法，在某种程度上是正确的。现代宇宙学的许多内容，尤其是由粒子物理学的统一理论和其他仅限于少数人的奇思怪想所产生的那些东西，是极为可笑的。更确切地说，它们是反讽的科学，那种即使在原则上也不可能由经验所检验或解决的科学。因此，从严格意义上说它们根本不是科学，其主要作用是使我们面对宇宙的神秘时保持敬畏之心。

滑稽之处在于，在同时代的众多杰出物理学家当中，正是霍金，第一个预言了物理学将很快得到关于自然的完备的统一理论，从而使物理学寿终正寝。他是在1980年提出的这个预言，就在他被任命为剑桥大学卢卡斯数学教授职位之后不久；而300年前的牛顿，也曾担任过这一职务。（几乎没有听众注意到，在名为"理论物理已经接近尾声了吗？"的演讲的结尾，霍金曾说：鉴于计算机的快速发展，它们很快会在智力上超过创造它们的人类，并且自己得出终极理论。）[2]霍金在《时间简史》一书中更详细地阐述了其预言，他在该书的结束语中宣称，得到终极理论会帮助我们"知道上帝的心智"。[3]这句话表明终极理论会给我们留下一个神秘的启示，我们将在这一启示的光环笼罩下打发此后的岁月。

但在这本书前面部分讨论他所谓的无边界宇宙假说时，霍金对终极理论可能解决的问题提出了一个完全不同的观点。无边界假说提出了一个古老的问题：大爆炸之前存在什么？我们宇宙边界之外存在什么？按照无边界假说，整个宇宙的历史，所有空间和时间的历史，形成一种四维球：时空。讨论宇宙的肇始或终结，就像讨论一个球的起点和终点一样，毫无意义。因此，霍金推测，物理学在被统一后也会形成一个完美无缺的整体，也许仅仅只有一种完全自洽的统一理论，才能产生我们所知道的时空，上帝在创造宇宙时可能别无选择。

"那么，造物主将置身何处？"霍金问。[4]他的回答是，没有造物主容身之处，一个终极理论将会把上帝以及伴着上帝的神秘排除出宇宙。像史蒂文·温伯格一样，霍金希望把神秘主义、活力论、神创论等，从其最后的避难所之一，也就是宇宙起源问题中彻底驱赶出去。根据一个传记作家所说，霍金和他的妻子简（Jane）之所以在1990年分道扬镳，部分原因就在于，作为一个虔诚基督徒的简逐渐被丈夫的无神论所触怒。[5]

在《时间简史》出版后，另外几本书继续讨论了物理学是否会达到一个完备的终极理论，一个能回答所有问题因而会使物理学终结的理论。那些宣称这样一个理论根本就不存在的人，更倾向于使用哥德尔定理和其他一些深奥的定理。但是在霍金的研究生涯中，他认为达到一个万物至理面临着一个更基本的障碍。只要仍存在着与霍金一样拥有超凡想象力的物理学家，他们就永远不能完全根除宇宙的神秘，永远不能找到终极答案。

霍金与其说是个真理的追求者，不如说是一个艺术家、一个幻想家、一个喜欢开宇宙玩笑的人。我怀疑，他一直清醒地认识到：找到并从经验上证实一个统一理论是极其困难的，甚至是不可能的。他宣称，物理学正处于找到终极答案的边缘，这可能是一段反话，与其说是一个断言，不如说是一种挑衅。1994年，霍金曾告诉一位记者，物理学可能永远也不会达到一个终极理论，这其实已承认了这一点。[6]霍金是反讽物理学和反讽宇宙学的大师级的践行者。

宇宙学的震惊

现代宇宙学中最引人注目的事实是，它的确不全是反讽的。宇宙学已经给予我们一些真实的、无可辩驳的惊奇。20世纪初的人们，曾认为太阳坐落于其中的银河系这样一个星群，就构成了整个宇宙。随后，天文学家认识到了那些叫作星云的小光斑（它们被认为是银河系中的星团），实际上却是恒星岛，银河系只不过是比任何人的想象都要大得多的宇宙所包含的众多星系之一。这个发现给人一个极大的、实证的、不可消除的震惊，甚至最坚定的相对主义者也难以对此加以否认。谢尔登·格拉肖的解释是，星系既不是想象出来的，也不是不可想象的；它们就是存在。

另一个很令人惊奇的事实出现了，天文学家发现，这些星系的光总是向可见光谱的红端移动，显然星系正远离地球而去，星系之间也在相互远离，这一后退速度使光线产生多普勒频移（同一种频移使得一辆远离听者的救护车上的汽笛声渐次降低）。红移现象支持一个建立在爱因斯坦相对论基础上的理论，该理论认为，宇宙肇始于远古发生的一场大爆炸。

20世纪50年代，理论家们预言：100多亿年前宇宙的炽热诞生，应留下很弱的微波辐射。1964年，贝尔实验室的两位无线电工程师偶然发现了所谓的宇宙背景辐射。物理学家还提出，那个创生火球可能起到了一个原子核反应堆的作用，火球内部的氢会聚变成氦和其他轻元素。过去几十年的仔细观察表明，银河系和其他星系中轻元素的丰度，与理论预言完全吻合。

来自费米实验室和芝加哥大学的戴维·施拉姆（David Schramm），更乐意把星系红移、微波背景辐射和元素丰度这三个证据，称为大爆炸理论的三大支柱。施拉姆身材高大，胸部厚实，并且热情奔放。他是一位飞行员，也是位登山运动员，还荣获过希腊罗马式摔跤赛冠军。他是大爆炸理论以及其个人在完善轻元素丰度计算中的丰功伟绩的不倦鼓吹者。我到会的时候，施拉姆让我坐下来，非常详细地重述了一遍大爆炸的现有论据，然后说："大爆炸理论确实很诱人，我们已有了基本框架，只要再做些查漏补缺的工作就行了。"

施拉姆承认，有些漏洞确实很大，理论家们还不能精确确定，早期宇宙的热等离子体究竟怎样凝结成恒星和星系的。观察表明，天文学家通过望远镜观察到的可见星云物质的质量，没有大到足以阻止一个星系四分五裂的程度；肯定有一些不可见的物质或暗物质，将星系束缚在一起。换句话说，我们能看到的所有物质可能都只不过是暗物质海洋上的泡沫。

另一个问题，则涉及宇宙学家称之为"大尺度结构"的东西。在宇宙学早期阶段，星系被认为是多少有些均匀地分散在整个宇宙中的，但随着观测技术的改进，天文学家发现星系似乎聚集成团，周围是巨大的真空。最后，还有一个宇宙在所谓量子引力阶段呈现什么行为的问题，这时的宇宙是如此之小，如此之热，以至于所有的力都被认为是统一的。这些就是瑞典诺贝尔专题研讨会所要讨论的主要议题。但是施拉姆认为，这些问题没有一个能对大爆炸理论的基本框架构成威胁，在他看来，"仅仅因为你不能预测龙卷风这一点，并不意味着地球不是圆的"。[7]

在诺贝尔专题会上，施拉姆不厌其烦地一再向其宇宙学同行们兜售着差不多同样的话，并不断声称宇宙学正处于"黄金时代"。他那商人式的热忱似乎激怒了一些同行，毕竟，他们之所以成为宇宙学家，并不是为了去填补先驱者留下的细节。在施拉姆第N次宣称"黄金时代"后，一个物理学家气呼呼地说："当你处于一个时代时你不可能知道这个时代是不是黄金时代，只有当你回顾时才能知道。"对施拉姆的戏谑被广为传诵，比如说，一位同行一本正经地推测，这个壮实的物理学家本身可能就代表着暗物质问题的解；另一位则建议，应把施拉姆用作阻止我们宇宙被吸入虫洞的塞子。

这次瑞典会议快结束时，霍金、施拉姆和其他宇宙学家都乘上公共汽车去附近的一个村庄听音乐会。当他们进入举办音乐会的路德教堂时，里面已经挤满了人。在教堂前面，管弦乐队已经就位，有头发淡黄、着装五颜六色的年轻人，也有形容枯槁的老人，手里拿着小提琴、单簧管及其他乐器；他们的邻里乡亲挤满了楼厅和教堂后部的座位。当科学家们由坐在自动轮椅中的霍金带头，排成队从中间通道走向前面为他们预留的座位时，镇上的居民们开始鼓掌，一开始有些踌躇，随后就热烈起来，一直持续了近一分钟。这是一个好的象征：至少在此时此刻此地，对于这些人来说，科学取代了宗教而成为宇宙真理的源泉。

然而，怀疑终究已经渗进了科学的卫士当中。在音乐会开始前，我无意中听到戴维·施拉姆和一个年轻的英国物理学家尼尔·图罗克（Neil Turok）之间的一段对话。图罗克向施拉姆吐露，解决与暗物质和星系分布有关的问题实在太难了，这令他心怀难释，以至于想放弃宇宙学，转向别的领域。他悲观地问："无论如何，谁说我们有权理解宇宙？"施拉姆摇头表示不同意。在乐队开始准备时，他坚定地低声说，宇宙学的基本框架，即大爆炸理论是绝对正确的，宇宙学家只需把尚未解决的问题解决就行。施拉姆说，"局面会好起来的"。

图罗克似乎感到施拉姆的话令人欣慰，但他可能仍然忧心忡忡。如果施拉姆是正确的怎么办？如果宇宙学家在大爆炸理论框架下已经探悉了宇宙之谜的答案怎么办？如果剩下的只是扫尾工作，却又偏偏尾大不掉怎么办？考虑到这种可能性，像霍金这样的"强者"科学家会跳过大爆炸理论，径直进入后实证科学，就不那么让人感到诧异了。如此有富于创造力和雄心的人，除此之外还能做些什么呢？

俄罗斯魔术师

作为反讽宇宙学的践行者，安德烈·林德（Andrei Linde）是少数几个堪与斯蒂芬·霍金相提并论的人之一。他是一位俄罗斯物理学家，1988年移民去了瑞士，两年后来到美国。林德也参加了瑞典的诺贝尔专题讨论会，其滑稽表演是这次会议中的精彩场面。在户外鸡尾酒会上喝了一两杯后，林德用空手道中的劈砍动作将一块石头敲成了两半，随后又用双手倒立起来，紧接着一个后空翻稳稳地立在地上。他从口袋里掏出一盒火柴，把其中两根呈十字形放在手上。当林德让手至少在表面上保持静止时，上面的那根火柴却抖动并跳了起来，仿佛被一根看不见的弦拉动着。这个魔术使他的同行们欣喜若狂。不久，由于有一打左右世界上最杰出的宇宙学家徒劳地试图重复林德的壮举，因而弄得火柴和咒语到处乱飞。当他们要求知道林德是怎样做到这一点的时候，他笑着喊道："量子涨落！"

林德更因其理论"变戏法"而出名。在20世纪80年代初期，他使暴涨理论（inflation）获得同行们的认可，这是一个从粒子物理学中推出的更为离奇的想法。暴涨的发明（在这儿用"发现"一词似乎并不合适），一般归功于麻省理工学院的艾伦·古思（Alan Guth），但林德帮助改进了这一理论，并使之得到公认。古思和林德提出，在我们宇宙历史的极早阶段——精确地说是T=10^{-43}秒时，那时的宇宙比一个质子更小——引力会变成斥力。因此，他们认为宇宙经历了一次惊人的、指数增长的膨胀；而时至今日，宇宙则以一个低得多的速率膨胀。

古思和林德的观点建立在未被检验的——几乎肯定是不可检验的——粒子物理学统一理论基础上。不过宇宙学家喜欢暴涨理论，因为它能解释一些由标准大爆炸模型产生的、扰人的问题。首先，为什么宇宙在所有方向上均表现出或多或少的相似性？答案是：与吹起一个气球时抹平了它的皱折类似，宇宙的指数膨胀使得它相对平滑。反过来，暴涨也解释了为什么宇宙不是一个完全均匀的、一锅汤似的发光体，而是以恒星和星系形式呈现的成团的物质。量子力学表明连真空也充满能量，这些能量不断地涨落，像风吹过湖面时湖面水波的起伏。按照暴涨理论，这些在宇宙极早期由量子涨落产生的波峰，在暴涨后会变得足够大，成为形成恒星和星系的引力种子。

暴涨有一些令人惊诧的含义，其中之一是我们通过望远镜所能看到的一切，都只代表在暴涨时产生的极大区域内的一个极微小部分。但林德并未就此止步，他进一步提出，甚至那个极大宇宙，也只不过是暴涨时产生的无限多宇宙中的一个。膨胀一旦开始，就永远不会结束，它不仅产生了我们置身其中的宇宙——我们依靠望远镜能探索到的嵌满星系的领域，还产生了无数的其他宇宙。这个超级宇宙具有所谓的分形结构：大宇宙生出小宇宙，小宇宙再生出更小的宇宙，如此继续下去。林德把他的模型称为混沌的、分形的、永远自复制的暴涨宇宙模型。[8]

林德既有公开逗趣和奇思妙想的一面，也有出人意料的冷峻一面。我去斯坦福大学采访他时，瞥见了其性格中的后一面。他和妻子丽娜塔·卡洛斯（Renata Kallosh）自1990年起供职于斯坦福，他妻子也是一位理论物理学家。当我走进他们租赁的灰色方形的房子时，林德带我草草地转了一圈。在后院，我们碰见了卡洛斯，她正愉快地在一个花坛上翻弄着什么。"看，安德烈！"她叫着，指着头顶上树枝中的一个鸟巢，满巢都是吱吱叫的鸟。林德只点了点头。当我问他是否发现加利福尼亚能使人放松时，他喃喃自语："可能是太放松了。"

在林德讲述他的经历时，很明显，焦虑乃至抑郁是激励他前行的重要因素。在他研究中的几个阶段，就在取得突破性进展前，他会对洞察事物的本质感到绝望。林德在20世纪70年代后期就已经偶然得出了暴涨的基本概念，当时他正在莫斯科，但是他认为这个想法缺点很多，以至于无法继续研究。艾伦·古思认为暴涨能解释宇宙几个使人困惑的特征，比如宇宙的平滑性。这使他的兴趣再次被激起，但是古思的看法也有毛病。林德在思考这个问题时是如此入迷，以至于得了胃溃疡；好在他终归还是厘清了该怎样修正古思的模型，才能消除其技术上的问题。

但即使这个新暴涨模型，也还是建立在林德深感怀疑的统一理论之上。最终，在陷入忧郁以至于缠绵病榻一段时日之后，他确信暴涨可能源于约翰·惠勒（John Wheeler）首先提出的更为普遍的量子过程。据惠勒所云，如果谁能拥有一台比任何现存显微镜的分辨率强大亿万倍的显微镜，他就能看到时空由于量子不确定性而剧烈地涨落。林德认为惠勒所说的"时间泡沫"会不可避免地产生暴涨所需的条件。

暴涨是一个自耗过程，即空间的膨胀使驱动暴涨的能量很快耗散。但是林德认为，一旦暴涨开始，由于量子不确定性，它将总是在某处继续进行（量子不确定性的一个特征）。在这个时刻，新的宇宙纷纷产生了，有些宇宙立即坍缩回去，另一些宇宙膨胀得如此之快，以至于物质没有机会聚合。一些类似于我们置身其中的宇宙安稳下来，以足够慢的速率膨胀，引力就使物质形成星系、恒星和行星。

林德有时将这种超宇宙比作无垠的大海。靠近看，这大海给人的印象是运动不息和变化不止的，波浪上下起伏。我们人类，由于生活在这引起起伏的波浪之一中，会认为整个宇宙正膨胀着。但是如果我们能升到海面之上，就会认识到膨胀的宇宙只是一个无限大的永恒的海洋中一个微不足道的局部。林德认为弗雷德·霍伊尔（Fred Hoyle）早期的稳恒态理论（该理论会在本章后面的部分予以讨论）在某些方面是对的；如果以上帝般的视角来看，超宇宙当然能表现出某种平衡。

林德并不是第一个假定存在其他宇宙的物理学家。虽然大多数理论家都将其他宇宙作为数学抽象对待，并对此感到困窘，但林德却喜欢推测它们的性质。例如，在说明其自复制宇宙理论时，他借用了遗传学话语，暴涨创造的每一个宇宙都生出另外的"婴孩宇宙"；这些后代中有一些会保持其先辈的"基因"，演化成类似的宇宙，有着类似的自然法则，也许还有着类似的土著居民。援引人择原理，林德提出，某种宇宙学版本的自然选择，会更倾向于让那些有可能产生智慧生命的宇宙永远存在。"在宇宙的某处存在着像我们一样的生命，这在我看来差不多就是板上钉钉的事实，"他说，"可惜我们永远也不会知道。"

像艾伦·古思和其他几个宇宙学家一样，林德也喜欢玄想在实验室中创造一个暴涨宇宙的可行性，但只有林德提出这样的疑问：为什么要创造另一个宇宙？它带有什么目的？根据林德计算，一旦某个宇宙工程师创造出一个新的宇宙，它会立即以超光速同其母体分离，不可能有进一步的通信。

林德猜测，或许这位工程师能以某种方式精心处理暴涨前的种子，使它演化成为一个有特定的维数、特定物理规律和自然常数的宇宙。这位工程师会以上述方式将某种信息嵌在新宇宙的结构上。林德认为，实际上我们的宇宙很有可能就是另一宇宙的生物创造的，而像他自己这样的物理学家，在摸索着试图揭示自然规律的过程中，实际上可能正在破译来自我们宇宙母体的信息。

林德抛出这些观点时显得相当谨慎，同时观察着我的反应，只是在最后，大概是对我吃惊地大张着嘴感到很满意，他才让自己露出了一丝笑意。然而，当我想知道嵌于我们宇宙的信息可能是什么的时候，他的笑容消失了，郁郁地说，"似乎我们还没有成熟到能知道这些信息的地步"。当我进而追问他是否担心其所有的工作可能只是——我竭力想要找到一个恰当的词语——胡说八道时，他的脸色阴沉得都快滴出水了。

"在我消沉的时候，我的确会感到自己是个彻头彻尾的白痴，"他回应道，"我思考的都是些相当原初的玩意儿。"他又补充说，他曾尽力让自己不要太沉迷于自己的想法。"有时这些模型相当奇怪，如果你对它们太认真，就有掉入陷阱的危险。我想这和在湖面薄冰上跑步相似，如果你跑得非常快，你可能不会沉下去并且能跑上一大段距离。可如果你只是站在那儿去思考是否跑对了方向，那无疑你就会掉下去。"

林德似乎想表明，他作为一名物理学家的目标并不是去追寻解，去追寻"终极答案"或仅仅是追寻某个"答案"，而是要不断前进，不断向前滑行。林德对终极理论的想法感到恐惧，其自复制宇宙论要这样解读才有意义：只有宇宙是无限且永恒的，科学作为对知识的探求，也才会是无限且永恒的。但林德认为，因为物理学受制于这个宇宙，所以它不可能趋近终极的解。"例如，你没有将意识包括进去。物理学研究物质，而意识并非物质。"林德同意约翰·惠勒的说法，现实在某种意义上是一种参与现象。"在你测量前，没有什么宇宙，没有你能称作是客观现实的东西。"林德说。

就像惠勒和戴维·玻姆一样，林德似乎对物理学永远不会十全十美的前景，既满怀着神秘的憧憬，又倍感煎熬。他说："理性的知识有一定局限。研究非理性的一条途径是深入其中思考，另一条途径是用理性工具研究非理性的边界。"林德选择了后者，因为物理学只是提供了一条研究世界运演的

"不能说完全无意义"的道路。但有时候他承认："当我一想到自己会像一个物理学家那样死去时，就会感到沮丧。"

暴涨理论的衰落

　　林德在学术界堪称威望素著，因为在他选择执教斯坦福之前，有好几所美国大学向他发出了邀请，这一事实既说明了他在夸夸其谈方面的确造诣高深，同时也显示出宇宙学界对新思想的渴望。不过，到20世纪90年代初期，暴涨和前十年在粒子物理学中出现的诸多奇思妙想一样，已经开始失去主流宇宙学家的支持。我在瑞典遇见戴维·施拉姆时，他对暴涨理论还是相当乐观的；几年后，当我再度跟他谈起时，他也有了疑虑。施拉姆说，"我喜欢暴涨理论"，但是它永远不能被完全证实，原因是它不能提出任何独特的、其他方法不能解释的预言。他又接着说："我们并不是由于暴涨而观察到这些现象，而就大爆炸本身来说，我们确实能观察到那些现象。美丽的宇宙微波背景辐射和轻元素丰度告诉你'这就是大爆炸'，没有其他途径能预见到这些观测结果。"

　　施拉姆承认，宇宙学家们越是冒险回溯到离时间起点更近，他们的理论就越变得带有猜测性。宇宙学需要一个粒子物理学的统一理论来描述极早期宇宙的过程，但是提出一个有效的统一理论是极其困难的。"即使有人提出一个非常漂亮的理论，如同超弦理论那样，也没有任何办法能检验它。所以你不是真正地在应用科学的方法，即作出预言，然后检验预言。没有进一步的经验检验，它只是数学自洽。"

　　这个领域会像量子力学的解释一样，以一种美学上的解释标准而宣告结束吗？施拉姆回答说："那正是我考虑的一个实际问题，除非有人提出检验方法，否则我们只能走向哲学而不是物理学的领地。现有的检验只能在已有观察结论的基础上，告诉我们宇宙是怎样形成的，这更多的是马后炮，而不是预言。"黑洞、超弦、惠勒的"万物源于比特"，以及其他一些奇异的东西，对它们的理论探索或许总有突破的可能。但是，除非有人提出了精确的检验，或者我们很幸运地找到了一个能小心探进的黑洞，否则就不能自认达到了那种"顿悟"（Eureka）程度，并真正确信自己知道了答案。

　　似乎意识到其正在讲述的东西意义十分重大，施拉姆忽然间就又恢复到其惯常的吹鼓手和公共关系代言人模式。他坚称，宇宙学家在大爆炸模型之外取得进展是如此困难，这是一个好兆头，由此又回到了一个大家都耳熟能详的话题上来。"例如，世纪之交的时候，物理学家们都在谈论物理学的大部分问题都已被解决了，虽然还有少数使人烦恼的小问题，但也已在攻克之中。后来我们发现根本不是那么回事。实际上，我们发现的往往是即将到来的另一个大飞跃的线索。当你认为尽头在望时，会发现那是一个虫洞，由此会进入一个对宇宙全新的认识天地。我认为那可能就是未来将要发生的，我们正接近并开始看到真相。我们会遇到某些过去一直未能解决但又一直反复纠缠我们的问题，我希望这些问题的突破，会导致一个全新的、激动人心的时代到来。宇宙学绝不会走上绝路。"⁹

但是，如果宇宙学已经过了其巅峰发展阶段，也就是说，它已不再可能给出如同大爆炸理论那般深刻的、经验性的意外发现的话，怎么办？哈佛大学粒子物理学家霍华德·乔治（Howard Georgi）认为，宇宙学家不会有确知每件事的好运。乔治的娃娃脸上总带着一种善意的嘲弄表情，他说："我想你不得不把宇宙学看作是像进化生物学那样的一门历史科学。你正试图观察今天的宇宙并推测过去，这是一件有趣但危险的事，因为可能一直存在有重大影响的偶然事件。宇宙学家十分努力地试图去理解哪种事物是偶然的、哪种特征是确定的。但是我发现很难较好地理解那些论据，因此并不真正感到信服。"乔治建议，宇宙学家应该读一些进化生物学家斯蒂芬·杰伊·古尔德（Stephen Jay Gould）的书，以获得应有的谦逊，古尔德讨论了根据我们现在的知识重建历史的潜在困难（可参阅本书第五章）。

乔治吃吃地笑出声来，似乎他自己也明白不可能有任何一个宇宙学家会接受他的建议。同谢尔登·格拉肖一样，乔治也曾一度是寻求物理学统一理论的领军人物；其与格拉肖另一个相似之处在于，他最终也指责超弦理论以及其他统一理论的"备胎"是不可检验的，因而是不科学的。他指出，粒子物理学和宇宙学的命运在某种程度上可以说唇齿相依，宇宙学家希望借助统一理论更清楚地认识宇宙起源；而粒子物理学家则希望能通过望远镜观察宇宙的边缘，以找到能证实其理论的事实，以此来代替地球上的实验。乔治轻声说："这给了我一点推动力。"当我问他对霍金、林德和其他人探索的量子宇宙学领域的看法时，他淘气地笑了："我能说些什么？一个像我这样普通的粒子物理学家，在未知领域很容易感到迷惑。"他翻出全是讨论虫洞、时间旅行和婴孩宇宙的关于量子宇宙学的论文，"相当有趣，就像读《创世纪》"，至于暴涨理论，它是"一种绝妙的科学神话，至少和我听过的其他任何创世神话相比，都毫不逊色"[10]。

叛逆者中的叛逆

总有一些人，他们不但不接受暴涨说、婴孩宇宙等猜测性的假说，甚至对大爆炸理论本身也持怀疑态度。大爆炸理论的异端之首，当属弗雷德·霍伊尔（Fred Hoyle），一位英国天文学家，同时也是物理学家。无论你就两者中的哪一领域去衡量霍伊尔的履历，都只能得出这是位真正的业内精英的印象。就读于剑桥大学时，他师从诺贝尔奖得主保罗·狄拉克（Paul Dirac），后者曾成功地预言了反物质的存在；1945年，霍伊尔被聘为剑桥大学讲师，并且在20世纪50年代协助阐明了恒星是如何缔造出行星和人体必不可少的重元素的；20世纪60年代初，霍伊尔在剑桥创建了著名的天文学研究所，并出任第一任所长。由于他所取得的诸多成就，1972年霍伊尔被册封为爵士。是的，霍伊尔就是弗雷德爵士。但是，由于霍伊尔顽固地拒不接受大爆炸理论，且在其他一些领域也死不悔改地支持非主流思想，导致他在自己曾倾心协助创建的领域里，已成了被放逐的异端。[11]

在英国南部有一个名叫伯恩摩斯（Bournemouth）的小城，自1988年以来，霍伊尔就一直住在那儿的一座高层公寓里。当我去那儿拜访他时，他的妻子巴巴拉（Barbara）迎接了我，并把我引进起居室，霍伊尔正坐在椅子上，津津有味地看着电视里播放的板球比赛。我进去时，他站起来和我握了握手，眼睛仍盯着电视。他的妻子有礼貌地提醒了他，然后过去把电视机关上了。这时候，霍伊尔才把注意力转向我，就像是刚从某一魔咒中清醒过来。

我原以为霍伊尔会很古怪，并且愤世嫉俗，但事实上，他本人非常和蔼。他的鼻子有些扁平，颚骨突出，并且嗜好使用俚语：把同事称为"伙计"，把那些似是而非的理论称为"破马桶"。所有这些使他看起来有种蓝领阶层所特有的正直与亲切。他似乎非常沉迷于局外人的形象。"我年轻的时候，老人们总认为我惹人讨厌。现在我老了，年轻人当然也就把我看作一个讨人嫌的老家伙了。"他笑了笑，接着说，"如果有人认为我总在唠叨自己说过的话，那么可以说没有什么比这更让我闹心了"，就像许多天文学家那样。"我担心的是有人走过来说：'你一直重复的那些东西在技术上根本就不成立。'这会使我很难受。"（事实上，已经有人在指责霍伊尔既重复啰唆又犯有技术错误了。）[12]

　　霍伊尔似乎掌握着某种能使事情听起来更加合理的诀窍，比如说，在其辩称生命的种子肯定是从外太空流落到地球上的时候。霍伊尔认为，生命在地球上的自发产生，就好比垃圾堆在龙卷风的作用下会自发组装成一架波音747飞机一样，都是绝不可能的事情。在我们那次会谈时，霍伊尔对这点作了详细说明。他指出，由于小行星的频繁冲撞，至少在38亿年以前地球上都不可能有生命存在；而细胞生命的出现几乎确定无疑是在37亿年以前。霍伊尔进一步发挥，如果把地球整个45亿年的历史比作是一天24小时，那么，生命是在大约半个小时内出现的。"你必须发现DNA分子；并且在这半个小时内你还得合成成千上万种酶，"他解释说，"而这一切都是在一个非常恶劣的环境下发生的。我发现，当你把所有这些加以综合考虑时，就会明白生命在地球上自发产生的说法不能让人信服。"在霍伊尔长篇大论时，我发现自己竟然在暗暗赞同他的观点。的确，生命自然不可能在这里产生，而更明显不过的又是什么呢？直到后来，我才认识到，按照霍伊尔的时刻表，类人猿是在大约20秒前演化成人类的，而现代文明的存在还不到0.1秒。或许有点儿可笑，但事实就是这样。

　　霍伊尔对宇宙起源问题的严肃思考始自二战之后不久，也是在同另外两位物理学家托马斯·戈尔德（Thomas Gold）和赫尔曼·邦迪（Hermann Bondi）长期讨论的过程中酝酿而产生的。他回忆道："邦迪在什么地方的一个亲戚——印象中好像哪儿都有他的亲戚——送了他一箱酒"，三位物理学家一起喝酒时，不知怎么就把话题转向了长期以来一直困扰着年轻人的问题：宇宙是如何存在至今的？

　　宇宙中所有的星系都正在相互远离这一发现，使得许多天文学家相信：在过去的某一特定时间点曾发生了一次爆炸，宇宙也由此产生，并且至今仍在膨胀中。霍伊尔主要是出于哲学层面的考量而反对这一模型的。除非你对宇宙得以在其中创生的时间和空间有足够的认识，否则讨论宇宙创生问题就显得毫无意义，因为"你失去了物理规律的普遍性"。霍伊尔向我解释："物理学便不复存在了。"排除这种荒谬性的唯一方案，霍伊尔宣称，就是时空必定是一直存在的。就这样，霍伊尔、戈尔德和邦迪提出了恒稳态理论。该理论假定宇宙在时间上和空间上是无限的，并且通过某种至今尚不可知的机制不断产生着新物质。

　　20世纪60年代初微波背景辐射的发现，看来给了大爆炸理论一个强有力的支持。在这以后，霍伊尔便不再发展他的恒稳态理论了。但是到了20世纪

80年代，当他看到宇宙学家们努力去解释星系形成和其他一些难题时，他那些旧有的疑虑便又重冒了出来。"我愈益清晰地感觉到，肯定有某种相当严重的错误存在"——不仅关乎暴涨和暗物质这样一些新概念，而且也包括大爆炸本身。"我坚信。如果你的理论是正确的，那么就能由此得出许多正确的结论。而在我看来，到1985年为止，20多年过去了，却没有多少事实支持这个理论。如果它没错的话，就不应该出现这种情况。"

这样，在做了一些改进之后，霍伊尔以一种新的形式再次提出了恒稳态理论。他认为，在宇宙的演化过程中，不是仅存在着一次大爆炸，而是有许多次小爆炸发生。这些小爆炸产生了轻元素，并造成了星系红移。至于宇宙微波背景辐射，霍伊尔的最好解释是：这是星际间的金属尘埃发出的辐射。霍伊尔承认，其"准恒稳态理论"不过是用许多的小奇迹取代了一个大奇迹，还远远不够完美。但他坚持认为：大爆炸理论最近一些推论，例如暴涨的存在、暗物质以及其他一些怪论，存在着更严重的缺陷。"这就像是中世纪的神学"，他带着不常有的怒气大声嚷道。

但是，同他谈话越久，我就越怀疑他对大爆炸的质疑是否出自内心。从他的某些谈话中，你会发现他对这个理论有种特别的偏爱。"大爆炸"这个词，正是霍伊尔1950年在一次有关天文学的系列广播演说中首次创造出来的，这件事本身就算得上是对现代科学的一大嘲讽。他告诉我说，当时自己绝没有半点儿要贬低该理论的意思，正如许多报道所提到的那样，他这样做仅仅是为了更好地描述该理论。那时，他回忆道，一个名叫弗里德曼（Friedman）的物理学家证明，爱因斯坦的相对论必然会导致一个膨胀的宇宙。从那以后，天文学家就把这个理论称为"弗里德曼宇宙论"。

"这实在是个祸害，"霍伊尔说，"为了更形象生动地说明这一理论，我率先用了'大爆炸'一词。要是我将之申请专利、申请版权的话……"他又陷入了遐思冥想之中。1993年8月，《天空与望远镜》杂志发起了一场为该理论更名的比赛。然而，在考虑了成千上万个建议之后，大赛的裁决者宣称，他们没有发现比"大爆炸"更贴切的词语。[13] 霍伊尔说，他对此并不感到吃惊。"词语就像标枪一样，"他认为，"一旦扎进去，就再难拔出来了。"

事实上，霍伊尔差点就发现了宇宙微波背景辐射。看得出来，他为此感到非常懊恼。那是在1963年的一次天文学会议上，霍伊尔与罗伯特·迪克（Robert Dicke）曾有过一场深入的对话。迪克是一位来自普林斯顿的物理学家，正准备研究大爆炸模型所预言的宇宙微波背景。他告诉霍伊尔，说他预计背景微波在绝对零度以上20K左右，许多理论家也这样认为。霍伊尔提醒他说，有位名叫安德鲁·麦卡洛（Andrew McKellar）的加拿大射电天文学家，已经在1941年发现星际气体辐射出的微波在3K而不是20K。

让霍伊尔感到终身遗憾的是，在那次对话过程中，无论他还是迪克，都没能捅破最后那一层窗户纸，道出麦卡洛的发现背后的隐示：微波背景辐射大概就在3K。"我们只是坐在那儿喝咖啡，"霍伊尔回忆道，声音也随之大了起来，"只要我们俩无论谁说句：'大概就是在3K，'我们就会径直去检验它。要是那样的话，这项成果在1963年便属于我们了。"一年以后，就在迪克着手进行其微波实验之前，贝尔实验室的阿诺·彭齐亚斯（Arno Penzias）和罗伯特·威尔逊（Robert Wilson）发现了3K微波背景辐射。由于这一发现，他们后来获得了诺贝尔奖。"我一直觉得那是我一生中最大的失误。"霍伊尔叹了口气，又沉重地摇了摇头。

霍伊尔差一点儿就发现了背景辐射，虽然现在他对此现象的发现大加嘲讽，认为纯粹是捏造，但他还是对此耿耿于怀，为什么呢？我想，就像许多叛逆者那样，霍伊尔也曾经期望带着荣誉和光环打入科学界。在实现这个目标的路上，他踽踽而行。但是1972年，剑桥官方的一纸公文使他辞去了天文学研究所所长的职务——这仅仅是由于政治上而不是学术上的原因。就这样，霍伊尔和他的妻子离开了剑桥，在英国北部荒野小镇的一幢房子里一住就是15年，直到他们迁至伯恩摩斯。在此期间，曾经对他很有帮助的反权威主义也不再是建设性的，而成了反动的。他堕落成哈罗德·布鲁姆所奚落的"单纯的叛逆者"，虽然其内心深处还存有过去的梦想。

看起来，霍伊尔似乎还由于其他一些问题而受了苦。科学家的使命是去发现自然界中存在的模式（patterns），但在发现模式的过程中，总是难以避免南辕北辙的风险。霍伊尔的后半辈子，似乎便陷入了这一泥潭。他的确发现了模式——或者，毋宁说是阴谋——无论是关于宇宙的结构，还是关于那帮反对其激进观点的科学家。霍伊尔的思维定式，在他关于生物学的看法上体现得最明显不过了。自20世纪70年代初以来，他就认为宇宙中到处都弥漫着病毒、细菌和其他微生物（在1957年出版的《黑云》一书中，霍伊尔首次提出了这种可能性，这本书也是他最为著名的一部科幻小说）。可能正是这些布满太空的微生物，提供了地球生命的种子，并由此引发了生物进化；自然选择对生物多样性所作出的贡献微乎其微，甚至根本就不起什么作用。[14] 霍伊尔还宣称，像流感、哮喘和其他一些流行性疾病的爆发，就是由于地球经过这些病原体汇聚区域所导致的。

长期以来，生物医学界的传统观点认为疾病是以人际方式传播的。谈到这一点时，霍伊尔有些愤怒了，"他们根本不看事实就妄加判断："咳!这是错的。"并拒绝讲授这些看法。他们说的仅仅是一些类似的废话。这正是他们看病时老是出错的原因所在。如果他们碰巧治好了你的病，那你就太走运啦!"但是如果太空充满了微生物，我问道，那么为何没有探测到呢?他回答说，哦! 也许它们在曾经的过去的确是存在的。他怀疑20世纪60年代美国在高空气球和其他一些平台上所做的实验已经证实了太空中生命体的存在，但官方却秘而不宣。为什么呢? 霍伊尔认为这可能涉及国家安全或者是由于实验结果与现有理论相悖。"在今天，科学被限定在一些固定的框架内，"他以一种庄严的语气说，"每一条前进的道路都被一些错误的信仰封锁了。现在要是你企图在杂志上发表一些有悖于常规的观点，你会遭到编辑的断然拒绝。"

但是，霍伊尔强调，他并不认为艾滋病毒像某些报道讲的那样，是来自于外太空。他说："艾滋病毒是如此奇怪，我不得不相信它是在实验室里制造出来的。"难道霍伊尔在暗示这种病原体可能是由于某项生物战计划出了差错而引起的? "没错，我就是这么认为的。"他答道。

霍伊尔还推测，生命，甚至作为整体的天地万物，肯定都是按照某种设计好的宇宙蓝图而逐步展开的;天地万物都是"受到明显控制的"，他说，"有许多事情看来偶然，事实上却并非如此"。我问他是否认为有某种超自然的智慧安排了这一切，他严肃地点了点头："我就是这样看待上帝的。宇宙的一切都遵循着他的安排，但他究竟以何种方式安排，我并不知道。"当然，霍伊尔的许多同行，甚至还包括相当多的一部分公众，也像他一样认为宇宙受到了刻意安排。或许事实的确如此，但又有谁说得清呢? 霍伊尔声称科学家们故意隐瞒了太空微生物存在的证据，或者大爆炸理论的一些理所当然的缺陷，这也许是对他的同行们的根本误解。事实上，许多科学家正巴不得能有这样富于革命性的发现呢。

太阳原则

撇开霍伊尔的怪僻不论，就他对大爆炸理论的质疑而言，未来的观测或许会证明至少从部分上说他是有先见之明的。天文学家会发现:宇宙微波背景辐射并非是大爆炸的结果，而是由诸如银河系中的尘埃之类更实在的东西造成。核合成也并不像施拉姆及其他大爆炸理论的支持者声称的那样，是这个理论强有力的证据。但即使抛开大爆炸理论的这两个重要证据不论，红移仍然是这个理论的重要证据。甚至就连霍伊尔也承认，红移现象表明宇宙的确是在膨胀着。

对于天文学来说，大爆炸理论如同生物学中达尔文的自然选择一样重要：它是天文学的内聚力，使天文学成为一个统一的整体并赋予它意义。但这并不意味着该理论能够或者将能解释所有的现象。尽管宇宙学与粒子物理学有着紧密的联系，但它本身并不像粒子物理那样精确严密。比如，就哈勃常数的取值而言，天文学家在很长时间内都没有就此达成一致，而哈勃常数正是衡量宇宙的大小、年龄和膨胀速率的一个重要物理量。要推算出哈勃常数，必须测出星系的红移量和星系到地球的距离。前者可以直接测出，后者的测量却极其麻烦。天文学家并不能假定星系的表观亮度与它到地球的距离成正比。一个星系看上去很亮或许只是由于它距我们很近，但也可能它本身就很亮。还有，某些天文学家坚持认为我们宇宙的年龄是100亿年甚或更年轻些，而另外一些天文学家却深信它不小于200亿年。[15]

对哈勃常数取值的争议显然表明：甚至一个看来很简单的计算，宇宙学家们也必须作出许多假定；而这无疑会影响他们的计算结果。像进化生物学家和历史学家那样，宇宙学家必须对自己的数据作出解释。因此，我们在接受天文学的任何一个断言时，须采取谨慎的态度（例如，施拉姆宣称核合成的计算结果在5位有效数字内与理论预言保持一致）。

对宇宙更为细致的观测，并不一定就能澄清像哈勃常数这样一些有争议的问题。考虑一下吧：所有恒星中最为神秘的莫过于太阳了，没有人能确切知道，比如说，太阳黑子是如何产生的，为什么其数目以大约十年为周期增加或减少。我们之所以能够用一个简洁优美的模型来描述自己所存身的宇宙，很大程度上是由于缺乏足够的资料，是由于我们的无知。然而，我们对这个宇宙的细节知道得越多，也就越难于用一个简洁的理论去解释它是如何这样存在至今的。即便是研究人类史的学生都能认同这一悖论，偏偏宇宙学家们对此却难以接受。这个太阳原则表明：宇宙学中形形色色的奇谈怪论该适时终止了。近至20世纪70年代初，黑洞还被认为是理论上的怪谈而没有被严肃地接受。（根据弗里曼·戴森的说法，爱因斯坦本人也认为黑洞是"一个瑕疵，有待建立一个更好的数学模型将其从他的理论中剔除出去"。）[16]

然而由于约翰·惠勒等人的大力宣传，黑洞已经逐渐被认为是一个真实存在的客体。许多理论工作者现在也相信，包括银河系在内几乎所有的星系，在其核心处都隐匿着巨大的黑洞。形成这一共识的原因在于，没有一种更好的模型能够解释在星系中心物质的强烈旋转。

之所以会有对于黑洞这种说法，仅仅是由于我们的无知。天文学家们应当扪心自问：如果以某种方式突然将他们带到仙女系或我们银河系的中心，他们会发现什么呢？他们会发现那儿就像是现在理论所描述的黑洞，抑或会是另外一种截然不同的情况——一种没有人能够想到亦不曾想过情况呢？太阳原则表明，出现后一种情况的可能性似乎更大。甚至对于银河系本身，我们人类可能永远也无法看清其心脏部位的真实面目，更遑论其他星系了。但是，我们可能有足够的知识对黑洞假说产生怀疑，我们可能有足够的知识再次发现人类的所知是多么有限！

对宇宙学来说也大致如此。我们已经知道宇宙的一个令人震惊的基本事实：宇宙正在膨胀，它大概已有100亿年—200亿年的历史，正如进化生物学家们知道所有的生命都是从同一祖先通过自然选择进化而来。同样地，就像进化生物学家不可能超越达尔文理论一样，宇宙学家们也不可能超越这个基本事实。正如戴维·施拉姆所言，将来当这个理论在所知与未知之间取得一种完美的平衡时，再回过头来看，人们也许就会发现20世纪80年代末90年代初的时期是宇宙学的黄金时代。随着时光的流转，当我们积累的资料足够多的时候，宇宙学会变得像植物学——由理论松散地维系在一起的巨大的经验事实的集合。

发现的终结

无论如何，在发现宇宙中有趣的新事物方面，科学家的能力终归是有限的。马丁·哈威特（Martin Harwit），一位天文学家，同时也是一位科学史家，直到1995年，他还在担任华盛顿特区史密斯索尼安学院国家航空航天博物馆的馆长。他在1981年出版的著作《宇宙学的发现》中指出——

历史表明：我们在科学发现上所做的诸多努力都遵循着同一个模式，无论是各种昆虫的发现，还是为寻找新大陆和岛屿而进行的海洋探险，或者是探寻石油在地层中的储藏。当它们吸引越来越多的人去探索某一项发现时，发现的速率开始加速上升；但是用不了多久，新发现减少了，发现的速率降低了，即使采用一些高效的新方法也无济于事，研究会逐渐走向终结。人们可能会偶然发现一两个先前忽视了的细节，或者会遇到某种很罕见的特殊情况；但发现的速率依旧快速减小。接着，大河变成涓涓细流，昔日高涨的热情冷却了，研究者们会纷纷离开这个领域。事实上，它已不再富有活力了。[17]

哈威特指出，天文学不像科学中那些更具实验性的领域，从本质上来说它是被动的。我们只能通过那些自太空降临到我们身边的信息来探测天文现象，而这些信息大多以电磁辐射的方式存在。在如何改进观测技巧，使之更趋成熟方面，哈威特做了许多猜测，诸如光学望远镜，还包括像引力波探测器这样一些尚未完善的手段。他作了一幅曲线图粗略估计了一下过去和未来在宇宙学方面新发现增长速率的变化情况。这是一条钟形曲线，在2000年很快达到它的最高峰。根据这条曲线，到那时，我们已有的发现将占我们所能作出的发现总量的一半；到2200年，已有的发现将占全部可能发现总数的90%；在接下来的几百年内，发现的速率逐渐减小；与此同时，人们也逐渐完成其余10%的任务。

当然，哈威特也承认，各种各样的变化会加快或阻止发现的步伐，从而改变这个"进程"。"考虑到政治因素，将来天文学得到的资助会更少。一旦爆发战争，就会使得天文学研究几近中止。虽然战后大量废弃不用的军用设施用于天文学研究会再度促进它的发展。"[18]黑暗中总会有一线光明。

反讽的宇宙学当然还会继续下去，只要那些富于想象且雄心勃勃的"诗人"——如霍金、林德、惠勒，当然还有霍伊尔等人——还存在着。他们的理论，一方面暴露了人类在这个领域的经验知识是多么贫乏，另一方面又证实了人类无限的想象力。这既令人沮丧又振奋人心。反讽的宇宙学至多只能使我们感到敬畏，但它并不是科学。

当约翰·多恩（John Donne）写下下面这些诗句时，他大概是说给霍金也是说给我们听的："我的思想无所不及，包罗万象。而令人费解的是，作为造物主的我却困顿在封闭的囚室里，在病榻上，在我所创造的万事万物中。在世界的每一个角落，我的思想伴随着光环，又超越了光环；遭遇到光环，又抛弃了光环。"[19]

就让这段话成为宇宙学的墓志铭吧！

第五章
进化生物学的终结

绝不会有别的学科像进化生物学那样，背负着如此沉重的历史包袱，它已完全笼罩在文学批评家布鲁姆所称的"影响的焦虑"的阴影之下。这一学科的主旋律，在很大程度上可以说就是达尔文那些才气横溢的追随者，不断地向他那势不可挡的影响力妥协的过程。达尔文的自然选择理论（自然选择也是达尔文理论的核心成分）建立在两个观察事实的基础之上：首先，植物和动物繁衍的后代总数通常情况下都可能超出环境可承受的范围；其次，子代与亲代以及子代个体之间，总会有些微的差别。达尔文就此总结：每一生物个体，为了能长久地生存下去并繁殖自己的后代，总要与同种的其他个体展开直接或间接的竞争；机遇在任何个体生物的生存中都起一定的作用，大自然只偏爱（或选择）那些其变异特征更具适应性的个体，也就是说，它们更易存活下去，有更大的机会把这些适应性变异传给后代。

至于那些至关重要的代际变异是怎么产生的，达尔文只能付诸猜测了。在初版于1859年的《物种起源》一书中，曾提及法国生物学家拉马克（Jean-Baptiste Lamarck）的设想，即认为生物体不仅可以把遗传性状传递给后代，而且可以把获得性状传递给后代。比如说，因为长颈鹿经常伸长脖子去啃高处的树叶，这使得它们的精子或卵细胞发生了某种改变，所以就生出了脖子较长的后代。但达尔文显然并不满意这种认为生物体可以自己引导适应方向的观点，他宁愿相信代际变异是随机的，只是在自然选择的压力下，某些变异才成为适应性的，并导致生物的进化。[1]

达尔文一直懵无所知的是，在他的有生之年，有一位叫作孟德尔（Gregor Mendel）的奥地利修道士，正默默地从事着足以推翻拉马克的观点，并最终证实达尔文之天才直觉的实验研究。孟德尔是第一位认识到自然形态可进一步分化成不连续性状的科学家，这些性状在亲代和子代间的传递，是借助于一种他称为"遗传因子"（*hereditary particles*）、现在称作基因的物质；基因阻止了性状的融合，从而保持其特性不变；基因重组发生在有性生殖的过程中，同时偶尔产生基因表达的失误或者突变，从而提供出各种变异的后代，使得自然选择可以在此基础上尽情施展魔法。

孟德尔于1868年发表了一篇论文，公布了自己的豌豆杂交实验结果，但直至世纪之交，始终未能得到科学界的重视。甚至到了20世纪初期，孟德尔的遗传学理论也未被立即整合到达尔文理论中去。某些早期的遗传学家，甚至认为基因突变和性状重组可独立决定生物的进化方向，而不必依赖于自然选择。但到了20世纪三四十年代，哈佛大学的恩斯特·迈尔（Ernst Mayr）和其他几位进化生物学家，开始将达尔文进化论与遗传学融为一体，重新表述了达尔文的理论，形成一种强大的综合体系，并断言自然选择是生物形态以及生物多样性的首席建筑师。

1953年DNA结构的发现（DNA结构被看作是"建构"所有生物体的蓝图），进一步证明了达尔文关于一切生物都是相互联系的、都有着共同来源的直觉。沃森（Watson）和克里克（Crick）的这一发现，也揭示出使自然选择成为可能的遗传现象——连续性和变异性相统一的现象——的深层原因。同时，分子生物学也宣布，所有的生物现象均可用机械的、物理的术语来解释。

但这一结论在斯滕特看来，却根本谈不上有什么先见之明。他在《黄金时代的来临》一书中指出，早在发现DNA结构之前，某些著名科学家就已注意到：对于理解遗传现象等基本生物学问题，传统的科学方法和科学假说是无法胜任的，物理学家玻尔（Niels Bohr）就是这一观点的主要提倡者。玻尔认为，就像物理学家在理解电子的行为时只能满足于不确定性原理一样，生物学家们在试图探索更深层的生命奥秘时，也必然会碰到一个根本性的限度。

在研究生物现象时，就像我们对待物理状态一样，必须容忍某种不确定性，这也就意味着，必须允许生物体在某些细节上保有一定的自由，而这一点足以使生物体对我们隐藏起其最终的奥秘。照这种观点来看，生命的存在应被当作生物学的出发点，类似于量子（这在经典机械物理学看来是极为荒谬的）作为基本粒子的存在方式，构成了量子力学的基础一样。[2]

　　斯滕特指责玻尔试图复活陈腐过时的活力论观点，即认为生命起源于某种神秘的、不可能被还原为物理过程的精髓或生命力。事实上，玻尔的活力论设想并未得到证实，分子生物学却证实了玻尔的另一论断，即只要科学足够发达，它就能把神秘现象还原为平凡的事实（当然不可能是全然通俗易懂的）。[3]

　　对于每一位雄心万丈的年轻生物学家来说，身处于这个后达尔文、后DNA的时代里，要怎样做才能出人头地呢？可选择的道路之一，就是要比达尔文更加达尔文，把达尔文理论奉为理解自然的金科玉律，奉为无可超越的至上法则，这就是还原主义者道金斯所选的路径。理查德·道金斯（Richard Dawkins）是牛津大学的高级讲师，他坚持唯物的、非神秘论的生命观，并把达尔文理论打磨成一件可怖的武器，用以屠戮任何敢于向自己的生命观挑战的观点。在他看来，任何坚持神创论或其他反达尔文观点的主张，都是对他个人的侮辱。

　　在道金斯的著作经纪人于曼哈顿召开的一次集会上，我见到了道金斯本人。[4]他是个英俊的、冰雕一般的男人，有一双犀利的眼睛，配上刀削般的鼻梁，使他那红润的脸颊显得极不协调；穿一身定做的昂贵套装。当他举起青筋裸露的双手来强调自己的某一论点时，那双手总是轻轻地颤抖着，但这并不意味着他是个神经质的人。事实上，在短兵相接的观念交锋过程中，他是个技巧娴熟、表现优异的对手，因而被称为达尔文的猎犬。

　　就像其著作风格一样，道金斯本人也处处流露出绝对的自信，他的每句话仿佛都带着一句潜在的引子："任何一个傻瓜也能明白……"作为一个理直气壮的无神论者，道金斯宣称他不像某些科学家那样，认为科学和宗教探讨着互不相干的问题，因而能够和平共存。他认为，大多数宗教都坚信，只有上帝才能为生命的设计和意图作出解释，因此他决心把这种观点连根拔掉，声称："所有的意图最终都源于自然选择，这就是我所推崇的信条。"

　　道金斯用了大约45分钟的时间向听众们阐述了他那极端的还原主义进化观。他建议我们把基因看作是一小段、一小段的软件，其目标只有一个：拷贝出自己更多的副本来；不管是石竹花还是猎豹，所有的生物都只不过是这些"自我复制程序"创造出来的精巧装置，以帮助它们"扩大再生产"。文化也是基于这种自我复制程序之上的，道金斯称之为"拟子"（memes）。道金斯请我们想象一本书，书中传递着这样的信条：相信这本书并使你的孩子也相信它，否则的话，当你死去之后，你只能到一个被称作地狱的恶心地方去。"这就是一条十分有效的自我复制密码。当然，除非是蠢到了家，否则谁也不会接受如此直白露骨的指令：'相信这本书并使你的孩子也相信它'；你得更狡猾些，以某种更加雍容华美的装饰把它包装起来。大家当然都明白我正在谈论的是什么。"当然！而基督教，就像所有的宗教一样，正是由此类密码构成的极其成功的系列文件，此外还能怎么解释呢？

然后，道金斯开始回答听众提出的问题。听众的身份十分复杂，有记者、教师、编辑和别的一些准知识分子，其中有个叫巴洛（John Perry Barlow）的家伙，以前曾是个地地道道的嬉皮士，偶尔为感恩而死乐队①写写词，后来摇身一变，又成了"新时代"的赛博空间先知。他举止笨拙，脖子上紧箍着一条红色的印花大手帕。正是这个家伙，向道金斯提了个长长的问题，似乎是关于信息到底存在于何处之类的事。

道金斯的眼睛眯了起来，鼻翼轻轻翕动着，仿佛已嗅到了某种糊涂头脑的气味。"对不起"，他说，他不知道巴洛到底想问些什么。巴洛又费了约一分钟的口舌。"我觉得你正试图表述的玩意儿，也许对你来说很有趣，但我却丝毫不感兴趣。"道金斯说完就开始扫视会场，寻找下一位提问者。房间里的气温骤然之间仿佛冷了好几度。

接下来，在讨论有关地球之外的生命这一话题时，道金斯又抛出了他的信条，认为自然选择是宇宙的普适规律：生命在哪里出现，自然选择就在哪里发挥作用。他告诫说生命不可能在宇宙中普遍存在，因为不论是在太阳系别的行星上还是在宇宙中别的什么地方，我们至今尚未发现有任何生命的迹象。巴洛勇敢地打断他的话，指出我们之所以未能探测到地外生命形式，可能仅仅是因为我们的观察能力不足。他最后又意味深长地补充了一句："我们不知道究竟是谁发现了水，但我们敢肯定绝对不会是鱼。"道金斯转过身来，冷静地注视着巴洛："你的意思是说，我们一直都在注视着'他们'，但却一直未能认出'他们'来？"巴洛点了点头。"噫——！"道金斯长长地吁了一口气，似乎借此吐出了使这个榆木疙瘩脑袋开窍的一切希望。

道金斯也会同样尖刻地对待自己的生物学家同行，尤其是那些胆敢挑战达尔文理论基本范式的人。他以不容置疑的口气宣称，不论以何种方式出现，任何试图修改或超越达尔文的努力，都将以失败而告终。他打开一本自己1986年出版的《盲人钟表匠》，宣读了如下一段文字："我们的存在曾被

---

① 感恩而死乐队（Grateful Dead）：著名迷幻始祖级乐队，其第一场音乐会于1965年在旧金山的菲尔莫尔（Fillmore）举行，自此风靡世界，以迷幻风格和能使听众最贴近地感受到爱而著称——译者注。

视为所有奥妙现象中最大的奥秘，但……它已不再是什么奥秘了，因为谜底已经被揭开。达尔文和华莱士（Wallace）已经解开了生命之谜，尽管我们偶尔还能为他们的答案添加那么几条注释。"[5]

当我后来就"注释"云云请教道金斯时，他答道："这里面难免会有点虚饰浮夸的成分。但这样说也并不过分"，因为达尔文确实解开了"生命的由来以及它何以如此美丽、如此有适应性、如此复杂的谜底"。道金斯也赞同斯滕特的观点，认为自达尔文以后生物学上所有重大进展，包括孟德尔对基因表达连续性的论证、沃森和克里克关于DNA双螺旋结构的发现等，都支持了而不是削弱了达尔文的基本思想。

最近的分子生物学研究显示，DNA还与RNA之间存在着相互作用，并且蛋白质的结构远较我们以往的设想复杂，但遗传学的基本范式——基于DNA之上的基因遗传——却丝毫未尝被动摇。"怎样才能真正使之动摇呢？除非你拿一个完整的生物体，比如说塞伦盖蒂草原上的一匹斑马，让它获得一种新的性状，比如说学会一种寻找水源的新方法，并进而把这一性状反向编进基因组中去。如果这类事情真的发生了的话，我把脑袋揪下来给你。"

仍然存在着某些十分重大的生物学课题，如生命的起源、性的起源以及人类意识的起源等。发育生物学——这一学科试图揭示单个受精卵是怎样发育成一条大蛇或者一名福音传教士的——同样提出了某些重要的问题。"我们当然需要弄清楚事情到底是怎样发生的，并且这些问题将被证明是特别特别复杂的。"但道金斯坚持认为，发育生物学就像在它之前诞生的分子遗传学一样，将仅仅是在达尔文范式之内填充进更多的细节。

道金斯"烦透了"那些主张单凭科学不可能解答关于存在的终极问题的知识分子。"他们认为科学过于傲慢，认为有许多问题科学无权过问，因为它们在传统上是神职人员的专利，就像那些神职人员真的能给出什么答案。诚然，像宇宙是怎样开始的、引发大爆炸的原因是什么、意识究竟是什么等问题，的确是难以回答的。但如果连科学家都无法解答某件事情，那么别的见鬼的家伙就更加解答不出来。"道金斯兴致勃勃地援引了英国大生物学家彼得·梅达沃（Peter Medawar）的一句话，说某些人"'天生喜欢在不求甚解的泥潭里得过且过地打滚'，我却要去追求理解"。道金斯又热切地补充道："但理解对我来说就意味着科学的理解。"

我问道金斯，为什么他的信条——即关于生命问题，我们所知的一切和需要知道的一切，达尔文已经基本上都告诉我们了——不仅受到来自神创论者、新时代信徒以及哲学诡辩论者方面的反对，甚至也显然遭到大多数生物学家的反对。他答道："这也许是因为我没把自己的观点表述清楚吧。"实际情况当然正好与此相反，道金斯已清楚地表明了自己的观点，并且是如此清楚明白，以至于他未给神秘、意义、目的等留下任何存身之地——甚至，在达尔文本人已经给出的科学发现之外，他也未给别的科学发现留下丝毫余地。

古尔德的偶然性方案

自然会有这样一些现代生物学家，他们对于自己的工作仅仅被看作是在为达尔文的杰作添加注脚这一点感到怒不可遏。达尔文的强者型（在布鲁姆的意义上）追随者之一，是哈佛大学的斯蒂芬·杰伊·古尔德（Stephen Jay Gould）。古尔德试图通过贬低达尔文理论的解释力，来抵抗达尔文的影响。他声称有许多现象是达尔文理论所无法解释的。20世纪60年代，古尔德通过抨击历史悠久的均变论教条（均变论认为地球和生命是被某种稳定的地质物理力量持续不断地创造出来的），正式向世界表明了自己的哲学立场。[6]

1972年，古尔德和纽约美国自然史博物馆的埃尔德雷奇（Niles Eldredge）合作，通过引入断续平衡理论[①]，把对均变论的批评扩展到生物进化过程中。（这一理论，古尔德等自称为"勇往直前的朋客"，[②]而批评家们却称之为"抽风式的进化"。）[7]古尔德和埃尔德雷奇认为，新物种的产生，几乎不可能通过达尔文所描述的那种渐进的、线性的进化方式实现，更多情况下，物种的形成是一种相对迅速的事件，当一个生物群脱离其稳定的亲本种群、开始自己的进化旅程时，新的物种就出现了；决定物种形成的，不是像达尔文（以及道金斯）所描述的适应过程，而是某些更加独特、复杂、偶然的因素。

在其后续的系列文章中，古尔德不厌其烦地、不遗余力地攻击达尔文理论的许多阐释中，被他认为是含义模糊的地方，即进化及其必然性。在古尔德看来，进化并未呈现出任何前后一致的方向，也无法指出其进化产品（比

---

① 原文是"the theory of punctuated equilibrium"，目前学界对该理论的译法并未统一，还有"间断平衡理论""点断平衡说""中断平衡进化说""跃进—平衡理论"等不同译法。这里仍沿用了旧译本的说法——译者注。

② Punk eek，是"punctuated equilibrium"两词首音节的谐音——译者注。

如说人类）在任何意义上是必然的；即使把"生命进化的录音带"重复播放上百万次，这种有着超常的大脑的奇特猿猴也不可能出现。古尔德只要发现了遗传决定论的观点，一定会给予迎头痛击，不管它是出现在关于种族和智力的伪科学声言中，还是出现在受人尊敬的有关社会生物学理论中。古尔德把他的怀疑论主张用自己的美文加以包裹，文章里引用了大量雅俗共赏的东西，同时处处自谦地宣称自己的文章只是一种文化赝品——这倒是颇具自知之明。古尔德获得了令人目眩的成功，几乎他写的所有的书都上了畅销书排行榜，并且他本人也一直是世界上引用率最高的科学家之一。[8]

在见到古尔德本人之前，我对他思想中某些明显是相互矛盾的方面困惑不已。比如说，我就搞不清楚他的怀疑主义以及他对进化论的反感，到底达到了怎样的程度。他是和库恩一样，相信科学自身并未显示出任何持续的进步吗？或者，他也认为科学的发展就像生命的进化过程一样，是曲折的、毫无目的的吗？那么，他是怎样绕过那些使库恩难以自圆其说的矛盾的呢？再者，某些批评家（古尔德的成功证实了他当然也有众多的追随者）指控他是一位未公开的马克思主义者，但马克思所信奉的是高度决定论、进步论的历史观，这与古尔德的观点是相对立的。

我也为古尔德在有关断续平衡的争议上是否做了某种妥协而感到迷惑。在1972年发表的原始论文的页头标题中，古尔德和埃尔德雷奇旗帜鲜明地宣称，他们的理论是达尔文渐进主义的"替代理论"，将来肯定会取代达尔文理论；但在1993年发表于《自然》杂志上的一篇回顾性文章（标题是《作为时代产物的断续平衡论》）中，古尔德和埃尔德雷奇则暗示，他们的假说可能只是对达尔文基本模型的"有益的扩展"或"补充"。"勇往直前的朋客"变成了"委曲求全的小媳妇"（punk meek）。古尔德和埃尔德雷奇最后以一种坦诚的、和解的姿态结束了全文。他们指出，其理论只不过是众多强调随机性、不连续性甚于强调秩序和进化的现代科学观点之一。断续平衡理论，照此看来，只是古生物学观点在时代精神中的一种体现，而时代精神就像时间长河中飘忽不定的幽灵（用一种文学语言来表述），不足采信。因此，断续平衡理论的发展前景，最终被证明为只是附庸时尚的产物并被抛进历史垃圾堆呢，还是包含着顿悟自然法则的火花，只有等待那断续的、不可预见的未来裁决了。[9]

我怀疑，这种言不由衷的谦恭，只不过是20世纪70年代后期所种下的前因的必然后果，当时的论者们，把断续平衡论称为超越达尔文理论的一场

"革命"，大肆加以兜售；而那些神创论者，也难免会抓住"勇往直前的朋客"理论为证据，攻击进化论并非普遍接受的理论；某些生物学家由此指责古尔德和埃尔德雷奇两人，说正是他们的花言巧语才鼓动起这些反进化论的言论。1981年，古尔德在阿肯色州举行的关于"神创论究竟能否和进化论一起在公立中学里讲授"的诉讼中出庭作证，并试图借此机会澄清是非。结果却迫使古尔德不得不承认，断续平衡论并非什么革命性理论，它至多也不过是件颇具匠心的"小玩意儿"，仅可供某些专家品鉴把玩。

古尔德相貌平凡。他身材矮而胖，红润的脸上油光光的，点缀着一个蒜头鼻子和一抹卓别林式的短髭。我见到他的时候，他正穿着一条皱巴巴的卡其布裤子，套件牛津衫，看起来就像一位事事心不在焉的老派教授。但只要他一开口讲话，所有那些平凡的假象就立即消失得无影无踪了。当讨论到一些科学上的重大争端时，他的语调变得疾快而低沉，各种知识领域的证据，哪怕是那些极其复杂、极端技术化的证据，他也能根据需要随口道来。就像他的文章一样，他的谈话也点缀着各种格言警句，并且总是一成不变地以这样的前缀引出："当然，你一定知道这句名言，关于……"他谈话的时候，常给人一种心神不属的印象，仿佛并未将注意力集中在自己所讲的话上。我有一种强烈的印象，就是单纯的谈话绝对不足以占据他的全部心神，他头脑中的高水平的程序，总是在话题的前面悠闲地散步，从容地四下探测着，尝试着谈话可能出现的主题，搜寻着一系列新的论据、类比和格言。我觉得不论我的思想走到何处，古尔德总会在前面等着我。

古尔德承认，他之所以走向了进化生物学的道路，部分地是受到了库恩《科学革命的结构》一书的激励，他在1962年此书出版后不久，就拜读了这部大作。库恩的著作使他相信：他，作为一位"来自皇后区（那儿的青年人从来就没有上过大学的）中产阶级低层家庭的年轻人"，也许同样能在科学领域里出人头地；它也同样引导古尔德抛弃了"那种归纳主义的、改良主义的、进步论的、'一次只增加一个事实，不到年老时绝不要去构建理论体系'的做学问模式"。

我问古尔德是否认为科学并非在向着真理前进，就像库恩一样。他坚决地摇了摇头，并否认库恩持有类似的立场。"我很了解库恩"，虽然库恩是社会建构主义者和相对主义者的"思想之父"，但他仍然相信"在我们的意识之外存在着一个客观世界"，古尔德肯定地说。库恩只是觉得这个客观世界在某种意义上是很难定义的，但他确实认为"比起几个世纪之前，我们对它的认识是大大进步了"。

如此说来，曾经坚持不懈地要把进化观念从进化生物学中剔除出去的古尔德，倒是位科学进步的信奉者了？"噢，当然，"他温和地说道，"我相信所有的科学家都是进步论者。"没有任何一位真正的科学家会是文化上的相对主义者，古尔德进一步发挥，因为科学是极其乏味的。"科学的日常工作是紧张而乏味的，你不得不去清扫老鼠笼子，滴定各种溶液，还得清洗培养皿。"如果不是考虑到这一切可能会带来重大的科学进展的话，没有谁能容忍下去。作为补充，他又再度回到了库恩的话题上，"某些具有丰富思想的人，却往往以一种近乎奇怪的曲折方式把自己的观点表述出来，这只不过是为了达到一种强调的效果"。（当我后来回味这句话时，不能不感到怀疑：难道古尔德是在拐弯抹角地为自己在早些年的"花言巧语"表示歉意吗？）

古尔德对我有关马克思的质询，只是含含混混地一带而过。他承认马克思的某些原理是很有吸引力的，比如说，马克思关于社会意识是由社会存在决定的，并且通过矛盾、通过肯定方面和否定方面的冲突而变化发展的观点。"确实是合理而有趣的变化发展理论，"古尔德评论道，"你通过否定最初的前提而前进了一步，然后你又否定了第一次否定，但却并不回到最初的前提，实际上你已经前进到了别的地方。我认为这些都很有意义。"马克思关于社会变革和革命的观点，"其中系统内小的量变的积累，最终会导致系统的质变"的观点，与断续平衡论是十分相容的。

虽说已几乎没有必要，但我还是问出了第二个问题，古尔德是——或曾经是——一名马克思主义者吗？"你只需回忆一下马克思是怎么说的就已足够了。"我的话音未落，古尔德就接上了茬。古尔德进一步解释道，没有任何一个聪明人会表白自己与某种"主义"有紧密关系，更不用说一个容量如此之大的"主义"了。古尔德也不赞成马克思关于进步的观点。达尔文虽然是"维多利亚时代的伟人，难以完全摆脱进步论的影响"，但与马克思相比，却更多地批判了维多利亚时代的进步观念。

作为一个不屈的反进步主义者，古尔德也并未排除文化能够展现某种进步的可能性。"因为社会继承是拉马克主义的，它为文化的进步信念提供了坚实的理论基础。进步虽然总是会被战争以及诸如此类的事件所打断，因而被推离了正常的轨道，但至少我们所发明的一切都直接传递给了后代，因此，定向的积累才有了可能性。"

最后，当我问及古尔德有关断续平衡理论的问题时，他表现出了强烈的维护态度。他说，这一思想的真正意义在于，"在个体生存竞争的水平上，你不可能用达尔文主义的，或经典达尔文主义的术语，来解释物种的形成"；对于进化的趋向，只能通过在种系水平上发挥作用的机理加以解释。"之所以存在着进化的趋向，是因为某些物种的形成更频繁，因为某些物种的存活时间比其他物种更长。"他说道，"物种出现和灭亡的原因，与生物个体的出生和死亡原因全然不同，这是一种不同的理论。这才是真正有意义的，这才是'勇往直前的朋客'中的新理论成分之所在。"

古尔德拒不承认在断续平衡理论争议中作出过让步，或曾屈服于达尔文理论至高无上的权威。当我问他该怎样解释从1972年原始论文中的"替代理论"到1993年《自然》杂志上回顾文章中的"补充"说这一转变时，他宣称："我没那样写过！"古尔德把这一切归咎于《自然》杂志的编辑约翰·马多克斯（John Maddox），说他在既未与自己协商，也未征求埃尔德雷奇意见的情况下，就擅自把"补充"一词插入文章标题之中。"我为此对他非常恼火。"他气呼呼地说。但接着，他又争辩说，其实"替代"和"补充"在含义上并没有太大差别。

"听着，把它说成一种替代理论，并不意味着旧的渐进主义就不存在了。明白吗，我想这是人们所忽略的又一个事实。这个世界上充满了各种各样的替代物，对吧？比如说，世界上既有男人也有女人，但在人类的性别中，哪一个能替代另一个呢？我的意思是说，当你宣称某物是一种'替代'时，并不意味着它是唯一起作用的。在我们写作论文之前，渐进主义拥有绝对的霸权，我们提出了一种有待验证的替代理论，我认为断续平衡理论在化石记录中得到的支持率是占绝对压倒性优势的，这表明渐进主义虽仍然存在，但在整个事情的综合模型中已经不再是真正重要的了。"

在古尔德继续口若悬河地讲述的时候，我却开始怀疑他是否真的对解答关于断续平衡论或其他问题的争端感兴趣，于是我问他是否认为生物学能达到一种终极理论，他做了个鬼脸。持这种信念的生物学家都是"幼稚的归纳优越论者"。他说："他们确实认为我们总有一天能测定人类的所有基因组序列。好吧，就算我们能有这么一天吧！"他承认，甚至是某些古生物学家也认为，"如果我们有足够的耐心去探索，肯定会弄清历史的基本特征，然后我们就可以再现生命的历史"。但古尔德却并不这么看，达尔文"揭开了生物个体之间基本的相互关系的谜底，但对我来说这才仅仅是开始。探索之路远未结束，现在只是刚刚开始"。

　　那么，古尔德认为进化生物学最突出的理论问题是什么呢？"哦，太多了，我都不知道该从何说起。"他指出，理论工作者仍需确定进化背后的"一整套原因"，从小分子开始直到大的生物种群；然后是"所有那些偶然性因素"，比如被认为曾引起众多种群灭绝的小行星撞击。"这样说起来，关于进化生物学的主要问题，可归结为进化的动因、动因的强度、动因的水平，以及偶然因素。"他若有所思地沉默了一会儿，"这是个不错的说法。"他边说边从衬衣口袋里掏出个小记事本，并在上面记了起来。

　　然后，古尔德兴高采烈地列举出科学无法解答这些问题的各种各样的原因。作为一门历史性科学，进化生物学只能提供回顾性的解释，而无法作出预见；有时它不能给出任何结论，因为缺乏足够的证据。"如果你缺乏历史事件先后次序的证据，那么就无计可施，"他说，"这就是为什么我会认为我们永远无法知道语言的起源，因为这绝非一个理论问题，而是个历史偶然性问题。"

　　古尔德也赞同斯滕特的观点，即为了在前工业化社会中生存而产生的人类大脑，在解决某些特定问题时显然无能为力。研究已经表明，人类在处理涉及复杂变量的可能性和相互作用问题上力不从心，比如处理遗传—环境问题。"如果基因和文化之间是相互影响的——当然，它们肯定是相互影响的——人们就永远无法理解遗传—环境问题，你不能因此说它20%是出于基因的作用，80%则应归于环境影响。你当然不能这样做，因为这是毫无意义的。自然发生的特性就是自然发生的特性，这就是你对于它所能说的一切。"无论如何，古尔德不是那种乐意给生命或意识赋予某些神秘特性的人。"我是一个老派的唯物主义者，"他说，"我认为意识产生于神经组织的复杂性，但我们对此了解的确还很少。"

　　出乎我的意料的是，古尔德紧接着竟突然絮叨起关于无限和永恒的话题来。"这是两个我们所无法理解的问题，"他说，"但理论上却一直要求我们解决它。这也许是因为我们的思考方法不对头。在笛卡儿空间中，无限是一种似乎很矛盾的说法，对吧？当我只有八九岁的时候，我习惯于说：'喂，那儿有一堵砖墙。'但砖墙的那边又是什么？这就是笛卡儿空间。即使空间是弯曲的，你仍然情不自禁地要思考弯曲空间之外又是什么，哪怕这并非思索'无限'问题的正确方法。也许这一切全都错了！也许这个宇宙确是分形膨胀的！我不知道它到底是怎么回事，也许存在着这一宇宙得以建构出来的某些途径，只是我们无法预料到罢了。"考虑到科学家们总是根据一些预想的概念对事物进行分类的趋向，古尔德怀疑他们能否在任何学科达到终极理论。"任何有关终极理论的宣言，归根到底是否仅仅反映了我们对事物进行概念化的方式，我也同样感到困惑不解。"

即使抛开所有这些局限，那么生物学，甚至是作为整体的科学，会不会发展到它的极限并因而走向其终结呢？古尔德摇了摇头。"人们在1900年就曾认为科学正走向终结，但紧接着我们就取得了板块构造学说、生命的遗传基础等新进展。科学怎么会终结呢？"古尔德又补充道。再说，现有的理论可能只反映了我们作为真理的追求者自身的局限性，而不是实在的真正本质。我还没来得及对此作出反应，古尔德就又抢到了我的前面："当然，如果那些限度是科学所固有的，那么科学也许会在其限度之内完结。哦，对了。是的，这是一种合理的判断，尽管我不认为它是正确的，但我能理解其结构。"

古尔德认为，在生物学的未来岁月里可能依然存在着重大的观念上的革命。"发生在我们这个星球上的生命进化，可能会被证明仅仅是整个生命现象很小的一部分。"别的地方的生命现象，也许完全不符合达尔文主义的原则，完全不像道金斯所坚信的那样。事实上，地外生命的发现，可能会证明道金斯关于达尔文主义的宣言，即认为达尔文理论不仅适用于这个小小的地球，而且适用于整个宇宙，是完全错误的。

如此说来，古尔德是相信在宇宙的其他地方也存在着生命了？我问道。"你相信吗？"他顺口反问过来。我回答说，我个人以为这种争论只是个见仁见智的问题。古尔德在这个恼人的问题面前退缩了，也许只有这一次，他才真的被噎住了。是的，地外生命当然是件见仁见智的事情，他说，但人们仍然可以根据既有事实作出猜测。在地球上，生命的出现是自然而然的，因为能提出这方面证据的古老岩石记录，确实也证实了这一点。另外，"宇宙是无限的，其任何一部分都不可能是绝对独一无二的，这就可以合理地推出，宇宙中存在别的生命形式的可能性也是极大的，只是我们并不知道它们，当然我们也无法知道。同时，我也相信，宣称有其他生命形式存在，绝不是哲学家们的语无伦次"。

理解古尔德这个人的关键，可能并不在于他的被册封的马克思主义或自由主义或反独裁主义等标签，而在于他对自己学科领域潜在的终结可能性的深深恐惧。通过把进化生物学从达尔文手里——以及从被定义为探求普遍规律的、作为整体的科学之中——解放出来，他试图使对知识的追求成为开放的甚至无止境的事业。古尔德是如此的老于世故，以至于他并不像某些笨拙的相对主义者那样，去否认那些尚未被科学所揭示的基本法则的存在。与此相反，他只是非常有说服力地让你相信：现有法则并不具备太强的解释力，它们遗留下许多无法解答的问题。他是一名极其圆滑的反讽科学的实践者，充分发挥了反讽科学的否定性力量。他对于生命的看法只能用大选期间人们贴在汽车保险杠上的标语来概括："废话漫天。"①

当然，古尔德对自己观点的表述要文雅得多。在我们交谈的过程中，他曾指出，许多科学家都不把历史学——那处理特殊事件和偶然因素的学科——看作是科学的一个部分。"我认为这是一种错误的分类方法，历史学是一种类型不同的科学。"古尔德坦承，他发现历史学的模糊性及其对直来直去的分析方法的拒斥，对他有着极大的诱惑力，"我热爱历史学！因为我实质上就是一名历史学家。"通过把进化生物学转换成历史学——一门本质上是解释的、反讽的学科，就像文学批评一样——古尔德应用他那花言巧语的出众伎俩折服了许多人。如果说生命的历史是一座包含大量偶然性事件的深不可测的采石场，那么他就可以不停地挖掘下去，逐一把玩那一个个古怪的事实，永远也不必为自己的工作是否已变得失去价值或多余而担心。在大多数科学家们都忙于鉴别隐藏在自然背后的"信号"的时候，古尔德却一直把注意力放在"噪声"之上，断续平衡理论根本就算不上什么理论，它只是对"噪声"的描述。

古尔德最大的弱点就是缺乏独创性。达尔文在《物种起源》一书中，早已预言了断续平衡理论的基本概念："许多物种一旦形成，就不再经历任何进一步的演变……物种曾经经历过的渐变阶段，虽然长达若干年，但与它们保持自身性状的阶段比较起来，可能就极其短暂了。"[10]古尔德在哈佛的同事恩斯特·迈尔在20世纪40年代也曾提出，新物种能够如此迅速地产生（例如，通过小种群的地理隔离），以至于它们在化石记录中未留下任何中间步骤的痕迹。

道金斯认为从古尔德的作品中找不出多少有价值的东西。他说，可以肯定的是，物种在有些时候（或者是经常性地）是以迅速的爆发形式产生的，但这又能说明什么呢？"重要的是，在物种产生那段时间里，渐进的选择过程仍

① "Shit happens"：每当美国总统大选的时候，各政党的候选人都竞相发表演说，陈述自己的政见，废话与空头支票满天飞。许多美国公民为了表示自己对这种劳民伤财做法的不满，就在自家汽车的保险杠上悬挂此类标语牌，以示轻蔑——译者注。

然存在，只是被压缩得很短。"道金斯评论道，"因此，我不认为这一点有
多么重要，只是把它看作是新达尔文主义理论的一个有趣的环节。"

道金斯还认为，古尔德关于人类或其他任何形式的智慧生命在地球上的
出现缺乏必然性的观点，同样不值一哂。"在这一点上我绝不会赞同他！"
道金斯说，"我认为，也不会有任何别的什么人赞成他！这就是我的观点！
他只不过是在向风车挑战！"生命曾以单细胞的形式存在了30亿年，如果不
是因为产生出多细胞生物的话，它还会在地球上再持续30亿年。"这样，当
然就不存在什么必然性了。"

我问道金斯从长远来看，古尔德的进化生物学观点有取得优势地位的
可能吗？毕竟，道金斯以前曾提出过，生物学的基本问题可能会是十分有限
的；而古尔德所致力研究的历史问题，实际上却是无限的。"如果你所谓的
'长远'，指的是等到所有有意义的问题都已被解决，而你所能做的一切只
不过是去发现一些细节的那一天，"道金斯冷冰冰地回答，"那么，我想古
尔德的观点是会占上风的。"他又补充道，生物学家们永远无法确认哪一种
生物学原理是真正具有普遍意义的，"直到我们发现了别的具有生命的星
球"。道金斯的这一陈述十分含蓄地承认了这样一个事实：对于那些最深层
的生物学问题——地球生命在多大程度上是必然发生的？达尔文理论究竟是
宇宙普适的法则，还是仅适用于地球？——只要我们仍然只有一种生命形式
可供研究，就永远也不可能被真正地、经验地解答。

盖亚异端

达尔文轻松愉快地就吞噬了古尔德所提出的假说，甚至连个饱嗝都不必
打；他也同样轻松地消化掉了另一位自诩为强者科学家的观点，这个人就是
马萨诸塞大学阿姆斯特分校的林恩·马古利斯（Lynn Margulis）。马古利
斯凭借其个人的几点学术主张，向她所谓的"极端达尔文主义"的正统地位
发起了挑战。其首要的也是最为成功的观点就是共生概念，而达尔文及其追
随者所强调的，却是生物个体以及种群间的竞争在进化过程中所起的重要作
用。20世纪60年代，马古利斯就开始论证，在生命的进化过程中，共生是同
样重要的因素——甚至是更为重要的因素。在进化生物学中，最重要的谜团
之一，就是关于从原核生物到真核生物的进化，前者的细胞没有细胞核，是
所有生物中最简单的；而后者的细胞却具有完整的细胞核。所有多细胞生物

包括我们人类，都是由真核细胞组成的。

马古利斯认为，真核细胞的出现，可能是因为一种原核细胞吞噬了另一种较小的原核细胞，而后者就变成了细胞核。她认为，不应把这种细胞看成是单个生命，而应看作是组合体。在马古利斯成功地给出共生关系在活的微生物中存在的实例之后，她关于共生在早期进化中起着重要作用的观点也逐渐赢得了支持，然而她并未就此止步。就像古尔德和埃尔德雷奇一样，她也认为用传统的达尔文学说的原理无法解释从化石记录中所观察到的生物种群的时断时续现象，而共生则可解释物种何以会突然产生，以及它们何以在如此长久的时间里保持稳定。[11]

马古利斯对共生的强调，自然而然地导致一种十分激进的观点：盖亚（Gaia）。这一概念和术语（盖亚是希腊神话中的大地女神），最初是在1972年由詹姆斯·拉夫洛克（James Lovelock）提出的，他是一位自雇的英国化学家和发明家，对传统观念的攻击更甚于马古利斯。"盖亚"具有多重含义，但其核心观念却是指由地球上所有的生命构成的生物圈，与其环境（包括大气、海洋以及地球表面的其他方面）构成的一种稳定的共生关系。生物圈借助自身的化学作用，以一种更有利于自身生存的方式改变着环境。马古利斯直接采纳了这一思想，从一开始她就与拉夫洛克在普及这一思想上携手合作。[12]

1994年5月，我在纽约宾夕法尼亚火车站的头等休息室中遇见了马古利斯，她正在那里候车。虽然上了年纪，但她看起来仍像个野丫头：短短的头发，红润的皮肤，穿一件带条纹的短袖衬衫，着卡其布裤子。一开始，她称职地表现了作为一个激进分子的特色，对道金斯以及其他一些极端的达尔文主义者关于进化生物学可能已接近完结的观点大加嘲讽。"他们已经完蛋了，"马古利斯宣称，"但那只是20世纪生物学史上的一件鸡毛蒜皮的小事，无碍于羽翼丰满的、生机勃勃的科学的进一步发展。"

她强调，自己并不怀疑达尔文理论的基本前提："进化毫无疑问是存在的，我们已经观察到了过去进化发生的证据，它如今也仍在发生着，每一个具有科学头脑的人都会赞同这一点。问题是，它是怎样发生的？正是在这一点上，人们才分成了不同的派别。"极端的达尔文主义者，把自然选择的单元定为基因，因而无法解释物种形成是怎么发生的。根据马古利斯的观点，只有那种把共生与更高层面上的选择结合起来的更为宽泛的理论，才有可能解释化石记录以及现有生命的多样性。

　　马古利斯补充道，共生也不排斥拉马克理论或者获得性状的遗传。通过共生，一个生物体能够可遗传地吸收或渗透另一生物体，从而变得更加适应。举例来说，如果半透明的真菌吞噬了可进行光合作用的水藻，真菌就可能由此获得光合作用的能力，并把它遗传给后代。马古利斯认为拉马克一直被极不公正地看成是进化生物学的牺牲品，"这是个英国人歧视法国人的问题。达尔文的一切都正确，而拉马克却一无是处，这简直太不像话了"。马古利斯指出，共生发生学，即通过共生创造出新物种，并不是一种真正的原创性思想。这一概念最早是在19世纪末提出的，从那时起已被修正了许多次。

　　在会晤马古利斯之前，我曾读过她和她儿子多里昂·萨根（Dorion Sagan）合著的一本书的草稿，书名叫《生命是什么？》。这本书用混合着哲学、科学和抒情诗的语言去描述"生命：永恒之谜"，但实际上所探讨的却是通向生物学的一种新的综合化的途径，试图把古代的泛灵论信仰与后牛顿、后达尔文科学的机械论观点融汇在一起。[13]马古利斯承认，写作本书的目的，不在于给出可检验的科学论断，更多的却是在生物学家中提倡一种新的哲学观。但她坚持认为，她与道金斯之类的生物学家之间的唯一区别，就在于她公开承认自己的哲学观，而不是假装自己没有，"那些以不受文化的左右而自命清高的科学家，并不比别人更清白"。

　　这是不是意味着她相信科学不能达到绝对真理呢？马古利斯沉思了一会儿，然后才解释道，科学的实用力量和说服力来自这样一个事实，即它的结论可以在现实世界中得到检验，这一点确实与宗教、艺术和其他知识模式的主张不同。"但这与宣称绝对真理的存在是两码事。我认为不存在什么绝对真理，即使存在，我认为也没有谁能得到它。"

　　或许是意识到她已走到相对主义的边缘，已接近被布鲁姆称作"单纯的反叛者"的那类人物了，马古利斯赶紧又竭力把自己拉回到主流科学中来。她说，虽然自己常常被看作是女权主义者，但她却并不是，也极度反感被贴上一个女权主义者的标签。她承认，与"适者生存"和"弱肉强食"之类的概念比较起来，"盖亚"和"共生"看起来更女性化些，"确实存在着这类文化上的联想，但我认为那只是一种彻头彻尾的曲解"。

　　她拒斥那种经常被某些人与"盖亚"联系在一起的主张，即认为地球在某种意义上是一个活的机体。"地球显然并不是一个活的机体，"马古利斯说，"因为没有任何一个单独的机体能循环利用自身产生的废物。这太拟人化、太易招致误解了。"她宣称，拉夫洛克之所以会赞同这一隐喻，是因为他以为这将有助于环保理念的发展，同时也因为这符合他自己的准宗教倾向。"他说这是一个不错的隐喻，因为它比原有的那个更好；我认为它不好，是因为它只能招致科学家对你的恼怒，因为你竟怂恿非理性思想。"（实际上，拉夫洛克也公开表露了对自己早期关于"盖亚"的言论的怀疑，甚至打算放弃这一术语。）[14]

古尔德和道金斯两人都曾嘲笑过"盖亚"假说，认为那是"伪科学"，是自命为理论的诗，但马古利斯至少在这一点上要比他俩更实在，更具实证色彩。古尔德和道金斯凭借对地外生命的推测来支持他们关于地球生命现象的观点，马古利斯毫不留情地嘲笑了这类伎俩。任何关于地外生命存在方式的说法——不管它是达尔文性质的还是非达尔文性质的——都不过是纯粹的玄想。她这样评论道："这种玄想流行也罢，不流行也罢，你都不必认真去对待。我就不明白人们为什么会对那些臆测之言产生如此强烈的反响。这么说吧，那些玄想并不是科学，它根本就没什么科学基础！只不过是些臆测之言！"

马古利斯回忆起20世纪70年代初的一件事，她曾接到大导演史蒂文·斯皮尔伯格（Steven Spielberg）的一个电话，他当时正在编写电影《外星人》的剧本。斯皮尔伯格问马古利斯"外星人是否可能或似乎可能也具有两只手，并且每只手上也有五个手指头"，她毫不客气地回答道："你是在拍电影！只要拍得有趣就行！你管那么多屁事干吗？别自己昏了头，竟认为那是科学！"

在采访接近尾声的时候，我问马古利斯是否介意她本人时常被视为煽动者或惹人厌的家伙，或者像某个科学家所评论的那样"一无是处"。[15]她抿紧了嘴唇，仔细想了一番。"这是一种傲慢的说法，而不是严肃的评论，"她答道，"我的意思是，你不会如此评论一个真诚的科学家，对吧？"她紧盯着我，使我终于意识到她的反问并非一种说话技巧，她确实在期待着一个答复。我只好回答说，那些评述确是有些失于厚道。

"是啊，的确如此。"她沉默了。她仍坚持说，这类批评并未给她造成什么烦恼。"任何人作出此类有辱人格的批评，都只会暴露出自己人格的卑下，难道不是吗？我的意思是，如果他们的评论只是罗列出一些描述我的挑衅性形容词，而不真正关注问题的本质，那么……"她的声音沉寂了下去。就像众多的强者科学家一样，马古利斯有时也难免渴望能成为现实社会中一位受尊敬的人。

考夫曼追求秩序的热情

达尔文的现代挑战者中，若要举出最具雄心、最为激进的一位，可能非斯图亚特·考夫曼（Stuart Kauffman）莫属了。他是复杂现象研究的前线指挥部——圣菲研究所（参见第八章）的一位生物化学家。20世纪60年代，考夫曼还在读研究生，那时他就开始怀疑达尔文的进化论有严重缺陷，因为它在无法明确解释生命所表现出的神奇能力的情况下，自己却能神奇地久盛不衰。毕竟，热力学第二定律已明确宣布：宇宙中的一切事物都正在无可挽回地滑向"热寂"（heat death），或宇宙平衡态。

为了验证自己的设想，考夫曼设计出各种变量，代表着抽象的化学物质和生物物质，并在计算机上模拟它们之间的相互作用，并由此得出了几点结论。其中一个结论是：当一个由简单的化学物质构成的系统达到一定的复杂程度时，就会产生戏剧性的转变，类似于液态的水结冰时所发生的相变。分子开始自发地化合，创造出复杂性和催化能力不断增加的大分子。考夫曼论证说，导致生命产生的更可能是这种自组织或自催化的过程，而不是某个具有自复制和进化能力的分子的侥幸生成。

考夫曼的其他假说走得甚至更远，它们有可能挑战生物学的核心原则，也就是自然选择原理。根据考夫曼的观点，由相互作用的基因物质那复杂的排列顺序所产生的自发突变，其实并不是随机产生的；相反，它们倾向于向少数几种模式收敛，或者，用混沌理论家所钟爱的术语来表述，这些突变收敛于少量的吸引子。1993年，考夫曼出版了一本厚达709页的巨著《生物序的起源：进化中的自组织与选择》。他在书中强调，上述基因生序原则，他有时也称之为"反混沌"（antichaos），在引导生命进化中所起的作用可能远远大于自然选择的作用，特别是当生物进化到足够复杂的程度之后。[16]

我1994年5月访问圣菲的时候，第一次见到了考夫曼本人，一张泛着深棕红色的宽脸，波浪状的灰发，越接近"中央地带"越稀疏。其着装是标准的圣菲模式：粗斜纹棉布衬衣，卡其布裤，旅游鞋。其给人的第一印象，往往是怯懦、脆弱而又极端的自负。其谈话就像是爵士乐手的即兴表演，主题很短，枝节却很长。仿佛一名推销员极力要在自己和顾客间建立一种愉快的氛围，他一直称呼我"约翰"。显然他很喜欢谈论哲学，在我们对话的过程中，他不断地推出一些具体的小型演讲，不仅是关于"反混沌"理论的，还涉及还原论的局限性、证伪理论的困境以及科学事实的社会语境等。

我俩的会谈刚开始不久，考夫曼就提起我曾为《科学美国人》杂志写过的一篇介绍生命起源理论的文章。[17]在那篇文章里，我引用了一句考夫曼评论自己的生命起源学说的原话："我确信我是正确的。"他告诉我，看到这句话白纸黑字地呈现在自己眼前，他感到十分窘迫，并曾发誓今后尽量避免再口出狂言。考夫曼在很大程度上也做到了这一点，这倒使我多少有些后悔自己当年的孟浪，因为在采访的整个过程中，他总是费力地不断澄清自己的主张："我当然不会再说我确信自己是正确的，约翰，但……"

考夫曼刚写完一本书——《在宇宙这个家》，阐述了其生物进化学说的含义。"约翰，我相信本书的基本观点是正确的——我的意思是，在所有的观点都有根有据这一点上，我认为它是正确的；当然它们还有待实验上的证明。但至少就数学模型而言，你可以从中得到一整套用以描述生命作为自然现象是怎样出现的模型，就好像给你提供了一组复杂而有效的反应分子，你就可据以产生自催化反应。如果这些观点能够成立，就像我两年前曾豪情满怀地向你宣称的那样，"他开心地笑着，"那么，我们人类的出现就不是什么难以置信的意外了。"事实上，几乎可以肯定生命在宇宙其他地方也都会存在着，他补充道。"这样，我们对自己在宇宙这个家里就获得了一种全然不同的理解。过去曾把生命看作是似乎难以置信的事件，它偶然发生在一个且仅此一个星球上，但可能性是如此之小以至于你不会相信它真的能发生。"

考夫曼用同样绕圈儿似的表达方式，陈述了他关于基因网络倾向稳定于反复出现的特定模式的学说。"再次假定我是正确的"，考夫曼说，那么，生物系统展现出来的许多"序"（order）并非"优胜劣汰"法则的作用结果，而是这种普遍存在的"生序效应"（order-generating effects）的结果。"至关重要的一点是，这种'序'是自然发生的，是自主的序，对吧？再次重申，如果这一观点是正确的，那么我们不仅必须修正达尔文理论以适合这一点，而且还能以一种不同的方式去理解生命以及生命之序的突现（emergence）。"

考夫曼告诉我，他的计算机模拟还提供了一些别的更合理的结论。就像在沙堆上再增加一粒沙子有时就会引发崩塌一样，一个物种的适应性发生变化，也会导致生态系统中所有其他物种在适应性上产生突发性改变，这可能以灾难性的大灭绝而告终。"用一个比喻来表述就是，我们每个人所采取的最佳适应行为，都有可能触发使社会系统崩溃的开关，并导致我们的最终毁灭，对吧？因为我们是按照各种规则生活在一起的，社会系统是我们共同创造出来的，并且我们每个人都对它产生着微妙的影响，这也就预示着我们在生活中必须保持一份谦恭之心。"考夫曼把有关沙堆分析的观点归功于佩尔·贝克（Per Bak），一位同样加入了圣菲研究所的物理学家，他曾提出过一种叫作自组织临界性的理论。（这一理论将在第八章中加以讨论。）

我提出了自己关心的问题，即许多科学家，特别是圣菲研究所的科学家，似乎把计算机模拟混同于现实了。考夫曼对此点了点头。"我同意你的看法，我个人也对这种做法深为不安。"他答道。面对某些模拟实验，"我搞不清，我们所谈的世界——我的意思是，外在于我们的那个世界——和那些十分精巧的计算机游戏、艺术形式等相比，其间的界线究竟在何处"。当他自己进行计算机模拟时，无论如何，他总是"试图估计出世界上的相应事物是怎么行为的，或近似地是怎样行为的。有时，我也会单纯地为发现某些仅仅是看来十分有趣的东西而忙碌，同时又会怀疑它们是否有实用价值。但除非你正要鼓捣的玩意儿，在总体上能被证明是与外在世界的某种东西相对应的，否则我不会认为你正在从事科学研究。这也就意味着，计算机模拟出的东西最终必须是可检验的"。

他自己的基因网络模型所"提供的各种各样的预言"，很可能将在今后的15～20年里得到验证，考夫曼这样告诉我。"只要附加某些限制条件，它是可以检验的。当你检验的是一个由10万种要素组成的系统，并且你不可能把这一系统拆分成枝枝节节的部分，是的，那么适宜的检验结果是什么呢？将只能是一些统计结果，对吧？"

他的生命起源怎样才能得到验证呢？"你可能正在问两个不同的问题。"考夫曼答道。一方面，这个问题关注的是在大约40亿年之前，生命在地球上实际产生的方式，考夫曼不知道他的理论——或者别的任何一种理论——是否能够令人满意地回答这个历史问题；另一方面，人们可以通过试着在试验室里操纵自催化装置来验证他的理论。"说实话，我们可以打一个赌。不管是我还是别的什么人，只要有人能用分子聚合性自催化装置显示出相变和反应的迹象，你就输给我一顿晚餐，如何？"

在考夫曼和其他对进化生物学现状不满的挑战者之间，存在着一些相似之处。首先，考夫曼的观点是有其历史渊源的，正如断续平衡论或共生论一样。康德（Kant）、歌德（Goethe），以及其他一些前达尔文时代的思想家都曾猜测，可能有普遍的数学原理或法则潜藏在自然的模式之下。即使在达尔文之后，也仍有许多生物学家相信，除了自然选择之外，还存在着某种生序力，阻止宇宙向普遍的热力学平衡态滑落，并导致生物序的产生。到了20世纪，这种有时也被称为"理性形态学"（rational morphology）的观点的支持者，包括汤普生（D'Arcy Wentworth Thompson）、贝特森（William Bateson）和更近期的古德温（Brian Goodwin）。[18]

再者，看起来激励着考夫曼的动因，既来自关于事物必定怎样的哲学信仰，也来自事物实际怎样的科学好奇心，前者至少也与后者同样强烈。古尔德强调机遇（即偶然性）在进化中的重要性；马古利斯拒斥新达尔文主义的还原论观点，坚持一种更加整体化的思路；同样地，考夫曼认为偶然事件自身不足以导致生命的产生，我们的宇宙中一定隐匿着某种基本的生序趋势。

最后一点，就像古尔德和马古利斯一样，考夫曼一直纠结于该如何界定自己与达尔文之间的关系。在同我会晤期间，他说他把"反混沌"看作是对达尔文自然选择理论的补充；而在其他场合，他曾标榜"反混沌"是进化的首要因素，而自然选择的作用则是微不足道的，甚至根本就不存在。考夫曼在这一问题上持续的矛盾心态，在其1995年春送给我的一份打印的书稿《在宇宙这个家》中清楚地表现了出来。在这本书稿的第一页，考夫曼宣称达尔文学说是"错误的"，但他又划去了"错误的"一词，而代之以"不完善的"；在书的清样中，考夫曼又改回"错误的"字样，书终于在几个月后出版了。那么最终的、正式出版后的说法是什么呢？"不完善的。"

考夫曼有一位强有力的盟友，那就是古尔德。他曾在考夫曼的《生物序的起源》一书的封面上宣称，"在我们探索更加全面、更加合理的进化理论的进程上"，这本书将成为"一座里程碑和一部经典"。这真是一对奇怪的盟友，因为考夫曼坚持认为，他在计算机模型中揭示出的复杂性法则，已经为生命的进化提供了切实的必然性；然而古尔德却毕生都在致力于论证：在生命进化的历史上，绝不存在什么必然性的东西。在我采访他的时候，古尔德旗帜鲜明地反对生命史是按照数学规律展开的说法。"那的确是一个深刻的立场，"古尔德说，"但我仍然认为那只不过是一种深刻的谬误。"古尔德与考夫曼真正相同之处在于，他们都在挑战着道金斯等中坚达尔文主义者的主张，即认为进化论在某种程度上已经解释了生命进化的历史。在他为考夫曼著作所写的护封吹捧文字中，古尔德明确表示自己信奉那句古老的箴言："敌人的敌人就是朋友。"

总的说来，考夫曼在推销自己的思想观点方面成效不大，或许主要的症结在于他的理论在本质上说是统计性的，正如他自己所承认的那样。在人们还只能考察地球生命这唯一的数据来源的情况下，这种关于生命起源及其后续进化过程的统计学预言是无法得到证实的。对考夫曼研究成果最严厉的批评，来自英国的进化生物学家梅纳德·史密斯（John Maynard Smith），他和道金斯一样，以言辞犀利同时又是把数学引入进化生物学的先驱这两点而著名。考夫曼曾在史密斯手下做过研究工作，所以为了使这位从前的导师相信自己工作的重要性，他曾花费了无数的口舌，但显然是劳而无功。在1995年的一次公开辩论中，史密斯曾对自组织临界性理论，即由佩尔·贝克提出并为考夫曼所信奉的沙堆模型，作过这样的评论："这一整套玩意儿只会让我觉得齿冷。"在会后进餐时，正当酒酣耳热之际，史密斯告诉考夫曼说，他并不觉得考夫曼研究生物学的进路多么有趣。[19] 对于考夫曼这样的反讽科学的践行者而言，其最大的伤害也莫过于此了。

但考夫曼的反驳却更加雄辩，更具说服力。他暗示像道金斯之流的生物学家所散布的进化理论，是冷酷无情的、机械的，并未给予生命的尊严和神秘以公道的评价。考夫曼是正确的，达尔文理论正是在这一点上让人无法满意并显得语无伦次，即使是由像道金斯这样技巧娴熟的雄辩家来陈述，也无法遮掩。但道金斯至少还能分辨出什么是有生命的东西、什么是无生命的东西，而考夫曼却似乎把一切东西，从细菌到星系，都看成是不断进行排列组合的抽象数学形式的不同表现。他是一位数学美学家，他的观点，与那些公然把上帝称作几何学家的粒子物理学家的主张相比，简直如出一辙。考夫曼曾经声言，与道金斯的观点比较起来，他对生命的看法会让人觉得更有意思，接受起来更舒服。但我却觉得，对于我们大多数人而言，更乐于接受道金斯那细小的、进取心十足的复制基因，而不是考夫曼那存在于N维空间中的布尔函数。这些极度抽象冰冷的玩意儿，有什么"意思"和"舒服"可言呢？

科学之保守主义

对于科学以及所有人类事业的现状而言，存在着一种持续的威胁，那就是年轻的一代人总想给这个世界烙上自身印记的欲望；而社会对于新奇的事物，偏偏又有着永远无法餍足的胃口。正是这对孪生现象，招致了艺术领域中表现形式上的急剧翻覆，在这一领域中，单纯地为变化而求变，早已是司空见惯的现象。科学也难以完全摆脱这类影响。古尔德的断续平衡理论，马古利斯的"盖亚"，考夫曼的"反混沌"，都曾经红极一时；但与艺术比较起来，若想在科学上产生有持久影响的改变，往往要困难得多，原因显而易见。科学的成功，在很大程度上应归功于它的保守主义，即它坚持有效性的高标准。量子力学和广义相对论之新、之奇，曾经超出了所有人的想象，但它们最终之所以能得到认可，并不是因为它们带来了一场思想上的"地震"，而是因为它们确实是有效的：它们精确地预言了某些实验结果。某些老旧理论之所以能历久不衰，也有其充分的理由，它们强劲而又坚韧，并且与现实有着不可思议的一致性，甚至可以说：它们就是真理。

自诩的革命家们还面临着另外一个问题。科学文化一度曾是极其弱小的，因而易接纳急剧的改变；现在，它已发展成一个由智力因素、社会因素和政治因素混杂而成的庞大官僚机构，具有同样庞大的惯性。考夫曼有一次在与我聊天的过程中，曾把科学的保守主义比作生物进化中的保守性，因为生物进化的历史也极力抵御着变化。考夫曼指出，不仅仅是科学，许多其他的观念系统——特别是那些具有重要社会影响的观念系统，都有一种随时间推移而"坚定不移"的倾向。"回顾一下船或载人飞行器的标准操作程序的演变吧，那是一个保守得令人难以置信的过程。如果你涉足其间，并且试图从零开始设计，比如说，航空母舰的操作程序，你会把事情彻底搞砸。"

考夫曼向我身边凑了凑，说："这真的很有意思。让我们再来看看法律的情况，好吗？经过了1200年的岁月，英国普通法已成了什么样子？不过是连篇累牍的废话，充斥着一套一套的概念，告诫着我们如何如何才是合理的行为。要想改变这一切，真是难如登天！我甚至都怀疑你是否能搞清这张概念之网，尽管它已扩展到所有的领域，并在每个领域里指导着我们：在这个世界上应该怎样走我们的生活之路。你更无法搞清，为什么它的核心部分会越来越固执地抵制着一切改变。"十分古怪的是，考夫曼所表述的正是一个绝妙的论证，足以说明为什么他自己关于生命起源和生物序的激进理论，可能永远也无法被接受。如果说有哪种科学思想已证明了自己具有征服一切挑战者的威力的话，那肯定就是达尔文的进化理论。

生命的神秘起源

假如我是一位神创论者，我会停止攻击进化论，因为它毕竟已得到了化石记录的强有力支持，而是转变方向，将目标集中于生命的起源上。毫无疑问，这是现代生物学大厦最羸弱的一根支柱。生命的起源是所有科学写手的乐园，这里盛产各种奇特的科学家和各种奇特的理论，却又没有哪一个被彻底地摒弃或接受过，他们仅仅是风行一时。[20]

在生命起源问题上，最勤奋、最受尊敬的研究者之一是斯坦利·米勒（Stanley Miller）。1953年，他还是位年仅23岁的研究生，就已开始尝试着在实验室中再现生命的起源过程。他在一个密封的玻璃仪器中装入几升甲烷、氨和氢气（代表原始大气）和一部分水（代表着海洋），一个电火花装置模拟着闪电，不断地电击着"大气"，同时用加热圈使水保持沸腾。几天之后，水和气中都显出了一种红色的黏糊糊的东西。通过分析这些物质，米勒兴奋地发现，其中竟含有大量的氨基酸，这些有机化合物是构建蛋白质的"砖瓦"，而后者又是构筑生命的基本原料。

米勒的实验结果看来有力地证明，生命可以从被英国化学家霍尔丹（J. B. S. Haldane）所称的"原始汤"中产生出来。权威人士们就此推测，像雪莱（Mary Shelley）笔下的弗兰肯斯坦博士[①]那样的科学家，短期内就会在实验室中魔术般地"变出"活着的生物体来，并由此揭示出生命起源的细节，但直到如今仍是事与愿违。米勒告诉我，在他的初始实验之后的近40年时间里，事实已经证明，解开生命起源这一谜团要远比他或别的什么人所预想的更为困难。他回忆起这样一则预言，那是在他的实验之后不久作出的，说在25年之内科学家们"肯定"会弄清生命到底是怎样开始的。"可是，25年早已过去了。"米勒冷冰冰地说道。

在其1953年的实验之后，米勒一直致力于探索生命的奥秘。他声名鹊起，这既因为他是个严谨的实验科学家，也因为他是个有点乖戾的倔老头，对他觉得不满意的研究成果总是暴躁地予以批评。当我在加利福尼亚大学圣地亚哥分校米勒的办公室里见到他时（他是那里的一名生物化学教授），他烦躁地说，自己的领域作为一门边缘学科，虽然仍有那么点声誉，但已经不值得去认真追求了。"有些研究相对其余的垃圾的确要好一些，但糟糕的研究人员队伍情况，却有可能把这一学科领域推向深渊，我为这事儿烦得要命。人们好好地做着研究工作，结果呢，却只能看着这些垃圾大出风头。"米勒似乎对各种关于生命起源的时髦假说全无好感，一律斥之为"胡说八道"或"纸上谈兵的化学"（paper chemistry）。他对某些假说是如此的蔑视，以致当我请教他对于这些假说的看法时，他只是摇摇头，深深地叹口气，然后窃笑几声，仿佛

① Dr.Frankenstein，雪莱于1818年所著小说中的主人公，是一名年轻的医学研究者，创造出了一个最终毁灭了他的怪物——译者注。

对人类的愚蠢已经无能为力了。考夫曼的自催化理论就落入了这一类假说中。"在计算机上运行各种方程绝不等于做实验。"米勒嘲笑道。

米勒认为，科学家也许永远都无法精确地知道生命到底在何时、何地发生的。"我们是在试图讨论一个历史事件，这与科学通常的那种争论截然不同，因此判断的标准和讨论方法也是完全不一样的。"但当我暗示米勒他对于揭示生命奥秘的前景过于悲观时，他看来竟十分惊讶。悲观？绝非如此！他是乐观的！

总有一天，他发誓说，科学家们将会发现那引发了伟大的进化传奇的自复制分子。正如大爆炸的微波背景辐射的发现确立宇宙学的合法地位一样，原初遗传物质的发现也将会证明米勒的领域的合理性。"那时，这一领域就会像火箭一样起飞。"米勒透过咬紧的牙关轻声而又坚定地说道。这样的发现会即刻成为显而易见的吗？米勒点了点头，"那将是具有这样一种性质的某种东西，它会使你不由自主地喊出：'天啊，原来就是这个。你们怎么能在如此长的时间里一直对它视而不见呢？'于是每个人都会被彻底说服了。"

当米勒在1953年完成其里程碑式的实验时，许多科学家仍在分享着达尔文的信念，认为蛋白质是自我复制分子的最可能的候选者，因为蛋白质被认为能够复制和组装自身。在发现了DNA是遗传和蛋白质合成的基础之后，许多研究者倾向于认为核酸是原初分子。但这个方案有一个十分重大的障碍：如果离开了被称作酶的催化蛋白质的帮助，DNA既不能合成蛋白质也不能进行自我复制。这一事实将生命的起源变成了一个经典的"鸡先还是蛋先"的问题：谁最先出现，是蛋白质还是DNA？

在《黄金时代的来临》一书中，岗瑟·斯滕特表现出了其一贯的先见之明。他提示说，如果研究者们找到一种既可自我复制同时自身又是催化剂的大分子，那么这道难题就可迎刃而解了。[21] 20世纪80年代初期，研究人员的确鉴别出了这样一种分子：核糖核酸或者叫RNA。这是一种单链的分子，它在蛋白质的合成中起着DNA的助手的作用。实验表明，特定类型的RNA可以作为自身的酶，可将自身剪切为两段，并且还能再度恢复原状。如果RNA能起到酶的作用，那么离开蛋白质的参与仍然能复制自身。这样，RNA既可以作为基因又可作为催化剂，同时兼有鸡和蛋两种身份。

但所谓的RNA世界假说，仍然存在着几个问题。在实验中，即使是在最好的情况下，也难以合成RNA及其组合，更不用说是在可能的前生命条件下了。RNA一旦合成，也只是在科学家们大量的化学诱导之下，才能复制出自身的新版本，而生命的起源"必须发生在简单条件下，而不是在某种特设条件下"，米勒说。他坚信，某种更简单的也可能十分不同的分子，必定已为RNA廓清了道路。

林恩·马古利斯作为反对者之一，怀疑对于生命起源的研究，是否真能得出米勒所梦想的那种简单、自洽的答案。"我认为那对于癌症的病因来说，可能是正确的，但对于生命的起源却又不同。"马古利斯指出，生命出现在复杂的环境条件之下，"存在着日与夜、冬与夏，温度的变化，湿度的变化。初始分子正是这些情况长期累积的结果；生化系统也是日积月累的结果。因此，我认为永远也无法为生命开出一个一揽子的处方：加水，加各种混合物，然后可获得生命。这不是个一步登天的过程，而是个包括大量变化的累积过程。"她进一步指出，即使是最小的细菌，"也比斯坦利·米勒的化学混合物更接近人类，因为细菌已经具有了高级生命系统的基本特征。然而，从细菌进化到人类这一步，还是相当遥远的；但比起从氨基酸混合物发展到细菌这一步来，反倒容易得多"。

弗朗西斯·克里克在其《生命本身》一书中写道："生命的起源似乎是个不可能的奇迹，因为要实现这一点，必须满足的条件是如此众多。"[22]（应该说明的是，克里克是一个倾向于无神论的不可知论者。）克里克认为外星人可能在数十亿年前乘宇宙飞船拜访了地球，并有意地播种了微生物。

或许斯坦利·米勒的希望最终能实现：科学家仍会发现某些"聪明的"化学物质或其混合物，能在似乎可信的前生命条件下进行复制、突变和进化。这种发现，无疑将会开创一个应用化学的新纪元。（绝大多数探索者都是着眼于"开创新纪元"这一目标，而不是去阐明生命的起源。）但考虑到我们对生命肇始条件的贫乏认识，那么，任何建立在上述发现基础之上的生命起源理论总是令人生疑的。米勒坚信：一旦生物学家们被告知生命起源之谜的答案，他们就会恍然大悟并一致赞同。但这一信念有一个必要的前提，即那一答案是绝对可信的，这样它就只能是回顾性的。又有谁敢保证"地球生命的起源"这一问题本身就是可信的呢？生命或许是由一系列不可能的甚至难以想象的事件的奇妙巧合造成的。

再者，对可能存在的原初分子的发现，一旦或假如真的实现了，似乎也不大可能告知我们那些真正想知道的事情：地球上的生命是必然发生的还是奇妙的巧合？它是随处皆可发生的呢，还是仅仅发生在这个偏僻的地方？这些问题，只有在我们发现了地外生命之后，才有可能解答。但社会对于资助这类研究，看来正变得越来越不情愿。1993年，国会砍掉了国家航空和航天管理局（NASA）的"探索地外智慧"（SETI）计划，这一计划就是为了监测太空中由其他文明发出的无线电信号而设定的。向火星，也就是太阳系最有可能存在其他文明的星球发射载人火箭的夙愿，也已经被无限期地推迟了。

即使这样，科学家们明天也许仍能找到关于地外生命的证据，这样的发现将改写科学、哲学以及人类思想的所有方面。古尔德和道金斯之间关于自然选择是宇宙普遍现象还是仅存于地球上的现象之争，可能会因此而平息下来（尽管毫无疑问地，他们仍会发掘出大量的证据以支持各自的观点）；考夫曼也会因而判明：自己从计算机模拟中得出的法则对真实的世界是否有效；如果地外生命已拥有了足够的智慧并已建立起自己的科学，那么威滕也就可能会最终弄明白：对于任何探索统治现实世界的基本规律的努力而言，超弦理论是否都是其必然的终结。科幻小说全都变成了现实文学，《纽约时报》将与现在的超市小报《世界要闻周报》一样，大印而特印各国政要们与外星人亲切交谈的照片。你总可以抱这样的指望。

第六章
社会科学的终结

对于爱德华·威尔逊（Edward O. Wilson）来说，只要能沉浸在蚂蚁的王国里，尘世的喧嚣对他来说就都无所谓了。他是在阿拉巴马州长大的，从孩提时代起，他就被蚂蚁诱惑进了生物学的世界。蚂蚁一直是他的灵感之源，他以这些小生灵为主题写出了大量的论文和好几本专著。威尔逊的办公室在哈佛大学比较动物学博物院，室外是成排的蚁丘，他向我炫耀这些蚁丘的时候，那份骄傲和激动的神气，简直就像是一个十来岁的小孩子。我问他现在是否已写尽了有关蚂蚁的话题，他却踌躇满志地说道："我们还只是刚刚开始。"他最近正着手进行一项对"大头蚁属"（*Pheidole*，蚂蚁王国中最大的一个属）的调查，据推测，大头蚁属包括2000多个蚁种，其中的大多数从未被描述过，甚至也未曾被命名过。"我想，促使自己承担这项任务的冲动，与许多中年汉子决定乘皮划艇横渡大西洋，或结队翻越乔戈里峰时的冲动是一样的。"威尔逊这样对我讲。[1]

威尔逊是保护地球生物多样性计划的领导人之一，他的主要目标就是使对"大头蚁属"的调查成为一项基准，以供生物学家们在监测不同地域的生物多样性时作参照。依靠哈佛大学对蚁种资源的丰富收藏（也是世界上最大的收藏），威尔逊精心绘制了大量大头蚁属中不同蚁种的铅笔草图，并标记上其生活习性和生态学特征。"也许在你看来这是一项极其枯燥乏味的工作，"他一边翻动着那些大头蚁种系草图（那是令人生畏的厚厚一摞），一边带点歉意地对我说，"但在我而言，这却是所能想象出的最佳消遣活动。"他承认，当自己透过显微镜鉴别出一个尚未认识的新蚁种时，有一种"似乎看到了——我不想被误解为诗人——造物主的真面目的感觉"。一只小小的蚂蚁，就足以使威尔逊对宇宙的玄妙保持敬畏之心了。

　　当我俩走到他办公室的蚁田（那是一只爬满蚂蚁的长木箱）前面时，我才第一次发现，在威尔逊那孩子气的举止背后，竟跃动着一种好斗的气质。这些是南美切叶蚁，他解释道，分布在从南美洲一直到路易斯安那州的广阔地域上。那些在海绵状蚁巢表面奔来奔去的小家伙是工蚁，而兵蚁却埋伏在里面。威尔逊拔掉蚁巢顶部的一个塞子，并向露出的蚁穴中吹了口气，转眼间，就有几只肥硕的庞然大物怒冲冲地奔了出来，气枪子弹状的脑袋来回摆动着，恶狠狠地大张开上颚。"它们能咬穿皮鞋帮，"威尔逊用有点过分的赞赏语气评论道，"如果你妄想挖开南美切叶蚁的窝的话，它们会一口一口地啃光了你，就像中国历史上那千刀万剐的凌迟极刑一样。"说到这里，他竟咯咯地笑了起来。

　　威尔逊这种好斗的气质究竟是与生俱来的，还是后天养成的？在我俩随后的讨论中，其好斗性表现得更加明显。话题转向为什么美国公众一直不愿正视基因在塑造人类行为上的重要作用，"人人平等，美好的未来决定于公民的美好愿望，诸如此类的平等主义信仰，已经牢牢控制了这个国家，以至于公民们对任何看来会有损于这一中心伦理观念的东西，都会不屑一顾"。在威尔逊布道般宣讲的时候，他那张瘦骨嶙峋的、就像南方农夫佬一样的长脸上，再也见不到惯常的和蔼与亲切，反而像清教传教士一样冷峻。

　　存在着两个——至少两个——威尔逊，一个是群居昆虫中的诗人，以及地球上生物多样性的热情维护者；另一个却是凶猛、好胜而又野心勃勃的斗士，与自己内心深处的这样一种感觉顽强地搏斗着：他只是一位迟来者，他的学术研究从某种意义上说已经彻底完成了。威尔逊对于"影响的焦虑"的反应，与另外一些也在与达尔文抗争的人截然不同，如古尔德、马古利斯、考夫曼等人。最大差别在于：古尔德等人论辩说，达尔文理论的解释力是有限的，实际的进化过程比达尔文及其现代信徒们所设想的要复杂得多；威尔逊则走上了一条相反的道路，他试图拓展达尔文主义，以证明它所能解释的现象，比任何人（甚至包括理查德·道金斯）所料想的都更多。

　　威尔逊之所以会承担起社会生物学的发言人角色，可以追溯到他在20世纪50年代末所经历的一场信仰危机，那还是他刚到哈佛时的事。虽然他那时已是世界上群居昆虫研究领域的权威之一，但这一研究领域至少在其他科学家眼中仍是无足轻重的，他一直对此耿耿于怀。分子生物学家们在欢欣鼓舞于他们对DNA结构（即遗传的基因基础）发现的同时，开始怀疑对整体生物（如蚂蚁）的研究是否有价值。当时沃森也在哈佛，威尔逊总难忘记仍沉浸在发现双螺旋结构的激动中的沃森那扬扬自得的神气，以及他对进化生物学"流露出的明显的轻蔑"，认为那只不过是被吹捧过了头的"集邮式描述"，应彻底让位给分子生物学。[2]

为了回击这一挑衅，威尔逊扩大了自己的视野去探求那些不仅决定着蚂蚁的行为，而且决定着所有群居动物行为的普适规律。这些努力的成果，就体现在《社会生物学——新的综合》一书中。这本出版于1975年的著作，在很大程度上可以说是对非人类的社会动物进行考察的扛鼎之作，涉及从蚂蚁、白蚁到羚羊、狒狒的众多社会动物，在生态学、群体遗传学以及其他一些学科的理论基础上，威尔逊详细阐明了交配行为和劳动分工怎样对进化压力作出反应。

只是在书的最后一章，威尔逊才转向人类社会。他把人们的注意力引向这样一个明显的事实：作为研究人类社会行为的社会学，迫切需要一种统一的理论。"在构建统一的理论框架方面，已经有了许多的尝试，但……收效甚微。今日社会学中所谓的理论，其实只不过是以一种想当然的博物学方式，对各种现象和术语加以罗列。真正的社会过程是难以进行分析的，因为其基本单位难以把握，也许根本就不存在。面对想象力丰富的社会学家们炮制出的大量定义和隐喻，所谓综合性理论只不过是在它们之间进行冗长而乏味的前后引证。"[3]

威尔逊认为，只有屈从于达尔文的范式之下，社会学才能变成一门真正科学的学科。他曾指出，类似于战争、排外、男权等现象，甚至我们偶尔爆发出的利他行为等，都可以看作是源自人类繁衍自身基因的原始冲动的适应性行为。威尔逊预言，在未来，随着进化论以及遗传学、神经科学的进一步发展，将会使社会生物学在更宽阔的领域里解释各种人类行为。不仅是社会学，甚至连心理学、人类学以及所有的"软"社会科学，最终都将被纳入社会生物学体系之内。

这本书受到了普遍的好评，然而一些科学家，包括威尔逊的同事古尔德，却指责他在提倡宿命论。批评家们指出，威尔逊的观点只不过是社会达尔文主义的一种现代翻版，把这一声名狼藉的维多利亚时代教条与一些自以为是的想象搅和在一起，为种族主义、性别歧视和帝国主义提供科学的辩护。对威尔逊的攻击，在1978年美国科学促进会（AAAS）的一次会议上达到了高潮。名为"国际反种族主义委员会"的激进团体的一名成员，把一满

罐水劈头盖脑地泼到威尔逊身上，并喊着："你在胡说八道①！"⁴

这一切并未吓倒威尔逊，他与多伦多大学的物理学家查尔斯·拉姆斯登（Charles Lumsden）合作，陆续写出了两本人类社会生物学方面的著作：《基因、意识与文化》（1981）和《普罗米修斯之火》（1983）。在后一本书中，威尔逊和拉姆斯登承认："对于基因和文化两者间的相互影响，要想构思出一幅精确的图像是很难的。"但他们宣称，妥善处理这一难题的途径，不能沿袭"用文学批评的方式描写社会理论的'光荣'传统"，而是要就基因和文化之间的相互影响，建立一种严格的、数学化的理论。威尔逊和拉姆斯登写道："我们所希望建立的理论，是一个由相互联系的抽象过程构成的系统，这些过程将尽可能清晰地用数学结构表示出来，并能被准确地翻译成感觉经验世界的过程。"⁵

但这后两本书，却并不像《社会生物学》那么受欢迎。一位评论家最近把他们关于人类本性的观点斥为"冷冰冰的、机械论的"和"过分简单化的"。⁶在采访的过程中，我发现威尔逊对社会生物学的前景竟然比以往更加乐观。在承认自己的理论在20世纪70年代很少得到支持的同时，他坚持认为，"现在已有足够的证据表明"，许多的人类习性，从同性恋到羞怯，都具有其遗传基础；医学遗传学的进展，也正使得人类行为的遗传学解释更易被科学家和公众所接受。人类社会生物学不仅在欧洲蓬勃发展并在那里组建了社会生物学协会，而且在美国也正逐渐深入人心，尽管许多美国科学家因为"社会生物学"一词的政治内涵而极力回避这一术语。那些冠以生物文化研究、进化心理学以及人类行为的进化论研究等名目的学科，都是从社会生物学这一主干中生长出来的分支。⁷

威尔逊仍然坚信，不仅社会科学，甚至连哲学也都将最终统一到社会生物学中来。他正在撰写一部新书，书名暂定为《自然哲学》，描写社会生物学的新发现将怎样有助于解决政治和道德问题。他打算提出，宗教信条可以而且应该"接受经验的检验"，如果它们与科学真理不符，就要坚决抛弃。比如，他建议天主教会应审查一下：其反堕胎禁律作为一条可导致人口过度膨胀的教条，是否与保护地球生物多样性这一更大目标相冲突。我一边听着威尔逊的诉说，一边不由自主地想起了一位同事的评论：威尔逊是个极端矛

---

① "You're all wet"，在美国俚语中，"all wet"是"胡说八道"的意思，而"wet"的本意是"湿"，所以这句话还可从字面理解为"你全身湿透了"。那位激进分子之所以泼威尔逊一身水，一方面是为了发泄不满，另一方面也是为造成这种双关效果——译者注。

盾的人物，在他身上，非凡的智慧和渊博的学识竟同一种幼稚——或简直就
是愚蠢——结合在一起。

就连那些赞赏威尔逊的尝试（即为一种解释人类本性的详细理论奠定基
础）的进化生物学家，也怀疑这样一种努力能否成功。举例来说，道金斯就
非常讨厌古尔德等"左倾"科学家对社会生物学理论所表现出的"不假思索
的敌意"。他曾说过："我认为威尔逊受到了不公正的对待，并且不仅仅来
自其哈佛的同事们。因此，只要一有机会，我就会站出来为威尔逊辩护。"
然而，对于"人类生活的混乱"能完全用科学术语解释这一点，道金斯却不
像威尔逊那样充满自信。道金斯认为科学不打算解释"能产生庞杂细节的高
度复杂系统。解释社会学现象，就好比用科学理论去解释或预言一个水分子
在通过尼亚加拉瀑布时的确切过程，你永远也无法做到，但这并不意味着其
中存在着什么基本原理上的困难，只不过是因为这一过程太过复杂"。

我猜想，威尔逊本人也怀疑社会生物学能否发展到他所相信的那种全能
地步。在《社会生物学》的结尾部分，他暗示这一领域将以一种关于人类本
性的完备的、终极的理论抵达其发展的终点。他写道："为了维护物种的多
样性，我们被迫去追寻能解释这一切的整体知识，一直探究到神经元和基因
的层次。一旦我们的知识发展到足以用这些机械的术语来解释自身的地步，
社会科学也就迎来了其全盛时期，但这一结局却是人们所难以接受的。"他
引用加缪①的一段话结束了全书："在一个被剥夺了幻想和光明的世界上，
人们会觉得自己是个天外来客，是个异乡人。这是一种永无大赦之望的流
放，因为他已被剥夺了故土的记忆，永远丧失了希望的福地。"[8]

当我提起这个悲观的结尾时，威尔逊承认自己在《社会生物学》一书
收尾时确实有点消沉。"我曾思索了很长时间，随着我们越来越多地了解了
诸如'我们从哪里来？'以及'我们忙忙碌碌为哪般？'等问题，并用精确
的术语表述出来，这无疑将会贬低……我搜肠刮肚要找的字眼是什么来着？
对了，贬低我们自鸣得意的自我形象，同时也会挫伤我们对未来无限增长的
希望。"威尔逊还认为，这样的理论将会把生物学这个维系他本人的生活意
义的学科彻底地带向终点。"到那时，我会说服自己放弃这一领域。"威尔
逊断定，人的意识作为文化和基因之间的复杂相互作用的产物，至今仍在这
种相互作用下发展着，它代表着科学的一个永无止境的前沿。"我发现这是
科学和人类历史中的一块尚未勘探的广阔领地，可供我们永远探索下去，这

----

① Camus，全称是Albert Camus，1913—1960，法国作家，生于阿尔及利亚——译者注。

使我感到非常愉快。"威尔逊摆脱窘境的方法是承认自己的批评者们是正确的：科学不可能解释所有那些变幻莫测的人类思想和文化，也不存在什么人类本性的终极理论，即足以解答我们自身所有问题的理论。

社会生物学到底有多大的革命性？并不太多，这是威尔逊自己的说法。尽管有着巨大的创造力和同样巨大的雄心，威尔逊仍只是一个极其传统的达尔文主义者。这一点在我请教他有关"亲生命性"（biophilia）这一概念的含义时，就显得更加清楚了。亲生命性意指人类与大自然（至少是大自然的某些方面）的亲和性是与生俱来的，是自然选择的产物。威尔逊用这一概念找到他所钟情的两大领域（社会生物学和生物多样性）的共同基础。威尔逊曾就亲生命性撰写过一本专著（出版于1984年），后来他又编辑了一部他和别的学者论述同一主题的论文集。在采访威尔逊的时候，我犯了个战术上的错误，不该就此妄加评论，说亲生命性这一术语让我联想起"盖亚"，因为两者都能唤起人们的利他主义激情，使人们去拥抱一切生命，而不仅是自己的亲属或者同类。

"事实上，这全是胡说八道！"威尔逊反驳道，语气是如此尖刻，以至于我竟被吓了一跳。亲生命性与那大而不当的利他主义根本就沾不上边，"对于人类本性从何而来我持有一种极强硬的机械论观点。我们对其他生物的关注，绝对是一种达尔文式自然选择的结果"。亲生命性之所以在进化中产生出来，不是为着所有生命的利益，而是为了人类个体的利益。"我的观点完全是以人为中心的，因为我从进化现象中所观察到、理解到的一切，都支持这一观点，而不是别的观点。"

我问威尔逊是否赞同他的哈佛同行恩斯特·迈尔的观点，即现代生物学已降至只能解难题的境地，其答案只能巩固占优势地位的新达尔文主义的范式。[9]"呵呵，把各种常数精确到小数点之后的下一位，这并不是什么新鲜的见解。"威尔逊以嘲笑的口吻说道，其措辞明显地易让人联想起传说中19世纪沾沾自喜的物理学家。但在文雅地奚落了一番迈尔的完成论观点之后，威尔逊又对其表示赞同。"我们不大可能推翻自然选择的进化观，或者推翻我们关于物种形成的基本理论框架。因而，我也同样怀疑：对于进化是怎样实现的、变异是如何发生的或生物的多样性是怎样形成的等基本问题，我们是否能在种系的水平上取得什么革命性的突破。"当然，关于胚胎发育，关于人体生物学与人类文化的相互作用，关于生态系统等复杂系统，仍然存在着许多有待研究的问题，但威尔逊断定，"根据我的判断"，生物学的基本原则"已经牢固地建立了起来，包括进化得以实现的法则、机理等决定进化过程的基本规律"。

威尔逊理应再补充的一点是，达尔文理论所蕴含的那种让人寒心的道德和哲学寓意，在很久以前也已被详细地阐明了。在1871年出版的《人类的由来》一书中，达尔文曾指出：如果人类是像蜜蜂那般进化的，那么，"毫无疑问，未婚姑娘们就会像工蜂一样，把杀死她们的兄弟看作是一项神圣的职责，而母亲们也会千方百计地杀死自己有生育能力的女儿，并且每个人都会认为这是天经地义的"。[10] 换种说法就是：我们就是动物，自然选择不仅塑造了我们的形体，而且也塑造了我们的信念，即判断对与错等基本价值观念的信仰。一位惊慌失措的维多利亚时代评论家，就此在《爱丁堡评论》上愠怒地写道："如果这些观点也能成立，那么一场思想上的革命也就迫在眉睫了，它将彻底推翻公众们圣洁的良知和信仰，从而把社会夷为一片瓦砾场。"[11] 这场革命早在19世纪结束以前就已经发生了，尼采宣布：并不存在支撑人类道德的神性基础，上帝死了。我们已不必劳驾社会生物学家们告知这一点。

诺姆·乔姆斯基的只言片语

批评社会生物学以及其他研究社会科学的达尔文式路线的学者中，一个最引人注目的人物是诺姆·乔姆斯基（Noam Chomsky），他既是语言学家，也是美国最强硬的社会批评家。我第一次见到乔姆斯基本人时，他正在就现代工会的实践问题发表演讲。他身材修长而强壮，因为长期伏案而稍有些驼背，戴一副铁框眼镜，穿着胶底运动鞋，一条丝光卡其军服布做的男裤，配一件开领衫。如果没有满脸的皱纹和那一头泛灰的长发，他倒更像个大学学生，尽管在大学生联谊会的晚会上，他更喜欢大谈黑格尔（Hegel）而不是狂饮啤酒。

乔姆斯基演讲的要点是：现代的工会领导人更关心的是维护自己的权力，而不是代表工人的利益。那么，他所面对的听众是谁呢？正是那些工会领导人。在自由问答时间里，正像任何人都可预料的那样，这些领导开始自卫反击，现场弥漫着明显的火药味。乔姆斯基面对四面八方的攻击，表情安详而冷静，在他那无可动摇的自信和大量无可辩驳的事实面前，批评者们不得不俯首承认：是的，他们确实可能已经变节，投向了资本家老板的怀抱。

当我后来向乔姆斯基表示自己对其演讲魅力的钦佩时，他告诉我，其实他对于"向人们布道"兴趣不大，他只是要反对一切专制制度。当然，通常情况下，他愤怒的火力并非指向工会，因为它已是失势的"权贵"；而是集中在美国政府、工业界和大众传媒上。他称美国为"恐怖主义的超级大国"，而大众传媒则是其"宣传工具"。他对我说，如果《纽约时报》（他最喜欢攻击的目标之一）哪一天开始评价他的政论书籍的话，那就表明他的政论出了问题。他把自己的世界观归结为一句话："不管是什么权威，我都要跟它对着干。"

我告诉他，我发现一个很有点讽刺意义的事实：他在政治观点上极力反对权威，但他自己又是语言学领域的权威。"错，我不是。"他厉声道。他的语调通常情况下是不愠不火的，并且总带着一种催眠师的镇定，甚至在他像医师般"解剖"论辩对手的时候也不失其平和，但这时却突然显出了锋芒。"我在语言学界的地位一直是微不足道的，现在也依然是这样"，他坚称自己"在学习语言方面全无特长"。并且，事实上，他甚至不是一位专业的语言学家。他说麻省理工学院只不过雇用了他，发给他薪水，其实院方并不懂得也不怎么关心人性问题，仅仅是要他去填补一个空缺的职位。[12]

我之所以提供以上背景，只不过是为增加些趣味性，但乔姆斯基确实是我所见过的最具反叛性的知识分子之一，也许只有无政府主义哲学家保罗·费耶阿本德可与之相提并论。他想把一切权威形象都从其宝座上打翻在地，甚至连他自己也不放过，身体力行地展示了什么叫"自我影响的焦虑"。因而，对乔姆斯基一切公开的言论，只能采取有所保留的态度去对待。尽管他自己极力否认，但乔姆斯基的确是一位当世最重要的语言学家。《不列颠百科全书》宣称："在当今的语言学领域，用于讨论所有重要理论问题的术语，都未超出他选择并定义的范围。"[13]乔姆斯基在思想史上的地位，已被认为几可与笛卡儿和达尔文比肩。[14]20世纪50年代，当乔姆斯基还在读研究生的时候，语言学与当时几乎所有的社会科学领域一样，处于行为主义的统治之下；而行为主义支持洛克（John Locke）的观点，认为心灵最初只是"白板"，像空白石板一样可供经验在其上任意刻画。乔姆斯基向这一主张发起了挑战，他认为：儿童单纯通过归纳和试错——像行为主义者所主张的那样——是不可能掌握语言的，肯定有某些基本的语言规则，或者说普适的语法，是我们大脑中所固有的。在人类语言和认知研究领域中，乔姆斯基的理论（他最早是在1957年出版的《句法结构》一书中提出的）促成了行为主义统治的彻底垮台，为一种更倾向康德主义、更重视遗传作用的观点扫平了道路。[15]

从某种意义上说，像威尔逊等试图用遗传学术语来解释人类本质的科学家们，都承蒙乔姆斯基之惠，但乔姆斯基却对人类行为的达尔文主义解释一向耿耿于怀。他承认自然选择在语言和其他人类品质的进化中肯定起过某种作用，但考虑到人类语言与其他动物那极简陋的交流系统之间所存在的难以逾越的鸿沟，以及我们对这段进化历史的贫乏认识，科学在解释语言是如何进化的这一问题上所能起到的作用其实是很有限的。乔姆斯基进一步发挥，仅仅因为语言在今天是一种适应性行为，并不足以说明它就是作为回应自然选择压力的结果而产生的；语言也许是某一次智力爆发的意外副产品，只是到了后来才派生出各种各样的用途，这一解释可能也适用于人类意识的其他方面。乔姆斯基指责说，达尔文式的社会科学根本不是真正的科学，只不过是"掺和进一点点科学因素的心灵哲学"，问题在于"达尔文理论就像一只宽松的大口袋，能把（科学家们）发现的任何货色都装进去"。[16]

乔姆斯基的进化观点使他坚信，我们理解人类或非人类的本质的能力是有限的。他抛弃了那种被大多数科学家所坚持的主张，即认为进化把大脑塑造成了一架多功能的学习和解题机器。与斯滕特和麦金一样，乔姆斯基相信，人类大脑中的固有结构为我们的认识能力设定了限度。（斯滕特和麦金之所以得出这样的结论，部分地是受到了乔姆斯基研究结果的影响。）

乔姆斯基把科学问题区分为疑难问题和神秘问题两类，前者至少是潜在的可解的，而后者却是不可解的。乔姆斯基向我解释道，在17世纪以前，真正现代意义上的科学尚不存在，那时几乎所有的问题都是神秘问题；后来，牛顿、笛卡儿等人开始提出并解答那些疑难问题，他们所应用的方法孕育出了现代科学。牛顿等人的研究，有一部分导致了"惊人的进展"，但其余的大部分都自消自灭了，比如，对于像意识和自由意志等问题的研究，根本就未取得什么进展。"我们甚至连不好的思想都没有。"乔姆斯基说。

他认为，所有的动物都具备由其进化史所形成的认知能力，比如说一只耗子，它可能学会走出要求它每隔一个岔路口就向左拐的迷宫，但如你指定一个素数，让它碰到每隔这一素数的岔路口再向左拐，那么它永远也学不会。如果你相信人是动物——并不是什么"天使"，乔姆斯基挖苦性地加了一句——那么，我们肯定也同样受制于这些生物学限度。语言能力使我们能以耗子所无法企及的方式提出并解决种种疑难问题，但最终，就像耗子陷入那座素数迷宫一样，我们也会面对那些完全无能为力的神秘问题，我们同样受制于我们提出疑难问题的能力。因而，乔姆斯基不相信物理学家或其他科学家能达成一种万物至理；充其量，物理学家们也只能创造出一种"就他们所知应如何系统地阐述万物的理论"。

在他自己的语言学领域，"关于人类语言何以或多或少地被注入了相同的铸具之中，使语言趋向一致的法则是什么，诸如此类的问题，现在已有了大量的认识"；但语言所提出的许多深奥的问题，仍然悬而未决。例如，笛卡儿就曾试图领悟何以人类能通过无尽的创造性途径去应用语言，而在这一问题上，"我们与笛卡儿同样的无知"，乔姆斯基说道。

在其1988年出版的《语言与知识问题》一书中，乔姆斯基提出，在探讨有关人类本性的许多问题时，言语的创造力比科学方法更富成效。他写道："我们对人类生活、对人的个性的认识，可能更多的是来自小说，而不是科学的心理学，有人甚至会认为多得简直无法比拟。科学形式的才能，只不过是我们智力禀赋的一个方面。但愿我们只把它用在能够应用的地方，而不是仅限于应用这一种能力。"[17]

乔姆斯基还谈道，科学的成功来源于"客观真理与我们认知空间结构的机缘巧合。之所以是一种机缘巧合，是因为发展科学并非出于自然选择的设计，并不存在什么遗传变异上的压力，使得我们非得发展出解决量子力学问题的能力不可，我们只是具备了这一能力。它的产生，与许多别的能力的产生，都出自某种同样没人能理解的原因"。

据乔姆斯基称，现代科学已经把人类的认知能力拉伸到接近断裂的地步。在19世纪，任何一位受过良好教育的人都能把握住当时的物理学全貌；但时至20世纪，"你若真的做到了这一点，反倒成了怪物"。机会来了！我立即接过话茬问道：科学持续增长的困难，是不是意味着科学正在趋近其极限？如果把科学定义为对自然界可理解的规律或模式的探求，那么它会有终结的一天吗？让我感到意外的是，乔姆斯基竟然把刚说过的话又都吞了回去。"科学的处境很艰难，这我承认。但如果你找那些孩子聊聊天，就会发现他们渴望理解自然。但这种渴望却被窒息了，被那些枯燥的教学方法和可恶的教育系统给窒息了。他们训斥那些孩子，说他们太蠢了，永远也不会理解自然。"把科学带入目前窘境的罪魁，陡然间又转换成教育机构等权威系统，而不是我们所固有的局限性了。

乔姆斯基坚持认为："重大的自然科学问题是存在的，我们可以界定它，并且能够解决它，这一前景是令人振奋的。"比如，科学家们仍需阐明——差不多肯定能阐明——受精卵怎样发育成复杂的生物，以及人的大脑是怎样产生语言的。科学仍然有着极大的发展余地，像物理学、生物学、化学等。

为了否认自己思想的寓意，乔姆斯基再次发作了那种跟自己过不去的古怪行为，但我怀疑，事实上他已屈从于一种愿望思维。就像许多别的科学家一样，他无法想象一个没有科学的世界。我曾问过乔姆斯基他对哪一方面的成就更满意，是他的政治活动，还是他的语言学研究。他看来对我连这一点都不明白感到相当诧异，但仍答道：他公然反对各种不公正现象，只是出于一种责任感，从中得不到什么智力上的愉悦。如果世上的不公正问题在某一夜全部消失了，他会欢天喜地地投入到对纯知识的追求中去。

与进步论唱反调的克利福德·格尔茨

反讽科学的实践者们大致可以分为两种类型：一类属天真型，他们相信或至少也是希望自己正在发掘关于自然的客观真理（超弦理论家爱德华·威滕便是其典型代表）；另一类是世故型，他们其实已经意识到，自己所从事的工作更像是艺术或文学批评，而不是传统意义上的科学。若要在后一类中挑选出一位代表人物，没有人会比人类学家克利福德·格尔茨（Clifford Geertz）更合适的了，他既是科学家，又是科学哲学家。如果说古尔德充当了进化生物学的否定性力量的话，那么格尔茨在社会科学中也担当着同样的角色。斯滕特在《黄金时代的来临》一书中关于社会科学的预言，即"它作为一个学科，在相当长的时期内只能满足于模糊而空泛的状况，就像它现在所表现的那样"，就是在格尔茨观点的基础上作出的。[18]

我第一次接触到格尔茨的作品还是在大学期间，当时我曾选修过一门文学批评课，导师让我去看格尔茨1973年的一篇文章《重彩描画：向着文化的解释理论前进》。[19]文章的基本观点认为，作为一位人类学家，不能仅仅通过"记录事实"来刻画一种文化，他或她必须去解释现象，必须尝试着去猜测现象背后的意义。格尔茨写道：考虑一下"某一只眼的某一次眨动"[这一例子应归功于英国哲学家吉尔伯特·赖尔（Gilbert Ryle）]，它可能代表着一次无意识的肌肉抽动，引发的原因或许是一次神经活动障碍，或许只是出于疲倦或厌烦；它也可能代表某种眼色，某种有意的信号，具有许多可能的寓意。文化实际上是由难以计数的此类信息、符号组成的，人类学家的任务就是去解释它们。从原则上说，人类学家对一种文化的解释就应该像文化本身一样复杂和富于想象，但就像文学批评家们永远不可能一劳永逸地构建出《哈姆雷特》的意义一样，人类学家也不要奢望能发现绝对真理。"人类学——或至少是解释人类学——作为一门学科，其进步的标志与其说是达成共识，不如说是调和争论，较好的情况是把握好我们彼此争论的分寸。"[20]格尔茨认识到，他那一门"科学"的特点，并不是要平息众说纷纭的状态，而是通过越来越有趣的方式，使这一状态永远保持下去。

　　在后来的著作中，格尔茨不仅把人类学比作文学批评，更进一步把它比作文学。人类学中交织着"故事讲述、图像构设、象征符号的编排以及比喻的运用"，与文学的手法一样；他把人类学理论称为"纪实小说"，或对"真实时间真实地点中的真实人物所展开的富于想象力的描述"。[21]（当然，对于格尔茨之类的人来说，用艺术取代文学批评并不是什么过激行为，因为在大多数的后现代主义者看来，艺术作品也好，文学批评也罢，都不过是文本而已。）

　　格尔茨在《深奥的游戏：巴厘人斗鸡札记》一文中，展现了自己作为一名纪实小说作家的才能。这篇作于1972年的文章在开篇之初就确立了他那平铺直叙的风格："1958年4月初，我和妻子小心谨慎地来到一个巴厘人村庄，准备在那里做些人类学考察。"[22]格尔茨的散文堪与普鲁斯特（Marcel Proust）和詹姆斯（Henry James）相媲美①。但格尔茨却告诉我说，把他比作前者是愧不敢当的；至于后者，自认尚可比拟。

　　文章的第一部分，描述了这对年轻夫妇怎样赢得了那些素称冷漠的巴厘人的信任。在格尔茨和妻子及一群村民观看一场斗鸡时，警察突然袭击了现场，这对美国科学家夫妇跟着那些巴厘村民一起鼠窜而逃，并未向那些警察们寻求特权待遇，这一做法深深打动了那些村民，他们真心地接纳了这对夫妇。

　　这样，格尔茨就成了村民们的"自己人"。他进一步刻画了与村民们打交道的趣事，然后开始分析巴厘人对斗鸡的执着，最后归结道：这种血腥活动（斗鸡被锋利的刺武装起来，撕斗至死方休）反映出巴厘人的一种恐惧心理，以为在表面看来极平静的社会之下潜伏着某种邪恶的势力，他们用斗鸡

---

① 普鲁斯特（Marcel Proust，1871—1922），法国小说家；詹姆斯（Henry James，1811—1882），美国伦理与宗教哲学家、作家——译者注。

来祛除邪恶势力以及自己内心的恐惧。就像《李尔王》或《罪与罚》所揭示的主题一样，斗鸡活动也"诘问着这些同样的主题：死亡、男子汉气概、狂热、骄傲、毁灭、慈善、冒险，并把它们整合成完整的结构"。[23]

格尔茨身材雍肿，花白的头发和胡须总是乱糟糟的。我第一次采访他是在一个寒冷的春日里，在普林斯顿高等研究院。当时的他显得坐卧不安，一会儿扯扯耳朵，一会儿搓搓脖子，一会儿深深地偎进椅子中，一会儿又从椅子里突然挺起身来。[24] 在听我提问时，他时不时地把毛衣领子向上拽起，直到遮住鼻子，就像是一个试图蒙起自己真面目的劫匪。他在回答问题时也是躲躲闪闪的，就像他的文章一样：到处是逗号和句号，完整的判断被数不清的限制词分割得支离破碎，并且充满了膨胀的自我意识。

格尔茨下定决心要更正他所感觉到的一种普遍错觉，即他是一个彻头彻尾的怀疑论者，不相信科学会取得任何能经受时间检验的真理。格尔茨说，某些领域，特别是物理学，显然具有达致真理的能力；他还强调，与我可能已听到的传言相反，他并未主张人类学只是一种艺术形式，不包含任何经验性内容，因而不是一个正统的科学领域。格尔茨说，人类学是"经验的、受证据检验的，它创立理论"。人类学的实践者们有时也能形成一种并非绝对的但可证伪的观点，因而它是科学，一种能够取得某种进步的科学。

"人类学中不存在任何东西足以和硬科学的硬成就相提并论，我认为将来也不会有"。格尔茨继续说道："某些无稽之谈认为人类学家能轻而易举地解答所有这些疑问，并且告诉你为了取得那根本不存在的成就你该怎么去做，但没人会相信它。"他得意地大笑了起来："这也并不意味着不可能了解某个人，或者不可能从事人类学研究，我一点也没这个意思，只是不太容易罢了。"

在现代人类学中，百家争鸣才应是主导方针，而不应是某种思想的一统天下。"形形色色的理论正日渐增多，但并不收敛向某一点。它们通过繁多的渠道传播和散布开来，我没看出来有什么向整体综合发展的迹象，只看到它越来越趋向分化和多元化。"

随着格尔茨继续讲下去，他所展望的进步听来却成了某种大倒退，那时的人类学家们会一个一个地排除掉所有能形成统一理论的假设，坚定的信念会逐渐消解，疑惑日益增多。他注意到，个别人类学家仍然相信，通过研究那些所谓的"原始"部落，也就是那些据说保持着原始状态、未被现代文明打断过的部落，他们就能够推断出关于人性的普遍真理，但人类学家们不可能扮演纯粹的客观资料搜集者角色，他们无法摆脱偏见和固有观念的影响。

格尔茨发现威尔逊的预言是极其可笑的，竟然认为只要把社会科学建立在进化论、遗传学和神经科学的基础之上，它最终就会变得像物理学一样严密、精确。格尔茨回忆，自诩的革命家们常常会跳出来，提出某些据说能统一社会科学的新点子，在社会生物学之前，就曾经有过一般系统论、控制论和马克思主义。格尔茨指出："认为天将降某人于世，使他在一夜之间改变一切，这种念头是一种学究们的通病。"

在高等研究院，某些物理学家或数学家，在构思出一个关于种族关系或其他社会问题的复杂的数学模型之后，偶尔也会来请教格尔茨的意见。"但他们对于城市内部的真实状况一无所知！"格尔茨宣称，"他们仅仅是拥有了一个数学模型！"他由此而大发感慨，说物理学家们绝不会容忍一个缺乏经验基础的物理理论。"莫名其妙的是，对社会科学却又是另一种态度。如果你需要一种反映战争与和平的普适理论，你要做的一切只不过是坐到书桌前写出一个方程式，而对于历史和人民却可以毫无认识。"

格尔茨悲哀地意识到，他所倡导的那种内省的、文学的科学风格同样隐藏着陷阱，它会导致极度的自我意识，或导致其实践者的"认识论上的多疑症"。这一被格尔茨称为"眼见为实"的倾向，的确已产生了某些有意义的成就，但也带来了些糟糕透顶的结果。格尔茨指出，某些人类学家一意孤行地要把自己所有潜在的偏见（意识形态的或其他方面的）公之于众，以至于写出来的东西就像是忏悔录，你从中能了解得更多的只是作者本人，而不是他所研究的对象。

格尔茨最近重访了他早年曾经调研过的两个地区，一个在摩洛哥，另一个在印度尼西亚。这两个地区都已发生了很大变化，他自己也一样。访问的结果，使他更清楚地认识到，对于人类学家来说，要想洞悉那些超越于时间、地点和条件之上的真理，该有多么困难。他说："我总觉得这种努力将以彻底失败而告终。我之所以尚能保持合理的乐观，是因为我认为尚有可为，只要你别对它要求太高。是不是我过于悲观了？绝不是！但我确实有种压抑感。"人类学并非唯一一个在其自身局限性问题上苦苦挣扎的学科，格尔茨指出，"我在所有学科领域中都切实感觉到了与此相同的基调"，甚至在粒子物理学领域，它看来也已抵达了经验检验的限度。"科学中曾经存在的那种盲目的自信，在我看来，已经并不那么有市场了。这并不意味每个人都已放弃了希望或已痛不欲生或诸如此类，但科学的处境确实已变得特别艰难。"

就在我们会晤于普林斯顿那段时间，格尔茨正在撰写一部反映其故地重游见闻的书。此书于1995年出版，书名简练地概括了格尔茨那种焦虑的心态：《追寻事实》。在书的最后一段格尔茨剖析了书名的多重含义：毫无疑问，像他一样的科学家，当然一直都在追求事实，但如果说他们真的得到什么事实的话，也只能是事后性的；等到他们搞清楚究竟发生了些什么事的时候，世界已经向前发展了，又变得像从前一样神秘莫测。

格尔茨总结道，这一简短书名同样暗合"后实证主义者对经验实在论的批评，在抛弃了真理与知识的相互对应观点之后，'事实'这一术语已变成某种极其脆弱的东西，不再有什么关于结局的允诺和判断，甚至也不必关心开始时追寻的是什么，只有这场无尽的追寻，这些千奇百怪的男女，穿越于这些各具特色的时代。但这确实是一条奇妙的旅途：有趣、沮丧、实用且富有娱乐性，值得投入一生的时光"。[25] 反讽的社会科学也许不能带给我们任何结论，但至少可以让我们有事可做——永远，只要我们高兴。

第七章

神经科学的终结

科学固守的最后一块阵地，并不是太空领域，而是人的意识世界。即使那些膜拜科学之解决问题能力的最狂热信徒，也认为意识是潜在的、无止境的问题之源。关于意识的问题可以从许多不同的途径进行探讨；从历史的角度来看，人类是怎样以及为何变得如此精明的？达尔文在很久以前就提供了一个通用的答案：自然选择偏袒人类，因为人类能使用工具，能预测潜在的竞争对手的行为，能组织成狩猎的团体，能通过语言来共享信息，并能适应不断改变的环境。与现代遗传学相结合，达尔文理论对于我们人类心智的结构、进而对于我们人类的性行为和社会行为等，已能给出更详尽的解释。（尽管距离爱德华·威尔逊等社会生物学家的期望，尚有一定的差距。）

但现代神经科学家们，对于从历史的角度去探讨意识进化的方式和原因问题，却并不怎么热心，他们更感兴趣的是心智现在是怎样建构的、怎样工作的。这两者之间的差别，类似于宇宙学和粒子物理学之间的区别，前者试图解释物质的起源及后续的进化，而后者却探讨目前已发现的物质的结构；一个学科是历史性的，因而也必然是暂时的、臆测性的并且是悬而未决的；与之相比，另一个则更加经验化、精确化，更能经受消解和终结的考验。

即使神经科学家不研究发育中的大脑，而把自己的研究仅仅局限在成熟的大脑之上，所面临的问题仍然是大量的：我们到底是怎样学习和记忆的？视觉、嗅觉、味觉、听觉等的机理是什么？这类问题对于大多数研究人员来说，虽然也是深奥难解的，但通过逆推神经回路系统的传导过程，仍然有望解决；但意识作为知觉的主观感觉，却一向被看作是一类完全不同的难题，即它不是个实证科学的问题，而是个形而上学问题。在20世纪的大部分时间里，意识一直被排除在科学研究的范围之外，尽管行为主义早已消亡了，但其幽灵仍徘徊在科学家的心头，使他们不愿意去考察主观现象，尤其是意识。

这种心态一直等到弗朗西斯·克里克（Francis Crick）把注意力转向这一问题时，才有所改变。克里克是科学史上最坚决的还原主义者之一。在他和詹姆斯·沃森（James Watson）于1953年揭示了DNA双螺旋结构之后，克里克进一步揭示出遗传信息是怎样被编码进DNA的。这些成就为达尔文进化论和孟德尔的遗传理论提供了坚实的经验基础，而这正是上述两种理论所一向缺乏的。20世纪70年代中期，克里克离开已在那里度过了大部分研究生涯的英国剑桥大学，迁往索尔克生物学研究所，那是加利福尼亚圣迭戈北部的一座颇具立体派艺术特色的"城堡"，从中正可俯瞰浩瀚的太平洋。他继续从事发育生物学和生命起源的研究，并最终把注意力转向所有现象中最难以捉摸但又无法回避的现象：意识。只有尼克松（Nixon）才能打开与中国的外交僵局；同样地，也只有弗朗西斯·克里克才能使意识成为合法的科学研究对象。[1]

1990年，克里克与一位年轻的合作者，加州理工学院的德裔神经科学家克里斯托夫·科克（Christof Koch），在《神经科学研究》上撰文宣告：使意识成为经验研究对象的时刻已经到来。他们断言，如果继续把大脑看作一个黑箱，也就是说，看作一个内部结构不可知的甚或是不相关的客体，那么，人们就无望获得对意识或其他精神现象的真正认识；只有通过研究神经元以及它们之间的相互作用，科学家们才能积累起必要的、确凿的知识，以创建真正科学的意识模型——类似于用DNA术语解释遗传现象的那类模型。[2]

克里克和科克摒弃了绝大多数同行奉行的信条，即认为意识是不可定义的，更不用说进行研究了。他们提出，意识（consciousness）和知觉（awareness）是同义词，而所有形式的知觉——不管是直接反映客观世界的具体事物，还是反映高度抽象、内在的概念——似乎都包含着同样的基本规律，即一种把注意和短期记忆相结合的规律。[克里克和科克把作出这一定义的荣誉归于威廉·詹姆斯（William James）。]克里克和科克极力主张，作为研究意识的一种举隅法（synecdoche），研究者应把注意力集中在视知觉上，因为视觉系统已被透彻地研究过了。如果研究者能够发现这一机能的神经机理，他们就有可能揭示更加复杂和微妙的现象，诸如自我意识——这可能是人类所独具的现象，因而也就更难以在神经层次上进行研究。克里克和科克完成了一项似乎不可能的奇迹：他们把意识从一个哲学上的玄奥问题转变成一个经验问题。一种关于意识的理论无疑将代表神经科学的顶峰，或极致。

据说，行为主义的重要代表人物斯金纳（B. F. Skinner）的一些学生，在亲炙他那无情的、机械的人性观之后，曾陷入生存的绝望之中，所以，当我在索尔克研究所中克里克那间宽敞且通风良好的办公室里见到他时，不由得回想起这一活灵活现的传说。但克里克并非一个沮丧的、郁郁寡欢的人，恰恰相反，穿着便鞋、米色便裤和一件俗艳的夏威夷款衬衫，竟然异乎寻常的快乐。他的眼角和嘴角皆向上翘起，构成一个恒久不变的顽皮笑容，浓密的白眉触角一般向上、向外奓张。他时不时地就会爽朗地大笑起来，每当他笑的时候，红润的脸上总是闪动着更深的亮泽。尤其是当他把某些一厢情愿而又模糊的想法——比如我关于人类应有自由意志的空想——贯串在一起时，他看起来总是特别兴高采烈。[3]

克里克用他那干脆利落的、亨利·希金斯（Henry Higgins）式的语调[①]告诉我说，即使是像"看"这样一种简单的行为，实际上也包含着大量的神经活动。"你作出的任何一个动作，比如说捡起一支笔，都会牵动大量的神经活动，"他继续说道，并从桌子上捡起一支圆珠笔在我眼前晃动着，"为了实现你所要完成的动作，要进行大量的计算。你能觉知到的只是一个结论，但究竟是什么使你作出了这一结论，却是不可觉知的。这一结论在你看来也许是凭空产生的，但实际上却是你未觉察到的一些过程的产物。"我皱起了眉头，克里克却抿着嘴轻笑起来。

为了让我真正理解"注意"（attention）的含义，这是他和科克关于意识的定义中最关键的因素，克里克强调说，它包含的内容远远不止简单的信息加工过程。为了证明这一点，他递给我一张纸，上面印着一个常见的黑白图案：一会儿看到黑色背景上的一只白色花瓶，过一会儿则又会看到两个人脸的黑色侧面轮廓。克里克指出，虽然输入大脑的视觉信号一直未变，但我所真正觉知的内容，或注意到的内容，却在不断地改变。与这种注意的改变相对应的大脑内部变化是什么呢？如果神经科学家能够回答这一问题，克里克说，他们无疑已向揭示意识的奥秘前进了一大步。

在他们1990年发表的一篇讨论意识的论文中，克里克和科克针对上述

---

① 亨利·希金斯（Henry Higgins），音乐剧《窈窕淑女》（*My Fair Lady*）中的角色。在剧中，希金斯教授和别人打赌，说可以让卖花女（由奥黛丽·赫本饰演）在大使的茶话会上以贵族身份现身；于是用了几个小时来教卖花女如何改变自己的发音，教她模仿上流社会小姐的举止——译者注。

问题给出了一种尝试性的答案。他们的假说建立在这样的证据基础之上：当视觉皮层对刺激产生反应时，某几组神经元极其迅速地同步发放。克里克解释说，这些发生振荡的神经元，可能就对应着视野中注意力所指向的那些方面，如果把大脑中大量的神经元想象成正在七嘴八舌地议论纷纷的人群，则那引起振荡的神经元，就像是忽然开始高唱同一首歌的一小群人。再回到"花瓶—人像"的图案上去，则一组神经元唱的是"花瓶"，而另一组唱的是"头像"。

振荡理论（另外一些神经科学家也独立地发展了这一理论）也有其弱点，克里克很爽快地承认了这一点。"我认为这是一个很好的、大胆的初步尝试，但对于其正确性如何，我也抱着疑虑的态度。"但他指出，他和沃森对双螺旋的发现，是在走过了无数次的弯路之后才最终获得成功的。"探索性研究其实就像是在雾中行路一样，你不知道自己正走向何方，而只能摸索前行。人们往往事后才能搞清探索的过程应该是怎样的，并把它想象得特别简单直接。"克里克深信，意识问题的解决，不可能通过对心理学概念和定义的争论而实现，只能通过做"大量的实验，这才是科学的真谛"。

任何心智模型都必须建立在神经元的基础之上，克里克这样告诉我。过去，心理学家们一直把大脑当作黑箱来对待，只能在输入和输出的层次上研究它，而不是建立在对其内在机理的认识基础上。"不错，如果黑箱足够简单，这样做当然能达到认识它的目的；但如果黑箱十分复杂的话，你能得到正确答案的机会微乎其微。这与遗传学的情况一样，我们必须了解基因，了解基因是怎样发挥作用的。为了说明这一点，就必须深入了解基因的本质，去发现构成基因的分子等内在成分。"

克里克对于自己在把意识提升为一个科学问题的进程中所占据的有利地位极为自得。"我无须征求许可。"他说，因为他已在索尔克研究所取得了固定的职位。"我之所以从事有关意识的研究，其主要原因是我发现这一问题令人着迷，并且我认为自己有权利去做自己喜欢做的事情。"克里克并未期望研究者们能在一夜之间就解决所有这些问题，"我想强调的只是：意识问题是十分重要的，它在相当长的时间一直被忽视了"。

在与克里克谈话的过程中，我不由自主地想起了詹姆斯·沃森的《双螺旋》一书（他在这本著作中，回顾了自己与克里克如何破译DNA结构的往事）那句十分著名的开场白："我从不知道弗朗西斯·克里克有谦虚的时候。"[4] 有必要在这儿来点历史修正主义，克里克是经常谦虚的，在我们谈话的过程中，他就曾对自己关于意识的振荡理论表示过怀疑；在谈到自己正在写作的一本关于大脑的著作时，他说某些部分"糟透了"，需要重写。当我问克里克该如何解释沃森的妙论时，他大笑起来。沃森的意思，克里克提示道，并不是说他不谦虚，而是说他"充满了自信、激情以及诸如此类的东西"。如果说他还是不时地有点自负，或者对别人有点苛刻的话，那只不过是因为他有过于强烈地想弄清真相的欲望。"我可以保持约20分钟的耐性，"他说，"但也就是那么多了。"

克里克对自己的分析，一如他对许多事情的分析一样，同样地切中要害。他有作为一名科学家、一名实验科学家的完美个性，是那种能解答疑难问题，并能给我们带来某种启迪的人物；他已经超越了缺乏自信、一厢情愿以及痴迷自己理论的境界，或至少他表现的是这样；之所以他会给人一种不知谦虚的印象，是因为他只想到要弄清真相，从不顾及后果会怎样；他不能容忍含糊其词、愿望思维或无法验证的猜测等，而这些正是反讽科学的标志；他也同样渴望与人分享自己的认识，从而尽可能地揭示事情的真相。这种品质在杰出的科学家身上，并非如人们所期望的那样常见。

在其自传中，克里克曾披露过这样一件事：在他还是个少年的时候，就一心想当个科学家，但总担心等到他长大的时候，所有的答案都已被别人发现了。"我向妈妈吐露了自己的忧虑，她向我保证说：'别担心，宝贝，会留下许许多多答案让你去发现的。'"[5]忆及这段话，我就问克里克是否认为将永远有大量的答案留给科学家们去发现。这完全取决于你怎样去定义科学，他答道。物理学家们或许很快就会确定自然界的基本规律，但这之后，他们可以应用这些知识不断发明新东西；生物学看来会有一个更长久的未来，某些生物结构，比如说大脑，是如此复杂，以至于在相当长的一段时间里还难以阐明。另外一些问题，比如说生命的起源，或许永远都不会完全解答清楚，原因很简单，能得到的资料总是不充分的。生物学中"存在着大量令人感兴趣的问题"，克里克说，"这些问题足以让我们忙到至少是我们的孙子的时代"。同时，克里克也赞同理查德·道金斯的观点，即对于进化的基本过程，生物学家们已经有了令人满意的普遍认识。

克里克陪我走出办公室时，经过了一张桌子，上面正堆着厚厚的一叠稿纸，那是克里克关于大脑的论著的草稿，题为"令人震惊的假说"。"你是否乐意读一读这部草稿的开头部分？""当然乐意"，我回答。"所谓令人震惊的假说，"那部书稿开宗明义地写道，"就是指'你的一切'——

你的欢乐和忧伤，你的记忆和抱负，你的自我感觉和自由意志——其实都只不过是大量神经元集群及其相关分子的行为。正如刘易斯·卡罗尔（Lewis Caroll）笔下的艾丽丝所云：'你只不过是一大堆神经元'。"[6]我望向克里克，他正笑得嘴巴一直咧到了耳根上。

几个星期之后，我打电话给克里克，就我采访他所写的文章里所涉及的一些事实，请他核实一下，他也同时征求我对《令人震惊的假说》一书的意见。他坦白地告诉我说，编辑对《令人震惊的假说》这一书名一点儿也不觉得"震惊"，她认为"我们只不过是一大堆神经元"之类的说法，并没有什么让人大惊小怪的，克里克问我对此怎么看。我告诉克里克说，我不得不赞同编辑的看法；他关于心智的观点无论怎么说，也只不过是陈旧的还原主义和唯物主义观点。我认为《令人沮丧的假说》（*The Depressing Hypothesis*）也许是个更恰当的书名，但它可能会招致读者的反感。但无论如何，书名并不那么重要，我又补充道；因为单凭克里克这一鼎鼎大名的感召力，这本书肯定就会畅销。

克里克以他一贯的好脾气接受了这一切。当他的书在1994年面世时，仍然题为《令人震惊的假说》，只不过克里克，或者更可能是他的编辑，为其添加了一个副标题："对灵魂的科学探索"。见到这一副标题，我不由得哑然失笑；因为克里克的意图显然并不是要去发现什么灵魂——也就是说，独立存在于我们肉身之外的某种超自然的东西——而是要彻底清除灵魂存在的可能性。他关于DNA的发现，已经对根除活力论作出了极大的贡献；现在，通过其关于意识的研究工作，克里克希望能把那种浪漫世界观清除得干干净净。

杰拉尔德·埃德尔曼围绕难解之谜装腔作势

克里克探讨意识的进路，其前提之一是：迄今为止所提出的任何有关心智的理论都没有多大的价值。然而至少有一位著名的、同样也不缺少诺贝尔奖桂冠光环的科学家，却宣称在解决意识问题上取得了重大进展，他就是杰拉尔德·埃德尔曼（Gerald Edelman）。埃德尔曼的经历一如克里克般多姿多彩，并且也同样成就显著。在他还是个研究生的时候，埃德尔曼就协助确定了免疫球蛋白的结构，这种蛋白质对机体的免疫反应起着决定性的作用。因为这一工作，他于1972年分享了诺贝尔生理学或医学奖。继而，埃德尔曼转向发育生物学领域，这一领域研究的是单个受精卵如何成长为一个成熟的生物体。他发现了一组称为细胞黏附分子的蛋白质，被认为在胚胎的发育中起着至关重要的作用。

然而，这一切相对于埃德尔曼创立一种心智理论的宏伟工程来说，只不过是个小小的序幕。埃德尔曼在四部著作中阐述了他的心智理论：《神经达尔文主义》《拓扑生物学》《记忆中的现在》《明亮之气，灿烂之火》。[7]埃德尔曼理论的要旨是：正像环境压力选择一个物种中适应性最强的成员一样，输入大脑的信息通过强化神经元之间的联系，选择不同的神经元集群，以对应于，比如说，最有用的记忆。

埃德尔曼膨胀的野心和个性，使他成为一位颇受记者瞩目的人物。《纽约客》上的一篇人物专访称他为"一个动机明确、富有朝气而又天真的苦行僧"，认为他"既像爱因斯坦，又更像亨尼·扬曼①"；文章同时提到诋毁者们把他看作是"一个私心极度膨胀的利己主义者"。[8]在《纽约时报杂志》1988年的一篇封面报道中，埃德尔曼给自己颁赐了种种神一般的非凡力量。其中，在谈及其在免疫学方面的工作时，他说："在我进入这一领域之前，那里是彻底的黑暗；而我进入之后则是一片光明"；他把基于其神经模型基础上的机器人称为他的"造物"，并且说，"我只能像上帝一样观察它。我俯察着它的世界"。[9]

我1992年6月在洛克菲勒大学拜晤埃德尔曼时，才亲自领教了他的狂妄自大。（时隔不久，埃德尔曼就离开了洛克菲勒大学，到了加利福尼亚的拉加勒，在斯克里普研究所拥有了自己的实验室，位置就在克里克实验室的南边。）埃德尔曼是个大块头，穿一身黑色的宽肩西装，散发出一种慑人的优雅和亲切气息。就像他的书中所表现的那样，他对于科学问题的探讨，不时地被硬塞进去的轶事、笑话或警句所打断。这些题外话与主题之间的关系常常是莫名其妙的，似乎只是为了证明埃德尔曼是完美智者的化身，只是为了表明：不管是论机智还是论实干，也不论是做学问还是混世道，他都不仅仅是个实验科学家。

在解释他是"如何开始对心智感兴趣的"这一问题时，埃德尔曼说道："我被科学中的黑暗、耽于幻想和悬而未决的争端深深刺激着，我不介意研究细节，不过更喜欢为解决终极性问题而有所作为。"埃德尔曼只想去探求

---

① 亨尼·扬曼（Henry Yongman，原本的意第绪语的姓氏为Yungman）：1906年3月16日生，卒于1998年2月24日，英裔美国喜剧演员和小提琴演奏家，以其清新时尚而又玩世不恭的俏皮话，享誉表演界近70年，被称为"俏皮话之王"（King of the One-Liners）——译者注。

那些重大问题的答案，他获得诺贝尔奖的研究，即对抗体结构的研究，已经使免疫学变成一门"近乎完成的科学"；其核心问题，即关于免疫系统怎样对侵入病原作出反应的问题，已被解决了。他和另外一些人已经协助阐明，自我识别的发生是通过一个叫精选的过程实现的：免疫系统有着难以计数的、不同的抗体，当外来抗原出现时，就会刺激机体加速产生——或者精选——针对该抗原的特异性抗体，同时抑制其他抗体的生成。

埃德尔曼对悬而未决问题的探讨，将他最终引向了对大脑的发育和工作机理的研究。他认识到，建立一种关于人类心智的理论，将标志着科学的最终完结，因为到那时，科学已能够解释其自身的起源。考虑一下超弦理论吧，埃德尔曼说道，它能解释爱德华·威滕的存在吗？显然不能。埃德尔曼指出，许多物理学理论都把涉及心智的问题划归"哲学或纯粹的思辨"。"你读过我著作中的这样一段吗？其中马克斯·普朗克（Max Planck）说道，'我们永远也无法解答这个宇宙的奥秘，因为我们本身就是不可理解的'；还有伍迪·艾伦（Woody Allen）那一段，他说'如果我可以重新来过，我会住在熟食店里'。"

在描述他对心智的研究时，埃德尔曼的话乍一听也像克里克一样，是有根有据、理直气壮的。他强调，对于心智只能从生物学的立场来认识，而不能通过物理学的、计算机科学的或其他无视大脑结构的途径。"除非我们能拥有深刻的、令人满意的神经解剖学理论，否则我们就不会拥有关于大脑的深刻而又令人满意的理论，对吧？道理多么浅显。"诚然，那些"机能主义者"，例如人工智能专家马文·明斯基（Marvin Minsky）等人，确曾宣称过，说他们无须了解解剖学就能造出智能机，"我的回答是，'当你们把实物展示在我面前时，再夸口也不迟。'"

但随着埃德尔曼继续说下去，其"马脚"也就逐渐显露出来了。与克里克不同，埃德尔曼是透过他所特有的、无法摆脱的有色眼镜和自负来审视大脑的。他似乎认为自己所有的见解都是原创性的；在他把注意力转向这个领域之前，没有谁真正认识过大脑。埃德尔曼回忆说，当他开始研究大脑，或者更确切地说，是开始研究各种各样的大脑时，立刻被它们之间的多变性惊呆了。"令我感到奇怪的是，那些在神经科学领域从事研究的人，竟一直都把大脑当作彼此相同的东西来讨论。你可以翻开研究报告去查一查，每个人所讨论的大脑，仿佛只不过是一台可复制的机器；但如果你真正深入地观察一下大脑，不论是在哪个层次上——当然，存在着的层次数目是惊人的——

真正打动你心弦的却是多样性。"他指出，即使是同卵双胞胎，他们的神经元结构形式也表现出巨大的差异，这些差异远非无关紧要的"噪声"，而是具有极重要的意义。"这种差异相当让人吃惊，"埃德尔曼说，"而这正是你所难以理解的事情。"

大脑那巨大的多变性和复杂性，可能与从康德到维特根斯坦的哲学家们沉思的一个问题有关：我们是怎样对事物进行分类的？埃德尔曼进一步解释道。维特根斯坦从本质上突出了分类的困难，就在于不同的游戏之间，除了它们同是游戏之外，别无共同之处。"这是典型的维根特斯坦式语言，"埃德尔曼沉吟着说，"平实中肯之中，却透出一种卖弄的意味，我又无法说清他卖弄的究竟是什么。他的思维撩拨着你，他的话很有蛊惑力。有时候，他的话中透出一种自负，并且那绝不是矫揉造作。那是解不透的谜，是围绕这难解之谜的装腔作势。"

跳皮筋的小女孩，下国际象棋的棋手，以及正在进行海战演习的瑞典海军水兵，他们都是在做游戏，埃德尔曼又继续发挥道。对于大多数观察者而言，这些现象彼此之间似乎很少有或干脆就没有什么关系，然而它们都是一组可能的游戏中的成员。"这在研究中被定义为多形态组，是一种十分棘手的境况，意味着既非由充分条件决定，又非由必要条件决定的一组。我可以给你看一看在《神经达尔文主义》中关于它的示意图。"埃德尔从桌上抓起一本书，一直翻到有两组关于多样组的几何示意图的那一页，接着又把书一扔，再次说出一句令我目瞪口呆的话："让我诧异的是竟没有人着手把它们整合到一起。"

当然了，埃德尔曼确实已把这些东西整合在一起了：大脑各种各样的多样性，使它能够对自然的多样性作出反应；脑的多样性并非无关的"噪声"。"遇到外界的一组未知物理联系时，从中作出选择的基础，对吧？这是很有发展前途的思路。让我们再进一步，作出选择的基本单位是神经元吗？"绝不是，因为神经元只是二进制的，不够灵活；它要么是开（即发放），要么是关（即静止）；但彼此相连的、相互作用着的神经元集群，却可以完成选择工作。这些集群之间相互竞争，就可以产生出代表无限多样的外界刺激的有效表达或图像。能够形成有效图像的集群长得更加强壮，而其他的集群则萎缩了。

埃德尔曼继续自问自答着，他说得很慢，语气中充满自负，仿佛要把他所说的每个字都实实在在地刻入我的大脑之中。这些相互联系着的神经元集群怎样解决困扰维特根斯坦的多样组问题呢？通过折返；什么是折返呢？"折返就是产生图像的区域间不断往复循环的信号，"埃德尔曼说道，"因此，你能够通过大量平行的交互联结绘制出图像。折返不是反馈，反馈只适用于两条线路间，这一术语早已有了确定的用场和说明——输入正弦波，然后输出放大的正弦波。"他的表情很严肃，甚至有几分气愤填膺的模样，仿佛我突然变成了他那些呆笨且又善妒的批评家的象征——他们竟然敢说折返只不过是反馈。

他沉默了一会儿，好像是要让自己因此平静下来，然后又再度开讲，声高，语缓，一字一顿，就好像一个旅行者正竭力要使一个大概很迟钝的土著理解自己的意图。与他的批评者之所云相反，他的模型是独一无二的，与神经网络全无共同之处，他这样说道，语气中充满了对"神经网络"一词的轻蔑。为了取得他的信任——同时也因为这是事实——我坦陈自己一直觉得神经网络很难理解（神经网络由不同强度的联结在一起的人造神经元或者开关组成）。埃德尔曼得意扬扬地笑了起来，说道，"神经网络包含一个被曲解了的隐喻，它与真实的神经网络之间存在着一条难以逾越的鸿沟，面对它，你自然会想：'我难道就是这个样子吗？还是我哪里出了问题？'"埃德尔曼向我保证，他的模型绝不会遇到此类问题。

我正准备问有关"折返"的另一个问题，埃德尔曼却举手止住了我，他说已经到了向我宣布其最新创造物的消息的时候了，那就是"达尔文4号"。验证其理论的最佳途径，应是直接观察活动物的神经元活动，但这当然是办不到的，而唯一的解决办法，埃德尔曼说，就是造出一部能体现神经达尔文主义原则的自动机。埃德尔曼及其合作者已经造出了四个机器人，每一个都命名为"达尔文"，并且每一个都比前一个更为复杂，而事实上"达尔文4号"已经不再是机器人了，埃德尔曼向我保证，它是一个"真实的创造物"，是"第一个真正能够学习的无生命物体"，明白吗？

他再度停顿了一会儿，我感到他那传教士般的激情正向我冲击而来。他似乎想营造出一种戏剧化的效果，仿佛他正在拉开一层层的幕布，而每一层的后面都隐藏着一个更深奥的秘密。"我们去瞧一瞧吧！"他说。于是我们相继走出他的办公室，穿过大厅。他打开一扇门，门后的房间里安装着一台正嗡嗡作响的巨型计算机，埃德尔曼向我介绍说，那是"达尔文4号"的"大脑"；接着我们又走入另一个房间，那个"真正的创造物"正在那里等着我们。那是一堆装有轮子的机器，端居于一个胶合板平台上，周围散布着些红色和蓝色的积木。也许是觉察到了我的失望——对于任何一个已看过电影《星球大战》的人来说，现实的机器人都难免会让他失望的——埃德尔曼向我重申"达尔文4号"尽管"看起来像个机器人，其实不是"。

埃德尔曼指着一个装有光敏传感器和磁控制器的条状物，说那是"嘴"；墙上装着一个电视监控器，正闪烁着不同的图形，埃德尔曼告诉我，那代表着"达尔文"的大脑的状态。"当它确实发现了一个目标物之后，它就会靠上前去，把它抓起来，然后判断它是好东西还是坏东西……判断结果会改变各部件之间的广泛联系和突触状态，并能通过大脑图像表现出来，"他指着监控器，"从而使突触间的联系得到加强或减弱，并最终改变肌肉的活动情况。"

埃德尔曼紧盯着达尔文4号，它顽固地保持着静默。"哦，它需要相当长的启动时间，"埃德尔曼解释道，接着又补充说，"进行启动所需要的计算量是极其惊人的。"最后机器人终于开始慢慢移动起来，仿佛被埃德尔曼的诚心打动了，开始在平台上慢慢地移动，用它的"嘴"轻轻触碰着积木，放过蓝色的，"叼"起红色积木块，并把它们送进一个被埃德尔曼称为"家"的大盒子中。

埃德尔曼一直配合着机器人的动作向我作着同步解："嘿，它刚翻了翻眼睛。它又发现了一个目标。噢，它捡起了它，现在正搜索家的位置。"

它的最终目的是什么？我问道。"它没有最终目的，"埃德尔曼皱起眉头提醒我，"我们已赋予它价值观，蓝积木是坏东西，红积木是好东西。"与目的比起来，价值观更为普遍，因而更适于帮助我们有效地对付这个多样的世界，而目的就显得较为专门化了。埃德尔曼费力地继续解释，当他还是个十来岁的孩子时，他曾渴望能占有玛丽莲·梦露（Marilyn Monroe），但玛丽莲·梦露并不是他的目的；他拥有一套价值观，这导致了渴望拥有具有特定女性特征的异性，而玛丽莲·梦露不过是刚好符合这些特征。

生硬地压抑住脑海中涌起的那幅关于埃德尔曼和梦露如何如何的幻象，我转而请教这个机器人与其他科学家在过去几十年里所制造的那些有什么不同，它们中的大多数最低限度也具有"达尔文4号"所展现出来的能力。不同之处在于，埃德尔曼斩钉截铁地答道，达尔文4号拥有价值观或者说本能，而其他机器人完成任何工作都需要特殊的指令。我反驳道，所有的神经网络系统不是都已避开了特殊的指令，并具有了普遍的学习程序吗？埃德尔曼皱起了眉头："但所有那些系统，你都必须另外定义输入和输出，这是最大的差别。我说得对吧，朱里欧（Julio）？"他转头问一个郁郁寡欢的年轻博士后，他一直站在旁边默默地听着我们的对话。

犹豫了一会儿之后，朱里欧点了点头。埃德尔曼开心地大笑起来并指出，大多数人工智能的设计者，都试图用精确的指令把所有的知识都编入一套万能的程序中去，以适应每一种情况，而不是从价值观中自然地判断出适当的知识来。就拿狗来说吧，猎狗可能从几条最基本的本能中获取知识。"这比任何一帮编写沼泽地程序的哈佛小伙子都更灵验！"埃德尔曼大笑着说，同时瞟了朱里欧一眼，朱里欧也尴尬地陪着笑了几声。

但"达尔文4号"仍然不过是一台计算机，一个机器人罢了，仍然只有对外界作出有限反应的有限指令系统，我毫不放松地坚持着；当埃德尔曼称它为一个拥有"大脑"的"创造物"时，那只不过是一种拟人的说法。在我说这些话的时候，埃德尔曼在一旁不耐烦地嘟囔着，"噫，噫，好吧，好吧。"同时连连点着头。如果计算机被定义为由算法系统或有效程序所驱动的东西，那么，"达尔文4号"就不是一台计算机。诚然，计算机专家们也能为机器人编出一套程序来，使它足以完成"达尔文4号"所能完成的动作，但那只不过是冒充生物行为，而"达尔文4号"的行为却是名副其实的生物行为。如果出现了意外的电子故障从而扰乱了其"创造物"的一组密码，埃德尔曼告诉我说："它就会自动修复它，就像一个受伤的生物体一样，然后又可以四处走动了。我敢肯定，对于其他机器人来说，发生这种情况后只有死路一条。"

我没有进一步指明所有的神经网络和许多普通的计算机程序都有这种自我修复能力，反而问埃德尔曼，为什么其他的科学家抱怨说他们一点儿也不明白他的理论。他答道，多数真正的新科学概念都必须克服类似的阻力。他已经邀请那些指责其理论晦涩难懂的科学家——包括著名的岗瑟·斯滕特，他指责埃德尔曼理论难以理解的言论曾被《纽约时报》杂志所引用——来他这儿访问，以便向他们亲自讲解自己的理论。（其实斯滕特对埃德尔曼工作的评论，就是在两人比邻而坐共同飞越大西洋的一次旅程之后作出的。）但没有人接受埃德尔曼的邀请。"之所以出现这种不被理解的局面，我相信，原因在于接受者方面，而不是我的表述问题。"埃德尔曼说。

这时，埃德尔曼已不再试图掩饰自己的恼怒。当我问及他与克里克的关系怎样时，他粗鲁地宣称自己还要出席一个重要的会议，只好让能干的朱里欧陪我了。"我和克里克的关系要说起来话可就太长了，这不是在离开前的三言两语——哼！哼！——就能说清楚的。或者，用格劳乔·马克斯①的话来说，'走开，永远别再来烦我！'"撂下这句话后，他便离开了，只留下一阵空洞的笑声。

埃德尔曼也拥有一批崇拜者，但多是些对神经科学略知皮毛的人。他的最出名的崇拜者是神经病学家奥利弗·萨克斯（Oliver Sacks），而萨克斯最妙的一篇正式报告，描写的是他与脑损伤患者打交道的事，已确立了一种文学性的——即反讽性的——神经科学标准。当弗朗西斯·克里克指责埃德尔曼把一些"勉强可端上台面"、但并没有什么真正独创性的观点，隐藏在其"莫名其妙的语言烟幕"背后时，正道出了许多同辈科学家的心声。关于埃德尔曼的达尔文式术语，克里克补充道，与其说与达尔文的进化论相似，倒不如说更接近于华而不实的修辞学；他还提议，埃德尔曼的理论应改称为"神经埃德尔曼主义"。"杰里②的毛病，"克里克说，就在于"他过分热衷于提出标语口号并四处张扬，而又从不关心别人怎么评价。这才是他真正积习难改的弊病，也是真正惹人诟病的缘由"。[10]

塔夫茨大学的哲学家丹尼尔·丹尼特（Daniel Dennett）在参观了埃德尔曼的实验室之后，仍然未被点悟。在一篇关于埃德尔曼的《明亮之气，灿烂之火》一书的书评中，丹尼特指出，埃德尔曼的思想仅仅是一些老皇历的粗劣翻版。在丹尼特看来，埃德尔曼的巧辩是站不住脚的，他的模型就是一种神经网络，而折返也就是反馈；丹尼特又补充道，埃德尔曼还"从根本上误解了自己所探讨的哲学问题"；他或许表面上对那些认为大脑只是一台计算机的人尽情嘲笑了一番，但他用一个机器人来"证明"自己的理论，却表明他骨子里也持有相同的信念，丹尼特这样评价道。[11]

一些批评者指责埃德尔曼蓄意将别人的观点用自己怪异的术语包装起来，企图以此来沽名钓誉。我的看法，在某种程度上要更宽仁些，埃德尔曼具有一个经验主义者的大脑，同时又有一种浪漫情怀。当我问他，在他看来，科学本质上是有限的还是无限的时，他用自己那种独特的拐弯抹角的方式，这样答道："我不知道这个问题到底是什么意思。当我说数学中的级数是有限的或是无限的时，我清楚它的含义；但我不明白说科学是无限的到底指什么。举个例子吧，我想借用华莱士·史蒂文斯③的《遗著集》中的一句

① Groucho Marx，1895—1977，英国喜剧演员——译者注。

② Jerry，是杰拉尔德·埃德尔曼的昵称——译者注。

③ Wallace Stevens，1879—1955，美国诗人——译者注。

话，'从长远而言，真理并不重要，重要的是要去追求。'"埃德尔曼似乎
是在暗示，对真理的追求才是重要的，而不是真理本身。

埃德尔曼补充道，据说当爱因斯坦被问及科学能否被穷尽时，曾答道：
"可能。但用气压波动的术语来描述贝多芬的交响乐有什么用呢？"埃德尔
曼解释说，爱因斯坦所指的是这样一个事实，即单凭物理学是不能解释有关
价值、意义及其他一些主观现象的。那么，人们同样可以反问一句：用折返
神经回路的术语来描述贝多芬交响乐又有什么用呢？不论是用气压波动或是
原子还是任何其他物理现象作为神经元的替代物，怎么能合理地解释那神秘
的心智之谜呢？埃德尔曼说他无法接受克里克的观点，即认为我们"只不过
是一大堆神经元"，因此，埃德尔曼就使自己的基本神经理论变得朦胧含混
起来——把它和那些从进化生物学、免疫学以及哲学中借用过来的术语和概
念搅成一团——再给它添加些莫须有的高尚、共鸣和神秘性等色彩。他就像
是一位小说家，冒着不被理解——甚至是刻意追求着这种不被理解——的危
险，只是期盼着能获取一种更加深奥的真理。他是一位反讽的神经科学的践
行者，但遗憾的是，他缺乏必要的修辞技巧。

量子二元论

包括克里克、埃德尔曼在内，实际上几乎所有的神经科学家都赞同这
样一种论点：不论从何种严格的意义上说，心智的特性都不会依赖于量子力
学。物理学家、哲学家等对于量子力学与意识之间的联系的推测，至少自20
世纪30年代就已开始了。当时，某些具有哲学倾向的物理学家就已开始论
证，测量行为——甚至意识本身——在判定含有量子效应的实验结果时起着
十分重要的作用。但这类思辨的结果，除了引人注目之外，并没有什么更多
的实质性内容，并且其支持者无一例外地都抱有隐秘的哲学的甚至宗教的动
机。克里克的搭档克里斯托夫·科克曾用一个三段论来概括量子—意识关系
的论点：量子力学是玄奥难解的，意识也是玄奥难解的，所以，量子力学与
意识必然是相关的。[12]

提倡意识的量子理论最厉害的人物之一，是英国的神经科学家约翰·埃
克尔斯（John Eccles），他曾因其神经方面的研究工作荣获了1963年度的
诺贝尔奖。埃克尔斯可能也是信奉二元论（即认为精神独立于物质基础而存
在）的最杰出的现代科学家，他与卡尔·波普尔合著了一本为二元论辩护的

书，书名是《自我及其脑》，出版于1977年。他们摒弃了物质决定论，赞成自由意志；在脑和身容许的限度内，精神可以在不同的思想和行为过程中作出选择。[13]

对二元论最常见的异议，就是它违背了能量守恒定律：如果精神没有任何物质实在性的话，那么它怎能引发大脑中的物质变化呢？埃克尔斯与德国物理学家弗里德里克·贝克（Friedrich Beck）一起给出了这样的答案：带电的分子（或离子）在突触部位的积聚，使脑神经细胞激活，并导致被激活的细胞释放神经递质；但给定数量的离子在突触的积聚，并不总能引发神经元发放，其原因，据埃克尔斯解释，在于那些离子至少在某一瞬间是以量子叠加态存在的；在某些状态下神经元发放，在另外的状态下则不发放。

精神通过"决定"哪些神经元发放、哪些不发放，从而对大脑施加其影响，只要整个大脑维持这种可能性，自由意志这种行为就不违背能量守恒定律。"我们对此没有丝毫证据"，埃克尔斯愉快地交了底，那是在一次电话采访中，他在向我介绍了自己的理论之后说出这番话的。不过他仍把这种假说称为"一个巨大进展"，可以激励二元论的复兴。唯物主义及其所有那些微不足道的后裔——逻辑实证主义、行为主义和同一性理论（它把精神状态与大脑的物质状态完全等同起来）——"都已终结了"，埃克尔斯宣称。

谈及自己在解释精神的特性时之所以转向量子力学的动因，埃克尔斯是很坦率的——若为他本人着想的话，他也许是过于坦率了。他说自己是个"有宗教信仰的人"，反对"廉价的唯物主义"。他相信"心智的真正本质与生命的本质一样，出自神性的创造"。埃克尔斯还坚信："就揭示存在的奥秘而言，我们还只是刚刚开始。"我问他，是否能有那么一天，我们能够探明存在的奥秘，并因此把科学带向终结呢？"我不这样认为。"他答道。沉默了一会儿，他又用激动的声音补充道："我不希望科学终结，唯一重要的事情就是接着干。"他赞同其二元论同伙、证伪主义者卡尔·波普尔的观点，即我们必须且能够"发现，发现，再发现，同时要进行思考。我们在任何事情上都不能宣称已获得了最后的结论"。

罗杰·彭罗斯的真正目的

罗杰·彭罗斯在掩饰自己的隐秘动机方面，比埃克尔斯要略胜一筹，或许是因为他自己对此也只有模糊的认识。彭罗斯最初是作为黑洞以及其他一

些奇异物理现象方面的权威而建立起自己的声望的；他还发明了彭罗斯镶嵌
（Penrose tiles），简单的几何形状拼在一起时，竟能产生出无限变化的、准
周期性的图案。1989年以来，他又因《皇帝的新脑》一书中所提出的论点而
声名大噪。这本书的主要目的是要驳斥那些人工智能支持者的狂言，即认为
计算机可以模拟人类的所有特性，甚至包括人的意识。

理解彭罗斯论点的关键在于哥德尔不完备性定理。这一定理的内容是：
任何一个自洽的公理系统，一旦超出某个基本的复杂性层次，就会产生出这
样一些命题，它们在这一公理系统内既不能被证明，也不能被否证；因而，
该公理系统总是不完备的。在彭罗斯看来，这一定理是在暗示着：任何"可
计算的"模型——不论是经典物理学的、计算机科学的，还是我们现在正讨
论着的神经科学的——都不足以模拟精神的创造能力，或者，更精确地说，
都不足以模拟意识的直觉能力。精神必定来自某种更为微妙的现象，或许就
是与量子力学有关的现象。

我第一次拜晤彭罗斯是在锡拉丘兹，正是那次采访引发了我对科学的限
度问题的兴趣；三年之后，我在他的"大本营"牛津大学再度拜访了他。彭
罗斯介绍说，他正在写作《皇帝的新脑》一书的续篇，以便更详细地阐述自
己的理论。他说他比以往更加相信：自己的准量子论的意识说所遵循的是一
条正确的路径。"这是一个你只能孤军奋战的领域，但我对这一切抱有强烈
的信念，除此之外我看不到还有什么别的路径。"[14]

我提示说，有些物理学家已经在思考：如何利用奇特的量子效应，比如
说叠加，去完成传统计算机所难以实现的计算。假如这种量子计算机真的成
为现实的话，彭罗斯会承认它们已经具备了思维能力吗？彭罗斯摇了摇头，
答道："一台能够思考的计算机，它所依据的工作机理，不可能基于目前形
式的量子力学，而应该以目前尚未发现的更为深刻的理论为基础。"据彭罗
斯吐露，他在《皇帝的新脑》一书中所真正要反驳的，是这样一种假设，即
认为意识的奥秘，或者普遍存在的奥秘，可以用现有的物理学规律来解释。
"我敢说这种假设是错误的，"他宣称，"支配世界的行为法则，我相信，
肯定比现有的物理学规律更为玄妙。"

当代物理学对此无能为力，他进一步解释道，特别是量子力学，它还存
在着缺陷，因为它与习以为常的宏观实在之间存在显著的不一致。电子怎么
能在一个实验中像粒子，而在另一个实验中却又像波呢？在同一时刻，它们
怎么能处于两个不同的位置呢？必然存在着某种更为深刻的理论，它能消除

量子力学的佯谬及其让人迷惑的主观因素。"当然，我们的理论最终是要为主观主义留有余地的，但我却不希望看到理论本身就是一种主观性的学说。"换种说法就是：理论能够容许意识的存在，但却不必诉诸意识而存在。

不论是超弦理论——它归根到底也不过是一种量子理论——还是其他任何一种统一理论的候选者，在彭罗斯看来，都不具备作为统一理论所必需的特征。"如果存在这样一种物理学的大统一理论，那么从某种意义上说，它必定具有已知的理论所不具备的特征，否则是难以令人相信的。"他说。这样一种理论，必须具备"某种不容置疑的自然主义色彩"，换句话说，大统一理论必须合乎情理。

然而，对于物理学到底能否成就一种真正完备的理论这一问题，彭罗斯的心里依然充满了矛盾，就像他在锡拉丘兹时所表现的那样。他认为，哥德尔定理暗示着在物理学中，就像在数学领域一样，将永远存在着无法解决的问题。"物理学探寻物质世界是怎样存在的，并用数学结构把它表示出来，"彭罗斯认为，"即使第一步探讨的工作真的终结了，物理学这一学科仍然不会终结，因为数学是无止境的。"他讲得十分谨慎，那种字斟句酌的劲头，比我俩在1989年初次讨论这个问题时尤过之。很显然，对于这一论题，他已倾注了更多的心神去思考。

我忆及理查德·费曼曾把物理学比作下国际象棋：一旦我们掌握了基本规则，就可以没完没了地探索其结果。"是的，这两者之间的确不无相似之处。"彭罗斯说。那么，这是不是就意味着，如果撇开应用规则的结果的话，他认为掌握所有那些基本规则是可能的？"我想，就乐观的一面而言，我相信这的确是可能的。"他又热切地补充道，"我当然不是那种固执的人，竟认为从物理学的角度认识世界的道路是没有尽头的。"在锡拉丘兹时，彭罗斯曾说过，相信"终极答案"的存在是件让人感到悲观的事；现在，他觉得这一观点是乐观的了。

彭罗斯说，对于人们对他的思想的理解和接受情况，他大体上是满意的；多数的批评者都保持起码的客气态度，唯一的例外是马文·明斯基，在加拿大召开的一次会议上，彭罗斯与明斯基曾有过一次不愉快的交锋。在那次会议上，他俩都被安排作大会发言，在明斯基的坚持下，彭罗斯首先发了言，而明斯基却紧接着就站起来提出反驳。在被告以"别忘了你是个穿着体面的绅士，请保持绅士风度"之后，他干脆脱掉外套，说："好吧，我不认为自己是什么绅士！"便继续攻击起《皇帝的新脑》来，而他的那些论据，据彭罗斯说，是极其愚蠢的。回想起这幕往事，彭罗斯看来仍觉得有些困窘，并且仍有些痛苦。他在处事态度上的温和谨慎，与其观点上的大胆激进所形成的强烈对比，使我不由得再次感到惊奇，就像我初次见到他时所感觉的那样。

1994年，就在我于牛津大学拜访彭罗斯两年之后，他的《心智的阴影》出版了。在《皇帝的新脑》中，彭罗斯对于准量子效应究竟应在何处施展其魔力，一直是很含混的；而在《心智的阴影》中，他的确提出了一种猜想：是在微管中。那是一种微细的蛋白质通道，它构成大多数的细胞（包括神经元）的骨架。彭罗斯的这一假说，是基于斯图亚特·哈默罗夫（Stuart Hameroff）的一项研究报告作出的，后者是亚利桑那大学的一位麻醉学家，他发现麻醉剂能抑制电子在微管中的运动。正是以这一脆弱的报告为基础，彭罗斯构筑起一座宏伟的理论大厦。他猜想微管执行着某种非决定性的准量子计算，这种计算在某种程度上产生了意识。这样，每个神经元就不再是个简单的开关，而是一台独立的复杂计算机。

彭罗斯的微管说势必成为画蛇添足之作。在第一本书中，他营造出了一种混合着悬念、期望和神秘感的氛围，就像是一部恐怖影片的导演，对于那奇异可怕的巨大海怪，只为你提供了惊鸿一瞥，惹得你心痒难搔；如今彭罗斯终于揭开了其"海怪"的面纱，但它看起来只不过是一位穿着一套粗劣的潜水服并且在不断摆动着四肢的肥胖演员。可以预料，某些本就抱着怀疑态度的观众，对这一"海怪"的反应，是嘲弄甚于敬畏的。他们指出，微管几乎在所有的细胞内都能发现，而不仅仅存在于神经元中，那么，这是不是意味着我们的肝脏也具有意识呢？我们的大脚趾又如何？草履虫呢？（1994年4月，当我就这一问题请教彭罗斯的搭档斯图亚特·哈默罗夫时，他答道："我不认为草履虫有意识，但它确实能表现出很聪明的行为。"）

彭罗斯的思想，还遭到反对自由意志的克里克的针砭。因为，通过单纯的内省，彭罗斯不可能追溯出他觉知数学真理的计算逻辑，他坚持认为知觉必定是源于某种神秘的、非计算性的现象。但是，正如克里克所指出的，仅仅因为我们未觉察到导致某一决策的神经过程，并不意味着那些过程就不存在。而人工智能的支持者则批驳了彭罗斯的哥德尔论据，他们坚信，人们总能设计出这样一台计算机，它为了解决新问题能自动扩展本身的基础公理系统；事实上，这种能学习的算法系统是极为常见的（尽管它们与人的心智比较起来，仍然是极其原始的）。[15]

某些彭罗斯的批评者指责他是位活力论者，在其内心深处，他并不希望心智之谜能被科学解开。但假如彭罗斯真是一位活力论者，他肯定会使自己的观点尽量保持一种模糊的、不可检验的状态，他将永远不会揭开其微管"海怪"的面纱。彭罗斯是一位真正的科学家，他想知道；他虔诚地相信我们目前对实在的认识是不完备的，存在着逻辑上的缺陷，并且它还具有某种神秘色彩。他正在寻找一把钥匙，一种顿悟，一些使所有问题豁然开朗的准量子把戏；他正在寻找那个"终极答案"。他最大的失误在于：他竟认为单凭物理学就足以理解整个世界，就足以赋予世界以意义。史蒂文·温伯格本应告诉他：物理学没那本事。

来自神秘论者的非难

尽管彭罗斯把意识理论远远地推到现代科学的地平线之外，但他至少还提供了一种希望，认为我们总有一天能够抵达那一理论；但某些哲学家却已经在质疑：是否任何纯粹的唯物主义模型——包括传统的神经过程模型，以及彭罗斯构想的奇异的、非决定论机理的模型——都不可能真正阐明意识的现象。哲学家欧文·弗拉纳根（Owen Flanagan）称这些怀疑者为"新神秘论者"，沿用的是20世纪60年代的摇滚乐组合"问号与神秘论者"（这一组合曾演出过主打歌曲"96滴眼泪"）的用法。（弗拉纳根本人却不是神秘论者，而是位地道的唯物主义者。）[16]

哲学家托马斯·耐格尔（Thomas Nagel）在其1974年发表的题为"作为一只蝙蝠的感受是什么？"的著名论文中，对神秘论者的观点给出了最清晰的表述（一种矛盾的表述？）。耐格尔假定，对于人类以及许多高等动物（比如蝙蝠）来说，主观经验是其最基本的特征。"毫无疑问，它[①]以我们所无法想象的、难以计数的形式发生着，它也发生于其他星球上、其他太阳系里，遍及整个宇宙。"耐格尔写道，"但不论其形式怎样变化，一个生物体具有意识经验这一事实，归根结底是意味着：在本质上，存在着某种东西使得这个生物体成其为这个生物体。"[17]耐格尔认为，不论我们掌握了多少关于蝙蝠的生理学知识，都无法真正了解作为一只蝙蝠的感受，因为科学无法洞悉主观经验的王国。

有人也许会称耐格尔为一名弱神秘论者（weak mysterian），因为他仍然维持着这样一种可能性，即认为哲学和／或科学也许有一天终能揭示出这样一条自然的途径，它可以跨越我们的唯物主义理论与主观经验之间的

---

① 指主观经验——译者注。

鸿沟。而科林·麦金却是一位强神秘论者（strong mysterian），同为哲学家，他也相信绝大多数重要的哲学问题是不可解答的，因为这已超出了我们的认知能力范围（参阅第二章）。就像老鼠有其认知限度一样，人类也不例外，我们的限度之一就是我们不可能解决心—身问题。麦金认为他关于心—身问题的主张（即它是不可解的），正是耐格尔在《作为一只蝙蝠的感受是什么？》一文中所做分析的必然逻辑结论。麦金认为，他的观点比被他称为"取消论者"的主张更优越，后者试图证明心—身问题根本就不是个问题。

麦金认为，对于科学家来说，发明一种能精确地预见实验结果，并能产生巨大的医学效益的关于心智的理论，是十分可能的，但一种有效的理论不一定就是可理解的理论。"并不存在什么真正的理由，使得我们的部分心智不能发展出一种具有这些显著预见性的形式体系，但我们却无法用理解事物的那部分心智来理解上述形式体系。因此，仅就意识来说，我们也许能够在这一方面提出一种类似于量子论的理论，一种确实优秀的意识理论，但我们却不能解释它，或者理解它。"[18]

这类论调激怒了塔夫茨大学的一位叫丹尼尔·丹尼特的哲学家。丹尼特身材高大，留着花白的胡须，总带着一副有感染力的、爽朗的快乐表情，简直就像是一位减过肥的圣诞老人。丹尼特是麦金所谓的取消论者主张的典型代表，在其1992年的著作《意识释义》中，丹尼特认为，意识以及我们所拥有的统一的自我感，都只不过是运行于大脑硬件中的诸多不同"子程序"彼此相互作用而产生的幻象。[19]当我请教丹尼特对麦金的神秘论观点的看法时，他称之为"滑稽可笑的"；他还诋毁麦金关于人与老鼠的类比，认为：老鼠与人不同，它们不能想出科学问题，所以，它们当然也就不能解决科学问题。丹尼特怀疑麦金以及其他神秘论者"并不希望意识被科学攻克，他们更乐意认为意识超出了科学的限度。除此之外，无法解释他们为什么会欢迎这样一种邋遢的观点"。

丹尼特尝试了另外一种策略，一种对于像他这样公认的唯物主义者来说，显得过于柏拉图化的策略，同时，对于一位作家来说，也是一种危险的策略。他忆及博尔赫斯在其小说《巴别图书馆》中，曾想象出一个庞大的图书馆，里面收藏着所有可能的思想，包括那些已经存在的、将要存在的以及可能存在的思想，从最荒谬的到最崇高的。丹尼特说，在"巴别图书馆"的某处，肯定存放着一个被完美表述的关于心—身问题的答案。他在提出这一论点时的态度是如此的自信，以至于我都怀疑他相信巴别图书馆确实存在。

丹尼特承认，神经科学永远也不可能产生出一种人人都满意的意识理论。"我们对于任何事情的解释都不可能令所有人满意。"他说。许多人对于科学给出的解释，比如说关于光合作用或者生物繁殖的解释，并不满意，但"对光合作用或繁殖的神秘感却消失了"，丹尼特说，"我认为，对于意识，我们最终也会有一种相似的解释"。

很突兀地，丹尼特一下子又转向一个完全不相干的方向。在现代科学中"存在着一个愈益清晰的悖论"。他说："使科学近来得以迅速发展的时尚之一，正是那种使科学变得越来越让人难以理解的倾向。当你从试图用优美的公式来模拟事物，转向从事大量的计算机模拟时……你可能以获得这样一种模型而告终：它能完美地模拟你所感兴趣的自然现象，但你却不理解这一模型；也就是说，你无法用从前理解模型的方式来理解现在这个模型。"

丹尼特指出，一个精确地模拟了人脑的计算机程序，对我们来说，可能会像大脑本身一样不可思议。他注意到："软件系统也接近了人类理解能力的边缘。即使像互联网（Internet）这样的系统，与人脑比起来仍是微不足道的，然而它已被逐渐增补和扩充到如此庞大的程度，以至于没有人能真正清楚它是怎样运行的，或者它是否会继续运行下去。因而，当你开始运行软件写作程序、软件修改程序和自我修复指令时，你创造了一些具有其自身生命的人工制品，它们变成不再受其创造者认识论支配的客观物，成了以光速运行的某种东西，变成科学将不断跌扑于其上的一道障碍。"

令人惊讶的是，丹尼特似乎正在暗示着他也具有神秘论者倾向。他认为心智理论虽然可能是高度有效的，并且有强大的预见能力，但对于单纯的人类来说，却又不能为人所理解。人理解其自身复杂性的唯一希望，可能就是不再成其为人。"每一个拥有必要的动机和才能的人，"他说，"都能有效地将自己融入这些巨大的软件系统中去。"丹尼特正在谈论的，正是由某些人工智能的狂热分子所提出的那种可能性，即有朝一日，我们人类会抛弃这副终有一死的血肉之躯的自我，变成机器。"我认为这在逻辑上是可能的，"丹尼特补充道，"我不敢保证这种可能性有多大。那将是一个和谐的未来，我认为它并不自相矛盾。"但丹尼特对于这种超级智能机是否有一天能够理解其自身，仍然犹豫难决。若试图去理解其自身，这些机器势必将变得更加复杂，从而陷入螺旋上升的不断增加的复杂性之中，永远只能追逐自己的尾巴。

我怎么知道你具有意识

1994年春，在亚利桑那大学曾举办过一次会议，题为"构建意识的科学基础"。就在这次会议上，我目睹了哲学世界观与科学世界观之间的一场奇特冲突。[20] 会议的头一天里，戴维·查默斯（David Chalmers），一位蓄着长发、气质绝似庚斯博罗①著名油画《蓝衣少年》中的人物的澳大利亚哲学家，用强有力的措辞陈述了神秘论者的观点。他宣称，研究神经元不可能揭示出为什么声波对我们耳朵的撞击能引起我们对贝多芬第五交响曲的主观体验。他认为，所有的物理学理论所描述的，仅是那些与大脑里特殊的物理过程相关的机能，诸如记忆、注意、意向、内省，但这些理论中的任何一种都不能解释：为什么在这些机能的执行过程中要伴随着主观体验。虽然人们完全可以构想出一个在任何方面都与人类相似的机器人世界，但有一点除外，它们不具备关于世界的自觉体验。在查默斯看来，不论神经科学家对大脑的了解有多么透彻，他们都不可能在物理世界与主观世界之间的"鸿沟"上架起一座理解的桥梁。

至此为止，查默斯表述的基本上是神秘论者的见解，与托马斯·耐格尔和科林·麦金的见解大体相同。但接下去，查默斯却又宣称，虽然科学无法解答心—身问题，哲学却仍能做到这一点。查默斯认为他已经找到了一种可能的答案：科学家们应该假定信息与物质和能量一样，是实在的一种基本特性。查默斯的见解，与约翰·惠勒的"万物源于比特"概念相差无几——事实上，查默斯曾公开声明过自己受惠于惠勒的思想——并且，它也同样难以摆脱"万物源于比特"的致命缺陷。除非存在着某个信息加工者——不论是一只变形虫，还是一位粒子物理学家——去接收并处理信息，否则信息概念就没有任何意义。物质和能量创世之初就已存在，但就我们所知，生命却并非如此，那么，信息怎能像物质和能量那么根本呢？但无论如何，查默斯的观点毕竟是打动了听众的心弦。在他的演讲结束后，听众们簇拥在查默斯身边，七嘴八舌地告诉他，自己有多么欣赏他的演讲内容。[21]

至少有一位听众，克里斯托弗·科克，克里克的合作伙伴，对查默斯的言论大为愤慨。科克是个穿着红色牛仔靴的高个头倔汉子，在那天晚上为与会者举行的鸡尾酒会上，科克余怒未息地找上了查默斯，强烈谴责了他的演讲。恰恰是因为对意识的哲学探索全都失败了，科学家们才把注意力集中

① Thomas Gainsborough, 1727—1788，英国画家——译者注。

到大脑之上，当好奇者聚拢起来后，科克用他那机关枪扫射一般且略带点德国味的声音，这样宣布道。查默斯那基于信息之上的意识理论，就像所有的哲学观点一样是无法验证的，因而是毫无用处的。科克继续攻击道："你为什么不直接说就在你有了大脑的时候圣灵便从天而降并使你具有了意识！"因为这样的理论不必复杂化，查默斯生硬地答道，并且它也不符合他自己的主观体验。"但是我怎么知道你的主观体验与我的一样？"科克气急败坏地说，"甚至于我怎么知道你具有意识？"

科克提出了一个唯我论的棘手问题，也是一个蕴含于神秘论者主张的核心之中的问题：没有人真的知道其他存在（人或非人）是否拥有对于世界的主观体验。通过提出这一古老而又费解的哲学难题，科克就像丹尼特一样，也展露出他自己的神秘论者面目。后来科克让我了解了更多的东西：科学所能做的一切，就是要提供一幅与不同的主观状态相关联的物理过程的详图，但科学无法真正解决心一身问题。没有任何经验的、神经病学的理论能够解释：为什么心理机能要伴随特定的主观状态。"我不认为有什么科学能够解释这种现象。"科克说。为着同样的原因，科克也怀疑机器是否能具有意识和主观经验这一问题，科学到底能不能给我们一个明确的答复。"这种争论或许永远也得不到裁决，"他这样告诉我，然后又莫明其妙地补充了一句，"我究竟怎么才能知道你具有意识呢？"

弗朗西斯·克里克虽然要比科克更乐观一些，但他也不得不承认，关于意识的解释可能不是直观易懂的。"在理解大脑时，我认为我们得到的不会是一个常识性的答案。"克里克说。毕竟，自然选择在粗劣地拼凑出各种生物体的时候，并未遵照什么逻辑的设计方案，只不过是应用了各种各样的花招和恶作剧，只要是能奏效的手段全用上了。克里克还提示说，心智之谜不会像遗传之谜那样，会被轻易地揭开谜底，比起基因组来，心智"是一个更加复杂的系统"，而关于心智的种种理论，也许只能具有更有限的解释力。

抓起一支钢笔，克里克解释道，科学家也许能够确定到底是哪一种神经活动与我对这支钢笔的感知有关，"但如果你要问：'你发怒和忧伤的感觉，与我发怒和忧伤的感觉一样吗？'那么，这就是某种你无法与我沟通的东西，所以，我认为我们不能解释觉察到的一切。"

克里克接着说道，仅仅因为心智是经由确定性的过程产生的，并不能意味着科学家们能够预测全部的繁复性状；这些性状也许是混沌的，因而是

不可预测的。"理解大脑或许还会受到其他一些情况的制约，谁知道呢？我
认为没人能预料那么多。"克里克也怀疑量子现象可能在意识中起着重要的
作用，就像罗杰·彭罗斯所主张的那样；克里克又补充说，某些与海森伯
（Heisenberg）测不准原理类似的规律，可能也存在于神经过程中，限制了
我们追踪大脑活动之精确、细节情况的能力；并且，意识的内在过程也许就
像量子力学一样是自相矛盾的，是我们所难以把握的。"要记住，"克里
克说，"当我们还是狩猎者的时候，或者再往前推，当我们还是猴子的时
候，我们的大脑就已进化到了足以处理日常事务的程度。"不错，这正是科
林·麦金、乔姆斯基以及斯滕特的观点。

马文·明斯基的多重心智

最不可能成为神秘论者的神秘论者，就是马文·明斯基，他是人工智能
（AI）的奠基人之一。根据人工智能的观点，大脑只不过是一台特别复杂的
机器，其特性可以用计算机加以模拟。在我去麻省理工学院采访他之前，同
事们曾忠告我说，明斯基是个脾气古怪的甚至极不友善的采访对象，假如我
不想在采访到一半的时候就被逐出门外的话，最好不要过于直接地提问有关
人工智能的黯淡前景，或者是有关其独特意识理论的凄凉境遇等问题。而明
斯基从前的一位同僚，也恳请我不要趁机利用明斯基那令人难以忍受的言辞
大做文章。"要问清他是否真是那么想的，如果他的言论未重复三次以上，
你就不要引用它。"那位前同僚这样劝我。

当我见到明斯基本人时，发现他的确十分急躁，但这种性格更可能是与
生俱来的，而不是习得的。他无休无止地做些小动作，不停地眨巴着眼睛，
不停地抖动着腿，并且不停地把桌面上的东西推来推去。与大多数科学名人
不同，他给人的印象似乎是正在即兴地构思出各种观点和语句来，而不是把
它们整个儿地从记忆中提取出来。他常常才思敏捷，但也并非总是如此，比
如，在一次关于如何证明心智模型的即兴演讲最终瓦解成一大堆支离破碎的
语句之后，他嘟哝道："我现在也不知所云了。"[22]

连他的外貌，也显示出某种即兴创作的模样：其硕大的圆脑袋上乍看
起来寸草未生，但实际上却点缀着一圈像光学纤维一样半透明的头发；腰系
一条镶边的皮带，除了用它系住裤子之外，上面还挂着一个腰包和一个小巧
的手枪皮套，里面插着一把折叠钳。他大腹便便，再加上那形似亚洲人面

相，使他看起来像极了一尊弥勒佛——作为一名患了多动症的黑客而转世的弥勒佛。

明斯基看来似乎不能，或者是不愿意，长时间保持某一种情绪。刚开始的时候，正如事先所预料的那样，他用自己的言行坐实了其作为一个"乖戾的家伙"的形象，一个"极端的还原论者"的名声。他表示了对于那些怀疑计算机能具有意识的人的轻蔑，认为意识只是个无关紧要的问题。"我已经解决了它，但我不理解人们为什么不愿听我解释。"意识仅仅是一种短期记忆，一种"保持记录的初级系统"。某些计算机程序，比如说LISP，它具有能回忆其运行步骤的特征，这类程序"绝对是有意识的"，甚至远比人的意识力更强，因为人只有可怜的、肤浅的记忆存储单元。

明斯基称不愿接受其自身物质性的罗杰·彭罗斯为"懦夫"，并嘲笑杰拉尔德·埃德尔曼的"折返回路"假说只是回锅的反馈理论。明斯基甚至指责麻省理工学院的人工智能实验室，而该实验室正是由他建立起来的，也正是我俩当时会面的地方。"我认为目前这个实验室还算不上是个严谨的研究机构。"他宣称。

然而，当我俩穿行于实验室中寻找关于会下国际象棋计算机的讲座时，他的情绪却大为改观。"下棋会议不是安排在这儿吗？"明斯基向一群正在休息室中闲聊的研究人员发问道。"那是昨天的事了。"有人回答。在问了几个有关演讲情况的问题后，明斯基讲述了一个关于弈棋程序的历史典故。这一微型演说最终变成了对他的朋友，刚刚过世的伊萨克·阿西莫夫（Isaac Asimov）的回忆。他回顾了阿西莫夫——正是他普及了"机器人"（robot）这一术语，并在其科幻小说中探讨了机器人的形而上学含义——是怎样拒绝了他的邀请，一直不愿意到麻省理工学院来参观正在制造中的机器人，因为阿西莫夫担心他的想象力"会被这些令人讨厌的现实所压抑"。

休息室中的一个人发现他自己和明斯基都带着同样的钳子，便猛然地从皮套中拔出了他自己的装备，手腕一抖，使折起的钳口弹出到工作位置。"咳！注意了！"他喊道。明斯基一边咧嘴笑着，一边也拔出了他的武器。于是，他和挑战者便不停地开合着折叠钳子去钳对方，就像朋客们正在练习玩弹簧小刀的技巧。明斯基在对抗中详细讲述了折叠钳的多功能性和局限性（后者对他来说十分重要）；他的对手乘机耍了些花招夹痛了他。"你能用折叠钳将其自身卸开吗？"有人这样问了一句。这句触及机器人学中的基本问题的妙语，使得明斯基和他的同伴会心地大笑了起来。

后来，在返回明斯基办公室的途中，我们碰到了一个因怀孕而显得大腹便便的韩国少妇，她是一位参加博士生复试的考生，正在为第二天的口试做准备。"紧张吗？"明斯基问道。"有点儿。"她回答。"别紧张。"他一边劝慰着，一边温和地将自己的额头贴在对方的前额上，仿佛要把自己的力量注入她的体内。看到这一幕，我才猛然意识到：明斯基其实也有许多个侧面。

事实上也理应如此，因为多重性正是明斯基的心智思想之核心。在其《心的社会》一书中，明斯基宣称大脑在进化中形成了许多不同的、高度特异性的结构，以解决不同的问题。[23] "我们的学习机制是由多重神经网络构成的，"他向我解释道，"其中的每一个层次都进化得足以修复自身出现的故障，足以在思考问题时与其他层次相配合。"因而，不大可能把大脑还原为一组特殊的规则或公理，"因为我们的大脑所处理的是一个真实的世界，而不是由公理定义出来的数学世界"。

如果说人工智能没能实现其早期的诺言，明斯基认为那只是因为现代的人工智能研究者已经屈从于"物理学的妒忌"——将大脑的错综复杂性还原成简单公式的奢望。令明斯基大为烦恼的是，这些现代的研究者竟然毫不理会他的指教：哪怕是处理一个单独的、相对简单的问题，人的心智也有许多不同的办法。举例来说，假如某人的电视机"罢工不干活了"，这首先可能会被看作是一个单纯的物理学问题，他或她会查看一下电视台是否在正常播放，或者电源插头是否已经插上；如果这些措施都无济于事，这个人会考虑把电视机送出去修理，从而把这个物理学问题变成了一个社会学问题——找谁修理这台电视机费时最少、费用最低。

"这是我无法向那些人讲清楚的经验之谈，"明斯基指的是他那些从事人工智能研究的同行，"在我看来，大脑所解决的问题，在某种程度上说就是在单独的方法难以奏效时，怎样把不同的方法组合起来，以解决特定的问题。"明斯基宣称，除了他以外，唯一一位真正把握住了心智的复杂性的理论家已经死了。"关于心智是怎样产生的，弗洛伊德（Freud）提供了迄今为止仅次于我的最好的解释。"

在明斯基接下来的谈话中，他对多重性的强调渐渐染上了一种形而上的甚至是道德的色彩。他把自己研究领域里的种种问题，以及在更普遍意义上科学的种种问题，统统归咎于他所谓的"投资原则"（the investment principle）。按照他给出的定义，"投资原则"是指人类只愿意做自己善于做的事情，而不愿意转而研究新问题的倾向。重复，或者更确切地说是执着于单一目的，似乎是件令明斯基所深恶痛绝的事情。"如果有某些事情使你特别喜欢，"他宣称，"那么你不要以为这是件好事，而应把它看成是大脑内的毒瘤，因为这意味着你的一小部分心智已经发现了你的弱点，它们可以随时关闭你意识中所有其余的部分。"

在其职业生涯中，明斯基之所以能掌握如此众多的技艺——他在数学、哲学、物理学、神经科学、机器人学以及计算机科学等诸多领域都堪称大家，还写过好几本科幻小说——是因为他已学会怎样在学习新东西所引发的"窘迫感"中体会到一种乐趣。"那种手足无措的感觉让人特别兴奋，就像人们发掘宝藏时的那种热切体验，并且这种感觉绝不会持续太久。"

明斯基还曾是个音乐神童，但后来他却断然宣称音乐只是催眠药。"我认为人们之所以喜欢音乐，是为了用它来压抑思想，尤其是压抑那些邪恶的思想，而不是为了用它来激发自己的思想。"明斯基仍然发现自己只要一不小心就能创作出"巴赫①式的曲子"——他的办公室里就摆放着一架电子琴——但他却总是抑制自己的音乐冲动。"在某些情况下，我不得不扼杀那个作为音乐家的我，他总要时不时地冒出头来，而我只好每次都予以迎头痛击。"

对于那些宣称心智太过微妙以致不可能被理解的人们，明斯基显得非常不耐烦："想想看，在巴斯德（Pasteur）之前，人们也曾说过，'生命是千差万别的，你不可能机械地解释它。'道理是一样的。"但是，关于心智的终极理论，明斯基强调，将比物理学的终极理论更加复杂，但他相信它也是可以达成的。他说，所有的粒子物理学内容可能会被压缩成一页纸的方程式，但要描绘出心智的所有成分，却需要占用更多的篇幅。毕竟，考虑一下要描述一辆汽车，甚至描述一个简单的火花塞，需要占用多长的篇幅就清楚了。"仅仅是解释怎样将油槽对准并焊接在陶瓷材料上，才能使汽车启动时火花塞不漏油这一点，就需要相当厚的一本书。"

明斯基认为，一个心智模型是否正确，可以通过几种途径来证明：首先，一台基于该模型原理之上的智能机，应该能够模拟人类的成长，"这台机器开始时应像个婴儿一样幼稚，随后，他会通过看电影和做游戏而逐渐成熟起来"；其次，随着成像技术的提高，科学家们应能确定活人的神经过程是否与模型一致。"在我看来，一旦拥有了1埃（一百亿分之一米）分辨率

---

① 　J. S. Bach，1685—1750，音乐史上最伟大的德国作曲家之一——译者注。

的（大脑）扫描仪，那么你就可以看清某人大脑中的每个神经元，这是完全合理的。你可以这样观察它1000年，然后说：'好了，我们已确切地知道每当这个人说"忧郁"这个词时，大脑里到底发生了怎样的变化。'然后，人们再用几代人的时间对此进行检查核实，理论确实是完美无瑕的。一切都正确无误，事情就此告一段落。"

如果人类获得了心智的终极理论，那么，还剩下什么前沿领域可供科学探索呢？我问道。"你为什么要问这个问题？"明斯基反问道。关于科学家们会穷尽所有可研究的事物，从而变得无事可做的顾虑，纯粹是杞人忧天。他说："有大量的事情可做。"我们人类作为科学家也许能够达到自己的限度，但有朝一日我们会创造出比我们更聪明的机器来，它们会将科学事业继续发展下去。但那将是机器的科学，而不是人类的科学，我争辩道。"也就是说，你是个种族主义分子"，明斯基亢声说道，同时，他那呈穹顶状的硕大前额竟然急成了酱紫色。我扫了一眼他脸上的表情，想找出开玩笑的迹象来，但没找到。"我以为对我们来说，最重要的事情是继续成长，"明斯基自顾自说道，"而不是要维持目前这种愚蠢的状态。"我们人类，他补充道，只不过是"披上了衣服的黑猩猩"，我们的任务不是要维护现状，而是要继续进化，去创造出比我们更优秀、更聪明的生命来。

但令人惊讶的是，明斯基却又难以确切说出这些伟大的机器将对何种问题感兴趣。与丹尼尔·丹尼特相呼应，明斯基兴致极为索然地提议说，机器们在向更加复杂的实在进化的过程中，也许会试着去理解它们自己。他似乎更热衷于讨论将人类的个性转变成计算机程序的可能性，那样的话，就可以将人类个性的程序装入机器。明斯基将个性程序的下载（down leading）看作是一种通常被他认为是十分危险的追求，就像服食LSD[①]或沉湎于宗教信仰一样。

明斯基坦白地说，他很想知道马友友[②]这位大提琴演奏家在演奏一首协奏曲时会有什么感觉，但他又怀疑这样一种体验到底是否可能实现。他

①　麦角酸酰二乙胺，一种致幻药，俗称"摇头丸"——译者注。

②　马友友（Yo-Yo Ma，1955.10.7—）：大提琴演奏家，出生于法国的华裔美国人，曾获得多个格莱美奖。马友友曾为多部电影配乐，其中包括布莱德·彼特主演的电影《西藏七年》，李安导演的《卧虎藏龙》；也曾在美国总统奥巴马的就职典礼中演出。1998年，马友友正式创建了音乐组织丝路计划（Not-For-Profit Silk Road Project），以其音乐家的浪漫情怀和使命感，致力于用音乐增进文化的融合和人类的相互理解。——译者注。

解释道，为了分享马友友的体验，他就必须拥有马友友全部的记忆，他就不得不变成马友友；但是一旦变成了马友友，明斯基怀疑，他可能就不再是明斯基了。

能够坦白地承认这一点，对于明斯基来说，是非常了不起的。像那些宣称文本的唯一正确阐释就是文本自身的文学批评家一样，明斯基是在暗示着：我们的人性是不可还原的；任何将个性转化成一种抽象的数字程序——一串由"0"和"1"组成的数字，可以被装进磁盘，被从一台机器传输到另一台机器，或者被合成为代表另一个人的另一个程序中——的企图，都会彻底地破坏个体的本质属性。明斯基用自己独特的方式，在提示着那个"我怎么知道你具有意识"问题是难以克服的。如果永远也不可能把两个人的个性融为一体，那么，所谓"装入机器"云云，也就是不可能实现的了。事实上，人工智能的整个前提——如果"智能"是在"人"的意义上被定义的——也就成了空中楼阁。

尽管作为一个狂热的还原主义者的明斯基早已声名在外，但他实际上却是个反还原主义者。他是一个独具特色的浪漫主义者，甚至比罗杰·彭罗斯更加浪漫。彭罗斯给出的希望是：心智可以被还原为简单的准量子恶作剧；明斯基却坚持认为，任何诸如此类的还原都是不可能的，因为多重性是心智——所有的心智，不论是人类的还是机器的——的本质属性。明斯基对于专一性、对于简单性所表现出的厌恶情绪，在我看来，从中折射出的不仅仅是一种科学的判断力，还有某种更深刻的内容。明斯基就像保罗·费耶阿本德、戴维·玻姆以及其他一些知名的浪漫主义者一样，似乎对于"终极答案"，对那终结一切启示的启示，怀有莫名的恐惧。对明斯基来说，幸运的是神经科学中似乎不可能产生这样的启示，因为正如他所认识到的，任何关于心智的有效学说都将是极繁复的；但不幸的是，考虑到这种复杂性，对于明斯基本人甚或是他的子孙来说，都不再有希望能目睹具有人类属性的机器人的诞生。万一我们真的造出了自主的智能机，它们肯定是与我们格格不入的异类，它们与我们不同，恰似一架波音747与一只燕子的不同，并且，我们永远也无法确信它们具有意识，就像我们任何人都无法确信别人具有意识一样。

培根解决了意识问题吗

征服意识问题是一项历时久远的任务，因为大脑是异常复杂的。但它是无限复杂的吗？根据神经科学家们现在研究大脑的速度，在几十年的时间里，他们就能给出关于大脑的高度有效的示意图，一个将特定的神经过程与特定的精神机能对应起来的示意图，包括像克里克和科克定义的意识在内。这些知识将带来许多实际效益，诸如用于精神病的治疗，以及可转用于计算机上的信息加工方法。在《黄金时代的来临》一书中，岗瑟·斯腾特曾预言，神经科学的进步或许有一天会赐予我们超越自身限度的力量，我们也许能够"有选择地向大脑输入可控制的电信号，这些输入信号可用以产生人为的感觉、情感和情绪……凡胎肉体的人们很快就会活得像神仙，无忧无虑，只要他们的愉快中枢被正确地接通电流"。[24]

但斯腾特先于耐格尔、麦金等人的神秘论一步，还预言："归根到底，大脑也许无力给出对其自身的解释。"[25]科学家和哲学家为了完成这不可能完成的事业，将会继续努力下去，他们将以一种后经验的、反讽的做派，确保神经科学持续下去。参与这一伟业的实践者们争论着各自物理模型的含义，一如物理学家们争论着量子力学的含义。时不时地，还会冒出一种特别富有号召力的理论，由某些沉浸于神经知识以及控制论知识中的当代弗洛伊德提出，能吸引大批的追随者，并气势汹汹地宣称自己将要发展成心智的终极理论；于是，又会有更新一代的神秘论者脱颖而出，指责这一理论所难以避免的缺陷：它能提供关于梦或者神秘体验的真正令人信服的解释吗？它能告诉我们变形虫是否具有意识吗？计算机呢？

有人也许会争辩说，只要有人能确定意识只不过是物质世界的附带现象，那么意识问题就被"解决"了。克里克那坦率的唯物主义观点，与英国哲学家吉尔伯特·赖尔（Gilbert Ryle）的主张正好同声相和，后者在20世纪30年代曾造出"机器中的幽灵"这一短语，用以讥讽二元论。[26]赖尔指出，二元论——主张心智是独立的现象，不依赖于肉体并能对肉体产生影响——违背了能量守恒定律，因而也就违背了所有的物理学规律。在赖尔看来，心智只是物质的一种属性，只有通过追踪大脑内错综复杂的物质变化过程，人们才能"解释"意识现象。

赖尔并非第一个提出这种既曾盛极一时，也曾湮灭无闻的唯物主义范式的人。早在16世纪以前，弗朗西斯·培根就曾力劝其同时代的哲学家们，要他们放弃证明宇宙怎样从精神中演变而出的企图，而应该去考虑精神是怎样从宇宙中演变的。[27]在这一点上说，培根早用现代进化论的术语，从更大的范围来说，用现代唯物主义的范式，给出了关于意识的现代解释。科学对意识问题的征服，将成为人类最终的一项祛魅的事业，并再度成为尼尔斯·玻尔的如下宣言的明证：科学的工作就是化神奇为平凡。但人类的科学将不会，也不可能会，解决"我怎么知道你具有意识"这一难题。也许只有一条途径能够解决它：将所有的心智铸成单一的心智。

第八章
混杂学的终结

我怀念里根时代。罗纳德·里根（Ronlad Reagan）使得道义的和政治的抉择变得特别简单，他赞成什么，我就反对什么。比如星球大战计划，正式的名称应为"战略防御计划"，这是里根的一个庞大计划，它的目标在于建立一套空间防卫体系，用于保护美国免遭苏联核弹的袭击。就此我写过许多文章，在这些文章中最让我感到困惑的是有关戈特弗里德·迈耶·克雷斯（Gottfried Mayer-Kress）的那一篇。这位曾经在原子弹的诞生地——洛斯阿拉莫斯国家实验室工作过的物理学家，利用"混沌"数学建立了一个计算机模型，对美苏两国的军备竞赛作了模拟。其模拟结果表明：星球大战将打破这两个超级大国间的力量平衡，并极有可能导致巨大的灾难——核战争。我曾写了一篇报道赞扬他的工作，这不仅因为我赞同迈耶·克雷斯得出的结论，更因为他的工作岗位本身为这件事增加了一点有趣的讽刺意味。当然，如果迈耶·克雷斯的模拟结果赞许星球大战这一设想，那么我会毫不犹豫地予以批判，因为显而易见这是在胡说。星球大战计划不可能不打破两个超级大国间的力量平衡，难道我们真的需要一些计算机模型来告诉我们这一点吗？

我并没有诋毁迈耶·克雷斯的意思，他是一个抱定美好愿望的科学家。（就在我报道迈耶·克雷斯关于星球大战研究工作几年之后的1993年，我看到一份来自伊利诺伊大学的材料，那时迈耶·克雷斯就在伊利诺伊大学工作，他在那份材料中宣布了根据其计算机模拟结果提出的解决波斯尼亚和索马里争端的方案。）[1] 我想说明的是，耕耘在混杂学领域的人们所做的工作包罗万象，无所不涉，而他的工作仅是其中之一。我之所以用混杂学（Chaoplexity）这个词，是它既指混沌，也指它的近亲——复杂性。这两个术语被定义时，往往都是界定清晰但又带有明显的定义者的个人特色，定义混沌尤其如此。然而，这两个术语被数不胜数的科学家和记者用许多相互交叉的方式加以定义，因此它们实际上是指同一个东西，而且这个定义并非是毫无意义的。混杂学便是这样一个术语。

混杂学这个领域成为风靡一时的文化现象，源于1987年《混沌——创建新科学》一书的出版，该书作者詹姆斯·格莱克是前《纽约时报》记者，他的这部力作一经出版即成为畅销书，以后又有数十位新闻记者和科学家追随他的成功，就类似的主题写了许多本类似的书。[2]关于混杂学有两点多少是有些矛盾的。许多现象是非线性的，从传统上来说无法预测它们，因为任意小的影响都有可能导致巨大的无法估量的后果。爱德华·洛仑兹（Edward Lorenz），这位麻省理工学院的气象学家，研究混沌—复杂性的先驱者之一，称这种现象为蝴蝶效应。因为它意味着如果一只蝴蝶在爱荷华州上空扇一下翅膀，原则上将有可能产生数不清的影响并最终在印度尼西亚引发一次季风。正是由于我们不能获得更多的关于天气系统的知识，人类预测天气变化的能力极其有限。

洛仑兹的这一见解，实际上不过是在拾人牙慧。昂利·庞加莱（Henri Poincaré）早在20世纪初就警告说："初始条件的微小差异，将会导致最终现象上的巨大不同；前者的一个很小的错误，将使后者的错误无法估量。因此，预测是不可能的。"[3]研究混杂学的专家——我称之为"混杂学家"——也喜欢指出自然界中的许多现象是"涌现的"（emergent），它们展示了事物的某些仅仅通过查明其所在系统的各个局部仍不能被预测和理解的特征。"涌现性"（emergence）也是一个古老的概念，它至少可以追溯到19世纪，与当时的整体论、活力论及其他一些反还原论的教条有关。当然，达尔文并不认为自然选择可以从牛顿力学中推导出来。

关于混杂学的负面论述到此为止。下面我们来看看混杂学的正面论述：计算机技术的长足发展和复杂的非线性数学计算技术的不断进步，将帮助现代科学家理解混沌的、复杂的、涌现的现象，而这些现象用过去的还原论方法是无法加以分析和解释的。《理性之梦》是关于复杂性这门"新科学"最有影响的著作之一，其作者海因茨·佩格（Heinz Pagels）在该书封底上这样评价这门新科学："正如望远镜为人类揭开宇宙奥秘，显微镜引导人类在微观世界探幽掠胜，计算机则正向我们开启一扇激动人心的新窗口，透过它人类将洞悉自然界的本质。利用计算机处理复杂现象的能力，人类第一次模拟实在，建立关于复杂系统的模型，如大分子、混沌系统、神经网络、人体和脑，以及进化模式和人口模型。"[4]

很大程度上，这一憧憬来源于人们对简单的数学指令集的观察：一经计算机运算执行，往往可以得到充满幻想、错综复杂但依旧非常有序的结果。约翰·冯·诺伊曼也许是第一个认识到计算机具有这种能力的科学家。在20世纪50年代，约翰·冯·诺伊曼发明了元胞自动机，它最简单的形式是将一个屏幕分割成许多元胞网格或者小四方块，并建立一套与颜色和状态有关的规则，约束每一个元胞以及与它相邻的元胞；这样，单个元胞的状态发生变化，便可引起整个系统的一连串改变。"生命"产生于20世纪70年代初，由英国数学家约翰·康威（John Conway）创造，是到目前为止最著名的一种元胞自动机。尽管大多数的元胞自动机都着力解决那些可预测的周期性行为，但是利用"生命"却造就了无限变化的模式——甚至包括那些卡通式的物体，也似乎参加了这些神秘的使命。受康威奇特的计算机世界所引发的灵感的刺激，大批科学家开始利用元胞自动机来模拟各种物理和生物过程。

计算机科学派生出的另一个产物是芒德勃罗集，它同样紧紧抓住了科学界的神思。芒德勃罗集以IBM的应用数学家伯努瓦·芒德勃罗（Benoit Mandelbrot）的姓氏命名，而芒德勃罗正是格莱克《混沌》一书的重要人物之一。（也正是基于芒德勃罗关于不确定现象的工作，冈瑟·斯滕特得出结论：社会科学将永远达不到它所期望的目标。）芒德勃罗发明了分形，用来描述一类在数学上具有分数维特征的对象：它们比直线模糊，具有更多的分叉，但从来不能真正填满平面。分形揭示了这类自然现象在愈来愈小的尺度上都具有自相似性的特征。芒德勃罗在创造了"分形"这一概念后，随即指出许多真实世界的现象都具有分形特征，比如云彩、雪花、海岸线、股市涨落和树木，等等。

实际上芒德勃罗集本身也是分形的一个例子。这个集相当于一个简单的数学函数被反复迭代，每一次在得到该方程的一个解以后，就代回方程再对它求解，如此无穷反复。可以用计算机将由这一函数簇生成的数绘制成著名的芒德勃罗图：它既像一个布满芽苞的心脏，又像一只烧焦的小鸡，或者有八个小肿瘤分布在其各个边上的一个肿瘤状的物体。如果你用计算机放大该图，就可发现它的边界并不是光滑的线条，而是像火焰的边缘一样在闪动。不断地放大这些边界将使你置身于巴洛克幻象艺术的无穷无尽和变幻莫测之中。芒德勃罗图中的某些模式，如基本的心形，总是重复出现，但每一次出现又都表现出一些细小的差别。

芒德勃罗集这个"数学中最复杂的对象"，现在已经成为数学家的实验工具，用来检验与非线性系统（或混沌系统，或复杂系统）的行为有关的设想。但是芒德勃罗的这些发现与真实世界有什么关联呢？芒德勃罗在其1977年发表的杰作《大自然的分形几何》一书中警告说：我们在观察自然界中的分形模式的同时，切不可忘了要尽量去确定产生那个模式的原因。芒德勃罗指出：虽然对自相似性结果的探索"显得惊心动魄，而它也确实在帮助我理解自然界的精细结构"，但是要想揭示自相似性的原因则"希望渺茫"。[5]

芒德勃罗似乎是在暗指隐藏于混沌—复杂性含义下的一个诱人的三段论：当用一套简单的数学规则在计算机上产生一个极端复杂的模式时，模式的样式从不自我重复；而自然界也包含许多极端复杂的模式，它们也从不在样式上自我重复；因此，在自然界的许多极端复杂的现象之下，必然有某种暗含的简单规律在起作用，而混杂学家能够在计算机的强有力帮助下，挖掘出这些暗藏的规律。当然，自然现象之下确实存在简单的规律，只不过这些规律已经体现在量子力学、相对论、自然选择和孟德尔遗传学中。然而，混杂学却认为仍有大量威力强大的规律有待发现。

31味复杂性

红色和蓝色的斑点在计算机显示屏上飘忽来去。然而，它们不仅仅是一些彩色斑点，它们代表被模拟的人正在做着一些真实的人最起码要做的事情：觅食、求偶、竞争、合作，等等。这一计算机模拟程序的创造者乔舒亚·爱泼斯坦（Joshua Epstein）是如此宣称的。爱泼斯坦是一位来自布鲁金斯学院的社会学家，他在自己做访问学者的圣菲研究所，向我和另外两位记者展示了他的模拟。圣菲研究所始建于20世纪80年代中期，并迅速成为研究复杂性的中枢。他们标榜自己是研究混沌这门新科学的成功典范，而混沌，在他们看来，或许会最终超越牛顿、达尔文及爱因斯坦等还原论者的"陈词滥调"。

当我和我的同行们注视着爱泼斯坦的彩色斑点，并聆听着他对这些斑点的运动所做的更丰富多彩的解说时，我们有礼貌地连声喃喃着，以表示自己对其工作的兴趣；但是在其背后，我们却只好相视苦笑，因为我们当中没谁会把这类玩意儿当回事。我们全都明白，含蓄一点讲，这只是反讽的科学。而爱泼斯坦自己在逼问下则公然声明他的模型无论如何都不是在预测，他称之为一个实验室、一个工具，或者一个人工神经装置，用于探索人类社会进化的过程。（而这些，正是圣菲人所津津乐道的术语。）但在公开展示其工作的时候，爱泼斯坦却曾经宣称，类似于他的这类模拟，必将引发社会科学的革命，并且有助于解决这些学科中众多久悬未决的问题。[6]

另一个崇尚计算机威力的人是约翰·霍兰（John Holland），他是同时供职于密歇根大学和圣菲研究所的计算机科学家。霍兰是遗传算法的发明者，而所谓遗传算法，是计算机代码中的片段，它们能自身重组以生成一个新的程序，以便更有效地解决问题。据霍兰自称，该算法实际上就是进化的，与活体生物体内的基因在自然选择压力下的进化是一个道理。

霍兰认为，以诸如体现在其遗传算法中的那类数学技巧为基础，完全可以建立一个"复杂适应系统的统一理论"。他在1993年的一次演说中这样描述自己的憧憬——

许多令我们颇感棘手的长周期问题，比如说贸易不平衡、可持续发展、艾滋病、遗传病、精神卫生和计算机病毒等，都居于其各自所属的复杂巨系统的中心。而产生这些问题的系统，诸如经济、生态、免疫系统、胚胎、神经系统或计算机网络，也与这些问题本身一样呈现出多样化。然而，尽管存在着表面上的诸多差异，这些系统却共享一些显著的特征，以至于我们可以按照圣菲研究所的观点将它们归为单独的一类，称之为复杂适应系统。这不仅仅是一个术语，它标志着我们的一种认识，即存在某些一般性的规律，它控制着这些复杂适应系统的行为；而这些一般性的规律，则指明了解决那些伴随问题的方向。我们目前所做的大量工作，其目标就是要将我们的这种认识变为现实。[7]

这段陈述所表现出来的雄心壮志，的确令人感到惊心动魄。混杂学家们常常嘲笑粒子物理学家的傲慢，原因是他们竟幻想着创造一种能够解释一切的理论。但事实上，粒子物理学家在实现其抱负方面却表现得相当谨慎，他们仅希望能够用一个小小的包裹装下自然界所有存在的力，这样他们或许就能描绘出宇宙的起源。但是很少有人像霍兰等人这样，如此大胆地奢望其统一理论能够一箭双雕，既揭示真理（即洞悉自然）又获得幸福（解决现存世界的诸多问题）。即便如此，霍兰仍被认为是在复杂性领域从事研究的最谨慎的科学家之一。

但是，如果科学家们在复杂性的定义上都不能达成一致，那么，他们能够达成一个关于复杂性的统一理论吗？学习复杂性的学生试图将自己区别于那些专攻混沌的学生，但这类努力却收效甚微。根据马里兰大学的数学家詹姆斯·约克（James Yorke）的定义，混沌指一类特定的现象，这些现象由一些显然不可预测的方式产生。比如它们都表现出对初始条件的高度敏感、非周期行为，并以一定的模式在不同的空间和时间尺度上再现，等等。（约克应该知道混沌和复杂性的区别，因为正是他在1975年发表的一篇论文中引入了"混沌"这一术语。）而在约克看来，复杂性则似乎是指"任何你想要的东西"。[8]

一个被广为兜售的复杂性定义与"混沌边缘"有关。这个生动的短语，被用在1992年出版的两本书的副标题中：《复杂性：混沌边缘的生命》，作者罗杰·卢因（Roger Lewin）；以及《复杂性：有序和混沌边缘的新兴科学》，作者米切尔·沃尔德罗普（M. Mitchell Waldrop）。[9]（无疑两书的作者都试图用这个短语来表明一种风格，而这种风格似乎正来自复杂性这个领域所研究的实际内容。）混沌边缘的基本概念是：在高度有序和稳定的系统（比如晶体）内，不可能诞生新生事物；完全混沌的或非周期的系统，比如处于湍乱状态的流体或受热气体，则将趋于更加无形。真实的复杂事物，如变形虫、契约贸易者以及其他一些类似的东西，则恰好处于严格的有序和无序之间。

大多数通行的看法都将这一概念记在圣菲的研究人员诺曼·帕卡德（Norman Packard）和克里斯托弗·兰顿（Christopher Langton）的账上。帕卡德作为混沌理论领头人物的经历，使他意识到对概念进行包装的重要性，于是在20世纪80年代末期创造了"混沌边缘"这个重要的术语。在元胞自动机的实验中，他和兰顿得出结论：一个系统的计算潜力，也就是它贮存和处理信息的能力，在其状态介于高度的周期性和混沌性行为之间时达到顶峰。但是同样在圣菲研究所工作的梅拉尼·米切尔（Melanie Mitchell）和詹姆斯·克拉齐菲尔德（James Crutchfield）则报告说，他们自己的实验并不支持帕卡德和兰顿的结论。他们甚至怀疑是否"任何促进全方位计算能力的事物，都是生物有机体进化的重要因素"。[10]虽然仍有少数几个圣菲人在使用"混沌边缘"这一术语（其中值得一提的是斯图亚特·考夫曼），其他大多数人目前拒绝接受它。

复杂性还有其他许多种定义，根据麻省理工学院的物理学家塞思·劳埃德（Seth Lloyd）在20世纪90年代初提供的清单，至少有31种（劳埃德也参与了圣菲研究所的工作）。[11] 这些定义主要来源于热力学、信息论和计算机科学，并往往涉及熵、随机性和信息等概念，而这些概念每一个都被证明是出了名的难以捉摸。所有这些复杂性的概念都有缺陷。举例而言，由IBM的数学家格雷高里·蔡汀（Gregory Chaitin）提出的算法信息理论认为：用描述一个系统的最简洁的计算机程序可以表征该系统的复杂性。但是根据这个判据，一篇由一群猴子敲出的文章将比轮船推进器留下的航迹更复杂——原因是前者更随机，因而更不具有可压缩性——比《为芬尼根守灵》更甚。

这些问题揭示了一个令人尴尬的事实，即从某种媚俗的意义来说，复杂性存在于观察者的眼光里（一如情人眼里出西施）。[12] 研究人员曾多次争论，复杂性是否已经成为鸡肋因而应予以彻底抛弃；而他们同时又不改初衷地认为，这一术语具有太多的公共关系价值。圣菲人经常使用"有趣"作为"复杂"的同义语，但是又有哪个政府机构愿意为建立针对"有趣事物的统一理论"提供研究资助呢？

人工生命之诗

圣菲研究所的成员或许不会在他们研究的内容上有一致意见，但是他们却使用同一种方法，即借助于计算机进行研究。克里斯托弗·兰顿为了表现他对计算机的忠诚，发起了一场旨在提高混沌和复杂性地位的运动。他认为，在计算机上运行的对生命的模拟就是——既不是某种类型的，也不是某种程度的，更不是比喻意义上的，而是实实在在的——鲜活的生命。兰顿是"人工生命之父"，人工生命是混杂学的分支领域，它吸引了混杂学领域里许多人的注意。兰顿曾经组织了几次人工生命会议，其中第一次会议于1987年在洛斯阿拉莫斯召开；而这几次会议的参与者中，既有生物学家，也有计算机科学家和数学家，他们都和兰顿一样表现出了对计算机动画片的强烈兴趣。[13]

虽然对人工智能的研究早于人工生命几十年，人工生命却是对人工智能的突破。尽管人工智能的研究者追求的是通过计算机模拟思维来认识思维，但是人工生命的倡导者，则希望能在更广的范围内通过他们的模拟洞悉生命现象。然而，正如人工智能更多的只是产生华丽的语言而不是真实的结果一样，人工生命也是如此。1994年，兰顿在一篇为《人工生命》季刊创刊而作的导言文章中写道：

人工生命会教给我们许多生物学知识，许多我们单凭研究生命的自然产物不可能获得的知识，但最终，人工生命一定会超越生物学，进入一个我们至今尚无以名之的王国，在这一王国里的文化和技术，一定会比以往自然发生的那些更广阔。我不想用玫瑰花般的图景来描绘人工生命的未来，它不会解决我们的所有问题，但是不管怎样，它会带领我们前进……或许强调这一点的最简单方式，就是只需指出：当初玛丽·雪莱关于弗兰肯斯坦博士的预言，在今天已不再被认为是科学幻想了。[14]

兰顿的大名，在我与他会面以前即如雷贯耳了，他在好几本关于混杂学的报告文学中都扮演着杰出的角色。无疑，他是一个典型的年轻嬉皮士科学家：既热情开朗又老成持重，长发，穿一条牛仔裤，皮外套，滑雪靴，佩印度珠宝。他有着堪称辉煌的生活经历，其中最耀眼的一点是他在一次滑翔事故中因昏迷而导致顿悟。[15]

1994年5月，我终于在圣菲研究所与兰顿见了面，我们决定在当地一家他偏爱有加的饭店里作一次午餐会晤。兰顿的轿车——难道你不曾在关于他的某一本书里读到过吗？——是一辆伤痕累累的老爷车，塞满了林林总总的杂物，从录音磁带、钳子，到盛着调味汁的塑料容器，所有这些东西都蒙着一层棕色的沙漠灰尘。在我们开车去饭店的路上，兰顿尽职尽责地谈起了混杂学那早已模板化的老一套。自牛顿以来的大多数科学家已经研究了具有周期性、稳定性和平衡态的系统，但是兰顿及其圣菲同事们则试图去理解暗藏在许多生命现象下的"无常的王国"。总之，他说："一旦你达到一个生命体的平衡点，你就完了。"

他咧嘴一笑。这时，外面开始下起雨来。他启动了雨刷，雨刷抹过之处，挡风玻璃立即由透明变得模糊不清。兰顿透过玻璃没有被雨刷抹到的一角瞥向前方，并继续着我们的对话，似乎对挡风玻璃这隐喻信息无动于衷。他说，很明显，科学通过将许多事物打成碎片并进而研究这些碎片，已经取得了巨大的进步。但是这种方法论仅能提供对高层次现象的有限理解，而这些现象实际上已经通过历史事件被放大到很大的程度。人们可以通过一个综合的方法论来超越这些限制，而它要求将现存事物的基本部分以新的方式在计算机中综合起来加以考虑，从而探求可能会发生什么或者将要发生什么。

"结果你将得到一个非常大的可能性的集，"兰顿说，"你能够探测到的不是这个集中已经存在的化合物，而是那些可能存在的化合物。同时也只有以这些可能存在的化合物为基础，你才可能看到规律性的东西。而这些规律性的东西不可能来自对自然界初始提供的很小的集所做的观察。"生物学家利用计算机，通过模拟地球上的生命起源，并通过改变各种条件并进而观察结果来研究机遇在生命进化中的作用。"因此，人工生命的部分内容，以及我刚称之为综合生物学的那个庞大计划的一部分，打个比方说，是在从一个装满自然发生的事物的大信封中，取出人们想要知道的东西。"兰顿设想，用这种方法人工生命或许能揭示出，在历史上发生的事件中，哪些方面是必然的，哪些方面是由偶然造成的。

在饭店里，兰顿一边嚼着辣味鸡肉卷，一边言之凿凿地再度声明，他确实赞同计算机所模拟的生命正是那些活生生的东西本身，他认定这个观点"有着强大的生命力"。他形容自己是一个机能主义者，相信生命由其机能本身而不是由其构成来刻画的。如果一个程序员依一定的规则创造了一个分子式的结构，而这个结构能自发地整合自己成为一个整体，能够摄食、繁殖、进化，兰顿即认为这个整体是有生命的——"即使它们只是存在于计算机中"。

兰顿说他的信念有道德上的后果。"我常常想，如果我看见某人坐在我身旁的计算机终端前折磨这些生命，比如向它们输送一些地狱般的数字，或者只付酬给那些能在屏幕上拼出他姓名的少数幸运儿，我会送这个家伙去接受心理治疗。"

我告诉兰顿，他似乎将隐喻或类比与真实混同起来了。"事实上我正在做的是比这个更具煽动性的事情。"兰顿微笑着回答。他期望人们认识到生命或许是一个过程，能够由物质的任何一种排列生成，甚至计算机内的电子涨落也参与了该过程。"在某个层面上，真实物体的生成与功能性质是无关的。"他说。"当然会有差异"，他接着补充道，"只要有不同的基质，就会有差异，但是这些差异对于活的生命来说是不是根本的呢？"

兰顿并不支持通常由那些人工智能的狂热分子所持有的观点，即计算机模拟本身也具有主观经验。"这就是为什么我更喜欢人工生命而不是人工智能，"他说，与绝大多数生命现象不同，主观状态不能被归结为机械功能，"没有哪一种机械解释可以向你说明此时此地我这个人的心理意识和自我感觉。"换句话说，兰顿是一个神秘主义者，他相信对意识的解释超越于科学之外。他最后承认计算机模拟是不是真的活的生命终究还是个哲学问题，因而是一个不可解的问题。"但是，对于人工生命来说，它们只要能解决自身的问题，帮助扩大生命科学的经验性的基础数据，并增加生物学的感性材料，这就足够了，它们没必要去解决哲学上的问题。事实上生物学家也从来没有真的解决这个问题。"

兰顿说得越多，似乎也越认识到了一个事实，一个他甚至感到庆幸的事实，即人工生命永远也不可能成为真正经验性的科学的基石。他说："人工生命的模拟，迫使我不得不回头重新审视我所做的关于真实世界的假设。"换句话说，人工生命模拟会放大我们的负面能力，模拟本身实际上是在挑战而不是支持关于真实世界的理论。况且，对于那些从古老的还原论方法中得来的东西，从事人工生命研究的科学家们不得不满足于不能"完全理解"的事实。"对于某些特定种类的自然现象，我们除了给出某种解释之外，还能做的至多也不过摇摇头，说句：'嗜，这就是历史。'"

然后他坦白承认，这样的结果对他而言的确很称心如意，他甚至期望宇宙在某种基本意义上是"非理性的"。"理性与科学的传统相结合已经长达300年之久，这时你跳了出来，以关于某物的某种可理解的解释将之终结；如果事情的确就是那样，我会很失望。"

兰顿抱怨他被科学语言的线性所困扰。他说："诗歌，是在非线性地使用语言，诗意绝不是每一语意单位的简单总和。而科学同样要求不仅仅把整体看作部分之和。进一步的事实表明系统确实不是部分之和，这意味着传统的方法，比如仅仅刻画部分和它们之间关系的方法，将不足以抓住系统的本质。这不是说就没有一个比诗歌更科学化的方法来做到它，相反，我总感到从文明发展的角度来讲，在科学的未来将会涌现出更多诗化的东西。"

模拟的限度

1994年2月，《科学》杂志发表了一篇题为"地球科学中数值模型的验证、确认和证实"的文章，论述了计算机模拟所带来的问题。这篇带着明显后现代色彩的文章有三位作者，分别是：达特茅斯学院的历史学家兼地球物理学家内奥米·奥雷斯克（Naomi Oreskes），同属达特茅斯学院的地球物理学家肯尼思·贝利茨（Kenneth Belitz），以及南佛罗里达大学的哲学家克里斯丁·施雷德·弗雷谢特（Kristin Shrader Frechette）。虽然他们在文章中将重点放在地球物理模拟上，但是他们的警告实际上适用于所有的数值模型（比如他们在数周后于《科学》上发表的一封来信中所提到的那些）。[16]

作者注意到数值模型正在迅速成为对许多事情极富影响力的东西，比如用于解决关于全球变暖的争论，探讨石油储备的耗竭和核废料堆置场所的适宜性，等等。他们的论文被视为一个警告："对自然系统数值模型进行验证，并使得模型本身富有效力是不可能的。"能够被验证的，即能够被证明为真的，只能是那些纯粹的逻辑和数学问题；而这些纯粹的逻辑和数学系统都是封闭的系统，在这些系统中所有的成分都基于一个被定义为真的公理。人们公认"2+2=4"，并不是因为这一等式对应于一些外在的真实，奥雷斯克及其同伴指出：与纯粹的逻辑和数学系统相反，自然系统却总是开放的，人类关于自然系统的知识总是不完备的，至多是近似的，我们从来都不敢确定一定没有忽略某些相关的因素。

他们解释说："我们称之为数据的东西，其实只是我们用来对那些不能完全逼近的自然现象进行推理的工具，它们隐含真实的信息。许多推理和假设能由经验判定为正确（某些不确定的东西能被估算出来），但是，我们却不能指望事物会按照我们事先所做的假设发展。正是过多的假设本身使得系统更加难以把握。"换句话说，我们的模型总是太过理想化，太过近似，带着太多的猜测成分。

三位作者强调，即便一个模拟精确地拟合了甚至预测了某一真实现象的运行过程，我们仍然不能说模型已经得到了验证。我们无法确定这种吻合是来源于模型与实在之间的真实对应，还是一种单纯的巧合。实际上，建立在完全不同假设基础之上的其他模型，也总有可能得到相同的结论。

奥雷斯克及其合作者在讨论了一番哲学家南希·卡特赖特（Nancy Cartwright）的观点——称数值模拟为"一类虚幻的工作"——之后，继续这样写道——

我们不一定接受她的观点。但是我们应该考虑她所提到的这个方面：一个模型，就像一部小说，它与自然界相契合，但它不是"真实的"事情本身。如果一个模型能够做到与我们对自然界的经验一致的话，它也会像小说一样使人相信，它像那么回事。但是，正如我们总想知道小说中有多少人物来自真实生活，又有多少来自艺术加工一样，我们也会对模型提同样的问题：模型中有多少是建立在对真实现象所做可靠观察和测量的基础上，又有多少是建立在有见地的判断基础之上，还有多少只是图一时之便所做的假定……（我们）必须承认，模型很有可能会加深我们的偏见，支持错误的直觉。因此模型最大的用处只在于证伪现存的模式，而不是证实它们或者为它们提供充足的证据。

数值模型在某些场合比在其他一些情况下更有用些。它们尤其在天文学和粒子物理学中贡献卓著，因为这些学科所考虑的物体和各种力都精确地适用于其数学定义。实际上，数学帮助物理学家定义那些舍此就无法定义的东西，夸克便纯粹是一个数学构造，离开数学定义它便没有任何意义。夸克的特性——粲数、颜色和奇异性——都是数学的特性，在我们生活的宏观世界里并没有与之相对应的事物。当数学理论被运用于具体的、复杂的现象时，比如被应用于生物领域的任何现象，它便不如在天文学和粒子物理学领域所表现出来的那么强有力。正如进化生物学家恩斯特·迈尔曾经指出的那样，每一个生物体都是独一无二的，而每一个这种独特的生物体又时时刻刻在发生着变化。[17]这就是为什么描述生物系统的数学模型比物理学模型的预测能力要差得多，而我们同样应该怀疑这些模型揭示自然界真理的能力。

佩尔·贝克的自组织临界性

这种"乏味空洞"的哲学怀疑论让佩尔·贝克感觉很不爽。这位20世纪70年代来到美国的丹麦物理学家，是一个与哈罗德·布鲁姆笔下的强者诗人类似的人物。他高大，肥胖，时而严肃，时而好斗，言辞间总是充满"深邃"的观点。他力图使我相信，复杂性研究优于其他任何形式的科学。他无情鞭笞了粒子物理学家的妄想，即通过探索更小尺度上的物质就能揭开客观存在的秘密。"秘密本身并非来自对系统更深层次的下行挖掘，"贝克以其明显的丹麦口音断言，"而是来自对其他方向的探索。"[18]

贝克宣告，粒子物理学已寿终正寝，被其自身的成功所扼杀。他指出，粒子物理学的庆功晚会已经结束，晚会现场也将打扫干净，"然而大多数粒子物理学家却认为他们仍然在从事科学"。同样的悲剧发生在固体物理学领域，这是贝克开始其职业生涯的领域。成千上万的物理学家在从事高温超导研究，其中大部分都只不过是白费力气，这一事实足以说明该领域已变得多么缺乏生机。"肉已所剩无几，却仍有众多的凶兽要吃。"而在混沌（贝克的混沌定义与詹姆斯·约克的一样狭窄）领域，早在1985年，即格莱克出版其《混沌》一书以前两年，物理学家就已对导致混沌行为的那些程序达成了基本的理解。"事情就是这样！"贝克厉声说道，"不管什么事，一旦到了一哄而上的时候，就已经完事儿了！"（当然，复杂性是贝克规则的一个例外。）

贝克非常瞧不起那些满足于仅仅在先驱者的工作基础上进行修补和拓展工作的科学家。"完全没有必要做那个！我们这儿不需要清洁工。"幸运的是许多神秘现象是目前的科学所无法解释的，比如说像物种进化、人类认知以及经济之类。贝克说："这些事物的共同点是它们都是具有许多自由度的巨系统，我们称之为复杂巨系统，这些事物将会带来一场科学革命。未来若干年内，对复杂系统的研究将造就一门硬科学，这正如过去20年内粒子物理学和固体物理学成长为硬科学一样。"贝克反对将这些问题视为我们人类贫乏的大脑所认识不了的禁区，认为这是"伪哲学的、悲观的、空洞无聊的"说法。"如果我也这么认为，我将不会愿意再继续这项事业！"贝克解释说，"我们应该乐观地面对挑战，踏踏实实地干，这样我们才能不断前进。我相信科学将在50年以后焕然一新。"

贝克和他的两位同事在20世纪80年代末提出自组织临界性理论，人们迅速看好这一理论，认为它很有可能发展成复杂性的统一理论。被他作为范例讨论的系统就是沙堆，当你在沙堆的顶部增加沙粒时，如果沙堆达到贝克所说的临界状态，那么，即使在其顶部增加哪怕仅仅一粒沙，也会在沙堆的周边引起一次"雪崩"。如果将在临界状态发生的雪崩大小和频率绘制成图，其结果符合幂律：雪崩发生频率与沙堆大小的幂成反比。

贝克认为混沌的先驱者芒德勃罗早已指出，地震、股市涨落、物种灭绝和其他许多现象都表现出符合幂律的行为模式。换言之，贝克所定义的复杂现象也全都是混沌的。"既然经济学、地球物理学、宇宙学和生物学全都具有这些奇异的特征，那么，其背后必定存在着某种理论。"贝克希望他的这个理论能够解释为什么小地震常见而大地震罕见，为什么许多物种存在数百万年而后突然消失，以及为什么股市狂泻。"我们不可能解释所有事物的所有方面，但却有可能对所有事物的某些方面作出解释。"

贝克认为像他的那一类模型甚至会引起经济学的革命。"传统的经济学不是一门真正的科学。他们在数学的教条之下谈论的是完美的市场、完美的推理和完美的平衡。"这种方式只是一种"怪诞的近似"，它不可能被用来解释真实世界的经济行为。"任何工作在华尔街并注视着股市变化的真实的人，他们都明白股市涨落来自经济系统本身的一连串反应，来自各种因素的干预，包括银行贸易家、顾客、小贩、强盗、政府及经济形势等，几乎无所不包。而传统经济学则根本没有描述这些现象。"

数学理论能够帮助人类洞悉文化现象吗？贝克喃喃地念叨着这个问题。"我不明白意义是什么，"他说，"在科学里任何事物都没有意义。科学不问原子受磁场作用时为什么向左，它只观察和描述。因此社会科学家应该走出去观察人们的行为，然后描述出这种行为会对社会产生什么后果。"

贝克认为这些科学理论提供的只是统计描述，而不是特定的预测。"我们无法预测。但是不管怎样我们能够理解那些我们预测不了的系统，并且能够理解为什么它们不能够被预测。"热力学和量子力学提供了这方面的典范，它们都是关于概率的理论。贝克说："我认为做成一个特定的和注重细节的模型是失败之举，这样的模型并不带来洞察力。模型应该允许被修改，模型应该具有普适性。"贝克嘲笑特定的东西只不过是工程。

当我问及他是否认为众多的研究者最终会走到一起，并得到一个单一、真实的关于复杂系统的理论时，贝克显得有些信心不足。"这很难说，"他举例说，"我怀疑科学家是否能够得到一个关于大脑的简单而又独特的理论，他们或许会发现一些控制大脑行为的规律，但是却不能期望太多。"他沉默了一会儿，接着补充道："我认为复杂性统一理论的获得，将是一个非常长期的过程，它甚至比得到混沌理论更难。"

贝克也担心联邦政府对发展基础科学漠不关心。他认为政府增加对应用科学的重视或许会阻碍对复杂性的研究。基础科学的日子正愈益艰难，因为科学必须有用，大多数科学家正被迫做那些他们并不真正感兴趣甚至令人生厌的东西。贝克的主要雇主——布鲁克海文国家实验室，正在强迫人们做"可怕的事情，那些令人难以置信的垃圾"。即使是像贝克这样一个始终保持强烈乐观情绪的人，也不得不承认现代科学陷入了令人忧虑的困境。

自组织临界性曾被许多人大加赞扬，比如阿尔·戈尔（Al Gore）在其1992年的畅销书《平衡中的地球》中指出，自组织临界性不仅帮助他明白了环境对潜在分裂的敏感性，也帮助他理解了"自身生活的改变"。[19]斯图亚特·考夫曼发现在自组织临界性和混沌边缘，以及他在对生物进化进行计算机模拟时发现的那些复杂性定律之间，存在某种亲缘关系。但是其他研究者则指责贝克的模型甚至都没有为他的沙堆范例系统提供一个令人信服的描述。芝加哥大学的物理学家用实验证明沙粒的大小和形状不同，沙堆的行为方式也不同，几乎没有沙堆像贝克所预测的那样表现出符合幂律的行为。[20]

更进一步的批评认为，贝克的模型也许太空泛，本质上又是统计性的，因而实际上不能说明它所描述的任何系统。毕竟，虽然许多现象都能用高斯曲线（俗称钟形曲线）描述，但是很少会有科学家敢于声明，人类的智商和星系的亮度也一定是从共同的机制中派生出来。

自组织临界性根本不是一个理论。像断续平衡理论一样，自组织临界性仅仅是针对遍布自然界的随机涨落和随机噪声的诸多描述之一。贝克自己也承认，他的模型既不能对自然作特定预测，也不能带来有意义的见解。那么，贝克模型又有什么用呢？

控制论及其他思想所带来的震荡

自古以来，人类无数次地试图寻求一种适于预测和解释包括社会现象在内的诸多现象的数学理论，不幸的是，所有这些努力最后都以失败而告终。17世纪，莱布尼茨就曾着迷于创立一套不但能解决所有数学问题，而且也能解决哲学、道德和政治问题的逻辑体系。[21] 而今，在这个怀疑的世纪，莱布尼茨的这种梦想却依然延续着。自第二次世界大战以来，科学家们就一度被至少三个这种类型的理论所吸引，它们是控制论、信息论和突变论。

控制论的创立几乎可谓是一人之功，其主要部分是由麻省理工学院的数学家诺伯特·维纳（Norbert Wiener）建立的。维纳1948年出版的著作《控制论》的副标题"关于在动物和机器中控制和通信的科学"，将其勃勃雄心表露无遗。[22] 他宣称，完全可能建立一个单一的、包罗万象的理论，用这样一个理论不但可以解释机器的各种运行机制，而且还能解释小至单细胞生物、大至国民经济系统的复杂行为。所有这些实体的行为和过程，从本质上看都基于信息之上，它们的运行机制无非是各种正负反馈以及用以分辨信号和噪声的滤波机制。

到了20世纪60年代，控制论渐渐失去了其魅力。1960年，杰出的电子工程师约翰·R. 皮尔斯（John R. Pierce）曾硬邦邦地指出："在这个国度，'控制论'一词已被广泛地用在各种新闻媒体、大众刊物，或是一些虽不能被称为'半文盲'、至少也可称之为'半文学'性的杂志中。"[23] 不过，控制论在一些相对隔离的国家仍不乏其追随者，其中最显著的是俄罗斯（在苏联时期，这个国家痴迷于这样一种幻想，即社会完全可以像机器那样按照控制论的准则进行精细的调控）。如果不是因为他对科学本身所产生的影响，维纳在美国大众文化中所具有的影响也就不可能如此持久：我们不能否认，所有诸如"赛博空间"（cyberspace）、"赛博朋客"（cyberpunk）以及"半机器人"（cyborg）等词汇的出现，都要归功于维纳。

信息论是与控制论关系十分密切的另一个理论。1948年，贝尔实验室的数学家克劳德·香农分两部分发表了题为"通信的数学理论"的论文[24]，这标志着信息论的诞生。香农的巨大成就在于他创立了基于热力学中"熵"概念的关于信息的数学定义。与控制论不同，信息论至今仍是热门学科。香农理论的目的在于改善电话或电报中信息的传输受电子干扰（或噪声）影响的问题。迄今为止，信息论仍是编码、压缩、加密及其他信息处理方式的理论基础。

到了20世纪60年代，信息论的影响已渗透到了通信以外的其他领域，包括语言学、心理学、经济学、生物学乃至艺术（例如，许多智者就曾致力于寻找能够表述音乐质量与其信息含量之间关系的公式）。尽管在约翰·惠勒等人的影响下，信息论在物理学领域正经历着变革，但不可否认它在各个具体方面仍对物理学有不可忽视的贡献。尽管如此，香农自己也怀疑他的理论的某些应用是否真的会产生什么成果。他曾跟我谈道："不知为什么，人们总是认为它能告诉你关于'意义'是什么，但事实上，它根本就不能也不打算这样做。"[25]

突变论或许是类似的形而上理论中被吹捧得最玄乎的理论，它是由法国数学家雷内·托姆（René Thom）在20世纪60年代提出的。托姆是以纯数学形式提出突变论的，但他和其他许多人都宣称，该理论能帮助人们洞悉隐藏于客观世界广泛存在的、呈现不连续行为现象背后的本质。托姆最杰出的著作是他在1972年出版的《结构稳定性和形态发生学》，这部书在欧洲和美国产生了轰动性的影响。伦敦《泰晤士报》的一位书评家断言："这部书的影响绝对无法用三言两语加以描述。从某种意义上说，能与之相提并论的也许只有牛顿的《原理》一书，它们都为认识自然界设计了一套概念框架，同时，它们又都引发了进一步的无尽的思索。"[26]

托姆的方程式揭示了一个貌似有序的系统是如何发生骤然的、突变性的状态变化的。托姆及其追随者指出，这些方程式不仅可以解释一些纯物理现象，如地震等，而且也能解释生物和社会现象，例如生命的发生、毛毛虫向蝴蝶的形态变化以及文明的瓦解等。对这个理论的抨击始于20世纪70年代末。有两位数学家在《自然》杂志上发表文章指出，突变论不过是"又一次试图通过独自思辨来推演世界的尝试"，他们称之为"一个诱人的、但无法成真的梦想"。其他一些批评家指责托姆的工作"并没有提供关于任何东西的新信息"，而且是"夸大其词，并非完全真实的"。[27]

由詹姆斯·约克定义的混沌，也经历了相似的由兴盛到衰落的过程。1991年，混沌理论的先驱之一，法国数学家大卫·吕埃尔（David Ruelle）也开始怀疑他自己的研究领域是否已迈过了其巅峰时期。吕埃尔是"奇异吸引子"这一概念的首创人。"奇异吸引子"是一类具有分形特征的数学对象，它是用来描述具有非周期特性的系统行为的。在其著作《机遇与混沌》中，吕埃尔提到："混沌吸引了一大批追名逐利、梦想成功的人，他们感兴趣的并不是这一思想本身，这样一来就使得学术气氛江河日下……在混沌物理学这个领域，尽管频频有人宣称取得了'新'突破，但事实上真正令人感兴趣的发现却越来越少。不过，令人欣慰的是，这种狂热一过，对这个领域之难度的冷静评价将会导致另一个高水平研究浪潮的到来。"[28]

重要的是差异

约翰·霍兰、佩尔·贝克及斯图亚特·考夫曼等人，都曾梦想能有一种超然的、统一的理论对复杂现象作出解释。对于科学是否能达到这一境界，即使是一些与圣菲研究所有密切联系的研究人员似乎也持怀疑态度，包括圣菲研究所的创立人之一菲利普·安德森（Philip Anderson），一位以倔强著称的物理学家，曾因其在超导领域的卓越成就荣获1977年诺贝尔物理学奖。他是反还原论的先驱者之一。在其1972年发表于《科学》杂志上的《重要的是差异》一文中，安德森指出，不仅仅是粒子物理学，事实上几乎所有的还原论方法在解释世界时都存在严重的局限性。客观世界有着层次结构，上下层结构之间存在着或多或少的独立性。"在每个阶段，创立全新的定律和概念都是必要的，进行这些工作所需的灵感和创造性都不逊于其前一阶段，"安德森指出，"心理学并不是应用生物学；同样，生物学也不是应用化学。"[29]

"重要的是差异"成了混沌和复杂性运动的战斗口号。然而具有讽刺意味的是，这个原则同时也暗示着：所有这些所谓的反还原论的努力，根本就不可能创立一种关于复杂性和混沌系统的统一理论，一种能够解释从免疫系统到经济系统这样广阔范围的复杂系统行为的理论，正如贝克之类的混杂学家所认为的那样（这个原则同时还暗示着，罗杰·彭罗斯用准量子力学的方法解释意识问题的尝试，完全是误入歧途）。当我在其大本营普林斯顿大学拜晤安德森时，他似乎也意识到了这一点。"我认为根本就不存在万物至理，"他说，"但我认为一定存在着具有广泛适用性的基本原理，例如量子力学、统计力学、热力学及对称性破缺等。但是当你获得了在一个层次上适用的原理时，千万不要认为它将适用于所有层次。"（关于量子力学，安德森说："我个人认为在可预见的将来是不大可能对之进行修正的。"）安德森很赞同进化生物学家斯蒂芬·杰伊·古尔德的看法，认为生命的形成更多的是由偶然性、不可预见的环境所决定的。他说："我猜想我所表述的或许是种偏见，但这种偏见却是被博物学所支持的。"

对于计算机模型揭示复杂系统行为本质的能力，安德森并不像他的某些圣菲研究所同事那样笃信不疑。"因为我对全球经济模型并非一无所知，"他解释道，"据我所知，这些模型毫无用处！我时常怀疑是不是就连全球气候模型、海洋环流模型及其他类似的东西，也都同样充斥着虚假的统计和虚假的测量。"安德森指出，进行更细致、更真实的模拟不一定是解决问题的办法。例如，我们可以用计算机模拟液体变成玻璃的相变过程，"但是，你从中又能了解到些什么呢？你比模拟前多懂了些什么？为何不干脆拿一片玻璃并说它正在进行玻璃相变？为什么你一定要通过计算机来观察玻璃相变？这样做实际上是一种归谬法。从某种角度说，计算机并不能告诉你系统本身正在做什么"。

我对他说，不过你的一些同事似乎仍然坚信，总有一天他们将会发现一个崭新的理论，从而将一切神秘现象的本质昭然于天下。"是这样。"他摇着头答道。突然，他向空中伸出双臂，像一个获得再生的教民一样大声说道："我终于见到了曙光！我明白了一切！"紧接着，他放下手臂作懊丧状，然后微笑着说："你永远别指望会明白一切。只有疯子才会明白所有事情！"

夸克大师逐走"别的东西"

默里·盖尔曼似乎不像是一位圣菲研究所的领导人，因为他是一位还原论的大师。由于发现了从加速器射出的各种粒子流背后的统一秩序，他获得了1969年的诺贝尔物理学奖。他称自己的粒子分类系统为"八正道"，一种佛教追寻智慧的方法。（盖尔曼常常指出这是一种开玩笑的说法，并不意味着他就是那些"新时代"疯癫人物中的一员，认为物理学与东方神秘主义有相通之处。）他显示出洞察复杂现象背后统一本质的天赋，并且有创造新术语的非凡才能。他指出中子和质子及其他一些短命粒子，都是由三个一组的更基本的粒子——夸克所组成。盖尔曼的夸克理论在加速器实验中已得到了充分的证实，至今仍然是粒子物理学标准模型的基石。

盖尔曼总是喜欢回忆，他是怎样在阅读詹姆斯·乔伊斯的小说《为芬尼根守灵》时发明了"夸克"这个术语的（这部小说里的原文是"Three quarks for Muster Mark"①）。这则轶事使人们注意到盖尔曼的思想是如此博大与活跃，单单粒子物理学是无法令他满足的。正如他在一份分发给记者们的个人声明中所言，他的兴趣不但包括粒子物理和现代文学，而且还涵盖了宇宙学、核裁军政策、博物学、人类史、人口增长、人类可持续发展、考古学及语言演化等领域。盖尔曼似乎在一定程度上通晓世界上大多数主要语言和方言，他总是乐于告诉别人有关他们姓名的词源及其正确的方言发音。他是最早赶复杂性研究浪头的著名科学家之一。他帮助建立了圣菲研究所，并于1993年成为该所的首位专职教授。（在此之前，他在加州理工学院当了大约40年的教授。）

毫无疑问，盖尔曼是20世纪最杰出的科学家之一——他的出版经纪人约翰·布劳克曼（John Brockman）说，盖尔曼"有五个大脑，而且其中任何一个都比常人聪明得多"[30]——同时，他又可以说是一个最令人讨厌的人。事实上，几乎每个认识盖尔曼的人都能现身说法地告诉你，他有强烈的扬己抑人的癖好。1991年，当我在纽约一家餐馆与盖尔曼第一次见面时，他几乎立刻就显示出了这种迹象。盖尔曼身材矮小，戴一副很大的墨镜，满头短短的白发，双眼总是充满怀疑色彩地斜睨着。我刚拿出录音机和记录本，还没来得及坐下，就听见他说："科学记者全都是些白痴！是一种'很差劲的人'，总是把事情搞错。只有科学家才有资格将他们的研究成果公之于世。"这些话令我产生了很强烈的受伤害的感觉。不过，随着谈话的继续，我的这种感觉慢慢消失了，因为我发觉盖尔曼对他的大多数科学界同事都是持轻视态度的。在对他的一些物理学同行进行一番大肆贬低之后，盖尔曼说道："我希望你不要将我这些言论写出来，这不大好，因为在这些人中有一些是我的朋友。"

为了延长会面时间，我特意安排了一辆轿车，以便能陪同盖尔曼一块去机场，而后又陪他通过行李检查，最后和他一起进了贵宾候机室。他突然想起自己身上没带足够的现金，到达加州后恐怕没钱叫出租车（盖尔曼至今仍没有搬到圣菲研究所定居）。于是他开口向我借钱，我给了他40美元，他签

---

① 意为"向麦克老人三呼夸克"。"夸克"是德语中借自斯拉夫语的一个词，有多种含义，其中之一是指"一种海鸟的叫声"。而之所以用"夸克"来命名其所发现的基本粒子，盖尔曼称也是因为乔伊斯这句话里的"三"字暗合了这种粒子三个一组的存在方式——译者注。

了一张支票给我。当他将支票递给我的时候，建议我不妨考虑别去兑换这张支票，因为兴许有朝一日他的签名会变得异常珍贵。（我最终还是将这张支票兑现了，不过保留了一份影印件。）[31]

我感到盖尔曼十分怀疑他在圣菲的同事们是否能发现任何真正深奥的东西，换句话说，任何接近他的夸克理论的东西。不过，如果奇迹真的发生，也就是说，如果混杂学家真能作出重大成就的话，盖尔曼希望他自己能享此殊荣。这样一来，他的研究领域就几乎完全涵盖了整个现代科学，从粒子物理到混沌与复杂性。

作为一位著名的混杂学研究的领军人物，盖尔曼与极端还原主义者史蒂文·温伯格有着极其相似的世界观。当然，他的表现方式却与温伯格迥然有别。1995年当我在一次会面中问他是否同意温伯格在《终极理论之梦》中关于还原论的观点时，盖尔曼答道："我不知道温伯格在他的书中说了什么，不过，如果你读过我的书的话，你会知道我是怎样谈这个问题的。"接着，盖尔曼开始复述他1994年出版的《夸克和美洲虎》[32]一书中的几个主题。在盖尔曼看来，科学本身也有一个层次结构，在其顶端的是那些在已知宇宙中普适的理论，诸如热力学第二定律和他自己的夸克理论；其他理论，比如说有关遗传的理论，则仅适用于地球范围，而且这些理论所描述的现象总是伴随着大量的偶然性，并受其所处历史环境的局限。

"从生物演化中，我们可以看到大量历史因素的影响。无数连续地、紊乱地发生的偶然事件所形成的物种类型，完全可能比仅由自然选择压力所形成的物种类型要多得多，由此我们可以确定，人类的产生是由极大量的历史条件决定的。但尽管如此，基本原理与历史进程之间，或基本原理与特定条件之间，肯定存在着某种确定性的关系，这一点是显而易见的。"

盖尔曼曾试图让其圣菲同事们用"*plectics*"（混一性）一词来代替"complexity"（复杂性），由这件事也可看出其强烈的还原论倾向。"这个词源于印欧语系'*plec*'一词，其含义为'简单性与复杂性的共同基础'，因此，'plectics'就具有'探寻简单与复杂之间的关系，尤其是探寻具有复杂结构的事物行为背后的简单原理'的含义，"他说道，"我们试图创立的是关于这些过程在通常情形或特殊情形下如何运作，以及特殊情形如何与通常情形联系起来的理论。"（与"夸克"不同的是，"混一性"没能风行起来，我从未见过除盖尔曼以外的其他人使用过这一术语——除非是用这一术语来嘲讽盖尔曼对它的一往情深。）

盖尔曼认为，他的同事们不可能发现一种普适于所有复杂适应性系统的理论。"各种各样的系统，有些以硅为基础，有些由原生质构成，它们彼此之间千差万别，根本就不可能等同起来。"

当我问他是否赞同其同事菲利普·安德森提出的"重要的是差异"原则时，他轻蔑地答道："我完全不明白他究竟在说什么。"我向他解释说安德森认为还原主义方法在解释世界方面有很大缺陷，一个人无法循着粒子物理学推演到生物学，盖尔曼听了之后大声说道："能！当然能！你读过我就此问题写的书吗？我用了两至三章的篇幅来讨论这个问题。"

盖尔曼说尽管理论上可以完成这样一条解释链，但是在实际上却往往做不到，因为生物学现象总是受到非常多的偶然性、历史性因素影响。这并不表明生物学现象是由与物理规律无关的其他神秘规律支配的。涌现说的全部要点，按照盖尔曼的说法，可归结为"我们不需要以'别的东西'来获取'别的东西'，当你从这种角度看这个世界时，你会发现一切都变得可理解了！你再也不会被那些奇怪的现象所折磨了"。

斯图亚特·考夫曼等人曾提出，宇宙万物应该是由另外一种至今尚不为人所知的规律支配，否则，按照热力学第二定律，宇宙要远比我们所看到的混乱。显然，盖尔曼对此是持有异议的。他认为，这种现象其实不成其为问题，因为在宇宙起源时，宇宙处于一种远离热力学平衡的紧张状态，在宇宙松弛的过程中，系统从整体上出现熵的增加，但同时，在宇宙的许多局部，这种熵增趋势又会被破坏。"这是一种趋势，但在这个过程中却有无数的小涡旋，"他说，"这与复杂性增加了很不一样，只是复杂性的范围扩大了。显然，从这种角度考虑就可发现，无论如何也不需要另外一种定律！"

宇宙中的确会产生盖尔曼所谓的"冻结事故"（frozen accidents），比如说星系、恒星、行星、岩石及树木等。这些物体都具有复杂结构，同时它们又是构成更复杂结构的成分。"作为一种普遍规律，在非自适应的恒星和星系演化之类的过程中，生命形式、计算机程序和各种天体等总是会日趋复杂。但是！让我们想象一下，在很远很远很远的将来，到那时，这种趋势也许就不存在了。"千万年来的复杂性将一去不复返，宇宙很可能被分解为"光子、中子之类乱七八糟的东西，再也不复像现在这样富有个性"，在这种情况下，热力学定律就足以解释一切。

"我所反对的是那种蒙昧主义和神秘主义的倾向。"盖尔曼继续说道。他强调，关于复杂系统还有许多问题需要研究，否则他也就不会协助成立圣菲研究所了。"还有许多相当引人入胜的研究工作正等着我们去逐步展开，但是，我要说的是，没有丝毫迹象表明我们还需要——我不知道是否还有别的表达方式（*something else*）！"盖尔曼边说边冷笑，仿佛已不能抑制对那些可能会反对他的看法的人的嘲弄情绪。

盖尔曼指出，自我觉知和意识是蒙昧主义者和神秘主义者最后的避难所，人类显然比其他动物具有更高的智力水平和自我意识能力，但是，从本质上它们却又没有什么不同。"重申一遍，这些只不过是在不同复杂性层次上的现象。可以推测，这些现象的出现只不过是基本原理加上各种各样的历史条件所共同决定的。罗杰·彭罗斯写过两本可笑的书，在书中他将所有观点都建立在一个错误基础上，即认为要使哥德尔定理与意识问题挂上钩，还需要——别的东西。"

如果科学家想要发现新的基本定律，盖尔曼说，那么他就必须勇敢地沿着超弦理论的方向，向着微观世界奋勇前进。盖尔曼觉得在21世纪的早期，将很可能证明超弦理论是终极的物理学基本理论。"但是，像这样一个具有额外维数的牵强的理论，究竟是否有可能被接受？"我这样问道。盖尔曼听后不禁睁大眼睛瞪着我，仿佛我刚刚表述的是一个关于轮回转世的信仰。"你用这种怪异方式看科学，好像科学是民意测验，"他说道，"世界有它确定的运行方式，民意测验是无济于事的！它也许能对科学事业产生一定的压力，但是，最终的选择压力还是来源于与客观世界的符合程度。"那么，量子力学又如何呢？我们是否要继续忍受其怪异性呢？"啊，不，我认为量子力学没有什么怪异之处！量子力学就是量子力学！像量子力学那样起作用！这就是全部！"对盖尔曼而言，世界完全是可以理解的，他已手握"终极答案"。

科学是有限的还是无限的？对这个问题，盖尔曼第一次没有了现成的答案。"这是个很难回答的问题，"他严肃地答道，"我没法说。不过，尽管关于整个科学事业是否有终点尚难以回答，但是，可以肯定的是，总会有许许多多、各种各样的细节问题等待科学家去解答。"

盖尔曼最让人难以忍受的一点就是他几乎总是对的，当考夫曼、贝克及彭罗斯等人狂热地寻求超出现代科学地平线的"别的东西"——即能够比现有科学理论更好地解释生命、人类意识及存在本身之谜的新理论——的时候，盖尔曼断言了他们必将失败，这个断言很可能仍会被证明是正确的。也许，只有在认为拥有所有那些额外的维度和闭合弦的超弦理论将成为物理学基石这一点上，他可能会是错误的，但是，天晓得！

伊利亚·普利高津与确定性的终结

1994年，史密斯大学一位名叫阿图罗·埃斯科巴（Arturo Escobar）的人类学家，在《现代人类学》杂志上发表了一篇有关科学技术中派生的新观念和隐喻的文章。作者认为，混沌和复杂性赋予我们一个与传统科学完全不同的世界观；"混沌和复杂性强调流动性、多样性、多元性、关联性、片断性、异质性和弹性；不是'科学'，而是具体和局部的知识；不是定律，而是有关无机现象、有机现象和社会现象的自组织动力学及有关问题的知识"。请注意，这段引文中的科学一词是加了引号的。[33]

事实上，不仅仅是埃斯科巴这样的后现代主义者把混沌和复杂性看作是一种"反讽的努力"，人工生命专家克里斯托弗·兰顿在预言科学的未来将更富有"诗意"的同时，也表明了类似的观点。兰顿的观点实际上是对化学家伊利亚·普利高津（Ilya Prigogine）观点的响应。普利高津曾因其在耗散结构理论方面的成就，获得了1977年的诺贝尔奖。所谓耗散系统，是指由一些特殊化学混合物组成的、在多种状态之间保持涨落并且从不达到平衡态的系统。普利高津在比利时自由大学和美国德州大学奥斯汀分校自己的研究所之间来回奔波，进行了大量的实验。在这些实验的基础上，他构造了一套关于自组织、涌现以及有序和无序关联的思想，简言之，也就是混杂学的理论。

普利高津总是念念不忘时间这一概念。数十年来，他一直抱怨物理学对时间只按一个方向流逝这一明显的事实没有给予足够的重视。20世纪90年代初，普利高津宣布创立了一个新的、能正确反映客观世界之不可逆本质的物理学理论。这个基于或然性的新理论被认为可以消除那些长期以来困扰着量子力学的哲学悖论，并能调和量子力学与经典力学、非线性动力学以及热力学之间的矛盾一样，普利高津断言他的新理论将有助于在自然科学和人文科学的鸿沟之间架起一座桥梁，从而引起对自然的"返魅"。

普利高津有其自己的追随者，至少许多非自然科学家就热烈支持他。未来学家阿尔温·托夫勒（Alvin Toffler）在普利高津1984年出版的《从混沌到有序》一书的前言中，将普利高津比作牛顿，并预言未来的第三次科学浪潮将是普利高津的时代。[34] 然而，那些熟悉普利高津著作的自然科学家，包括那些吸取过普利高津观点和思路的年轻的混沌与复杂性研究者，却极少对普利高津表示赞扬。他们指责普利高津过于自高自大，其实对自然科学并没有什么具体的贡献，他只不过是重复了别人的实验并夸大了其哲学意义；因此，同其他诺贝尔奖获得者相比，普利高津应该是最不够格的一个。

这种指责也许是对的。但换个角度，普利高津之所以受到科学家们的敌视，很可能是因为他揭示了20世纪后期自然科学的阴暗面，甚至从某种意义上说，自掘了科学的坟墓。在《从混沌到有序》[与依莎贝尔·斯唐热（Isabelle Stengers）合著]一书中，普利高津指出，20世纪的几个重大科学发现已突破了科学的限度。普利高津和斯唐热指出："无论是在相对论、量子力学还是热力学中，对不可能性的证明告诉我们，若想像旁观者那样'从外界'描述自然是不可能的。"现代科学用概率来描述自然，从而导致了一种与传统科学的"透明"相对立的"模糊"。[35]

1995年3月，我在奥斯汀见到了普利高津，那时他刚从比利时回到美国。已是79岁高龄的他未显示出丝毫时差反应的迹象，非常机警和富有朝气。他体形瘦小，举止高雅，对于自己辉煌的成就淡然处之，并没有显示出丝毫傲慢。当我提议与他讨论一下那些问题时，他略带不安地点头同意，并连声说"好吧，好吧……"不过我马上就意识到，他其实是非常愿意就自然本质给我上点启蒙课的。

我们坐下不久，该中心的其他两位研究人员就加入进来。一位秘书后来告诉我，这两位人员来的目的是，如果普利高津过于健谈而不容我提问时，他们可帮助我在适当时候打断他。不过，尽管他们怀着这种初衷而来，最终还是没能达到目的。普利高津的话匣子一打开，就口若悬河谈起来，像一股不可阻挡的急流，词、句、段落从他的口中奔涌而出。有时，他的口音有点重——这使我想起《粉红豹》中的检察官克鲁索——当然这并不妨碍我理解他的谈话。

普利高津简要讲述了自己的年轻时代，1907年，他出生于俄国一个资产阶级家庭；1917年，全家逃亡到比利时。他的兴趣非常广泛：弹奏钢琴，研究文学、艺术、哲学，当然还有自然科学。他认为年轻时动荡的生活激发了他对时间这一概念的持久兴趣。"给我印象很深的是，科学长期以来都忽视了时间、历史和演化。也许，正是这种印象使我开始思考热力学问题。因为热力学中最重要的量就是熵，而熵意味着演化。"

20世纪40年代，普利高津提出：由热力学第二定律决定的熵增加，并不意味着总是产生无序；在某些系统中，比如他自己在实验研究中发现的，在放置混合化学物质的容器中，熵的变化会产生奇妙的模式。他开始意识到：

"结构根植于不可逆的时间流向，时间箭头在宇宙结构中是一个重要因子。从某种意义上说，正是因为这些认识使我与大物理学家爱因斯坦发生了分歧。在爱因斯坦看来，时间只是一种幻觉。"

按照普利高津的观点，大多数物理学家认为不可逆是源于观察者观察手段不足的幻觉。普利高津宣称："我不同意这种观点，因为照这样说，在某种意义上就意味着仅仅是因为我们的测量和近似过程，便在一个时间可逆的宇宙中引入了不可逆性！然而事实上，我们不是时间之父，而是时间之子，我们是由进化过程产生的。我们必须在我们的描述中包括进化模式。我们需要在物理学中包含达尔文的观点、进化的观点、生物的观点。"

普利高津和他的战友一直致力于建立这样一种新的物理学，普利高津告诉我："新模型的产生将使物理学获得新生，这与史蒂文·温伯格（和普利高津在同一大楼工作）等还原决定论者的悲观预测完全相反。新物理学将弥合总是把自然描述成确定性实体之结果的自然科学与强调人性自由和责任的人文科学之间的鸿沟。按传统的观点，自然科学坚持将自然描述为确定规律作用的结果，而人文科学则强调人类的自由和责任。"普利高津讲道："从这种观点出发，一个人就不能一方面认为自己是自动机的组成部分，而同时又带有人文主义色彩。"

普利高津强调说，这种统一当然是隐喻的，而不是字面的，它不可能解决科学的所有问题。普利高津在驳斥圣菲研究所和其他一些地方的抱有这种幻想的研究人员时说："我们不应该夸大和幻想这样一种统一理论，认为它将能解决政治、经济、免疫系统、物理、化学等所有问题，也不应该设想化学非平衡反应的进展会成为解决人类政治学的关键。当然不会！但是，这一模型将引入统一因子，引入分叉的因素，引入历史维数，引入进化模式，这些是在所有层次都会发现的。从这种意义上说，它是我们宇宙观的统一因素。"

普利高津的秘书从门外探头进来，提醒他已经在教员俱乐部预订了午餐。在秘书的三次提醒后，于中午12点5分，普利高津兴奋地结束了他的发言，宣布现在该去吃午饭了。在教员休息室，我们和该中心的其他工作人员聚到一起，他们都是些普利高津主义者。大家围坐在一个长方形桌子周围，普利高津高坐在桌子一边的中间位置，就像《最后的晚餐》中耶稣的位置，我坐在他的旁边，在犹大的位置上，与其他人一道听他滔滔不绝地说教。

普利高津偶尔也让他的弟子谈上一两句，这足以显示他与弟子们在雄辩能力方面的巨大差异。有一次，他让一个坐在我对面的脸色苍白的高个子同事（当时，他的同胞兄弟也在桌边，同样苍白，同样紧张不安）谈一下是如何用非线性和概率的观点看宇宙的。这位弟子放松了一下，带着西欧口音侃侃而谈，讲起了令人费解的泡沫、不稳定和量子涨落。普利高津很快就插话了。他解释道，他同事的工作意义在于说明时空范围内不存在稳定的基态，没有平衡条件；因而宇宙没有开始，也没有终结。

在吃鱼的间隙，普利高津再次强调他反对决定论。（早些时候，普利高津承认卡尔·波普尔对他影响巨大。）他认为"笛卡儿、爱因斯坦以及其他著名的决定论者都是悲观论者，他们试图进入另一世界，一个具有终极美丽的世界"。但在他看来，这样的决定论世界并不是理想社会，而是罪恶社会。奥尔德斯·赫胥黎（Aldous Huxley）在《美丽的新世界》中，乔治·奥威尔（George Orwell）在《1984》里，以及米兰·昆德拉（Milan Kundera）在《生命中不能承受之轻》等书中，所表述的都是这种观点。普利高津解释道，当一个国家企图用暴力压制进化、变动和流动时，它就会摧毁生命的意义，产生一个"不受时间影响的机器人"社会。

一个完全非理性的、不可预测的世界也是可怕的。"我们寻求的是折中的方法，必须找到一种概率描述的方法，这种方法能解决一些问题，但不解决一切问题，也不是什么问题都不解决。"普利高津认为他的观点能对认识社会现象提供哲学基础。同时强调：人类的行为不可能用科学的数学模型来确定。"人类生活中没有简单的基本方程！当你决定是否喝一杯咖啡时，就已经是一个非常复杂的决策。结果取决于某日某时你是否想喝咖啡等因素。"

普利高津一直致力于作出伟大的发现，现在他终于公布了成果。混沌、不稳定性、非线性动力学以及相关概念，这些概念不仅为自然科学家所接受，而且也为普通公众所接受，因为社会总是处于变动不安状态之中。无论是对宗教、政治、艺术还是科学，人们的高度统一的信念在逐步解体。

"今天，即使是非常虔诚的天主教徒也已不像其父辈那样忠于上帝。人们对于马克思主义或自由主义的信仰也发生了改变，对于传统科学的信仰已完全动摇。"对于艺术、音乐、文学，情形也差不多；社会已开始学会接受行为形式和世界观的多样性。普利高津总结道：人类已经到达了"确定性的终点"。

普利高津稍作停顿，让我们掂量他的话的分量。我打破沉默，提出一个问题：像极端主义者这样一些人，他们似乎比以往更坚持确定性。普利高津耐心地听完了我的提问，然后说，极端主义者是这一规则的例外。突然，他盯住一位拘谨的金发妇女，她是该研究所的副主任，正坐在我们对面，"你认为怎么样？"他问道。"我完全同意。"她回答道。或许感到同事的暗笑声，她赶忙补充说，极端主义"仿佛是对纷乱世界的反响"。

普利高津慈祥地点点头，承认他对确定性的终结的断言已在知识界引起了"猛烈反应"。《纽约时报》婉拒对《从混沌到有序》进行评论，普利高津听说这是因为编辑们认为讨论必然性的终结太"危险"。普利高津理解这种担心。"如果科学不能给出确定性，你该相信什么呢？在以前，这简单得多，要么信仰耶稣，要么信仰牛顿。但是现在，科学只能给你可能性而不是确定性，那么它就是一本危险的书！"

但是，普利高津认为他的观点正确反映了无限深奥的世界和我们的存在。这正是他的习语"自然的返魅"所蕴含的内容。比如，我们正在进午餐，有什么理论可以预测这件事！"宇宙是奇妙的，"普利高津谈道，将声音提高了一度，"我想我们都赞成这一点。"当他用平和而犀利的眼光扫过房间时，同事们抬起头，脸上露出略带紧张的微笑。他们的不安是有原因的，因为他们正受雇于一个相信经验的、严格的科学行将寿终正寝的人，尽管这些科学能解决某些问题，能帮助我们理解世界，还能使我们内心变得祥和。

作为对确定性的回报，普利高津同克里斯托弗·兰顿、斯图亚特·考夫曼及其他混杂学家一样，寄希望于"自然的返魅"（尽管佩尔·贝克非常狂妄自大，却至少避开了这种伪心灵学辞令）。普利高津这一阐述的意义在于，同牛顿、爱因斯坦或现代粒子物理学的精确、有力的理论相比，模糊、无力的理论可能更有意义、更合乎事实。这不禁使人疑惑，为什么不确定的、模糊的宇宙反而并不比一个确定的、透明的宇宙更冷酷，更骇人？更具体说，如何能用非线性的概率动力学来揭示世界，来安慰一位亲眼看到自己独生女儿被奸杀的波斯尼亚妇女？

米切尔·费根鲍姆和混沌之解体

正是与米切尔·费根鲍姆（Mitchell Feigenbaum）的会面，使我最终认识到混杂学的前景是黯淡的。费根鲍姆很可能是格莱克的《混沌》一书中最引人注目的角色，同样，在混沌领域也是如此。费根鲍姆本来是一名粒子物理学家，但不久他所思考的问题就超出了那个领域研究的范围，对湍流、混沌、有序与无序的关系提出了质疑。20世纪70年代中期，他在洛斯阿拉莫斯国家实验室做博士后时发现了一种隐秩序，称为周期倍化律，这一规律是许多非线性数学系统行为的基础。系统的周期是指系统返回初始状态所经历的时间。费根鲍姆发现，某些非线性系统在其演化过程中周期一直倍增，因而很快达到无穷大。实验证明，现实中许多简单系统都会表现出周期倍化（尽管不如预计的多）。例如，一个缓缓打开的水龙头，流水显示了周期倍增，从一滴、一滴到形成一股急流。数学家大卫·吕埃尔称周期倍化律为"极其优雅并具有重要意义"的工作，"在混沌理论中占有突出地位"。[36]

1994年3月，我在洛克菲勒大学采访了费根鲍姆，他在那儿有一间宽敞的办公室，可以俯视曼哈顿东河。一眼看上去就可发现，他处处显示着传说中的智慧，异乎寻常的大脑袋和后梳的头发，使他看上去很像贝多芬，当然更漂亮些，少了些野性。费根鲍姆发音清晰准确，不带方言口音，但的确总是透出一种特殊的味道，仿佛英语是他的第二语言，完全是靠他的聪明才掌握的一样。（超弦理论家爱德华·威滕的嗓音也具有同样的特点。）当被逗得发笑时，费根鲍姆的表情更多的是痛苦般的扭曲，本已突起的眼珠更加从眼眶中突出来；嘴唇后缩，露出了两排销子般的牙齿。他的牙齿由于长期吸无过滤嘴香烟与喝浓咖啡而变成了深色（在采访过程中，他一直在享受着这两种嗜好）。由于这些有害物的长期作用，他的声带常发出类似于男低音歌手那种低沉而丰富的声音，仿佛是在深沉而又阴险地窃笑。

像大多数混杂学家一样，费根鲍姆对于粒子物理学家敢于设想得到一个普适理论嗤之以鼻。他认为也许有朝一日真能找到可以诠释自然界所有基本力（包括引力）的理论，但是将此作为终极理论却又是另外一回事了。"我的许多同事相信终极理论，因为他们是虔诚的。上帝已难以令人信服，因而只能用这一理论取代上帝的位置，他们只不过是创立了上帝的一个新化身。"费根鲍姆认为，一个统一的理论也不可能解决所有的问题。"如果你真的相信这是理解现实世界的途径，我马上就可反问：在这一形式系统中，我怎样才能表述出你是个什么样子，难道要数出你脑袋上头发的根数吗？"他盯住我，使我感到有些头皮发麻。"一个回答是'这个问题没意思'，这种回答违背了我的意愿，有些受辱的感觉；另一个回答是'它挺好，但我们做不到'。正确的回答是将二者结合起来，我们的手段有限，不可能完全解决这类问题。"

　　而且，粒子物理学家过度关注于发现"真理"，用它来诠释得到的数据；科学的目标应该是"在你脑中产生一个崭新的、令人激动的思想"，这才是我们所企盼的，费根鲍姆解释道。接着，他补充道："就我所知，没有什么手段可以确保认识的正确性。我根本不关心认识的对错，我只想知道我是否拥有一种思维方式。"我开始怀疑：费根鲍姆与戴维·玻姆一样，有着艺术家、诗人甚至神秘主义者的灵魂，他在探求自然的启示，而不是探求真理。

　　费根鲍姆提到，粒子物理学的方法论（也是物理学的通用方法）只是力图看到事物的最简单的方面，即"可揭示一切事物的方面"。最极端的还原论者，曾认为研究复杂现象仅仅是个"工程问题"。随着混沌和复杂性的研究进展，费根鲍姆说："曾经归结为工程的问题现在需要从理论的角度来重新思考。不应该局限于得到正确结果，还应知道事物的运作方式，你甚至于可以从那最后的议论中悟出一个即将完成的理论的意义。"

　　同时，混沌也带来太多的滥竽充数者。"把这门学科命名为'混沌'就是一个骗局，"费根鲍姆说，"请想一下，我的一个同事（粒子物理学家）参加过一个酒会，结识了一个人，这人滔滔不绝地谈起混沌，并告诉他所有这些还原论者的货色全是胡说。哼，这真令人气愤，因为这人的言谈是愚蠢的。我想，这些人是如此轻率，真为他们感到惭愧，他们根本没有资格来代表混沌研究者。"

　　费根鲍姆补充道，在圣菲研究所，他的一些同事过于天真地相信计算机的威力。"酸甜苦辣，不尝不知，"他说，稍微停顿了一下，似乎在考虑怎样才能说得圆滑一些，"在许多数值试验中，要想观察出一些名堂是非常困难的，因而人们总想用越来越高级的计算机来模拟流体，这就需要研究被模拟流体的性质。如果你不知道看什么，那么你能看到什么呢？因为如果我从窗户看出去，看到的景色比任何计算机模拟都要逼真。"

　　他对着窗户连连点头，窗外青灰色的东河在缓缓流淌。"我不能那么尖刻地质问，但如果对此不闻不问的话，在数值模拟中太多的东西就会让我一无所获。"由于这些原因，非线性现象的近期研究"还没有结果。原因在于这些问题非常深奥，而且缺乏适当的工具。这项工作需要富有创造性的计算，同时还要有信心和运气，人们还不知道如何开始着手这项工作"。

　　我承认我常常为人们在对待混沌和复杂性上的浮夸不实而困惑。有时，他们似乎勾勒出了科学的限度，比如蝴蝶效应；而有时，他们又认为可以超越这种限度。费根鲍姆大声说道："我们正构造工具！我们不知道如何处理这些问题，它们太难了。我们偶尔弄到一笔钱，投到我们知道怎么办的地方，然后就把它吹嘘得尽可能远。它达到所能达到的边界时，人们会迟疑片刻，然后停顿一会儿，指望一些新点子。但用中听一点的话来说，它仍然扩大了科学领地的疆域，这不单是从工程角度考虑的结果，也不仅仅是给出一些近似答案。"

"我想知道为什么，"他继续说，仍紧盯着我，"事情为什么会这样？"这项事业可能会或将要失败吗？"当然！"费根鲍姆喊道，并发出了狂笑。费根鲍姆承认他后来有些压抑自己。20世纪80年代后期，他关注于改进一种描述分形体的方法，比如云在边界各种力的扰动下的变化过程。他的两篇有关这一主题的长文，发表在1988年和1989年的专业物理学杂志上。[37] 费根鲍姆带着蔑视的口吻说："我不在意谁读过它们，实际上我从没有机会就此作个演讲。"他认为或许根本没有人理解他的想法。（费根鲍姆就是因其观点让人难以理解和睿智而出名的。）随后，他补充道："我还没有更好的思路来推动这一研究。"

与此同时，费根鲍姆开始转向应用科学和工程学。他帮助一家地图绘制公司开发了一种软件，该软件能以极小的空间扭曲和以最好的美学方式自动生成地图。他同时参加了一个重新设计美国钞票的委员会，该委员会是为了提高美元钞票的防伪性而成立的（费根鲍姆提出利用分形图案被复印后变模糊这一想法）。我觉得这些工作对于大多数科学家来说，都是非常值得的工作。但人们心中的费根鲍姆的形象是混沌学理论的领导者，如果人们知道他在从事地图与钞票工作，会作何感想呢？

"他不再干正经事。"费根鲍姆说，仿佛在自言自语。我赶忙说，不仅如此，当混沌理论的最具才华的开拓者不再继续这项工作时，人们会认为这一领域已走到了尽头。"这听起来颇有道理。"他答道，并承认从1989年起，一直没有更好的思路来拓展混沌理论。"当一个人专注于实实在在的事物时，却突然……"他又作了停顿，"没有了思路，找不到……"他那双大大的闪亮的眼睛再次转向窗外的河流，似乎在寻找某种启示。

我略带愧疚地告诉费根鲍姆，希望能读一下他关于混沌学的最后一篇论文，问他是否有复印件。费根鲍姆马上从椅子上站起来，侧身走向办公室最远处的一排文件柜。途中，低矮的咖啡桌碰到他的胫骨，他缩了一下，咬紧了牙，然后继续前进，显然是撞伤了。这一幕倒好像是塞缪尔·约翰逊[①]的著名的踢石一幕的翻版。那张"恶毒"的桌子似乎在暗暗高兴："我驳倒了费根鲍姆。"

制造隐喻

对混沌、复杂性及人工生命等领域的研究将持续下去。某些参与者将满足于在纯数学和计算机理论方面进行研究，而其他大多数参与者则将为了工程开发新的数学和计算技术。他们将逐步积累经验，比如扩展天气预报的范围，或提高其他工程技术人员模拟喷气机或其他技术的能力。但是，他们不可能在洞悉自然方面取得突破——当然也无法与达尔文进化论及量子力学相提并论：他们不可能在了解客观世界和描述创生过程方面作出重大进步；他们也不会找到盖尔曼所说的"别的东西"。

迄今为止，混杂学家创造了一些有力的隐喻：蝴蝶效应、分形、人工生命、混沌边界、自组织临界性。但无论从正面或负面意义来看，这些东西在帮助我们理解具体的世界和令人惊奇的事物方面，并没带给我们任何助益。它们只是略微扩展了某些领域中知识的边界，清晰地描述了一些学科的轮廓。

计算机模拟表达的是一种现实世界的衍生物，我们可以利用它们来摆弄科学理论，甚至在一定程度上对科学理论进行检验，但它们不是现实本身（尽管许多沉迷者已看不到这一区别）。而且当科学家使用不同符号、以不同方式模拟自然现象的能力提高时，计算机将会使科学家怀疑他们理论的真实性，不仅怀疑结果的真实性，而且还怀疑绝对真实性的存在。计算机极有可能加速经验科学的终结。克里斯托弗·兰顿是对的：未来的科学将更富有诗意。

---

① Samuel Johnson，1709—1784，英国词典编纂者、作家、文学批评家，以Dr. Johnson闻名。曾创办《漫游者》杂志；编纂第一部《英语词典》；编注《莎士比亚戏剧集》并作序言。这里所谓的"踢石"典故，出自James Boswell 的《约翰逊的一生》（*Life of Johnson*），说的是约翰逊为了批驳贝克莱"心外无物"的主观唯心主义观点，就踢了一脚岩石，并说"这样我就驳倒了贝克莱"——译者注。

第九章
限度学的终结

正如恋人们只在他们的关系恶化之后才开始谈论它一样，当科学家们的努力得到越来越少的回报时，他们也会变得较以前更为清醒和怀疑。科学行将步文学、艺术、音乐的后尘，变得更为内省、主观、发散，并且为自身所使用的方法而困扰。1994年春，我在参加一个由圣菲研究所举办的题为"科学知识的限度"的讨论会时，似乎看到了科学未来的缩影。在三天的会期里，一大群的思想家，包括数学家、物理学家、生物学家和经济学家，思考了这样一个问题：科学是否有限度？如果有，那么科学能否通过自身力量知道其限度何在？会议是由两位与圣菲研究所有关的人组织的，其中一位组织者是约翰·凯斯蒂（John Casti），曾写过无数本数学和科普读物的数学家；另一位是约瑟夫·特鲁伯（Joseph Traub），一个计算机理论方面的科学家，哥伦比亚大学的教授。[1]

我之所以参加这次会议，有很大一部分原因是为了会晤格里高利·蔡汀，他是供职于IBM的数学家和计算机学家，从20世纪60年代初开始，一直致力于用其所谓的算法信息理论来解释和扩展哥德尔定理。就我所知，蔡汀已经接近于给出对这样一种思想的证明，即一个关于复杂性的数学理论是不可能的。在见到蔡汀之前，我把他想象成一个性格怪僻、外貌寒酸、长着一对毛茸茸的大耳朵且具有东欧口音的人，最为根本的是，某种古旧的哲学焦虑充斥在其关于数学限度的研究之中。但是，蔡汀和我的这种想象毫无共同之处，他粗壮、秃顶而且孩子气，穿着一身新颓废派装束：肥大的白色长裤，印有马蒂斯速写的黑色T恤衫，穿着袜子和凉鞋。他比我预想的要年轻，后来我才知道，在1965年，年仅18岁的他就已发表了第一篇论文，而其过人的活力使他看起来更显得年轻。蔡汀说话总是要么就越说越快，仿佛他已被自己的言辞打动了；要么就越说越慢，好像是因为他已意识到自己正在接近人类理解力的限度，应该放慢速度。他说话的速度和音量构成了相互交叠的正弦曲线，在费力地组织某个想法时，他会紧闭双眼，头部倾斜，似乎要从那不合作的大脑中把语言赶出来。[2]

　　与会者围坐在一个很长的长方形桌子周围，而桌子则放在一个很长的长方形房间里，房间的一端有一块黑板。凯斯蒂用一个问题拉开了会议序幕："现实世界是不是复杂得我们已无法理解了？"凯斯蒂解释说，哥德尔不完备性定理蕴含着这样一个结论，即数学描述总是不完备的，世界的某些方面总是抗拒对它的描述。类似地，图灵（Alan Turing）也表明了很多数学命题是"不可判定的"，也就是说，一个人不可能在有限的时间内判定这个命题是对还是错。特鲁伯试图用一种更为积极的态度来重述这个问题："我们能知道什么是我们不能知道的吗？我们能证明数学计算是有限度的，就像哥德尔和图灵证明数学和计算是有限度的一样吗？"

　　E.阿特立·杰克逊（E. Atlee Jackson），一位来自伊利诺伊州立大学的物理学家，宣称构造这种证明的唯一方法，就是构造出一种科学的形式表征。为了说明这一任务是如何之困难，杰克逊跳到黑板前面，画了一幅似乎是代表科学的极为复杂的流程图。当听众们一脸茫然地看着他的时候，他求助于一句格言来解释他的思想。他说："要想确定科学是否有限度，你就必须给科学下定义；而一旦你给出一个科学的定义，实际上就已经为科学设定了限度。"接着，他又补充道："我虽然不能定义我的妻子，但我却能认出她来。"在得到一阵礼貌的笑声作为奖励之后，杰克逊又回到了自己的座位上。

　　反混沌理论家斯图亚特·考夫曼不断地溜进会场发表一些禅语般的简短讲话，然后又溜出会场。在某一次露面的时候，他提醒我们说，我们能否生存下来，依赖于我们对世界进行分类的能力，我们能以不同的方式"切开"或者说区分它。为了对现象进行分类，我们就必须抛弃一些信息。考夫曼用下面这段咒语式的东西作了一个总结："存在就是去分类，就是去行动，这两者都意味着抛开一些信息。所以，恰恰是求知的行为本身却要求着无知。"他的听众看起来一齐被激怒了。

　　接着拉尔夫·高曼瑞（Ralph Gomory）说了几句。高曼瑞是IBM研究部门的前任副主管，他现在正领导着斯洛恩基金会，一个为推动与科学有关的各种计划而建立的慈善机构，这次圣菲会议就是由该基金会资助的。在听别人说话的时候，甚至在自己说话时，高曼瑞总带着一种不轻易相信任何事情的表情。他把头向前伸出，就好像正透过两个隐形的双焦点透镜看着什么，同时皱紧又粗又黑的眉毛，把它们拧成一个疙瘩。

高曼瑞声明，他之所以决定资助这个讨论会，是因为他一直感觉到现行的教育系统过分强调已知的东西，而对未知甚至不可知的东西强调得不够。大多数人根本没有意识到我们已知的东西是多么少。高曼瑞说，这是因为现行教育系统所教授的关于实在的观点，是一种无衣无缝、毫不矛盾的观点。比如，我们关于古代波斯战争的所有知识都来自于同一个来源：希罗多德（Herodotus）。我们怎么知道希罗多德是一个忠实的记录者呢？也许他只获得了不全面的或不准确的信息！也许他心怀偏见，也许整个事情根本就是他自己编的，我们永远也无法知道！

过了一会儿，高曼瑞又谈道，通过观察地球人玩国际象棋，一个火星人也许能正确地推出它的规则。但是火星人能确定这些规则是正确的规则，或者，是唯一可能的规则吗？每个人都对高曼瑞的疑问沉思了一会儿。随后，考夫曼推测了维特根斯坦将会如何回答这个问题。考夫曼说，维特根斯坦在思考棋手走出一步破坏规则的棋——且无论其是否故意为之——的可能性时，会"极端痛苦"；毕竟，火星人怎么能知道这一步棋究竟仅仅是个错误，还是另一个规则的结果？"明白了吗？"考夫曼问高曼瑞。

"我不知道谁是维特根斯坦，没半点儿印象。"高曼瑞有点不耐烦地回答。

考夫曼扬起了眉毛："他是个非常有名的哲学家。"

他和高曼瑞两人斗鸡般相互盯着，直到有人说了声："别管维特根斯坦了！"

派屈克·萨皮斯（Patrick Suppes），一位来自斯坦福大学的哲学家，不断地打断讨论并指出康德在其关于二律背反的讨论中，实际上已经预见到了所有他们在讨论会上思考的问题。最后，当萨皮斯提出另一个二律背反的时候，有一个声音高叫道："别再提康德了！"萨皮斯则抗议说，他只想再提一个二律背反，而且非常重要。但是，他的同伴们驳回了他的建议（显然，他们并不想被人告知，说他们仅仅是在用一些时髦行话和隐喻，重新陈述着在很久以前——不仅是康德，甚至古希腊人——就已提出过的论点）。

说话像打机关枪的蔡汀终于把话题拉回到了哥德尔身上。蔡汀说，不完备性定理只是从数学中提出来的一组深刻问题中的一个，而不是像大多数数学家所认为的那样，只是一种与数学和科学的进步毫无关系的悖论摆设。

"有些人把哥德尔定理当成是一种荒唐的、病态的、从自指悖论中得到的东西，"蔡汀说道，"哥德尔本人有时也担心这只是我们玩弄文字游戏时制造出来的悖论。而现在，不完备性看起来如此自然，你可以去问一下我们的数学家对此有什么办法！"

蔡汀关于算法信息理论的研究表明，当数学家试图解决复杂性不断增加的问题时，他们必须不断地增加公理；换句话说，为了知道得更多，人们必须假设得更多。蔡汀认为，其后果是数学必然越来越像一种实验科学，越来越疏离绝对的真理性。蔡汀还论证说，就像自然蕴含着基本的不确定性和随机性一样，数学也同样如此。他最近发现一个数学方程，该方程根据变量的数值不同，可能有有限或无限多个解。

"一般而言，你都会承认如果一个人认为某件事情是真的，他是由于某个理由才这样认为的。在数学中，一个理由称为一个证明，而数学家的工作就是从公理或预设的原则出发，去寻找证明和理由。而现在我所发现的是一种根本没有任何理由能证明其为真的数学真理，它们是偶然地或随机地为真的。而这就是我们永远无法找到真理的原因：因为没有什么真理，没有任何理由能说明它们是真的。"

蔡汀还证明，人们不可能确定一个给定的计算机程序，究竟是不是关于某一问题的各种解法中最简练的一种：我们总是可以给出一种更为简练的程序。（这一发现的后果是，其他人插言指出，物理学家永远也无法肯定他们是否找到了一个真正的终极理论，一个能给自然一种可能的最为紧凑简洁的论述。）很明显，蔡汀为自己是这一可怕潮流将要到来的预报者而扬扬自得，看起来他正在为推倒数学和科学神殿的想法而陶醉。

凯斯蒂反驳道，数学家可以通过使用简单的形式系统来避开哥德尔效应——比如仅含加法和减法的算术系统（不含乘法和除法）。凯斯蒂又说，非形式的演绎推理系统同样也可能避开这个问题。在讨论自然科学问题时，哥德尔定理可能最后被证明仅仅是一种障眼法。

弗朗西斯科·安东尼奥·"奇科"·多利亚（Francisco Antonio "Chico" Doria），一位巴西数学家，同样认为蔡汀的分析过于悲观了。他说，由哥德尔定理标志出的数学的限度，根本不能将数学带向终结，而只会丰富它。比如，多利亚建议说，当数学家遇到一个明显的不可判定的命题时，他们就可能创造出两个新的数学体系，其中的一个假设这一命题为真，而另一个则假设它为假。"和得到一个知识的限度正相反,"多利亚总结道,"我们可以得到大量的知识。"

在听多利亚说话时,蔡汀不断地转动着他的眼睛,萨皮斯看起来也是满腹狐疑。萨皮斯慢腾腾地说道,任意地设定不可判定的数学陈述的真假,具有"盗窃相比诚实劳动而言所具有的所有好处"。他把他的这一讥讽追溯到某个大名鼎鼎的人物那里。

谈话的方向在不断地改变,就好像趋向于某个奇怪吸引子一样,趋向那些具有哲学倾向的数学家和物理学家们最喜欢的话题之一——连续统问题。实在是光滑的还是凹凸不平的?是模拟的还是数字的?是所谓的实数(那种能被无限分割的数)还是整数能够最好地描述世界?从牛顿到爱因斯坦的物理学家都依赖于实数,但量子力学表明物质和能量,甚至还有时间和空间(在极小的尺度下)都是离散的,不可分的整体。计算机同样把所有的东西都表征成整数:"0"和"1"。

蔡汀痛斥实数为"废话",考虑到噪音和模糊性,关于世界的实数精确性只不过是个摆设。他宣称:"物理学家都知道,每一个方程都是一个谎言。"

有人用毕加索的名言予以回击:"艺术是帮助我们洞察真理的谎言。"

特鲁伯附和道,实数当然是一种抽象物,但却是一种极为有力、有效的抽象物。当然总会有噪音,但是我们可以用实数系统去处理噪音。数学模型能够抓住某些事物的本质,而且也并没有谁假称它把握住了整个现象。

萨皮斯大踏步地走到黑板前面,写下了一些方程,这些方程被萨皮斯认为可以一劳永逸地取消连续统问题,但听众们看起来却并不为之所动。(在我看来,这是哲学的主要问题所在:没有人真的愿意看到哲学问题被解决了,因为那样的话他们就再也不会有什么谈资了。)

其他与会者提出科学家所实际面对的知识障碍,其抽象性要比不完备性问题、不可判定性问题、连续统问题等种种问题的抽象性小得多,这些人中的一个是比耶特·哈特(Piet Hut),他是一位来自丹麦高等研究院的天体物理学家。他说,借助于强有力的统计工具和计算机的帮助,他和他的天体物理学同事学会了怎样克服著名的N体问题。而这一问题,即预测三个或多个通过引力相互作用的天体的运动规律,原本被认为是不可能的。计算机现在可以模拟包括亿万颗星星的整个银河系的演化过程,甚至星系团的演化过程。

但是哈特补充说,天文学家还面对着其他看起来不可能逾越的障碍。他们只有一个宇宙可供研究,所以不可能在它上面作受控实验。宇宙学家只能从现在往后追溯宇宙的历史，但是却永远也不会知道大爆炸之前的情况，或宇宙边界之外的情况——如果真有什么边界的话。另外，粒子物理学家在检验那些关于统一引力和自然界其他力的理论（比如超弦理论）时可能会一筹莫展，因为只有在大尺度和超过任何可想象的加速器的能力范围之外的能量状态下，这些效应才会变得明显。

一个类似的悲观论调出自罗尔夫·兰道尔（Ralf Landauer），他是IBM的物理学家，也是计算之物理极限研究的先驱之一。兰道尔带有德国口音的咆哮声，使他那带有讽刺意味的幽默变得更加锋芒毕露。当某位发言者一遍又一遍地表述自己的观点时，兰道尔发话了："也许你的观点很清楚，但你这个人却夹缠不清。"

兰道尔说科学家不能依赖于计算机来保持其能力的无限增长。他承认，很多曾经被认为是由热力学第二定律或量子力学加之于计算能力上的物理限制，现在已经被证明并不存在。但是，计算机制造厂的生产成本增长得如此之快，以至于他们威胁要中止长达几十年的计算价格不断下降的趋势。兰道尔同时还怀疑，计算机设计者是否能够把像叠加——也就是量子实体同时出现在一个以上位置的能力——这样奇异的量子效应很快地推向实用，并由此突破现有的计算能力，就像某些理论家所建议的那样。兰道尔论证说，这样的系统将会变得对微小的量子级的扰动也十分敏感，以至于实际上变得毫无用处。

布利恩·阿瑟（Brian Arthur）来自圣菲研究所，是一位红光满面的经济学家，说话带着轻快的爱尔兰口音。他将话题引向经济学的限度。他说，在试图预测股票市场的表现时，一个投资者必须对其他人将会怎样猜测进行猜测，以此类推，以至无穷。经济学领域本质上是主观的、精神性的，因此是不可预测的；不确定性只是"被系统过滤了"。一旦经济学家们决定简化其模型，比如说通过假设投资者可以拥有对市场或代表着某种真正价值之价格的良好知识，模型就变得不切实际了。两个具有无限智力的经济学家，可能会对同一个系统作出不同的结论，所有的经济学家所真正能干的，就是说一些诸如"可能是这样，可能是那样"的话。阿瑟补充道："如果你在股票市场赚了大钱，所有的经济学家都会听你的。"

然后考夫曼以另一种更为抽象的方式重述了阿瑟所说的东西。人类是这样一种"因子"，为了回应其他因子的"内在模型"的改变，必须不断地改变自己的内在模型，由此构成了一种"复杂的、相互适应的境域"。

　　兰道尔沉着脸，打断了他的话，说有一些比这类主观因素更为明显的理由使经济现象不可预测。艾滋病，第三次世界大战，甚至大投资公司的首席分析员拉肚子，都能对经济产生深刻的影响。有什么模型能预测这样一些事件呢？

　　罗杰·色帕德（Roger Shepard）是来自斯坦福大学的心理学家。他一直默默地听着，现在终于发言了。色帕德看上去有一点轻度的忧郁症，但这也可能只是由于他那象牙色的、下垂的胡子所产生的错觉，或者是由于他沉迷于那些无法回答的问题而产生的副产品。色帕德说他到这里来的一部分原因，是弄明白科学真理或数学真理到底是被发明的还是被发现的。他最近还对科学知识到底存在于何处思考了很多，而且还得出了一个结论，即它不可能独立于人类的心智而存在。比如一本没有人去读的物理教科书，就仅仅是纸和墨迹。但是，这就提出了一个色帕德认为令人困扰的问题：科学显得越来越复杂和越来越难以让人理解。看起来，将来有可能出现这种情况：即使最聪明的科学家也无法理解某些科学理论，比如关于人类心智的理论。色帕德说道："也许我太老派了。"但是，如果一个理论如此复杂以至于没有人能明白它，那么我们从中又能得到什么益处呢？

　　特鲁伯也被同样的问题困扰着。我们人类可能相信奥卡姆剃刀，即最好的理论总是最简单的理论，因为只有这样的理论才是我们可怜的大脑所能理解的。但是计算机没有这种限度，他补充说，也许计算机会成为未来的科学家。

　　在生物学中，"奥卡姆剃刀会置人于死地的"，有人绝望地补充道。

　　高曼瑞认为，如果世界根本上是无法理解的话，科学的任务就是找到实在中那些可以理解的部分。高曼瑞建议说，一种把世界变得更容易理解的方法，是把它变得人工化，因为人工系统总是要比自然系统更加容易理解，更加容易预测，比如，为了使天气预报更加容易，社会应该把世界放入一个透明圆罩中。

　　所有的人都盯着高曼瑞看了一会，然后特鲁伯评论道："我想拉尔夫的意思是说制造未来要比预测未来更容易一些。"

　　随着会议的进行，奥托·罗塞勒（Otto Rössler）的发言显得越来越有意义，要不就是其他人的发言变得越来越没意义了？罗塞勒是理论生物化学家和混沌理论家，来自德国的图宾根大学。他在20世纪70年代中期发现了一个数学怪物，即罗塞勒吸引子。他的满头白发看起来总是非常乱，就好像他刚从昏睡中苏醒过来一样。他有一副夸张的、木偶式的相貌：一双总带着些吃惊神情的眼睛，隆起的下嘴唇，球状的下巴上布满深深的皱褶。我和其他人都不太懂他的话（而且我怀疑是否有任何人能懂）。但是当他结结巴巴地低语时，每个人都向他前倾了身子，就好像他是位先知。

罗塞勒看到了两个基本的知识限度。一个是不可通达性，比如，我们永远也不能确切地知道宇宙的起源，因为无论是从空间还是从时间上说，那都离我们太远了。另外一个限度是歪曲，这个限度更糟：世界会欺骗我们，让我们觉得理解了它，可实际上却没有。如果我们能站在宇宙之外（罗塞勒这样建议道），那么我们就能知道我们知识的限度。但是我们却置身于宇宙之中，所以关于我们自身限度的知识只能是不完整的。

罗塞勒提了几个问题，据他说，这些问题是由18世纪的一位名叫罗杰·博斯科维奇（Roger Boscovich）的物理学家首先提出来的：如果一个人站在一个有着绝对黑暗的天空的星球上，他能确切地知道这个星球是否在旋转吗？如果地球和我们一起呼吸，我们能觉察到它在呼吸吗？也许不能。按照罗塞勒的说法："存在着这样一些情况，即你不可能从内部发现事情的真相。"他补充道，只是通过一些思想实验，我们就能找到一个超越这些知觉限制的方法。

罗塞勒说得越多，就越使我感觉到一种与他的想法的共鸣。在一次休息时，我问他是否认为智能计算机会超越人类科学的限度。他非常坚定地摇了摇头："不，不可能。"他用一阵激烈的低语答复道："我可以用海豚或抹香鲸为例来回答这一问题，它们有地球上最大的大脑。"罗塞勒告诉我说，当一条抹香鲸被捕鲸者击中时，其他的鲸有时候会聚集在它的周围，围成一个星形图案，然后自己也被捕杀。"人们往往会认为这只是一种盲目的本能，"罗塞勒说，"但实际上它们只是在向人类证明它们要比人类进化得更高级。"我只能点头。

在会议即将结束的时候，特鲁伯建议大家分成专题小组来讨论各个具体领域中的限度：物理、数学、生物、社会科学。一个社会学家直言他不想参加社会科学组，他到这里来就是为了和他的专业领域之外的人交流，并向他们学习。他的话激起了其他人的仿效。有人指出，如果所有的人都和这位社会科学家一样，那么情况就会变成社会科学组中没有社会科学家、生物组里没有生物学家了。特鲁伯说，他只是提一个建议，他的同伴们当然可以按照他们自己的选择去参加讨论。下一个问题是，这些不同的组在什么地方进行讨论。有些人建议说应该分散到不同的房间中去，这样一些大嗓门的发言者就不会影响到其他组的讨论。这时所有的人都盯着蔡汀，他那降低自己发言音量的许诺受到了嘲弄。又是一阵讨论。兰道尔对此说道，对一个简单问题投入过多的智力也不是一件好事。正当一切看起来已经没有指望的时候，各小组自发地组织起来了，并大致遵照着特鲁伯最初的建议，而且分散到了不同的地方。我想，这就是圣菲研究所的成员们所说的自组织或从混沌走向有序现象的令人印象深刻的体现吧。也许，这就是生命起源的方式。

我跟在数学组的后面，这个组里有蔡汀、兰道尔、色帕德、多利亚和罗塞勒。我们找到了一个空闲的、有黑板的休息室。有那么几分钟，每个人都说了一些应该说的话；之后，罗塞勒走到黑板之前，很潦草地写了一个最近发现的式子，这个式子给出了一个极为有趣的、复杂的数学对象——"一切分形之母"。兰道尔很礼貌地问罗塞勒，这个分形有什么意义。它"能安抚你的大脑"，罗塞勒答道，而且它还满足了他个人的一个愿望，即物理学家也许能利用这类混沌的但又是经典的方程来描述实在，这样就能取消量子力学那可恶的不确定性。

色帕德插进来说，他之所以参加数学小组，是因为他想让数学家来告诉他到底数学真理是被发现的还是被发明的。每个人都对此说了一会儿，但是没有得到最终的结论。蔡汀说，大多数数学家都倾向于发现的观点，但爱因斯坦却是一个发明论者。

在一次休息的时候，蔡汀又一次说数学已经死亡了。未来的数学家将只能依赖于大量的计算机计算来解决问题，而这些计算由于过分复杂而无人能懂。

看起来，每个人都烦透了蔡汀。数学确实在起作用，兰道尔咆哮道，它能帮助科学家解决问题。显然它并没有死。其他人也添油加醋地指责蔡汀言过其实。

蔡汀第一次缓和了下来。他补充说，他的悲观主义也许和那天早上吃了太多的硬面包圈有关。他说德国哲学家叔本华的悲观主义可以溯因到他那不太健康的肝——他宣扬自杀是对生存自由的最高表达。

斯蒂恩·拉斯缪森（Steen Rasmussen），一位物理学家，也是圣菲的正式成员，重新表述了为人所熟知的混杂学家的观点，即传统的还原主义方法不能解决复杂问题。他说科学需要一个"新牛顿"，一个能发明出全新的、解决复杂性问题的概念方法和数学方法的人。

兰道尔大声斥责拉斯缪森也沾染上了危害着很多圣菲研究人员的"痼疾"，即一种对"伟大的宗教式洞见"的迷信，认为这种洞见能一下子解决所有的问题。科学不是以这种方式进行的，不同的问题需要不同的工具和方法来解决。

当罗塞勒把自己从一段很长的、混乱的独白中解脱出来的时候，似乎表达了这样一个意思：我们的大脑只表达了世界所提出的许多问题的一种解决方案，进化本应创造出表达别种解答的另一种大脑来。

兰道尔似乎对罗塞勒怀有某种奇怪的戒备心理，他礼貌地问罗塞勒是否认为我们可能会为了获得更多的知识而换一种大脑。"只有一条路，"罗塞勒双眼盯着他面前桌子上某件隐形的东西回答道："发疯。"出现了一阵令人尴尬的沉默。接着，一个关于"复杂性"究竟是一个有用的术语，还是由于过分宽泛的定义已经变得毫无意义因而应该被抛弃掉的讨论热烈展开了。蔡汀说，就算像"混沌"和"复杂性"这样的术语，也已经没有多少科学的意义了，但是从公关角度来说它们还是有用的。特鲁伯补充说，理论物理学家塞思·劳埃德（Seth Lloyd）曾经至少列举出31种关于复杂性的不同定义。

"我们从复杂性走向困惑性。"多利亚吟起了诗。每个人都在点头，而且为了这句小诗而夸奖他。

当专题小组重新集合起来的时候，特鲁伯请每个人都来思考这样两个问题：我们已经学到了什么？还有什么尚未解决？

蔡汀则滔滔不绝地提了一大串问题：什么是元数学的限度？什么又是元元数学的限度？什么是我们能够知道的限度的限度？这种知识有限度吗？我们能模拟整个宇宙吗？如果能，我们能造出一个比上帝所造的还要好的宇宙吗？

"还有，我们能搬到那儿去住吗？"有人开玩笑地补充道。

李·西格尔（Lee Segel），一位以色列生物学家，警告他们在公开场合谈论这些问题要小心谨慎，以免助长社会上的反科学情绪。他继续说道，毕竟，已有太多的人认为爱因斯坦已经证明了一切都是相对的，而哥德尔则证明了没有任何东西能被证明。每个人都非常严肃地点了点头。西格尔自信地补充道，科学具有一种分形结构，而且很明显我们所能研究的事物是没有限度的。每个人又都点了点头。

罗塞勒提出了一个新词来描述他和伙伴们正在干的事情：限度学（limitology）。限度学是一项后现代事业，罗塞勒说，它是这个世纪不断解构实在的努力的产物。当然，康德早就思考过知识的限度问题；伟大的苏格兰物理学家麦克斯韦（Maxwell）也是。麦克斯韦设想了一个微型人，或者说一个小精灵，来帮助我们突破热力学第二定律的限制。但是，罗塞勒说，我们从麦克斯韦小妖中学到的唯一的东西就是：我们是在一个热力学牢笼中，这是一个我们永远不能逃离的牢笼。当我们从世界收集信息的时候，我们也增加了世界上的熵，从而增加了它的不可知性。我们正无可避免地走向热寂。"整个关于科学限度的讨论都是关于精灵的讨论，"罗塞勒嘶嘶地说，"我们是在和精灵搏斗。"

哈德逊河畔的一次小聚

每个人都同意这次讨论会是富有成果的，有些与会者告诉作为组织者之一的约瑟夫·特鲁伯，说这是他们所参加过的最好的会议。一年多以后，拉尔夫·高曼瑞同意从斯洛恩基金会中为将来在圣菲和别的地方举行的会议提供资金。比耶特·哈特、奥托·罗塞勒、罗杰·色帕德和罗伯特·罗森（Robert Rosen），一位同样参加了讨论会的加拿大生物学家，聚集在一起要写一本关于科学的限度的书。在听到他们准备论证说科学有一个十分美好的未来时，我并不觉得十分惊异。"失败主义的态度对我们没有任何好处。"色帕德坚定地告诉我。

在我看来，圣菲会议实际上是以一种杂乱的形式重演了岗瑟·斯滕特在四分之一世纪之前以优美的形式提出过的许多论证。和斯滕特一样，那些与会者承认科学面临着物理的、社会的和认知的限度。但是这些真理的追求者看起来没能像斯滕特一样，把他们的论证引向其逻辑结论。没有一个人能够接受科学——被定义为对那些关于自然的可理解的、可被经验证实的真理的研究——会很快结束或已经结束的结论。我当时以为，除了格雷高里·蔡汀之外，没有一个人愿意承认这一点。在所有的会议发言人中，他看起来是最乐于承认科学与数学可能正在超越人类的认知限度的人。

因此，圣菲会议几个月之后，我满怀希望地安排了一次和蔡汀的会晤，地点是离我们各自的家都很近的哈德逊河畔的一个乡村：纽约州的冷泉镇，我们在位于小镇的微型主干道上的一家咖啡店里共享了咖啡和烤饼之后，散步到河边的一个码头上。河对岸是暴风王山和西点军校那雄伟的要塞，鸥鸟在我们的头上盘旋。[3]

当我告诉蔡汀我正在写一本关于科学可能已进入一个收益递减时代的书时，我期待着一种赞同，但他却怀疑地大笑起来："真的？我希望这不是真的，因为如果真的是这样，那就太无聊了。每个时代都在这么想，那是谁来着？是开尔文勋爵吗？他说我们需要做的唯一事情就是把小数点后面的位数提高。"当我提到历史学家从来没有找到证据说开尔文勋爵确实说过这种话时，蔡汀耸了耸肩："瞧瞧，我们还有多少事情不知道！我们不知道大脑是如何工作的，不知道什么是记忆，也不知道什么是衰老。"如果我们能解决我们为什么会变老，也许我们就能找到一种延缓衰老过程的方法，蔡汀这样说道。

我提醒蔡汀，在圣菲他曾经表示数学甚至整个科学都可能已接近它们的终极限度。"我只是想唤醒人们，"他答复道，"当时听众们死气沉沉的。"他自己的著作，他强调说，代表了一个开端，而不是完结。"也许这产生了负面的效应，但我把它理解为告诉人们如何去寻找一种新型的数学真理的方法：像一个物理学家那样去做，更加经验地研究数学，增加新的公理。"

蔡汀说，如果他不是一个乐观主义者，他就不可能去研究数学的限度问题。悲观主义者会看着哥德尔定理，然后他们就会喝着苏格兰威士忌直到死于肝硬化。尽管现在的人类处境可能和几千年前一样"一团糟"，但是我们在科学技术中获得了巨大的进步，这一点却是无可置疑的。"当我还是一个小孩的时候，每个人都怀着一种神秘的尊敬之情谈论哥德尔。他又几乎是不可理解的、但又肯定是非常深刻的思想。于是我想知道他到底在说些什么，以及为什么是对的，而且我成功了！这使我成为一个乐观主义者。我想我们知道的东西很少，我希望我们知道的东西很少，因为这样生活会更有意思。"

蔡汀回忆起有一次他和物理学家理查德·费曼争论有关科学的限度问题，那是在20世纪80年代召开的一次关于计算的会议上，在费曼去世前不久。当蔡汀提出科学才刚刚开始时，费曼变得非常愤怒。"他说我们已经知道了关于日常生活中一切事物的物理学知识，剩下的东西是无关紧要的。"

费曼的态度令蔡汀十分困惑，直到他得知费曼正为癌症所苦时才不再困惑了。"对于费曼来说，像他这样能够完成如此伟大工作的人，不可能有这样悲观的态度。但是在他生命的尽头，当这个可怜的人知道他已经来日无多的时候，我就能够理解他为什么会有这种观点了。"蔡汀说，"如果一个人自知死期将至，他就不想错过所有的乐趣，他不想让自己有这种感觉，即还有某些关于物理世界的美妙的理论与知识，是他还不知道的，而且再也不可能知道了。"

我问蔡汀是否听说过一本题为《黄金时代的来临》的书，蔡汀摇了摇头。我就把斯滕特关于科学的终结的论证简要地复述了一遍。蔡汀转了转眼珠，问我斯滕特写这本书时的年龄。三十左右吧，我回答道。"也许他的肝脏有毛病，"蔡汀如此说道，"也许他的女朋友抛弃了他。一般说来，当男人们发现他们不能像以往那样和妻子做爱时，他们就会开始写这样的东西。"实际上，我回答道，斯滕特是在20世纪60年代在伯克利写的这本书。"噢，原来如此。"蔡汀欢呼道。

蔡汀并不为斯滕特的观点（即大多数人并不是为了自己才去研究科学）所打动。"从来不是这样，"蔡汀回击道，"那些作出优异科学工作的人总是一群狂人，其他人则只关心生存的问题，偿还他们的贷款，孩子生病了，妻子需要钱，或者她正准备和什么人私奔。"他大笑着说："想想量子力学，这个杰作是被一些人当作业余爱好，在没有任何资金的情况下完成的。量子力学和核物理学就像希腊诗歌一样。"

蔡汀说，幸亏只有少数几个人把自己献身于对伟大问题的探求上。"如果人人都想理解数学的限度，或作出伟大的作品来，那才是巨大的灾难呢。抽水马桶会堵住，电器会坏掉，建筑物会倒塌！我的意思是，如果每个人都想去搞伟大的艺术或深刻的科学，世界就会乱了套！只有我们几个人做这种事是件好事！"

蔡汀承认，粒子物理学看起来因为加速器所需的巨额费用而停滞不前了，但是他相信射电望远镜仍然可以通过揭示由中子星、黑洞和其他奇异的东西产生的猛烈过程，为物理学带来突破。但是，我问道，是否有可能出现这样的情况，即所有这些新的发现不是把物理学、宇宙学带向更为精确或更为一致的理论，而是表明建构这些理论根本是徒劳无益的呢？他自己的数学研究结论也表明，当人处理越来越复杂的现象时，必须不断地扩张其公理的基础，这看起来就起到了类似的作用。"啊哈，那么它们会变得更像生物学？也许你是对的。但是我们仍然是对世界有了更多的了解。"蔡汀答道。

蔡汀断言，科学技术的进步同时还在很多领域里降低了设备的费用，"如今在很多领域里，用很少一点钱所能买到的东西，简直是令人震惊"。计算机对他的工作而言是至关重要的。蔡汀还发明了一种新的计算机程序语言，用来使他的关于数学限度的想法变得更为具体。他把自己的著作《数学的限度》在国际互联网上传播。"国际互联网正在把人们联系起来，并让某些以前从来未发生的事情变得有可能发生。"

到了将来，蔡汀预言道，人类也许能够通过基因工程来提高他们的智力，或者，通过把自己输入到计算机中而做到这一点。"我们的后代的智力和我们相比，可能就像我们的智力和蚂蚁的智力相比一样。""假如每个人都吸海洛因，萎靡不振，整天看电视，你知道我们是不会达到这些目标的。"蔡汀停了一会，大声说道，"人类会有一个未来，如果他们配得到一个未来！如果他们消沉了，就没有未来！"

当然，科学总是会因为文明的终结而终结，蔡汀补充道。向着河对面嶙峋的山峰挥了挥手，他指出，这条河道是在上一个冰期由冰川切割出来的，仅仅是在一万年以前，冰雪还覆盖着这里整个区域。下一个冰期也许会摧毁整个人类文明，但就算是这样，宇宙中的其他生命也会继续追求知识。"我不知道是否存在其他的生命，我希望有，因为看起来他们不会把事情弄得一团糟。"

我张开嘴想要说，在未来科学研究也许会由智能机来进行，这是有可能的。但是蔡汀开始说得越来越快，越来越快，而且进入了某种狂乱的状态之中，让我无法开口。"你是一个悲观主义者！你是一个悲观主义者！"他大叫道，并用一些我在此之前告诉他的事情来提醒我，就是我的妻子已经怀上了我们的第二个孩子了。"你就要有一个孩子了！你得乐观起来，你应该乐观起来！我才应该悲观！我比你老，我没有孩子！IBM的业绩很不好！"一架飞机掠过，鸥鸟四散而飞，蔡汀嗥嗥的笑声越过浩渺的哈德逊河，消散在远方，没有一点回声。

历史的终结

实际上，蔡汀自己的工作经历和岗瑟·斯滕特的收益递减图景符合得相当好。算法信息理论并不是一个真正的新发展，而是对哥德尔洞见的一个扩展。蔡汀的工作同样支持斯滕特的另一观点，即科学在探测那些愈益复杂的现象的过程中，即将用尽我们先天的公理。不过，斯滕特为他的悲观预言留了一些透气孔：社会可能会变得非常富有，以至于可以为最异想天开的科学实验，比如说环绕地球的粒子加速器之类，提供经费而不考虑代价；科学家也可能取得一些突破性的进展，比如超光速传输系统或提高智能的基因工程技术，从而能使我们超越物理的和认知的限度。我们可以在这张单子上再加上一种可能性，科学家也许能发现地外生命，创造出比较生物学中的一个壮丽的新时代。如果剔除这些成就，科学就会逐渐以收益递减的方式走向停顿。

那么人类会变成什么样子呢？在《黄金时代的来临》中，斯滕特提出，科学在它终结之前，也许至少能把我们从压力最大的社会问题中解放出来，比如贫困和疾病甚至国家之间的冲突。未来将会变得和平而又安逸，如果不嫌它过于沉闷的话。大多数人将会终生以追逐享乐为己任。1992年，弗朗西斯·福山（Francis Fukuyama）在其《历史的终结》中提出了一种颇为不同的未来图景。⁴福山，一个于布什政府期间曾在国务院任职的政治理论家，把历史定义为人类为了寻找最明智的——至少是危害最小的——政治体系而做的斗争。按照福山的说法，截至20世纪，资本主义的自由民主制度一直是最好的选择。它只有一个真正的挑战者，那就是马克思主义者的社会主义。

福山接着考虑了由他的理论所产生的深刻问题：政治斗争时代结束了，我们下面该做些什么？我们为什么而活着？人性的本质是什么？除了象征性地耸了耸肩膀之外，福山并没有给我们提供任何答案。他发愁地说道，自由与繁荣并不足以满足我们尼采式的权力意志和我们不断进行"自我超越"的需要。没有了大的意识形态斗争，我们人类可能会仅仅为了给自己找点事做而制造战争。

福山并未忽略科学在人类历史中的作用，恰恰相反，他的理论要求历史有一个进步的方向，而他认为科学能提供这种方向。科学对于现代民族国家的进步是至关重要的，因为科学是获得军事和经济实力的手段，但是，福山却连考虑都没考虑这样一种可能性：科学可以为后历史的人类提供一种共同的目标，这种目标是鼓励合作而不是冲突。

为了了解福山这一忽略的原因，我在1994年1月给在兰德公司的他打了一个电话，当他的《历史的终结》成为一本畅销书之后，他在那里得到了一份工作。他带着那种经常和怪人打交道但又并不以为有趣的人所特有的谨慎小心，回答了我的问题。一开始，他误解了我的问题：他以为我是在问科学是否能够帮助我们在后历史时代作出政治上和道德上的选择，而不是走向终结。福山严厉地给我上了一课：当代哲学的课程，科学至多是道德中立的，事实上，如果科学进步与社会和个体之中的道德进步不相适配的话，"能把我们带入一种比没有这种进步还要糟糕的境地中去"。

当福山最终明白了我的想法——科学可能为文明提供一种使大家联合起来的主题或目标——之后，他的语气就变得甚至更有优越感了。是的，有不少人曾经写信给他提出过这种观点。"我想他们是空间旅行的爱好者，"他窃笑道，"他们说，你知道，如果我们没有了意识形态斗争，我们可以在某种意义上和自然开战，就是说把科学前沿不断向前推进并且征服太阳系。"

他发出了另一阵轻蔑的笑声。"那么你并不把这些预言当真啦？"我问道。"不，一点也不。"他不耐烦地说。为了从他那里得到更多的情况，我向他指出，不仅是那些《星际旅行》的爱好者，很多杰出的科学家和哲学家都相信科学（即对纯知识的追求）代表着人类的命运。"嗯。"福山回答道。看来他并没有在听我说话，而是重新回到了我给他打电话之前正在阅读的引人入胜的黑格尔的小册子上去了。我挂断了电话。

甚至没有经过太多的考虑，福山就得出了斯滕特在《黄金时代的来临》中提出的同样的结论。从非常不同的角度出发，他们都看出了科学与其说是求知欲的副产品，倒还不如说是权力欲的副产品。福山以厌倦的方式表达出来的对把未来奉献给科学这一态度的反对，是意味深长的。人类的绝大多数，不仅仅是无知的群众，也包括福山这样有学问的人，都认为科学充其量是有点有趣，而且肯定不值得拿来作为全人类的目标。无论人类长远的命运是什么——是福山那永恒的战争，还是斯滕特那永恒的和平，或者，更有可能的，两者的某种混合——人类的目标都不太可能是追求科学知识。

星际旅行的因素

科学早已遗留给我们一笔非同寻常的遗产，它使我们得以描绘出整个宇宙（从夸克到类星体）和了解统治着物理和生物领域的基本规律，它产生了一个真正的创世神话。通过应用科学知识，我们获得了驾驭自然的令人畏惧的力量。但是科学仍然未能让我们摆脱贫困、疾病、暴力、仇恨，仍然留下许多尚未回答的问题，其中最重要的问题是：我们是必然的，还是仅仅是一种侥幸？还有，科学知识并没有令我们的生活更有意义，正好相反，科学迫使我们去面对我们存在的无意义（正如斯滕特所说的那样）。

科学的终结肯定会加剧我们的精神危机。那句老话是对的，在科学中，正如在其他事情中一样，真正重要的东西不是结果而是过程。当科学揭示着新的、可理解的复杂事物的时候，它起先唤醒了我们的惊异感，但是每一次发现，最终却都在平淡中结束。让我们来假设一个奇迹已经发生了，物理学家们用某种方式证实了所有的实在都是来自十维超空间中能量环的扭曲。物理学家们，或我们其余的人，又能为这一发现震惊多久呢？如果这就是终极真理了，就是说它已经蕴含了所有其他的可能性，那么这一窘迫境地反而会带来更多的麻烦，这一点也许能够解释为什么甚至像格雷高里·蔡汀这样的研究者——他的成就能引出别人的更多的研究成果——都发现自己难以接受这样一种观点：纯科学（即对知识的不懈追求）是有限的，更不要说已经结束了。但是对科学将永远继续下去的信念终究只是一个信念，一个源自我们那与生俱来的自负的信念。我们情不自禁地相信，我们自己是某个宇宙剧作家幻想出来的叙事剧中的演员，这个剧作家爱好悬念片、悲剧、喜剧和——最终的也是我们最希望拥有的——皆大欢喜的结局。而最为令人欢喜的结局，就是没有结局。

如果我的经验可以算数的话，甚至那些对科学仅仅偶尔怀着一些兴趣的人，也会觉得十分难以接受这样一种观点：科学已时日无多。原因不难理解，我们满脑子想的都是进步，无论是真正的还是虚假的。每年我们都会有更小更快的电脑、更豪华的汽车、更多的电视频道。我们关于进步的观点被那种可以被称为星际旅行的因素进一步歪曲了。在我们还没发明出能以超光速航行的太空飞船之前，或者，当我们还不能通过基因工程或电子修复术获得真正奇异的"特异功能"（那些在赛博朋克小说中描述的东西）之前，科学怎么能接近它的完结呢？科学自身——或者，不如说是反讽的科学——助长了这些幻想的传播和增殖。人们会在有声誉的、众人瞩目的物理学期刊上，发现关于时间旅行、远距离传输和平行宇宙的讨论，而且至少有一个诺贝尔物理学奖的获得者，布瑞安·约瑟夫森（Brian Josephson）曾经宣称，物理学在能够解释超感官知觉和心灵致动（telekinesis）之前，是不会完结的。[5]

但是，约瑟夫森在很早以前就为了神秘主义和超自然说而抛弃了真正的物理学。如果你真的相信现代物理学，你就不可能给予ESP（超感官知觉）或超光速飞船以太多的信任。你也就不可能相信（就像罗杰·彭罗斯和超弦理论家那样）物理学会最终找到一种在经验上有效的统一理论——一种将广义相对论和量子力学熔于一炉的理论。由统一理论所断定的在微观领域中展示出的现象，与我们可以想象的经验的距离，甚至比宇宙边缘和我们的距离还要远。只有一个科学奇想看起来有可能实现：也许有一天，我们会制造出某种机器，这种机器能超越我们物理的、社会的和认知的限度，离开了我们，它们仍然能继续进行对知识的追求。

第十章

科学神学，或机械科学的终结

尼 采告诉我们，人类只不过是通往"超人"的一块垫脚石、一座桥梁。
如果尼采能活到今天，他一定会赞同这样的观点，即超人并非有血
有肉的人，而是用硅造出来的。随着人类科学事业的没落，那些希望对知
识的探索将永远持续下去的人们，肯定不会再把希望寄托在智人（*homo
sapiens*）身上，而是寄希望于智能机：只有机器才能克服我们体能上、认识
能力上的弱点——克服我们的平庸。

在科学中有一个小小的、奇特的亚文化群，其成员们冥想着一旦或假如
智力挣脱了尘世的束缚，它将会怎样发展。这些人当然并非在搞科学，而是
在搞反讽的科学，或痴心妄想。他们所关心的并不是世界究竟是什么，而是
在几个世纪或千万年后，世界可能是或应该是什么。这一领域或可称为科学
神学，它所提供的可能只不过是某些古老的哲学问题，甚至神学问题的新视
角，这些问题包括：如果我们什么都能做得到，我们会做些什么？生命的目
的何在？知识的终极限度是什么？苦难是存在的必要部分吗？我们能达到永
恒的极乐吗？

科学神学的第一位现代实践者，是英国的化学家（也是一位马克思主义
者）贝尔纳（J. D. Bernal）。在其1929年的著作《世界、肉体和魔鬼》中，
贝尔纳提出，科学很快就会给我们带来能引导人类自身进化的力量。他设
想：首先，人类会尝试着通过基因工程完善自身；但最终我们将因为有了更
有效的设计，而抛弃这副由自然选择赐给我们的臭皮囊。

一点一点地，人的外形——地球上萌发的原始生命留下的遗产——将会逐
渐退化，到最后基本上完全消失了，可能保留下来的只是某些奇形怪状的遗

迹。新的生命（基本上没有什么物质成分，差不多全是原来的精神）将会取而代之，继续发展。这样一种变化，将会与生命在地球上的诞生同样重要，并且同样是渐进的，不可觉察的。最终，意识本身也会在一类变得特别稀薄（etherealized）的'人'中终结或消失，不再拥有紧密结合在一起的有机体，变成一团团太空中的原子，被宇宙中的射线传到四面八方，并被彻底分解。那可能标志着一种结局，抑或是一种开端，但在此时此刻却是无法预见了。[1]

汉斯·莫拉维克招惹口舌的"特殊智力儿童"

贝尔纳就像他的许多同类人一样，一旦考虑到智力进化的结束阶段，就开始苦于自己想象力的极度贫乏，苦于自己的缺乏耐性。贝尔纳的追随者们，如卡内基梅隆大学的机器人学工程师汉斯·莫拉维克（Hans Moravec），试图解决这一问题，结论却是一团糟。莫拉维克是个快乐的甚至有些轻狂的人，他看起来好像的确也被自己的观点陶醉了。当他在一次电话访谈中揭去其未来图景的面纱时，竟发出一阵持久的、几乎上气不接下气的笑声，其强度可与他所说的话的荒诞程度成正比。

莫拉维克在给出自己的评论之前，首先宣称科学迫切需要有新的目标。"20世纪所取得的大部分成就，其实都是19世纪思想的衍生物，"他说，"现在是刷新观念的时候了。"还有什么别的目标比创造出"特殊智力儿童"（mind children）——能作出我们所无法想象的壮举的智能机——更令人激动的呢？"你造出他们，然后施以某种教育，再往后就是他们自己的事了。你尽了最大努力，但你仍无法预见他们的生活。"

莫拉维克在《特殊智力儿童》中首次向公众披露了这一特殊物种可能怎样形成。这本书出版于1988年，那时私营公司与联邦政府正把大把的钞票投向人工智能和机器人的研制，[2] 尽管至今这些领域也并未兴旺起来，但莫拉维克仍然深信未来是属于机器人的。他向我保证，在20世纪末，工程师们就会创造出能做家务的机器人。"在20世纪的最后十年里，制造出一种真空吸尘机器人是完全可能的，我对此确信不疑，这再也不仅仅是停留在争论阶段的话题了。"（事实上，随着20世纪末的临近，家用机器人问世的希望看来仍十分渺茫。但不必担心，科学神学家总会有他们的说法。）

莫拉维克指出，到21世纪中叶，机器人的智能将会与人不相上下，并会基本上接管经济活动。"到那时，我们是真正的失业了。"莫拉维克纵声大笑起来。人们可能会继续"某些像写诗之类的古怪行为"，因为它们源自超出机器人能力之外的古怪心理活动，但机器人将从事所有那些重要的工作。"把一个大活人放在公司里是毫无益处的，他只会在那里帮倒忙。"

就乐观的一面来说，机器人将生产出足够多的财富，所以人们可以不必再去工作。机器人将消除前机器时代存在的贫困、战争以及其他一些灾难，"这不过是小菜一碟"。通过购买力，人们仍然能对机器人经营的公司加以管理。"我们可以选择从哪家公司买东西，以及拒斥哪家公司的产品。比方说，对于那些生产家用机器人的公司，我们会把钞票投给产品质量最优的那一家；人类也会联合起来抵制那些其产品或生产政策对人类有害的公司。"

莫拉维克继续说，毫无疑问，为了获得新资源，机器人会向外层空间扩张，他们会全方位开发宇宙，把太空原材料转化成信息处理机。在这一人类所无法达到的新前沿领域，借助于纯粹的计算机模拟，机器人会越来越有效地利用一切可得到的资源。"最终，每一个小小的量子的行为都具有了物理意义，人类最终进入了电脑空间，一个效率越来越高的计算机世界。"当生活在电脑空间中的人们试图更迅速地加工信息的时候，传递信息的速度看来会慢得让人难以忍受，因为这些信息竟然仍只能以光速传送。"因而，编码技术上的重大改进的效果，将使这个高效率的世界在范围上越来越大。"从某种意义上说，电脑空间将会比真实的物理世界更广大、更密集、更复杂、更有趣。

大多数人将最终抛弃其短命的、有血有肉的自我，以换取在电脑空间中更大的自由和永生。但莫拉维克猜测，仍然可能存在着这样一些冥顽不化的"原始"人，他们会拒绝："'不，我们绝不想变成机器的一部分'，从而形成某种未来社会中的阿米什人①。"机器人们会容忍这些"有恋祖癖的怪物们"的存在，让他们生活在类似于伊甸园或保护区这样的环境里。但地球"毕竟只是新世界系统中一粒小小的灰尘，尽管它确实具有无限丰富的历史意义"，那些垂涎地球上的原材料的机器人们，可能最终会强迫最后一位地球人搬入电脑空间的新家。

但是机器人要那么大的威力、那么多的资源干什么？他们会出于自己的利益而去追求科学吗？我问道。绝对正确，"这正是我的未来构想的精髓所在：我们的没有生物血缘关系的后代们，他们基本上没有人类的局限性，可

---

① Amish，美国东部山区原始荷兰移民的后裔，至今仍保存原有风格，衣着黑色，生活朴素——译者注。

以重新设计自己，他们肯定会追求关于事物的基本知识。"事实上，对于智能机器来说，追求科学将是他们唯一有价值的动机。"像艺术之类人们常引以为傲的东西，似乎并不十分深奥。对于机器人来说，那只不过是他们用以调节自我情绪、消愁解闷的基本方式。"说到这里，他的大笑不由自主地升了级，变成了一阵狂笑。

莫拉维克说，他自己坚信科学或任何程度上的应用科学都是无止境的，"即使基本规律是有限的，但对它们进行组合的方式却是无限的"。哥德尔定理以及格雷高里·蔡汀对算法信息理论的研究都表明：通过增大其公理基础系统，机器将能不断创造出愈益复杂的数学问题。"人们也许最终需要接受庞大得如天文数字般的公理系统，到那时，你就可以从中得出在较小公理系统中无法得出的新东西。"当然，机器人的科学绝对不会模仿人类的科学，两者间的差别恐怕要比量子力学与亚里士多德物理学之间的差别更大。"我敢保证，描述自然性质的基本框架和其内部细目都将发生改变。"比如机器人将把人类关于意识的认识看作是一无是处的原始思想，从而把它与古希腊人的原始物理学归为一类。

说到这儿，莫拉维克突然转变了话题，他强调机器人将会比生物有机体更加复杂多样，所以猜测他们的兴趣是什么，肯定是犯傻；其兴趣显然将决定于其"生态位"①。莫拉维克接着谈道，他就像福山一样，是抱着严谨的达尔文主义者的观点来审视未来的。在他看来，科学其实只不过是进化着的智能机之间永无休止的生存竞争的副产品。他指出，知识本身从未曾有过什么尽头，大多数生物体都在被迫寻求"知识"以帮助它们渡过即将到来的难关。"如果谁能借助某种控制手段安然生存下来，就意味着它可以在更大的范围和更长的时间里猎食。所以许多看似与觅食无关的行为，其实正是探求知识的过程，并很可能帮助它们渡过将来的难关。"就连莫拉维克所养的猫都表现出类似的行为："在饱食无事的时候，它到处溜达着，考察着各种事物，也许会在无意之中发现一个耗子洞呢，那它将来就多了条食物来源。"换种说法就是：好奇心就是一种适应行为——"只要你能消受它"。

莫拉维克由此怀疑：机器人们在追求纯知识的过程中，或追求其他任何目标的时候，到底能不能学会竞争与合作。因为没有竞争就没有选择，而没有选择就不会有进化。"因此需要有某些选择原则，否则一切都是废话。"最终，世界也许会超越所有这类竞争，"但必须有某种推动力量使我们能够

① Ecological niche，指机器人个体"生活"于其中的小环境——译者注。

达到那一境界。好了，我们漫游未来世界的旅程就此打住，这种漫游本来就是半开玩笑性质的"。然后，他就像着了魔般疯狂地大笑起来。

我忍不住要再提一个与这次采访相关的小插曲——不是在我实际采访莫拉维克的过程中，而是在我重放那次会谈的录音带时发生的：莫拉维克讲得越来越快，音调钢丝般笔直地越拔越高。一开始我认为录音机只不过是如实地录下了莫拉维克那发作得越来越重的歇斯底里，但当他听起来开始像花栗鼠阿尔文①一样的时候，我才意识到自己听到的不过是一种听觉幻象。显然，在我们的谈话快结束的时候，录音机的电池也快耗尽了。当我继续听下去的时候，莫拉维克那尖厉的声音逐渐超出了正常智力所能理解的范围，最终变得无法分辨——仿佛他的声音已滑入了通向未来的时间隧道中。

弗里曼·戴森的多样性

汉斯·莫拉维克作为人工智能的狂热支持者，坚决抵制这样的观点，即认为机器人将以集体的力量追求自己的目标，可能会彼此组织成一种元意识（metamind）。但他并不是唯一的反对者，对单一意识（single-mindedness）怀有深深恐惧的马文·明斯基，也同样持有相似的观点。"合作只在进化的晚期才是必要的。"明斯基告诉我说。"从那时开始，大规模的变化已失去了必要。"明斯基又轻蔑地补充道。当然，总是存在着这样的可能性，即超级智能机感染了某种宗教信仰而放弃自己的个性，结成单一的元意识。

另一位反对这种终极统一论的未来学家是弗里曼·戴森（Freeman Dyson），在其文集《永无止境》中，他考察了这个世界为什么会充满暴力和苦难，得出的结论是：这与他所谓的"极度多样性原则"有关，这一原则——

在肉体和精神两个层次面上都发挥着作用，其含义是：自然法则和初始条件决定了这个世界只能是异彩纷呈的。这一原则决定了生命是可能的，却又

---

① Alvin the chipmunk，花栗鼠阿尔文是美国家喻户晓的卡通形象，是由罗斯·巴达塞里安（Ross Bagdasarian，Sr.）于1958年首创的动画音乐组合，以歌声的尖锐急促著称。它包括三只活泼的拟人化花栗鼠歌手：淘气的惹祸精阿尔文，高个子眼镜博士西蒙（Simon），以及肥胖而又冲动的西奥多（Theodore），其中的阿尔文人气最高。该组合的成功，催生出后续的系列花栗鼠动画产品，并被搬上了银屏——译者注。

是大不易的。每当生活变得枯燥乏味的时候，就会发生某些意料之外的事情来为难我们，阻止我们陷入生活的俗套之中。危及生命的事例在我们身边随处可见：彗星碰撞、冰河期、战争、瘟疫、核裂变、计算机、性犯罪、罪孽和死亡。并非所有这些难关我们都能安然渡过，因此悲剧时有发生。极度多样性常常导致极度的压力。我们最终得以生存下来，只不过是侥幸。[3]

在我看来，戴森这是在暗示着：我们不可能解决所有的问题，不可能造就一个天堂，不可能找到终极答案。生命是——而且必定是——一场永恒的抗争。

是不是我曲解了戴森书中的观点呢？带着这一疑问，我在1993年8月到他家拜访了他。他的家在普林斯顿高等研究院，自20世纪40年代以来，他就一直住在那儿。他身材矮小，瘦得似乎只剩下了一把骨头，高而尖的鼻子，深陷的双眸中透出锐利的目光，酷似一只被驯服了的猛禽。他的举止冷淡而沉默，除了在他大笑的时候。他的笑声似乎全是通过鼻腔发出来的，同时双肩剧烈耸动，就像是一个12岁的小学生刚刚听到一个下流的笑话；那是典型的颠覆分子式的笑声，只有发出这种笑声的人，才会把太空看作是"宗教狂热分子"和"难以管教的不良少年们"的天堂，才会坚持认为：科学充其量也只不过是"一种对权威的反叛"。[4]

我并未直接向戴森提问有关极度多样性的问题，而是首先请他就影响其一生经历的几次抉择谈谈他自己的看法。戴森曾在寻求物理学统一理论的最前沿领域中从事研究。20世纪50年代初，这位出生于英国的物理学家，曾在构造电磁学的量子理论的研究道路上，与费曼以及其他一些巨头展开过激烈的角逐。曾有过这样一种说法，认为戴森的成就理应获得诺贝尔奖，至少他也应该获得比现在更高的荣誉。他的一些同事也曾怀疑；可能正是因为失望以及由此而来的对立情绪，才导致戴森后来去追求那些与其非凡才能极不相称的研究工作。

当我向戴森提起这些议论时，他只抿着嘴冲我一乐，然后就以惯常的回答方式向我讲起了轶事。他说英国物理学家劳伦斯·布拉格（Lawrence Bragg）就是个类似的典型，在他1938年成为剑桥大学那历史悠久的卡文迪许实验室的主任之后，就扭转了实验室的研究方向，把它从得以建立盛名的核物理研究转向新的领域。"人人都认为布拉格使卡文迪许实验室偏离了主流研究方向会毁了它，但后来的事实证明，这当然是个英明的决定，正是由此产生了分子生物学和射电天文学这两项贡献，使得剑桥大学在随后50年左右的岁月里能够盛名不衰。"

纵观戴森的经历，他也在不断地转向新的未知领域。大学时代他主攻数学，后来转向粒子物理学的研究，并由此进入固态物理学、核工程、军备控制、气象学研究，以及我所称的科学神学等不同的领域。1979年，一向以严谨著称的《现代物理学评论》杂志发表了戴森的一篇文章，就宇宙中智慧生命的长远前景作了猜测。[5]戴森写作此文，是为了反驳史蒂文·温伯格的断言，即"对宇宙了解得越多，就会发现它越没意义"。戴森反驳说，存在着智慧生命的宇宙绝不是毫无意义的。他试图证明，在一个开放的、永远膨胀着的宇宙中，通过对能量的巧妙储存，智慧生命能永远持续下去——或许正是像贝尔纳所说的那样，以一种带电粒子云的方式。

与莫拉维克或明斯基等计算机的狂热支持者不同，戴森认为智慧生命不会很快被人工智能（更不用说玄妙的气体）所取代。在《永无止境》一书中，戴森设想基因工程会在将来"培育出"某种航天器，它"只有一只鸡那么大，但比鸡更聪明"，能借助太阳能动力的翅膀穿行于太阳系中，有时也能飞到太阳系之外，充当人类的"侦察兵"（戴森给它们取名叫"太空鸡"[6]）；更遥远的宇宙文明，也许正为能量的枯竭而困扰，他们会在恒星周围建起球形建筑（被人们戏谑地称为"戴森球"），以吸收恒星的全部辐射加以利用。他还预言，智慧生命最终会散布到整个宇宙之中，从而把整个宇宙变成一个巨大的意识载体。他坚持认为："不论我们向未来推进多么远，都永远会遇到新的事物、得到新的信息、发现有待探索的新世界，也仍然会不断拓展生命、意识和记忆的疆域。"[7]对知识的追求应该是——也必定是——永无止境的。

在这一预言中，戴森实际上已经触及了一个至关重要的问题："智慧生命在认识并控制了整个宇宙之后，又将做些什么？"戴森认为这是个神学问题而不是科学问题。"我看不出在智慧生命和上帝之间存在着什么泾渭分明的界线，当智慧生命超出了我们所能理解的范围时，就变成了上帝。上帝既可被认为是一种世界灵魂，也可被看作是世界上所有灵魂的集合体，我们只不过是上帝在其自身发展的这一阶段安插在这一星球上的一颗颗棋子，我们也许会随着他今后的成长而成长，或者，我们会被远远地抛在后边。"[8]戴森最后终于又回到了其前辈贝尔纳的观点上来，认为我们不要奢望自己能解决这样的问题，即这个超级存在（superbeing）——这个上帝——将会做些什么，或想些什么。

戴森承认，他关于智慧生命之未来的观点，其实是虚妄的想象。当我问他科学是否会永远持续发展下去时，他答道："我希望会是这样！我喜欢生活于其中的世界，就应该是这样的世界。"如果人的思想能赋予这个世界以丰富的意义，世界才会向他提供有意义的问题去思考；这样，科学必然是永恒的。他驾轻就熟地罗列了许多可支持自己预言的论据，并解释说："考察这一问题的唯一途径，就是回顾历史。"两千年以前，某些"极睿智的人们"发明出某些在当时显然极为"先锋"的东西，但若用现代的思想来衡量，它们当然不是科学。"如果我们真正走入未来，就会发现今天所谓的科学同样全不是那么回事，但这并不意味着那时就不存在有意义的问题了。"

像莫拉维克（还有罗杰·彭罗斯，以及许许多多持类似观点的人）一样，戴森也希望哥德尔定理除了适用于数学之外，同样也能适用于物理学。"既然我们知道物理学定律都是数学规律，我们还知道数学是一个不自洽的系统，于是推出物理学也将是不自洽的就顺理成章了。"也就是说，物理学是个开放的系统。"我想，从长远来看，那些预言物理学将会终结的人可能是正确的，物理学也许会过时，但我仍然相信，物理学会被看作是某种类似于希腊科学的东西：一个有意义的开端，但未抓住要点。所以，物理学的终结同样也是另一个新的开端。"

最后，我问戴森他自己是怎样看待其"极度多样性"观点的。他耸了耸肩，说他无意让人们把它看得太认真，并且对那种"宽银幕电影式的"未来图景兴趣不大。他说自己最爱引用的一句话就是"上帝存在于细节之中"。既然他强调多样性和思想的开放性是存在的根本，那么他是不是觉得许多科学家把一切都分解成单一的、最终的单位这一做法是难以忍受的？这样的尝试难道不是一种危险的游戏吗？"是的，在某种程度上确实是这样。"戴森答道，同时微微一乐，仿佛在暗示着他已窥破我那么关注"极度多样性"的真实目的。"我从不认为这是个多么深奥的哲学问题，在我看来，那只不过是诗意的幻想。"戴森当然是在坚持着在他自己和他的观念之间保持一种适当的反讽的距离。但这种态度却显然并不怎么坦诚，毕竟，纵观他自己的不拘一格的经历，他一生的奋斗都在坚持着极度多样性原则。

戴森、明斯基以及莫拉维克，他们在内心深处都把神学倾向的达尔文主义、资本主义以及共和主义奉为圭臬；就像弗朗西斯·福山一样，他们把竞争、奋斗、分工看成是存在的基本要素——即使是对后人类智慧生命

（posthuman intelligence）而言也依然如此。但某些倾向于更为"开明"观点的科学神学家，却以为竞争将被证明只不过是一个暂时阶段，机器人很快就会超越它。爱德华·弗里德金（Edward Fredkin）就是这样一位"开明人士"。弗里德金以前与明斯基是麻省理工学院的老同事，现在是一位富庶的计算机企业家和波士顿大学的物理学教授。他毫不怀疑未来社会是属于机器的，属于那些"比我们聪明数百万倍"的机器，但却认为，明斯基和莫拉维克所设想的那种竞争，在未来的智能机器们看来，只不过是种返祖现象，一种不合时宜的行为。弗里德金解释说，智能计算机们在追求其共同的目标时，会进行无与伦比的合作。无论什么新知识，只要被一台计算机所掌握，就等于所有计算机都掌握了；同样，任何一台计算机的进化，也就意味着所有计算机的进化。合作将创造出一种"一荣俱荣"的新局面。

就那些超越了达尔文主义者的"老鼠赛跑"的超级智能机而言，它们会想些什么，做些什么呢？"计算机们当然会发展它们自己的科学，"弗里德金答道，"我觉得这是再明显不过的事。"机器的科学与人类的科学有什么显著差异吗？弗里德金犹疑着说，他认为不同是肯定会有的，但他无法事先说出到底不同在哪些地方；如果我真想得到此类问题的答案的话，最好是去请教科幻小说作家。说到底，又有谁能真的知道呢？[9]

弗兰克·蒂普勒和欧米加点

弗兰克·蒂普勒（Frank Tipler）认为他知道。蒂普勒是图兰大学的物理学家，曾提出过一种名为"欧米加点"的理论[①]，该理论把整个宇宙描绘成一台全知、全能的计算机。与探索遥远未来世界的其他人不同，蒂普勒似乎并未意识到，他所从事的并非经验科学，而是反讽的科学。他实际上也说不出两者间有什么区别。但也可能正因如此吧，他才会那么肆无忌惮地去构想一台有着无穷的智慧和能力的计算机会做些什么。

我在1994年9月采访蒂普勒时，他正在为《不朽的物理学》一书作巡回宣传途中。（这本528页的巨著以精心推敲的细节描写，详细探讨了其欧米加点理论的影响。）[10] 蒂普勒身材壮硕，有一张肥而阔的脸盘，灰须、灰发，戴一副角质框的眼镜。在我们交谈的过程中，他一直用南方人的那种让

---

① Omega Point，即"Ω"点。"Ω"是希腊语24个字母中的最后一个，常用以意指终止或结局；"欧米加点"也就是"终点"的意思——译者注。

人无法忍受的拖泥带水风格长篇大论地宣讲着，表现出一种浅薄的自鸣得意的神气。我曾问他是否嗜好LSD或其他迷幻药，否则的话，为什么他会在20世纪70年代于伯克利大学做博士后的时候，就开始思索欧米加点呢？"没有！绝对没有！"他僵硬地摇着头说，"我甚至从不饮酒！我可以自豪地宣称，我是世界上所有姓蒂普勒的人中唯一一个绝对戒酒主义者。"

他从小就被教养成一个极端主义浸礼会教友，但到青年时期，他开始相信科学是通向知识和"改善人类境况"的唯一途径。他曾在马里兰大学所做的博士论文中，探讨了建造时间机器是否可能的问题。这一研究与他改善人类境况的目标之间有什么联系吗？"是的，时间机器显然可以增强人之为人的功能，"蒂普勒答道，"当然，它也可以被用于作恶。"

欧米加点理论是蒂普勒和英国物理学家约翰·巴罗（John Barrow）两人合作的成果。在其1986年出版的厚达706页的大作《人择宇宙学原理》中，蒂普勒和巴罗考察了这样一个问题：如果智能机把整个宇宙变成一个巨大的信息加工装置，那将会出现怎样的局面？[11] 他们设想：作为一个封闭的宇宙，它终将停止膨胀，并开始自己坍缩；而随着宇宙逐渐向终极奇点收缩，其加工信息的能力将趋向无穷大。"欧米加点"这个术语，是蒂普勒从耶稣会会士、科学家皮埃尔·泰哈迪·德·查丁（Pierre Teihard de Chardin，中文名德日进）那里借用过来的，查丁曾设想过一种人类未来图景，其中所有的生物都被并吞进一个体现基督精神的单一、神圣的存在中。（这一命题要求他必须深思另一个更大的问题：上帝除了向地球派遣作为其替身的拯救使者之外，是否也向其他承载生命的星球派遣使者呢？）[12]

蒂普勒起初认为，没有人能想象有无限智慧的存在会想些什么、做些什么，直到有一天他读了德国神学家沃夫哈特·潘宁伯格（Wolfhart Pannenberg）的一篇文章，认为所有人将来都会在上帝的心智中再生。这篇文章触发了蒂普勒"尤里卡"式的顿悟，使他产生了写作《不朽的物理学》的灵感。他意识到，对于每一个曾为追求永恒的天国之乐而生活的人来说，欧米加点都具有使他们再生——或复活——的力量。欧米加点不仅能使死者再生，而且会使他们更趋完美，并且使我们不会有什么放弃尘世的遗憾。举例来说，每个男人不仅能够拥有他所见过的（或虽未见过但曾存在过的）最美的女人，女人也同样能享有她们所钟情的伴侣。

蒂普勒说他起先对自己的观点并不怎么认真，"但你在思考这些事情的时候，不得不对这样的问题作出抉择：你真的相信这些是自己基于物理学规律而创造出的学说，还是违心地认为它们仅仅是与现实没有任何关系的玩笑呢？"后来他接受了自己的理论，孰料这竟带给他极大的满足。"我说服了自己——当然啦，或许只是愚弄了自己——使自己相信那将是一个多么美妙的世界。"然后他像背诵赞美诗一样说道："美国大哲学家伍迪·艾伦曾说过，'我不想借助于自己的著作得以永生，我想通过不死而永生。'我认为，这句话很好地表达了我开始设想计算机的可能性时所受到的心灵震撼。"

他接着又向我讲述了1991年的一件事，当时有一位英国广播公司（BBC）的记者约他拍摄一段介绍"欧米加点"的电视节目。后来，他六岁的女儿在看了演播之后问他，她那刚刚过世的外婆是不是在某一天也会作为计算机模拟的产物而得以复活呢？"我能怎么说？"蒂普勒这样问我，同时耸了耸肩膀，"当然！"蒂普勒陷入了沉默，脸上闪现出一丝迷惘，然后又消失了。

对于很少——无限接近于零的那种很少——有物理学家把他的理论当回事这一点，蒂普勒自云并不介意。毕竟，哥白尼的日心说在他死后的一个多世纪里，仍未得到公认。"私下里，"蒂普勒向我凑过来，瞪大的眼睛闪着暧昧的光亮，"我总以哥白尼自居，我俩最大的差别——让我再强调一遍——就在于人们已普遍承认他是对的，却还没有人承认我！"

科学家与工程师间的差别，在于前者探求何为真，后者寻求何为善；蒂普勒的神学表明，他本质上是一个工程师。与戴森不同，蒂普勒认为探求纯知识——他定义为支配世界的基本规律——的道路是有限的，并且已接近终点；但科学在终结之前，还有一项最伟大的任务：建构天堂！"我们怎样达到欧米加点，这仍然是个问题。"蒂普勒评论道。

蒂普勒献身于善，而非诉之于真，这至少提出了两个问题。其中一个已是众所周知的，对于像但丁（Dante）以及其他一些胆敢构想天堂的人们来说这一问题都普遍存在，即怎样逃避终极的无聊？正是这一问题，迫使弗里曼·戴森提出了他的"极度多样性原则"——他认为，极度的多样性导致极度的压力。蒂普勒也赞同戴森的说法："除非存在着失败的可能性，否则我们不可能真正享受成功的喜悦；成功与失败是不可分割的。"蒂普勒只是太懒了，不愿意仅仅为了使"欧米加点"中的公民们不至于感到无聊，而费心地在自己精心堆砌的天堂里引入真正的痛苦；他以为只要"欧米加点"能给予其臣民们变得"更聪明、更有学问"的机会，这就足够了。但在"欧米加点"的臣民们变得越来越聪明之后，假如对真理的追求完结了，怎么办？他们要这份"愈益"的聪明有什么用呢？用来与那"最美丽""最钟情"的超级伴侣们谈情说爱吗？

蒂普勒对苦难的嫌恶，导致他陷入了另一个悖论之中。他曾在其著作里宣称，即使"欧米加点"自身尚未被创造出来，它也仍然可以创造世界。当我向他指出这个悖论时，他喊道："噢！我有我的理由！"于是就给出了一个冗长而烦琐的解释，其要点无非是：因为未来占据着宇宙历史的最高点，所以我们的思考应以未来为参照系——正如我们思考天文学问题时，以恒星为参照系，而不以地球和太阳为参照系一样。按照这一思路，假定宇宙的终点——欧米加点——在一定意义上也是其起点，就是自然而然的了。但这是纯粹的目的论，我反驳道。蒂普勒点了点头："我们通常是按照从过去到未来的思路去认识宇宙的，但这只不过是我们人为规定的思路，宇宙没有任何理由也非得那样认识事物。"

为了支持这一论点，蒂普勒引证了《圣经》中摩西质询燃烧的荆棘那一节。在钦定本《圣经》[①]中荆棘答道："我就是我。"但据蒂普勒称，希伯来语原文应该译作"我将是我"。他用毋庸置疑的口气总结道：这一段文字揭示出，尽管《圣经》中的上帝已经劳神费力地创造了整个宇宙，并降尊纡贵与其手下的先知们聊过天，但是，他却只会在未来现身。

蒂普勒终于泄露出自己为什么会异想天开地要这样一个花招。如果"欧米加点"已经存在，那么我们都必然是它的再生物——或是其模拟物——之一；但我们的历史却又不可能是一种模拟物，它只能是本真的。为什么呢？因为"欧米加点"太完美了，它不可能创造出一个有着如此之多苦难的世界。像所有崇拜神明的信徒一样，蒂普勒也在邪恶与苦难这一绊脚石上摔了跟头，他不敢直面"欧米加点"可能对我们世界的一切恐怖负有责任的可能性，只好顽固地坚持着自己的悖论：欧米加点创造了我们，即使它本身尚未存在。

1984年，英国生物学家彼特·梅达沃（Peter Medawar）出版了一本名为《科学的限度》的著作，在很大程度上不过是反刍着波普尔主义的观点。例如：梅达沃在书中反复强调，"科学解答自己能够解答的问题的能力是无限的"，就好像这是什么含义隽永的至理名言，其实不过是毫无意义的同语反复罢了。但梅达沃也确实提出了几点卓见。他在一段讨论"怪力乱神"

---

① King James Version，1611年由英王詹姆斯一世核准发行的英译《圣经》版本；下述故事出自《旧约·出埃及记》——译者注。

（bunk，他用来指神话、迷信及其他一些缺乏经验基础的信仰）的文字中评论道："对怪力乱神着迷，有时候极为有趣。"[13]

蒂普勒可能是我所见过的对怪力乱神最为着迷的一位科学家；还必须补充的一点是：我发现欧米加点理论，特别是剥去其基督教的伪装之后，是我所遇见过的反讽科学中最吸引人的一个。弗里曼·戴森设想出一个有限的智慧生命，它漂泊于一个开放的、不断膨胀的宇宙之中，竭力抗拒着"热寂"；而蒂普勒的"欧米加点"却伸开双臂去拥抱那人类的大限、那永恒的大赦，只为能获得一种"无限智慧"的幻觉。在我看来，显然是蒂普勒的未来图景会更加令人神往。

我只是对那拥有无与伦比能力的欧米加点究竟想做些什么持有疑义。它会为是否复活希特勒的"精确"（nice）版本而烦恼吗？（由此引出的问题之一肯定会令蒂普勒烦恼不堪！）它会充当一个月下老人的角色，成天忙着为撮合那些呆头鹅般的男女与其"最钟情的"电脑超级伴侣而疲于奔命吗？我想肯定不会是这样。正如戴维·玻姆曾告诉我的，"世事不会尽如人意，真的"。我相信，欧米加点试图达到的绝不是善——也不是天堂或新波利尼西亚或任何别种永恒的至福——而是真；它将努力去发现"我是谁？我是怎么来的？"之类问题的答案，就像那些能力远低于它的人类祖先所做的那样。它会去寻求"终极答案"。难道还有什么别的目标更值得它去追求吗？

尾声
上帝的恐惧

物理学家保罗·戴维斯在其出版于1992年的著作《上帝的心智》中，曾深入思索了这样一个问题：人类是否能借助科学而达到绝对知识——终极答案。他的结论是否定的，因为量子不确定性、哥德尔定理、混沌等为理性知识设置了限度。据他猜测，达到绝对真理的唯一通道，或许只能由某种神秘体验提供。但他马上又补充道，他无法保证这一通道的可能性，因为他自己从未经历过什么神秘体验。[1]

在我成为专职科学记者之前某一年，曾有过那么一次特殊的体验，我认为就可以称之为神秘体验——当然，精神病学家也许会称之为精神失常者的幻想曲。随他们怎么称呼吧，在此我只是想描述一下当时真实发生的事情。客观上，当时我正张开四肢，仰卧在郊区的一片草坪上，对身边的一切都无动于衷。而主观上，那时我正深深坠入一片令人眩晕的黑暗之中，似乎正在走近生命的终极奥秘，它所带来的震撼潮水般地袭击着我，一浪高过一浪；同时，一种不可抗拒的唯我论紧紧攫住了我，使我相信——或许，更确切的说法是我知道——自己是宇宙中唯一有意识的生灵。没有什么将来，也不存在什么过去，甚至连现在也已不复存在，只有我的想象才是真实的，想什么就是什么，平生第一次觉得自己充满了无限的快乐和力量。然后我突然意识到：如果自己更深一步地陷入这一幻境中，它可能真的就此吞没了我。如果只有我一个人存在，那么谁能从无我中把我带回现实？谁能拯救我？意识到这一点，极度的幸福立刻变成了极度的恐怖，驱使我逃离开曾一度使我陶醉的境界，觉得自己跌入无垠的黑暗之中，并在坠落的同时逐渐融解，消失在无限的自我之中。

在这次梦魇般的经历过去以后几个月时间里，我一直相信自己发现了存在的奥秘，那就是潜藏在万事万物背后的"上帝的恐惧"——上帝对自己之为上帝的恐惧，以及对自己潜在死亡可能性的恐惧。这一信念既使我感到深深的自得，又使我体验着深深的恐惧——我一日日地疏远了朋友、家庭和一切使生活有意义的平常事物，我不得不拼命地工作，以求能忘却这场噩梦，并重新回到正常的生活中去，在某种程度上我也确实做到了这一点。马文·明斯基也许会认为，我是把这一体验封闭进了意识世界的某一个相对孤立的角落，这样它就不再能淹没其他更有生活意义的部分——那些关注着现实的工作、伴侣等事物的部分。不管怎么说，事隔多年之后，我之所以再度把这段插曲从记忆深处挖掘出来，重新加以思索，原因之一就在于我遇到了一种稀奇古怪的、伪科学的理论，一种可以帮我发掘过去那一幻觉经历的隐喻感的理论——欧米加点。

把存在想象为上帝，这已并不新鲜，也不是最好的形式，但把存在想象为一台奇大无比、威力无穷的计算机，它无处不在——它就是——整个宇宙。随着欧米加点趋近于时间、空间的极点和存在本身的最终解体，它将会经历一种神秘的体验；它那空前的智慧的唯一作用，就是可以认识到自己的存在其实毫无意义；除了自身之外，没有造物主，也没有上帝，只有它存在着，此外一无所有。"欧米加点"将会意识到，它对于终极真理和终极大一统的追求，已把它带到永恒虚无的边缘；如果它完蛋了，一切都完了，存在本身也随之而消失。"欧米加点"对自己这种可怖处境的认识将迫使它逃离自己，逃离自己那可怕的孤独和自知。随着"欧米加点"在绝望和恐怖中逃离自身，创造将会带着它所有的痛苦、美丽和千奇百怪的多样性生发出一个全新的世界。

这一观念的苗头，已在某些领域里零星地闪现出来。在一篇名为《博尔赫斯和我》的散文中，这位阿根廷寓言家描述了对于这种存在消蚀感的深深恐惧——

我喜欢古代的沙漏和地图，钟情于18世纪的印刷术、词源学，喜爱咖啡的滋味以及罗伯特·路易斯·史蒂文森[①]的散文；那个人（博尔赫斯），也享有这些嗜好，但却带着一种炫耀的心态，并最终把这些心爱的事物变成其装腔作势的资本。如果说我俩的关系充满着敌意，那确是有些夸大其词；我活着，并将继续活下去，因而博尔赫斯可以继续炮制他的文学作品，并在其作品中对我品头论足……几年前，我试图从他的阴影中把自己解救出来，于

---

① Robert Louis Stevenson，1850—1894，苏格兰小说家、诗人和随笔作家——译者注。

是就离开了城郊的神话世界，走入与时间和无限竞赛的游戏中；但是现在，连这些游戏也都属于博尔赫斯了，我将不得不去构思些别的什么。这就是我的生活，疲于逃避的生活，并在逃避的过程中遗失了一切。一切都归于忘却的救主，或者属于博尔赫斯。我不知道我俩之中，究竟是谁正在写这些文字。[2]

博尔赫斯正在逃避的，恰恰是他自己！当然，那个在后面追击的也是他自己。另一个相似的自我追击形象，出现在威廉·詹姆斯[①]的《宗教体验的多样性》一书的脚注中。在这条注解文字里，詹姆斯引证了一位叫克拉克（Xenos Clark）的哲学家对麻醉体验的描述。当克拉克从那种体验中清醒过来后，终于确信——

通常所谓的哲学，就像是一只追逐自己尾巴的猎狗，它越是想达到目标，所需走的路途就越远，但它的鼻子永远也够不到自己的蹄踵，因为鼻子永远在蹄踵的前面。"现在"总结的永远是往昔，使我永远赶不及理解真正的现在。但当我从麻醉中醒来的时候，就在我恢复成平时的我之前那一刻，我瞥见了——可以这么说——自己的脚后跟，在开始的时候就已瞥见了整个的过程。事实上，我们所跋涉于其中的旅途，在我们出发前就已结束了，哲学真正的终结早已完成，不是完成于我们抵达目的地的时候，而是在我们尚未出发的时候；而在现实生活中，如果我们放弃一切理智的探究，同样可以达到这一目的。[3]

但是，我们绝不能放弃理智的探究，因为一旦放弃之后，一切都就不存在了，只剩下遗忘。"黑洞"的命名者，物理学家约翰·惠勒已直觉到了这一真理，静卧在现实的最深处的，不是什么答案，而是一个问题：为什么一定要存在些什么而不能一无所有？"终极答案"就是根本没有答案，只有一个问题。惠勒关于世界仅仅是"想象力的虚构"的怀疑，并非全无根据。世界是个难解的谜，上帝之所以要创造出它来，是为了使自己免受无边的寂寞和死亡的恐惧所施加给他的折磨。

---

① William James，1842—1910，美国心理学家、哲学家、实用主义的倡导者——译者注。

查尔斯·哈茨霍恩的不朽上帝

我曾劳而无功地寻找一位赞同"上帝的恐惧"观点的神学家，后来，弗里曼·戴森向我指点了一条希望之路。有一次，戴森在演讲中介绍了自己的神学观点——就是前面曾提到过的那个设想，认为上帝并非无所不知、无所不能，而是同我们人类一样要不断成长，不断学习。演讲结束后，一位耄耋老人迎上前来，自我介绍说他叫查尔斯·哈茨霍恩（Charles Hartshorne），戴森后来才知道，这位老先生竟是20世纪最为著名的神学家之一。哈茨霍恩告诉戴森，说他对上帝的理解，与16世纪一位名叫索齐尼①的意大利神职人员很相似，索齐尼正是因自己的观点而被烧死在异端火刑柱上。根据戴森的印象，哈茨霍恩正是一位索齐尼式的人物。我问戴森知不知道哈茨霍恩是否还活着。"我不敢肯定，"他答道，"即使他仍活着，也必定是个很老很老的家伙了。"

离开普林斯顿之后，我借了一本介绍哈茨霍恩的神学思想的文集，文章作者除他本人外，还有别的一些人。通过阅读这本书，我发现他确实是个索齐尼派教徒，[4]也许他能理解我的"上帝的恐怖"观点——如果他还活着的话。我查阅了《名人录》，发现哈茨霍恩最后的职位是在德克萨斯大学奥斯汀分校。我向那里的哲学系拨了个电话，秘书告诉我说，哈茨霍恩教授活得好好的，他每周都到学校来几次，但与他联系的最好方法是直接给他家里打电话。

哈茨霍恩接通电话后，我向他作了自我介绍，然后告诉他我是从戴森那里得知他的大名的，并问他是否还记得那次与戴森就索齐尼问题展开的对话。他确实还记得，并与我谈论了一会儿戴森的情况，然后就精力充沛地投入一场关于索齐尼的讨论之中。尽管他的声音沙哑而又颤抖，但谈话的语气却是绝对自信的。交谈不到一分钟，他的声音就开始暗哑、飘忽起来，变成《米老鼠》动画片中某个角色的假声，为我们的谈话平添了一种超现实主义的奇幻色彩。

哈茨霍恩告诉我，与许多中世纪的甚至是现代的神学家不同，索齐尼教徒相信上帝是随着时间的推移而变化、学习和演化的，正如我们人类一样。"你要明白，经典的中世纪神学传统认为上帝是不能改变的，但索齐尼教徒却说，'不，完全不是那么回事儿。'这些'异教徒'是完全正确的，这一

---

① Socinus，全称是Faustus Socinus，1539—1604，意大利宗教改革家。他所提倡的教义，否认基督的神性，反对三位一体论，主张用理性解释犯罪和救赎——译者注。

点对于我来说再明显不过了。"哈茨霍恩解释道。

如此说来，上帝并非是无所不知的了？"上帝知道存在着的每一件事，但诸如将来的事是根本不存在的，上帝也就无法知道这些，除非它们已经发生了。"他的语调似乎暗示着：这不过是每个笨蛋都能明白的事。

如果上帝不能预知未来，那他会不会由此产生对未来的恐惧呢？他会害怕自己死亡吗？"不！"哈茨霍恩高声嚷道，并且嘲笑这个观念的荒谬。"我们有生也有死，这正是我们与上帝不同的地方；如果上帝同样是有生也有死的，他就不成其为上帝了。上帝体验着我们的生，但只是作为我们的生，而不是作为上帝的生；同时，他也体验着我们的死。"

我试图向他解释，我的意思不是说上帝真的会死，而只是说他可能会恐惧死亡，即他可能会对自己的不朽产生怀疑。"噢，"哈茨霍恩唯唯应道，我似乎能看到他在电话线的那一端正大摇其头，"我对此丝毫不感兴趣。"

我问哈茨霍恩是否听说过欧米加点理论。"是不是泰哈迪·德·查丁的？""是的"，我答道，"泰哈迪·德·查丁正是诱发这一理论的思想根源。这一理论的要点是：人类创造出来的超级智能机扩张到整个宇宙……""得得得，"哈茨霍恩轻蔑地打断了我的话，"我对这些更是不感兴趣，那纯粹是想入非非。"

我真想回敬两句：那些是想入非非，但所有这些索齐尼教徒的昏话就不是吗？可我只是追问了哈茨霍恩一句：他是否认为上帝的进化和学习会有达到终点的时候。他第一次在回答之前陷入了沉默，最后才答道，上帝不是一种存在，而是一种"演变形式"（mode of becoming），这种演变无始亦无终——永远。

但愿如此。

上帝的指甲

我曾试图向各种各样的熟人描述自己"上帝的恐惧"观点，但成功的机会微乎其微，就像我与哈茨霍恩交流时的情况一样。一位科学记者同行——是那种极其理性化的类型——正襟危坐地耐心听完我的长篇大论，又耐心地听着我无话找话，直到我陷入再也无话可说的绝境之后，他才突然冒出一句："你是说所有的事都已沦落到这样的地步，以至于连上帝也只有啃指甲的份儿了？"我愣怔地思索了一会，然后点了点头。诚然，所有的事都已沦落到连上帝也无可奈何的地步了。

事实上，我认为"上帝的恐惧"这一假说有许多可取之处，它提示我们为什么人类既那么热切地追求真理，却又在面对真理时踌躇不前。对真理的恐惧，对"终极答案"的恐惧，充斥于人类文化的每一部经典之中，自《圣经》以降，直到最新上映的关于狂人科学家的电影。人们通常都以为科学家们对这种不安是有免疫力的，某些科学家也的确有此免疫功能，或至少表面看起来是这样。弗朗西斯·克里克，这位唯物主义的靡菲斯特①，首先就会闪现在人们脑海里；还有冷峻的无神论者理查德·道金斯，还有斯蒂芬·霍金——这位爱跟宇宙开玩笑的家伙。（是不是英国文化中存在着某种特殊的品格，使得它孕育出的科学家们对形而上学的焦虑有着如此强大的免疫力呢？）

但相对于每一位克里克或道金斯，都有更多位科学家存在，他们对绝对真理这一信仰心中深藏着难言的矛盾。就如罗杰·彭罗斯，竟无法判断自己对终极理论的信仰究竟是乐观的还是悲观的；或者像史蒂文·温伯格，在"易于理解"和"毫无意义"之间画上了等号；或戴维·玻姆，在这种心理的压迫下，既想澄清现实，又想把它搞成一团糟；还有爱德华·威尔逊，既想追求关于人类本性的终极理论，又对可能到达这一理论的想法感到惶惶不可终日；还有马文·明斯基，对于单一心智（Singlemindedness）的想法惊恐莫名；再比如，像弗里曼·戴森，坚持认为焦虑和疑惑本就是存在的基本要素。这些真理追求者对于终极认识的矛盾心理，反映出上帝——或"欧米加点"，如果你喜欢这样称呼它的话——对于其自身困境的绝对认识的矛盾心理。

维特根斯坦在其散文诗《逻辑哲学论》中强调："真正神秘的，不是世界'怎样？'，而是世界'就是如此！'"[5]维特根斯坦认识到，真正的启迪（enlightenment），只是在人们面对存在的无情事实时作出的那张令人解颐的呆瓜面孔。科学、哲学、宗教以及各种形式的知识，其表面的目的是把对神秘奇迹的巨大"？"转变成一个更大的认识之"！"。可是假如人们掌握了"终极答案"之后，事情将会怎样？一旦考虑到我们的奇迹感将会被我们的知识一劳永逸地彻底消解，总难免要产生一种深深的恐惧：真的到了那一天，存在的意义何在？恐怕将是一无所有。神秘奇迹的问号，永远也不会被彻底抻直，即使是上帝的心智也无法做到这一点。

---

① Mephistopheles，是歌德所著诗体剧《浮士德》中的一个重要角色，是魔鬼的化身——译者注。

　　这一观点将会给别人造成怎样的印象，我并非一无所知。我自认是个很理智的人，喜欢——也许有点出格——嘲弄那些对自己的形而上学空想过于认真的科学家。但是，请允许我再度借用马文·明斯基的观点，我们都有着多重意识世界，那个实干的、理性的"自我"对我说：关于"上帝的恐惧"这一套纯粹是心神错乱的胡话；但还有别的"自我"呢，其中一个时不时地瞥一眼占星术专栏，并偶尔怀疑：既然有那么多报道都在讲述着地球人与外星人发生性关系的轶事，这背后是不是真的有什么可信的东西？还有一个"自我"则坚信：所有的事情都已沦落到令上帝也只能啃手指甲的地步了。这一信念甚至带给我某种奇特的满足感，我们的困境竟然也正是上帝的困境。既然科学——求真的、纯粹的、经验的科学——貌似已经结束了，还有什么值得相信呢？

跋
未尽的终结

作为一名科学记者，我一直对科学的主流观点抱有十分的敬意。公然蔑视这一现状的独行其是者，虽有可能成为轰动一时的人物，但结局几乎毫无例外：他／她错了，大多数人的观点正确。自从《科学的终结》一书于1996年6月问世以来，它所引发的反响就一直把我置于一种十分尴尬的境地。我并非没有预见到——甚至曾一度期盼过——本书的某些预言会遭到驳斥，但它竟会在如此之广的范围内受到如此众口一词的谴责，诚非我始料之所及。

公开驳斥我的"科学终结论"的人物包括：比尔·克林顿总统的科学顾问、（美国）国家航空和航天管理局局长、一打左右诺贝尔桂冠获得者，以及许多声名稍逊的评论家。批评来自每一个大陆，只有南极洲除外。即使那些自称欣赏本书的评论家，往往也要煞费苦心地与它的前提保持距离。著名的科学记者纳塔利·安吉尔（Natalie Angier）在1996年6月30日的《纽约时报书评》上发表了一篇表述不同意见的评论文章，在文章的结尾部分就曾这样表白："我无法苟同（本书）关于限度与没落的中心论点。"

人们也许会认为，这类善意的指责足以促使我进行自我反省。但自打我的书出版以来，我甚至比以往更加坚信：我是正确的，别人差不多都错了——这通常被看作是发疯的先兆。这并不意味着我已为自己的假说构建出了什么天衣无缝的论据。我的书就像所有的书一样，是个人抱负与各种竞争着的需求——家庭的、出版商的、雇主的，等等——折中的产物；在我怀着满腹的不甘把最终清样送交编辑的时候，对于自己的书稿应从哪些方面进一步加以完善，心里是相当清楚的。在这篇跋中，我希望能弥补书中某些明显的疏漏，并就批评家们所提出的那些理智而又可笑的观点给予回答。

不过是又一本"终结论"的书罢了

对《科学的终结》一书最普遍的反应可能就是："这只不过是又一本宣告'某重大领域的终结'的书。"评论家们暗示说，我的小册子和另外一些类似的玩意儿——值得一提的是弗朗西斯·福山的《历史的终结》以及比尔·麦克吉本（Bill McKibben）的《自然的终结》——都是同一种21世纪前的悲观主义的表现，虽能风靡一时，但并不值得严肃待之；批评家们还指控我和类似的终结论同伴们具有自恋倾向，因为我们坚称自己置身其中的时代是独特的，充斥着危机和盛极而衰的先兆（culmination）。正如1996年7月9日的《西雅图时报》所云："我们都希望生活在一个独一无二的时代，于是像历史的终结、新时代、基督重降或科学的终结之类的宣言，就有了不可抗拒的魔力。"

但我们的时代的确是独一无二的，像苏联的解体、接近60亿的人口总数、工业化导致的全球变暖和臭氧层破坏等，都是史无前例的；还有，热核炸弹或月球登陆或便携式电脑或乳腺癌的基因检测，总而言之，对于作为20世纪标志的知识与技术的爆炸而言，肯定是史无前例的。因为我们都生于斯世、长于斯世，所以我们就想当然地认为：当前这种指数级增长式的进步是现实的永久特征，它将会并且一定会持续下去。但对历史的考察却揭示出：这种进步也许只是一种反常现象，它将会并且一定会走向终结。相信进步不朽，而不相信危机与盛极而衰的先兆，是我们的文化之主要误区。

1996年6月17日的《新闻周刊》提醒读者，我的未来观代表着一种"想象力的破产"。诚然，想象遥远将来的重大发现十分容易，我们的文化已经替我们做到了这一点——应用《星际旅行》之类的电视剧、《星球大战》之类的电影、汽车广告以及向我们保证明天将会与今天不同甚至更美好的政治辞令。科学家以及科学记者总是声称：革命、突破和梦寐以求的重大发现正呼之欲出。

我所期望人们思考的是：如果在遥远的将来并不存在什么重大发现怎么办？如果我们已基本拥有了可能拥有的一切怎么办？我们不大可能发明出可把我们载到其他星系甚至其他宇宙的超光速直达飞船，基因工程也不大可能使我们变得无限聪明与长生不老，正如无神论者斯蒂芬·霍金所云："我们也不大可能发现上帝的心智。"

那么，等待我们的宿命将是什么？我怀疑它既不会是无所用心的享乐主义，如岗瑟·斯滕特在《黄金时代的来临》中所预见的那样；也不会是无事生非的战争，如福山在《历史的终结》一书中所警示的，而应是这两者的某种混合。我们将一如既往地得过且过下去，在至福与不幸、启蒙与蒙昧、仁慈与残忍之间摇来摆去。我们的宿命不会是天堂，但同样也不会是地狱。换言之，后科学世界将不会与我们现在的世界有什么天壤之别。

就像我对任天堂游戏的偏爱，我觉得，一些批评家"以我之矛，攻我之盾"的做法是公平的。《经济学人》杂志在其1996年7月20日的一期上，就曾得意扬扬地宣称，说我的"科学终结"命题本身就是反讽推理的典型案例，因为该命题说到底也毕竟是不可检验、不可证实的。但正如当年我质疑其证伪说是否可证伪时，卡尔·波普尔所云："这是人们所能想出来的最愚蠢的批评之一！"与原子、星系、基因等纯正的科学探索的客体相比，人类文化却是朝生暮死的；一颗小行星就可以在任何时候毁灭我们，从而给我们带来的不仅仅是科学的终结，还有历史、政治、艺术（随便你列举）的终结。因此，显而易见，人类文化方面的预言，往好里说，都只不过是受过良好训练的猜测；与之形成鲜明对比的是核物理学、天文学或分子生物学等领域所给出的预言，因为这些领域处理的是实在的某些更为永恒的方面，并且能达成更为永恒的真理。就这一点而言，我的"科学终结说"的确是反讽的。

但是，仅仅因为我们不能确知未来，并不意味着我们就不能对某种未来趋势是否胜于另一种给出令人信服的说明。一如某些哲学的、文学批评的或其他反讽性事业的成果，的确要比另一些更胜一筹，某些关于人类文化之未来的预见也是如此。我认为我给出的脚本比自己所欲取代的那些看来更合理；而且在我的脚本中，我们仍可持续不断地发现深奥的事实，关于宇宙的新事实；或者，我们也许会达到一个终点，并在到达之后获得极度的智慧，且能够驾驭自然。

《科学的终结》反科学吗

在过去的几年里，科学家们一直在以前所未闻的刺耳术语，悲叹他们所谓的"正日渐高涨的潮流"，即对科学的非理性和敌意。"反科学"这一称号被横加到各种迥然不同的靶子上，如挑战科学能导致绝对真理这一断言的后现代哲学家、基督教神创论者，诸如美国通俗电视剧《X档案》之类的神秘性娱乐节目的提供者，还包括我——这毫不足怪。因其在凝聚态物理领域的成就而荣膺诺贝尔奖的菲利普·安德森（Philip Anderson），就曾在1996年12月27日的《伦敦泰晤士报高等教育副刊》上指控说，如此尖刻地对待某些科学家和理论，我已经"极端不负责任地在为我们正经受着的反科学浪潮提供火力支援"。

具有讽刺意义的是，粒子物理学家们早就把安德森打入了反科学的另册，因为他早在超级超导对撞机于1993年被最终取消之前，就曾对该项目横加指责。正如当我质问他为何总是急不可耐地跳出来评判其他科学家，那他自己的信誉何在时，安德森所给的答复一样："我只不过是看到了他们，并打了声招呼。"我在自己的书中力图去做的，不过是尽可能生动且如实地描画出自己所采访的那些科学家和哲学家的肖像，再加上点儿我个人对他们的反应。

我要重申自己在《科学的终结》一书引言部分曾说过的话：我之所以成为一名科学记者，是因为我认为科学，尤其是纯科学，是所有人类创造中最神奇、最有意义的。我也不是一名勒德派分子，我喜欢自己的便携式电脑、传真机、电话和汽车。尽管我也强烈反对某些科学所带来的副产品，如污染、核武器以及智力上的种族歧视理论等，但我相信，就总体而言，科学已经在精神上、物质上使我们的生活变得无比富裕。科学也并不需要增加一个公共关系方面的防御炮台，抛开科学近来所引发的种种不幸事件勿论，科学仍旧是我们文化中一种无比巨大的力量，其威力远较后现代主义、神创论或其他装腔作势的理论为甚。科学需要——并且一定能够经受得住——言之有物的批评，而这正是我不惮浅薄所力求提供的。

某些观察家担心，《科学的终结》一书将被用来为进一步削减——如果不是彻底取消的话——研究基金作辩护；如果在联邦政府官员、国会议员和公众之间出现了群情滔滔地支持我的论点的局面，我肯定会吃不了兜着走。但事实恰恰相反，白纸黑字可以证明，我并未支持进一步削减科学基金，不论是基础科学的还是应用科学的，尤其是在国防基金仍居高不下的可恶情况下。

也有人抱怨我的预言会使年轻人面对科学探索之路望而却步，对此我必须严肃对待。我书中观点的"必然推论"是由《萨克拉门托新闻》公之于众的（1996年7月18日），我自己就有两个年幼的孩子，假如十年之后她们问我是否认为科学之路是一条死胡同，我该怎么说呢？

我也许只能给出类似这样的回答：你们不应该因为我所写的东西而丧失做一个科学家的勇气，因为仍有许多重大而又激动人心的事情等待着科学家们去做：去发现治疗疟疾和艾滋病的更佳方法，以及环境公害更少的能源，对污染将如何影响气候作出更准确的预测，等等。但是，如果你们想作出某种类似于自然选择或广义相对论或大爆炸理论那样的重大发现，如果你们想超越达尔文或爱因斯坦，那么你们成功的机会将微乎其微。（考虑到她们的个性，我的孩子们很可能会用自己一生的行动来证明她们的父亲只是一个彻头彻尾的傻瓜。）

詹姆斯·格莱克曾在为理查德·费曼所撰写的传记《天才》一书中，叩问为什么科学似乎已不再能产生出像爱因斯坦和玻尔之类的天才。他给出了一个看似矛盾的答案，认为：存在着如此众多的爱因斯坦和玻尔，如此众多的天才水平的科学家，以至于对任何一位个体来说，要想从这一群体中脱颖而出都已变得极为艰难。我对此深有同感，但格莱克假说的严重缺失是：与爱因斯坦和玻尔相比，我们时代的天才们所面对的发现主题要少得多。

让我们再花些笔墨回到反科学的主题上来。科学界不足为外人道的一个小秘密，就是许多著名的科学家都怀抱着明显的后现代情绪。我在书中给出了大量与这一现象相关的证据，可在此稍加回顾：斯蒂芬·杰伊·古尔德曾坦白承认，他对影响巨大的后现代文本《科学革命的结构》情有独钟；林恩·马古利斯宣称，他不认为存在着什么绝对真理，即使存在，他也不认为有谁能拥有它；弗里曼·戴森则预言，现代物理学在未来的物理学家眼中，将会像亚里士多德物理学在我们眼中一样古老。

对既有知识的这种怀疑主义态度应如何解释？在这些科学家以及其余的诸多知识分子看来，使生活富有意义的是对真理的追求，而不是真理本身；但现有知识在它得以成立的范围内是极难超越的。通过强调现有知识可能被证明是暂时的，这些怀疑主义者才能维系这样一种幻觉，即发现的伟大时代尚未终结，更深入的发现即将莅临。而后现代主义则判定：所有将来的发现，最终会被证明同样是暂时的，将屈从于另外一些伪称的洞见，以至无穷。后现代主义者情愿接受这种西西弗斯般的存在状况①，他们牺牲了对绝对真理的信念，所以才能对真理永远追求下去。

只是定义问题

1996年7月23日，我应邀参加了"查理·罗斯访谈秀"（Charlie Rose Show）

---

① Sisyphean condition：在希腊神话中，Sisyphus是科林斯城邦的一位贪婪的君王，死后被罚永远在地狱推一块巨石上山，而该石推到山顶后必定滚下——译者注。

节目，同时受到邀请的还有耶利米·奥斯特里克（Jeremiah Ostriker），一位普林斯顿大学的天体物理学家，他担负着在节目中反驳我的论点的期许。奥斯特里克与我一度就暗物质问题展开了唇枪舌剑，其核心假设是认为星体以及其他发光物质在宇宙的总体构成中只占了很小的比例。奥斯特里克坚称，暗物质问题的解决必将使我关于宇宙学家再也做不出什么真正重大发现的断言不攻自破；我不同意，说答案即便找到了，其意义也是微不足道的。罗斯插话说，我们两人之间的分歧看起来"只是定义问题"。

我必须承认，罗斯已经触及了我的著作的一个严重缺陷。在给出有关科学家们将不再可能作出堪比达尔文进化论或量子力学般深刻的科学发现这一观点时，我本应花费更多笔墨解释清楚自己所谓"深刻的"究竟意指什么。一个事实或理论是否"深刻"，与其在空间和时间上所适用的范围成正比。无论量子电动力学还是广义相对论，就我们的目前所知，自其诞生之日起，就适用于整个宇宙和所有时间范围，这才成就了这些理论的真正深刻性。作为对比，一种关于高温超导的理论，只适用于可能存在的物质的某一特定类型，并且就我们目前所知，也只适用于地球的实验室里。

毋庸置疑，在对科学发现进行评价时，更多主观的标准也会掺和进来。严格说来，与物理学那些奠基性的理论相比，所有生物学理论在深刻性上都要逊色得多，因为生物学理论只适用于——再说一次，就我们目前所知——物质的某一种特定组合，它仅存在于我们这个孤独而又渺小的星球上，且仅存在了35亿年。但生物学却有一种比物理学更含义隽永的潜质，因为它更直接地涉及在我们看来更加令人着迷的现象：人类自身。

在《达尔文的危险思想》（*Darwin's Dangerous Idea*）中，丹尼尔·丹尼特曾雄辩地论证说，借助于自然选择的进化论是"迄今为止人们所能想到的绝无仅有的最佳思想"，因为它"将众多的领域整合于一身，诸如生命、意义、时空王国的存在目的、因果、生物机能与物理定律，等等"。的确，达尔文的成就，尤其是在它融合了孟德尔遗传机制而形成新的综合后，已经宣告了后续的生物学只能是锦上添花，至少从哲学的角度看是这样（虽然，就像我后面论证的那样，进化生物学对我们认识人类的本质只能提供有限度的洞见）。即使是沃森和克里克关于双螺旋的发现，虽然已经带来了可观的实践成果，也只不过揭示了遗传机制的基础，并没有为新的综合增添什么重大的洞见成分。

再回到我与耶利米·奥斯特里克的辩论上来，我的立场是：宇宙学家们永远也不可能超越大爆炸理论，它已经奠基性地说明了宇宙处于膨胀之中，并且与其今天的情形相比，一度曾经是尺度更小、温度更高、密度更大；该理论为宇宙演化的历史提供了一个自洽的叙事版本，并且具有深刻的神学意味。宇宙有其起点，也应该有其终点（尽管宇宙学家们也许永远也不可能拥有足够的证据，为后一观点的争论一劳永逸地画上句号）。还有什么能比这更深刻、更有意义呢？

与此相对照，有关暗物质问题的最有可能的答案，当然就显得没那么重要了。它不过是宣称：单一星系或星系团的运动的最佳解释就是假定星系中包含着星尘、死星以及其他一些惯常形式的物质，这些物质通过望远镜是无法观测到的。还有一些关于暗物质问题的更富戏剧色彩的版本，想当然地假定宇宙中高达99%的部分都是由某种与我们在地球上所熟悉的一切截然不同的异种物质构成。这类说法，不过是暴涨宇宙以及其他一些更没边儿的宇宙假说的推演产物，永远也不可能得到证实；至于理由，在本书的第四章里已给出了详细的阐述。

应用科学又如何

有几位批评者挑剔我忽视了——更严苛的说法是诋毁了——应用科学。其实，我认为有极好的证据表明，应用科学也正迅速地趋向其极限。举例来说，我们一度曾认为，物理学家们关于核聚变的知识除给我们带来氢弹之外，必然会为我们提供一种清洁、经济而又无穷尽的能源；聚变研究人员也曾夸口了几十年，说"只要继续投入资金，20年内我们就能给你们拿出便宜得近乎免费供应的动力"。但最近几年，美国却大幅削减了聚变研究的预算。现在，即使是最乐观的研究人员也预测说，要想建成经济可行的聚变反应堆，至少还需要50年之久；现实主义者则承认，聚变能源是一个永远难圆的梦，原因很简单：技术、经济和政治上的障碍过于巨大，根本无法克服。

再看看应用生物学，其终点完全可以用人类长生不老的实现来标志。科学家们能鉴别并掌握支配人类衰老机理的可能性，一直是科学记者们经久不衰的关注热点。公众对科学家们攻克衰老之谜的信心本可以更强一些，前提是他们在一个明显更为简单的问题即癌症问题上，取得的成就更多一些。但自从理查德·尼克松总统于1971年代表联邦政府正式宣布"对癌症开战"以来，虽然美国已经投入了大约300亿美元的研究经费，但就整体而言，癌症死亡率却比那时上升了6个百分点。治疗方法也只有很小的改善，医疗者们仍然通过手术切除癌变组织，用化疗方法抑制其转化，并用放射疗法杀死癌细胞。或许终有一天，我们能穷自己的研究之力给出一种"疗法"，使癌症变得像小小的水痘一样不足为害；或许，我们做不到这一点；或许，癌症——推而广之还包括长生不老——只不过是个过于复杂以至于无法解决的难题。

具有讽刺意味的是，生物学面对死亡的无能，也许正是它被寄予最大希望之所在。在1995年11／12月号的《技术评论》上，麻省理工学院的社会政策学教授哈维·萨波尔斯基（Harvey Sapolsky）指出：二战结束之后，为科学基金进行辩护的主要理由是为了国家安全，或者更具体地说，是冷战的需要。现在，"邪恶帝国"这一借口已不复存在，科学家们若要为其庞大的开销辩护，作为替代的其他目标是什么？萨波尔斯基给出的答案就是长生不老。他指出，许多人都认为活得更久些——甚至可能的话，活到永远——是值得追求的；并且，把长生不老作为科学的首要目标的最大好处是：它几乎可以肯定是无法达到的，因此科学家们就可以源源不断地得到资金，以进行更多的研究，直到永远。

关于人类心智呢

1996年7月的《电气和电子工程师协会总览》（IEEE Spectrum），曾发表了科学记者戴维·林德利的一篇综述，宣告物理学以及宇宙学都已达到了其发展的尽头。（这一傲慢的宣判并没有什么特别令人感到吃惊的，要记得林德利曾写过一本著作，书名赫然是《物理学的终结》。）但他依然坚持认为，对于人类心智的探索终将催生出某种强有力的范式，尽管在目前还处于"前科学状态"，探索于其中的科学家们严格说来连自己究竟在研究什么都无法达成共识。但愿如此！但科学在超越弗洛伊德范式方面所表现出的无能为力，的确很难激起太大的希望。

自弗洛伊德一个世纪前创建其心理分析理论以来，有关心智的科学，就某些特定方面来看，的确已具有了更多的实证性、更少的思辨色彩。我们已掌握了一些令人惊异的能力去探测人的大脑，用微电极、磁共振成像以及正电子发射断层扫描。但这些研究既没有导致任何认识上的更深刻洞见，也未带来治疗手段上的重大进展，正如我在1996年12月号的《科学美国人》杂志上发表的文章中所揭示的，文章题为"为什么弗洛伊德阴魂不散？"。心理

学家、哲学家以及其他学者之所以仍揪着弗洛伊德成就延伸出的问题争论不
休，原因很简单：无论是在心理学层面还是药理学层面，关于人类心智该如
何理解和救治，迄今尚未产生任何足以一劳永逸地超越心理分析的更好替代
品。

　　也有科学家认为，要达成有关人类心智的统一范式，最有希望的出路在
于达尔文理论，其最新的版本就叫作进化心理学。回顾一下本书第六章，我
曾引述过的诺姆·乔姆斯基的批评意见，即"达尔文理论就像一只宽松的大
口袋，能把（科学家们）发现的任何货色都装进去"。这一点很重要，我在
这一观点基础上进一步发挥写了篇文章，发表在《科学美国人》杂志1995年
10月号上，题为"新社会达尔文主义者"。进化心理学最主要的对立范式，
也许可被称为"文化决定论"（cultural determinism），其基本假设是：决
定人类行为的首要因素是文化，而不是基因禀赋。为了论证自己的观点，文
化决定论者们指出了生活于不同文化环境下的人们，在行为上所表现出的巨
大多样性——其中的绝大多数似乎都是非适应性的。

　　作为回应，某些进化理论家就理所当然地把从众，或"驯顺性"
（docility），视为一种适应性的、固有的品质。换言之，那些"顺之者
生"的个体，就会保持驯顺的品质。诺贝尔桂冠获得者司马贺（Herbert
Simon）曾在《科学》杂志上撰文（1990年12月21日），推测驯顺性可以
解释为什么人们会遵从宗教的信条而遏制自己的性冲动，或在战争中勇往
直前；而作为个体的他们，其所失去的往往远超其所得。司马贺的假说，
的确睿智地吸收了文化决定论者的立场，同时也削弱了进化心理学作为一
门科学的合法地位。如果某一给定的行为与达尔文宗旨相符，那当然很
好；如果不相符，该行为只不过是展现了我们的驯顺性。这样，该理论就
变成了对证伪完全免疫的货色，因此也就证实了乔姆斯基关于达尔文理论
什么都能解释的批评。

　　要知道，人类会遵从其文化的倾向，还会给达尔文主义的理论家们引出
另外一个问题。为了证明某一品质是固有的，达尔文主义者们所力图揭示的
是它在所有文化中都会发生。这样，达尔文主义者力图揭示——比如说——
雄性天生就比雌性更倾向于滥交；但考虑到现代文化的相互关联性，某些
已经被达尔文派学者证明是普适性的，因而也公认是天赋的态度和行为，
实际上也许仅仅是由驯顺性导致的。这正是文化决定论者们一直以来极力
主张的。

　　科学在把握人类心智上的无能为力，同样也反映在人工智能的进展上，也就是要创造出足以模拟人类思维的计算机的努力上。许多权威人士都把1996年2月在IBM公司的计算机"深蓝"和世界冠军卡斯帕罗夫（Gray Kasparov）之间进行的国际象棋比赛，看作是这一领域的辉煌成就。毕竟，在以4比2的比分最终输掉之前，"深蓝"曾在比赛的第一场中占尽优势。但以我的偏见看来，这场比赛强有力地证明了人工智能自从40余年前被马文·明斯基等人创立以来就已经彻底失败。有着直接的规则和极小的笛卡儿运动空间的国际象棋，正是那种为计算机所精心炮制的游戏；而"深蓝"的五位人类教练之中，包括当今世界最好的国际象棋大师，而且其本身又是一台威力惊人的机器，有32位并行处理器，每秒能够检校两亿个位点。如果在国际象棋比赛中，这一硅结构的庞然大物仍不能击败小小的人的话，那么，用计算机模拟人类更复杂的才能——比如说，在鸡尾酒会上认出你大学时代所钟爱的女朋友，并立即想出恰当的说辞，使她为15年前抛弃你的行为感到追悔莫及——还有什么希望呢？

关于混杂学花招

　　自我的书首度出版以来，事实上就已提出了一对额外的参数，用以说明混沌与复杂性研究的限度——我毫无贬义地把它们统合在了"混杂学"（*chaoplexity*）的术语之下。由斯图亚特·考夫曼、佩尔·贝克、约翰·霍兰等人所着力追求的混杂学领域，其最为深远的目标之一就是要阐明某种东西———一种规律，或一套准则，或某种统一的理论，它将会使我们理解种种表面看来全无相似之处的复杂系统并使预测其行为成为可能。与之紧密相关的一种说法是：宇宙中蕴藏着某种生成复杂性的力，它抵制着热力学第二定律，并创造出星系、生命，甚至是智慧程度足以深入思考这种力量本身的生命。

　　类似于这样的假说，如果希望自身变得有意义，其支持者们必须清楚明白地告诉我们复杂性是什么以及它可以被怎样度量。我们凭直觉就可以知道，今天的生命比之2000年前，或200万年前，或20亿年前，的确更为"复杂"，但这种直觉怎么才能用一种非主观的方式予以量化呢？直到或除非这一问题得到解决，所有关于复杂性规律或复杂性生成力的假说统统都属于废话。我对这一问题能得到解决深表怀疑（意料之外啊，意料之外），隐藏在大部分关于复杂性的定义之下的无非是这样一种看法，即某种现象的复杂性与其不可能性成正比，或者与其必然性成反比。假如我们摇动一口袋的分子，有多大可能会由此得到一个星系、一颗星球、一只草履虫、一只青蛙或者一名股票经纪人呢？解答此类问题的最佳途径莫过于找到另一个宇宙或另一种生物系统，并对它们进行统计分析，这显然是不可能的。

尽管如此，混杂学家们依然辩称自己可以解答关乎上述可能性的问题，通过在计算机中构建别样的宇宙和生命演化史，并断定哪些特征是稳健的、哪些是有条件的或短命的。我相信，这一希望源自于对计算机科学和数学领域某些特定进展的乐观解读。在过去的几十年里，研究人员已经发现，许多简单的规则，一旦用计算机来执行，就会产生一些图案，它们表现得就像是时间或变量的函数般随机变化，我们可以把这种虚假的随机性称为"伪噪音"（pseudo-noise）。最典型的伪噪音系统就是芒德勃罗集，它如今已变成混杂学运动的标志物。混沌与复杂性研究两个领域都怀有一种共同的期望，即几乎遍布自然界的噪音，其中绝大多数实际上都属于伪噪音，是某些更基本的、决定论的算法的作用结果。

但那使得地震、股票市场、天气以及其他现象的预测变得如此困难的噪音，在我看来，虽非那么显而易见，但却真实存在，并且永远也不可能被规约成任何一组简单的规则。可以肯定，更快的计算机以及更进步的数学技术，的确会提升我们预测某些特定复杂现象的能力。至少在一般公众的印象中，天气预报在过去的几十年里已经变得越来越准确了，这部分地可归功于计算机建模；更重要的却是数据采集技术的进步，尤其是卫星成像技术。气象学家们用了更大、更精确的数据库，就可以在此基础上建立模型并加以检验。正是通过模拟与数据采集之间的辩证关系，预报能力的提升才得以实现。

在某种程度上，计算机模型正在超越科学本身的界限，向工程学发展（想来令人不寒而栗）。模型有用或无用，所遵照的不过是某些有效性原则，而"事实"却变得无关紧要了。再说，混沌理论告诉我们，蝴蝶效应已经为预测设定了根本的限度。要想预测特定系统的演变轨迹，人们就必须近乎无限精确地掌握其初始条件。这正是混杂学家们令我一直感到困惑的地方：若根据作为其基本信条之一的蝴蝶效应，他们所追求的绝大多数目标似乎都是无法实现的。

在涉及实在的多方面属性的可能性问题上，混杂学家们的探究之路并不孤独。这些问题同样繁衍出各色具有浓厚反讽意味的假说，诸如人择原理、暴涨宇宙、多重宇宙说、间断平衡说，以及盖亚。不幸的是，你无法确定宇宙或地球生命的可能性，因为作为思考对象的宇宙以及生命演化史都是独一无二的，统计所需要的数据点远远不止一个。

缺乏经验数据并不能阻止科学家和哲学家们在这类问题上固执己见。一方面是必然论者（inevitabilists），让他们感到欣慰的理论，必须把实在描述成某些不变法则的高度可能性的甚或必然的产物；多数科学家都是必然论者，其最著名的代表人物当属爱因斯坦，他抵制量子力学的理由就是其隐喻着上帝在创造世界时玩骰子。但是，也有一些著名的反必然论者（anti-inevitabilists），特别是卡尔·波普尔、斯蒂芬·杰伊·古尔德、伊利亚·普里高津等，他们把科学的决定论视为对人类自由的威胁，因而去热情拥抱非确定性和随机性。我们或者是命运的棋子，或者是不可思议的侥幸，二者必居其一。随你挑选！

火星生命

1996年8月，就在我的书于美国出版两个月之后，一个由（美国）国家航空和航天管理局（NASA）以及别的什么地方的科学家构成的小组宣称，在一块由火星降落到南极上的陨石碎片中发现了微生物生命的痕迹。评论家们立即抓住这一发现，用以证明科学正走向终结的说法是多么荒谬。但火星上存在生命的叙述，非但无法证伪我的命题，反而证实了我的断言，即科学正处于一种巨大危机的阵痛中。我还没有愤世嫉俗到足以轻信某些观察家的暗示，即NASA官员们虽然明知其所谓发现不大可能站得住脚，但为了招徕更多的资金，却在拼命吹嘘。然而，对这一叙述的夸张反应——来自NASA、政客、媒体、公众以及部分科学家——却恰恰证明人们对获得真正重大的科学发现有多么绝望。

正如我在书中几次提到的，关于我们并非宇宙中绝无仅有的智慧生命的发现，将是人类历史上最震撼人心的事件，我希望自己能在有生之年目睹这一重大发现的来临。但NASA团队的发现与此根本就不沾边儿。自一开始，就有些在原始生物化学方面真正博识的科学家对火星生命的叙述能否成立提出了质疑。1996年12月，有两个科学家团队各自独立地在《地球化学与宇宙化学学报》（*Geochimica et Cosmochimica Acta*）上发表了研究报告，称所谓在火星陨石上发现的生命物质，很可能是由非生物过程产生的，或者是由陆地生物的污染所致。"火星生命的丧钟"，这是《新科学家》（*New*

Scientist）杂志1996年12月21/28号所发布的悼词。

只有我们对火星进行了彻底的探索之后，才能确实知道这一星球上是否真的存在生命；最好是能有一组科学家深入钻探到火星表面之下，因为正如我们所知的，只有那里才可能分布着足够的水和热量，足以适合微生物的生存。要想派出这样一组科学家，负责空间事务的官员们至少要花几十年的时间去集攒资金和技术资源——即使政客们和公众乐意支付其开销。

退一步讲，就算我们最终真的确定了火星上曾经存在过或现在仍然存在着微生物，这一发现将极大地推动生命起源研究的发展以及更普遍意义上的生物学的发展。但这是不是就意味着，科学被一下子从其诸多现实的约束条件中解放出来了呢？绝非如此。假如我们在火星上发现了生命，我们的确会知道，在太阳系里，生命也存在于地球之外。但对于太阳系之外是否存在生命，我们仍是一如既往的无知；要想确切回答这一问题，我们仍将面对重重障碍。

天文学家最近鉴别出一大批邻近的恒星，认为其周围环绕的行星可能提供了生命存在的条件，但物理学家弗兰克·德雷克（Frank Drake，他曾是SETI，即"地外生命探索"计划的奠基者之一）曾估算过，要想抵达其中最近的星系并确定是否有生命居住其上，现有的宇宙飞行器要用40万年的时间。或许终有一天，由SETI设置的无线电波接收装置，能够接收到由其他星球发射的类似于"我爱露西"（I Love Lucy）的电磁波信号。

但正如20世纪最著名的进化生物学家之一恩斯特·迈尔所指出的，大部分SETI成员都是像德雷克这样的物理学家，他们对实在持有一种极端决定论的观点。物理学家们认为，既然地球上存在高度技术化的文明，那么在地球外信号可达范围内也存在着类似文明就是很可能的。迈尔等生物学家却认为这一观点是荒谬可笑的，因为他们知道进化中包含着极大的偶然性，或者说简直就是幸运；即使生命进化的宏大实验重演数百万次，也不一定就能演化出哺乳动物，更不用说聪明到足以发明电视机的哺乳动物了。1995年剑桥大学出版社再版的《外星人：他们在哪里？》中，收录了迈尔的一篇文章，对SETI项目作了总结，认为该计划正顶着"天文学尺度"的困难挣扎前行。尽管我认为迈尔很可能是正确的，但在1993年，当国会真的终止了对SETI的资助时，我仍然难以抑制沮丧的心情。该项目目前正靠着私人基金苟延残喘。

想入非非部分

最后，拙著的结尾部分转向了神学和神秘论，或者如一位熟悉的朋友所称，是"想入非非部分"。我曾担心，某些评论家会因这部分内容而把我判作是无可救药的奇谈怪论者，并因而抛弃我关于科学之未来的整个论点。幸运的是，这种情况并未真的发生，大部分评论家要么完全忽略了尾声部分，要么就是简单地对它表示一番疑虑了事。

最敏锐的评论，或许该说是最获我心的，是由物理学家罗伯特·帕克（Robert Park）给出的，文章见于1996年8月11日的《华盛顿邮报图书世界》。他说，起初他对我用"幼稚的反讽科学走向疯狂"终结全书深为不满，但在进一步反思的过程中他认识到，这一结尾是"一种隐喻"霍根以此警示着：这正是科学的终结之所……科学正经历着抵抗种种后现代异端邪说的考验，不存在什么客观真理，在围墙之内只有后现代主义。

我自己也不能给出比这更好的表述，但我之所以这样终结全书，内心中还有别的动机。首先，我觉得，对于书中那些曾被我大加嘲笑的科学家们来说，揭示出我自己就像他们一样也具有形而上学幻想的倾向，是件十分公平的事。同时，我在尾声部分所描写的神秘体验插曲是我一生经历中最为重要的体验，在其后长达十余年的时间里，这一体验曾时时地刺激着我，使我终于决定要有所诉说，即使这一做法将有损我作为一名记者可能已经拥有的小小名誉，也再所不计。

其实，真正重要的神学问题只有一个：如果真的存在一位上帝，那么他为什么要创造出一个具有如此众多苦难的世界？我的经验提示着这样一种答案：如果存在着一位上帝，那么他不仅是从欢乐和爱中，而且也是从恐惧和绝望之中，创造了这个世界。这就是我给予存在之谜的解答，并且，我觉得有必要与别人分享这一答案。请允许我在这里做彻底的坦白，我写作《科学的终结》一书的真正目的是创立一种新的宗教——"神的恐惧教派"（The Church of the Holy Horror）。从一名科学记者摇身一变，过一过教派领袖的瘾，将是一种很好的调剂，更不用说其中尚有钱可赚呢。

1997年1月于纽约

致谢

　　首先要感谢《科学美国人》杂志的一贯支持和鼓励，使我能全面地发展自己的兴趣，否则，我永远也不可能写出此书；《科学美国人》杂志社还惠允引用我为杂志撰写的以下人物传略系列（版权归《科学美国人》杂志社所有，保留所有的权利）：《克利福德·格尔茨》（1989年第7期），《罗杰·彭罗斯》（1989年第11期），《诺姆·乔姆斯基》（1990年第5期），《在起跑线上》（1991年第2期），《托马斯·库恩》（1991年第5期），《约翰·惠勒》（1991年第6期），《爱德华·威滕》（1991年第11期），《弗朗西斯·克里克》（1992年第2期），《卡尔·波普尔》（1992年第11期），《保罗·费耶阿本德》（1993年第5期），《费里曼·戴森》（1993年第8期），《马文·明斯基》（1993年第11期），《爱德华·威尔逊》（1994年第4期），《科学能够解释意识吗？》（1994年第7期），《弗雷德·霍伊尔》（1995年第6期），《斯蒂芬·杰伊·古尔德》（1995年第8期）。我还荣幸地获准引用下列书籍的资料：《黄金时代的来临》，岗瑟·斯滕特著；《科学进步论》，尼古拉斯·里查著；《告别理性》，保罗·费耶阿本德著；以及《宇宙发现论》，马汀·哈威特著。

　　我对我的代理人Stuart Krichevsky深怀感激，是他帮助我把一个模糊的观念最终转化为一种连贯、清晰的观点；还有Addison-Wesley出版公司的Bill Patrick和Jeff Robbins，他们曾给予我中肯的批评和鼓励。在历时一年有余的写作过程中，我曾受惠于许多朋友、熟人和同事，有的是《科学美国人》杂志社的，也有社外的，他们向我提供了许多有价值的信息。这些人是（按姓氏字母顺序排列）：Tim Beardsley, Roger Bingham, Chris Bremser, Fred Guterl, George Johnson, John Rennie, Phil Rose, Russell Ruthen, Gary Stix, Paul Wallich, Karen Wright, Robert Wright, and Glen Zorpette。

2015年版前言　重启《科学的终结》之争

1. 2007年，著名文化批评家乔治·斯坦纳（George Steiner）在葡萄牙的里斯本组织了一次专题研讨会，并邀请我在会上发言，当然也邀请了一些大腕儿，如弗里曼·戴森、杰拉尔德·埃德尔曼、刘易斯·沃尔珀特（Lewis Wolpert）等，而这次会议至少是部分地受到了拙著的激发，题为"科学正趋近于其极限？"。斯坦纳在其开场白中讲道："科学理论及其实践是否正在撞向某种根本性的、无以逾越的秩序之墙？即使只是提出这一问题，就已经触及了某些特定的禁忌，某些已被写入我们文明根基中的教条。"甚得我心啊，乔治！此次研讨会的会议纪要，载于《科学正趋近于其极限？》（*Is Science Nearing Its Limits*? Lives and Letters, 2008）。诺贝尔物理学奖得主伊瓦尔·贾埃弗（Ivar Giaever）曾在2013年指出："或许，我们已经抵达了科学的终点；或许，科学不过是一项有限的事业，尽管源自这一有限事业的发明是无穷的。那么，这就是科学的终结吗？绝大多数科学家对此存而不论，但我却旗帜鲜明地赞同这一点。"参见《我们已抵达纯科学的终点：伊瓦尔·贾埃弗》（"We have come to the end of pure science: Ivar Giaever," by Nikita Mehta, *Live Mint*, December 24, 2013）（http://www.livemint.com/Politics/JVpA3xKlQGbE8vStlsepLI/We-have-come-to-the-end-of-pure-sciences-Ivar-Giaever.html）。

2.其最具代表性的案例，可参阅《我们正趋近于科学的终点吗？》（"Are we nearing the end of science?" by Joel Achenbach, the *Washington Post*, February 10, 2014）（http://www.washingtonpost.com/national/health-science/are-we-nearing-the-end-of-science/2014/02/07/5541b420-89c1-11e3-a5bd-844629433ba3_story.html）。

3.参见伊奥尼迪斯的文章《公开发表的研究成果为何大多都是虚假的》（"Why Most Published Research Findings Are False," by John Ioannidis,

PLOS Medicine, August 30, 2005）（http://www.plosmedicine.org/article/
info%3Adoi%2F10.1371%2Fjournal.pmed.0020124）。

4.参见伊奥尼迪斯的文章《虚假主张的流行》（"An Epidemic of False
Claims," by John Ioannidis, *Scientific American*, May 17, 2011）（http://www.
scientificamerican.com/article/an-epidemic-of-false-claims/）。

5.《不端行为是导致应撤科学出版物产生的罪魁祸首》（"Misconduct
accounts for the majority of retracted scientific publications," by Ferric Fang et al,
Proceedings of the National Academy of Sciences, October 1, 2012）（http://www.
pnas.org/content/early/2012/09/27/1212247109）。

6.《经济学人》，2013年10月19日（*The Economist*, October 19, 2013）
（http://www.economist.com/news/briefing/21588057-scientists-think-science-self-
correctingalarming-degree-it-not-trouble）。

7.可参阅文章《丹尼尔·卡尼曼从社会心理学角度解释"垃圾论文满天飞"现象》
（"Daniel Kahneman sees 'Train-Wreck Looming' for Social Psychology"），刊
于《高等教育年鉴》（*the Chronicle of Higher Education*）2012年10月4日（http://
chronicle.com/blogs/percolator/daniel-kahneman-sees-train-wreck-looming-forsocial-
psychology/31338）。

8.加来道雄和我打的这个赌，得到了久赌基金（the Long Bet Foundation）的赞
助（http://longbets.org/12/）。

9.可参阅伦纳德的《宇宙景观》（*The Cosmic Landscape*, by Leonard, Back Bay
Books, 2006）。坦白讲，该书最好别看，知道我说的那些就够了。

10.肖恩·卡洛尔在2014年的一篇文章里贬斥了证伪观，文章发表在边缘网
（Edge.org），一个由科学书籍代理商John Brockman监管的网站（http://edge.org/
response-detail/25332）。卡洛尔以此文回应Brockman所提出的问题："哪种科学观
念可以寿终正寝了？"

11.斯蒂芬·霍金对哲学的抨击，可见于其著作《大设计》（The Grand Design,
co-written with Leonard Mlodinow, Bantam, 2012）；而劳伦斯·克劳斯的，则是在
其与Ross Andersen的一次在线访谈节目中，见于2012年4月23日的大西洋网（http://
www.theatlantic.com/technology/archive/2012/04/has-physics-madephilosophy-and-
religion-obsolete/256203/）。

12.这一点是哲学家/物理学家大卫·阿尔伯特（David Albert）指出的，他曾就
克劳斯的《无中生有的宇宙》（*A Universe from Nothing*）一书，撰写过一篇尖刻的

（请从正面意义上理解）评论文章，刊于《纽约时报书评》（*New York Times Book Review*），2012年3月23日。

13.若欲寻一份关于系外行星探索之旅的精彩报告，推荐阅读《五十亿年的孤独》（*Five Billion Years of Solitude*, by Lee Billings, Current, 2013）。

14.《纽约时报》2011年的报告，题为"生命摇篮理论令人欢欣鼓舞的进展"（"A Romp into Theories of the Cradle of Life," by Dennis Overbye, *New York Times*, February 21, 2011）（http://www.nytimes.com/2011/02/22/science/22orgigins.html?_r=1&ref=science）。

15.弗朗西斯•柯林斯的这一声明，是在2003年4月14日美国国立卫生研究院的新闻稿中给出的（网址：http://www.nih.gov/news/pr/apr2003/nhgri-14.htm）。

16.参见由《基因医学杂志》（*Journal of Gene Medicine*）给出的在线表单（http://www.wiley.com/legacy/wileychi/genmed/clinical/）。

17.可参阅雷诺•斯佩克特的文章，《癌症战争：一份写给怀疑论者的进展报告》（"The War on Cancer: A Progress Report for Skeptics", by Reynold Spector, *Skeptical Inquirer*, February 2010）（http://www.nytimes.com/2009/04/24/health/policy/24cancer.html?_r=0）。

18.可参阅我的博文，《研究人员真的发现过什么能决定人类行为的基因吗？》（"Have researchers really discovered any genes for behavior?", *Scientific American*, May 2, 2011）（http://blogs.scientificamerican.com/cross-check/2011/05/02/have-researchers-really-discovered-any-genes-for-behavior-candidates-welcome/）。

19.参阅杰里•科因的文章，《你没有自由意志》（"You Don't have Free Will," by Jerry Coyne, the *Chronicle of Higher Education*, March 18, 2012）（http://chronicle.com/article/Jerry-A-Coyne/131165）。

20.人类学家理查德•兰厄姆的这一主张，出自其与新闻记者达尔•彼得森合著的大作《恶魔般的雄性》（*Demonic Males*, co-written with journalist Dale Peterson, Mariner Books, 1997）；而爱德华•威尔逊的言论，则出自其《地球的社会征服》（*The Social Conquest of the Earth*, Liverright, 2012）一书。我对这类主张的驳斥，可参阅《不，战争不是不可避免的》（"No, War Is Not Inevitable," Discover, June 12, 2012）（http://discovermagazine.com/2012/jun/02-no-war-is-not-inevitable）一文，以及本人2012年的书《战争的终结》（*The End of War*, McSweeney's, 2012）。

21.参阅卡哈尔·梅尔莫的文章，《震怒于DNA先驱者的理论：非洲人不如西方人聪明》（"Fury at DNA pioneer's theory: Africans are less intelligent than Westerners," by Cahal Milmo, *The Independent*, October 17, 2007）（http://independent.co.uk/news/science/fury-at-dna-pioneers-theory-africans-are-less-intelligent-than-westerners-394898.html）。其中引述了沃森评论非洲人的话："我们所有的社会政策都建立在这样的事实基础上，即他们的智力与我们的一样，但所有的检测结果却告诉我们，事实并非如此。"

22.我对神经编码的探讨，可参阅《意识之谜》（"The Consciousness Conundrum," *IEEE Spectrum*, June 2008）（http://spectrum.ieee.org/biomedical/imaging/the-consciousness-conundrum）。

23.克里斯托弗·科赫对心灵论的辩护，见于其文章《意识是普遍的吗？》（"Is Consciousness Universal?" *Scientific American*, December 19, 2013）（http://www.scientificamerican.com/article/is-consciousness-universal/）。

24.可参阅玛西娅·安吉尔的文章《精神病学的幻象》（"The Illusions of Psychiatry," by Marcia Angell, *The New York Review of Books*, July 14, 2011）（http://nybooks.com/articles/archives/2011/jul/14/illusions-of-psychiatry/）。

25.可参阅史蒂夫·洛尔的文章《"大数据"的起源：一个词源学的侦探故事》（"The Origins of 'Big Data': An Etymological Detective Story," by Steve Lohr,the *New York Times*, February 1, 2014）（http://bits.blogs.nytimes.com/2013/02/01/theorigins-of-big-data-an-etymological-detective-story/）。洛尔把优先权给了硅图公司的计算机科学家约翰·马什艾（John Mashey），他早在20世纪90年代就开始谈论大数据了。

26.可参阅克里斯·安德森的文章《理论的终结：海量数据使科学方法变得过时》（"The End of Theory: The Data Deluge Makes the Scientific Method Obsolete," by Chris Anderson, *WIRED*, June 23, 2008）（http://archive.wired.com/science/discoveries/magazine/16-07/pb_theory）。

27.可参阅雷·库兹韦尔的著作《奇点临近》（*The Singularity Is Near*, by Ray Kurzweil, Penguin, 2006）。

28.可参阅阿什利·万斯的文章，《仅仅是人类？昨天才是这样》（"Merely Human? That's So Yesterday," by Ashlee Vance, the *New York Times*, June 12, 2010）（http://www.nytimes.com/2010/06/13/business/13sing.html?pagewanted=all）。

29.若想更多地了解我关于科学的古怪思想，还可访问我的网站：johnhorgan.

org；以及我在《科学美国人》的博客："交叉检索"（Cross-check）。

引言　寻找"终极答案"

1.《皇帝的新脑》，罗杰·彭罗斯著（*The Emperor's New Mind*，Roger Penrose，Oxford University Press，New York，1989）。文中所提到的书评，作者为天文学家、作家Timothy Ferris，发表于《纽约时报书评》（*New York Times Book Review*），1989年11月19日，第3页。

2.我采访彭罗斯的文章发表在《科学美国人》（*Scientific American*）杂志1989年11期，第30–33页。

3. 我到锡拉丘兹采访彭罗斯是在1989年8月期间。

4.对"反讽"的这一定义，参考了Northrop Frye的文学理论经典著作《批评的解析》（*Anatomy of Criticism*，Princeton University Press，Princeton，N.J.，1957）。

5.《影响的焦虑》，布鲁姆著（*The Anxiety of Influence*，Harold Bloom，Oxford University Press，New York，1973）。

6.出处同上，第21页。

7.出处同上，第22页。

第一章　进步的终结

1.古斯塔夫·阿道夫研讨会的报告集已出版，题为《科学的终结？——攻击与辩护》（*The End of Science? Attack and Defense*，edited by Richard Q. Selve，University Press of America，Lanham，Md.，1992）。

2.《黄金时代的来临——进步之终结概论》，斯滕特著（*The Coming of the Golden Age：A View of the End of Progress*，Gunther S.Stent，Natural HistoryPress，Garden City，N. Y.，1969）。也可参阅"注释1"《科学的终结？——攻击与辩护》一书中收录的斯滕特的论文。

3.参阅《亨利·亚当斯教育学》（*The Education of Henry Adams*，Massachusetts Historical Society，Boston，1918；reprinted by Houghton Mifflin，Boston，1961）。亚当斯的加速度定律是在34章提出的，写于1904年。

4.出自斯滕特的《黄金时代的来临——进步之终结概论》，第94页。

5.出处同上，第111页。

6.波林对化学的议论，可参阅其著作《化学键的本质以及分子和晶体的结构》（*The Nature of the Chemical Bond and the Structure of Molecules and Crystals*，Linus Pauling，published in 1939 and reissued in 1960 by Cornell University Press，Ithaca，N. Y.），该著作一直是最具影响力的科学范本。波林曾告诉我，早在此书出版的十年之前，他就解决了化学的基本问题。1992年9月，我在加利福尼亚的斯坦福采访了他。他说："我认为截至1930年底，或者在年中，有机化学就已很圆满地建成了；无机化学和矿物学仍然存在着许多至今尚未完成的工作——硫矿学除外。"波林逝世于1994年8月19日。

7.出自斯滕特的《黄金时代的来临》，第74页。

8.出处同上，第115页。

9.出处同上，第138页。

10.我采访斯滕特的时间是1992年7月，地点在伯克利。

11.这一令人沮丧的事实，我是在《银河系时代的来临》（*Coming of Age in the Milky Way*，Timothy Ferris，Doubleday，New York，1988）一书第371页上看到的。若想更详尽回顾美国太空计划的历史，请参看《二十五年之后——登月竞赛的内幕》（"25 Years Later，Moon Race in Eclipse," by John Nobel Wilford）一文，载于1994年7月17日《纽约时报》第1页，此文为纪念登月计划25周年而作。

12.关于衰老的这一悲观（乐观？）的看法，可在《我们为什么会衰老》一书（*WhyWe Get Sick*：*The New Science of Darwinian Medicine*，Randolph M.Nesse and George C.Williams，Times Books，New York，1994）第8章找到。该章题为"年老是年轻之源"（Aging as the Fountain of Youth），其作者George C.Williams是现代进化生物学的老前辈。也可参考其经典文章《基因多效性、自然选择与衰老过程的演进》（"Pleiotropy，Natural Selection and the Evolution of Senescence，" *Evolution*，vol.11，1957：398–411）。

13.迈克尔逊这段话的出处有不同的版本，这里所引用的一段出自《今日物理》（*Physics Today*）1968年第4期，第9页。

14.错误地把迈克尔逊的"小数点说"归于开尔文的文字，可见于《超弦——万物至理？》（Superstring：*A Theory of Everything?* edited by Paul C. Davies and Julian Brown，Cambridge University Press，Cambridge，U.K.，1988）第3页。这本书还有一点引人瞩目的地方，就是披露了诺贝尔奖得主费曼（R. Feynman）曾对超弦理论抱着深深的怀疑。

15.布拉什对19世纪末物理学状况的分析，可见于其文章《六位数的浪漫情怀》

（ "Romance in Six Figures," *Physics Today*，January 1969：9）。

16.例如，加利福尼亚大学圣巴巴拉学院的科学史家巴达什（Badash），就不同意布拉什的观点，认为"完成论的症候的确并未到病入膏肓的地步……更准确地说应是一种'低度感染'，但它确实是存在的"。参阅其《十九世纪科学的完成论》（ "The Completeness of Nineteenth-Century Science," by Lawrence Badash，*Isis*，vol.63，1972：48–58）一文。

17.请参阅科什兰的文章《水晶球与呼唤的号角》（ "The Crystal Ball and theTrumpet Call" ），以及随后的一组预见性文章，载于《科学》（*Science*）杂志，1995年3月17日。这一短视的专利局长的故事，在软件大王比尔•盖茨1995年的畅销书《未来之路》（*The Road Ahead*，Nathan Myhrvold and Peter Rinearson，Viking，New York，1995）第Xiii页上被再度重述。

18.可参阅杰弗里的文章《无可发明》（ "Nothing left to Invent," by Eber Jeffery，*Journal of the Patent Office Society*，July 1940：479–481.）。在这里，我要感谢加利福尼亚大学戴维斯学院的科学史家舍伍德（Morgan Sherwood），正是他慷慨地向我提供了这篇文章。

19.《进步的理念》，伯里著（*The Idea of Progress*，J. B. Bury，Macmillan，New York，1932）。我对伯里观点的简短总结，借鉴了斯滕特《黄金时代的来临》中的分析。

20.《进步的悖论》，斯滕特等著（*The Paradoxes of Progress*，Gunther S. Stent，W. H. Freeman，San Francisco，1978），第27页。这本书中有几章的内容已在斯滕特早期著作中发表过，如《黄金时代的来临》；另增加了对于生物学、伦理学以及科学的认识局限的讨论。

21.《科学——无尽的前沿》，布什著（*Science：The Endless Frontier*，Vannevar Bush），1990年由华盛顿特区 "国家科学基金会" 重新刊行。

22.这段引文转引自《科学进步论》（*Scientific Progress*，by Nicholas Rescher，Basil Blackwell，Oxford，U. K.，1978：123–124）。作者里查是匹兹堡大学哲学教授，他在另外几处也曾谈到了恩格斯的科学能力无限论信条，究竟是怎样被现代马克思主义者所坚持的。也可参看前面提到的《进步的悖论》，斯滕特在书中曾提到，对《黄金时代的来临》一书最严厉的批评来自苏联哲学家V. Kelle，他认为科学是永恒的，斯滕特的科学终结论只是腐朽的资本主义的象征。

23.哈维尔的评论可参阅霍尔顿的《科学和反科学》（*Science and Anti-Science*，by Gerald Holton，Harvard University Press，Cambridge，1993），第

175–176页。作者霍尔顿是哈佛大学哲学教授。

24.这里对斯宾格勒的评述摘引自霍尔顿的《科学和反科学》。在该书以及另外一些出版物上（包括发表在《科学美国人》上的一篇文章，刊于1995年第10期第191页），霍尔顿试图否定科学终结论，其方法是诉诸爱因斯坦的权威，因为爱因斯坦常说追求真理之路是永恒的。霍尔顿似乎从未意识到爱因斯坦当时的话所表述的主要是一种希望，而不是对科学前景的实际估量。霍尔顿还认为，那些主张科学正走向终结的人，通常都站在了科学和理性的对立面。但明显的事实却是：预言科学将要结束的观点，主要并不是来自反理性主义者阵营，而恰恰是来自科学家阵营，如温伯格、道金斯、克里克等坚信科学是通向真理之最佳路径的人们。

25.参见《科学——无尽的地平线还是黄金时代？》（"Science: Endless Horizons or Golden Age?"）一文，作者格拉斯，载于《科学》（Science）杂志1971年1月8日，第23–29页。格拉斯是"美国科学促进会"（AAAS）的前任会长，该文是根据他1970年12月28日在AAAS年会上所做报告内容改写的。

26.参见格拉斯的文章《生物学发展的里程碑和增长速率》（"Milestones and Rates of Growth in the Development of Biology," Bentley Glass, *Quarterly Review of Biology*, March 1979：31–53）。

27.我电话采访格拉斯的时间在1994年6月。

28.参阅卡达诺夫的文章《艰难岁月》（"Hard Times," Leo Kadanoff, *Physics Today*, October 1992：9–11）。

29.我电话采访卡达诺夫的时间在1994年8月。

30.出自里查的《科学进步论》，第37页。

31.出处同上，第207页。尽管我在很多方面都不同意里查对科学的看法，但其两本著作《科学进步论》和《科学的限度》（*Scientific Progress* and *The Limits of Science*, University of California Press, Berkley, 1984），对于任何一位想要了解科学的限度问题的人来说，都是不可多得的资料来源。很不幸的是，这两本书目前已全部脱销了。

32.本特利•格拉斯对里查著作《科学进步论》的评价，载于《生物学评论季刊》（*Quarterly Review of Biology*），1979年12期，第417–419页。

33.康德的这一论述，转引自里查的《科学进步论》，第246页。

34.培根"不断超越"的含义，在梅达沃的《科学的限度》（*The Limits of Science*, Peter Medawar, Oxford University Press, New York, 1984）中有所讨论。梅达沃是著名的英国生物学家。

35.参见亚当斯编辑的《柏拉图以来的批评理论》（*Critical Theory Since Plato*, edited by Hazard Adams, Harcourt Brace Jovanovich, New York, 1971），第474页。

第二章　哲学的终结

1.见西奥查理斯和皮莫波洛斯（T.Theocharis and M.Psimopoulos）的文章，题为"科学哪儿错了？"（"Where Science Has Gone Wrong"），载于《自然》（*Nature*）杂志329卷，1987年10月15日，第595–598页。

2.皮尔斯关于科学和终极真理关系的思想，里查在《科学的限度》中有所讨论（参阅第一章注31）；也可参阅皮尔斯的《选集》（*Selected Writings*, edited by Philip Wiener, Dover Publications, New York, 1966）。

3.波普尔的主要著作包括：《科学发现的逻辑》（*The Logic of Scientific Discovery*, Springer, Berlin, 1934; reprinted by Basic Book, New York, 1959）；《开放社会及其敌人》（*The Open Society and Its Enemies*, Routledge, London, 1945; reprinted by Princeton University Press, Princeton, N.J., 1966）；《猜想与反驳》（*Conjectures and Refutations*, Routledge, London, 1963; reprinted by Harper and Row, New York, 1968）。波普尔的自传《无尽的探索》（*Unended Quest*, Open Court, La Salle, Ill., 1985）以及《波普尔选集》（*Popper Selections*, edited by David Miller, Princeton University Press, Princeton, N.J., 1985），对他的思想给予了很好的介绍。

4.参阅《无尽的探索》（*Unended Quest*）第17章：是谁埋葬了逻辑实证主义？（Who Killed Logical Positivism?）

5.出处同上，第116页。

6.我对波普尔的采访，是在1992年8月。

7.我的那篇关于量子力学的文章，标题是"量子哲学"（"Quantum Philosophy"），载于《科学美国人》杂志1992年7月，第94–103页。

8.参阅波普尔和埃科尔斯合著的《自我及其大脑》（*The Self and Its Brain*, Popper and John C. Eccles, Springer-Verlag, Berlin, 1977）。埃科尔斯因其对神经信号的研究而获1963年诺贝尔医学和生理学奖，我在第7章中讨论了他的观点。

9.岗赛·魏契特肖瑟的生命起源观点，发表在《国家科学院资料汇编》上（*Proceedings of the National Academy of Science*, vol.87, 1990），第200–204页。

10.波普尔对达尔文理论的批评，在"自然选择及其科学地位"（Natural

Selection and Its Scientific Status）中有详细的讨论，后收入《波普尔选集》第10章。

11. 喵夫人当时正在寻找的是波普尔的一本著作，题为《倾向性的世界》（*A World of Propensities*，Routledge，London，1990）。

12.参见物理学家彭迪（Hermann Bondi）为波普尔的90寿辰所发表的贺文，载于《自然》（*Nature*）杂志1992年7月30日，第363页。

13.波普尔的这则讣闻，刊于《经济学人》（*The Economist*）杂志，1994年9月24日，第92页。波普尔逝世于同年的9月17日。

14.出自波普尔《无尽的探索》，第105页。

15.《科学革命的结构》，库恩著（*The Structure of Scientific Revolutions*，Thomas Kuhn，University of Chicago Press，Chicago，1962）。本书引文页码，系指1970年版。我采访库恩的时间是1991年2月。

16.见《科学美国人》，1964年第5期，第142–144页。

17.库恩把科学家比作"瘾君子"以及《1984》中被洗过脑的角色的详情，可在《结构》的第38页和第167页中找到。

18.我对布什政府的所谓"新范式"说的批评，最初是在一篇关于库恩的人物小传中提出的，发表在《科学美国人》（*Scientific American*）杂志1991年第5期，第40–49页；后来，我接到詹姆斯•平克敦（James Pinkerton）的一封抱怨信，因为他正是当时布什总统的策略规划副助理，"新范式"说正是他炮制出来的。平克敦坚称其"新范式"绝对"不是里根经济政策的改头换面，而是有内在联系的一整套观点和原则，侧重于选择、下放权力以及用更少的中央控制取得更大的利益"。

19.关于库恩曾给范式下过21种定义的说法，出自马斯特曼（Margaret Masterman）的文章《范式的本质》（"The Nature of a Paradigm"），收入《批评与知识的增长》（*Criticism and The Growth of Knowledge*，edited by Imre Lakatosand Alan Musgrave，Cambridge University Press，New York，1970）。

20.《反对方法》，保罗•费耶阿本德著（*Against Method*，Paul Feyerabend，Verso，London，1975，1993年重印）。

21."实证主义者的茶杯"云云，见《告别理性》（Farewell to Reason，Feyerabend，Verso，London，1987），第282页。

22.费耶阿本德关于犯罪集团的说法，请参阅其短文《致一位专家的安慰信》（Consolations for a specialist），收入《批评与知识的增长》中。

23.费耶阿本德这些怪论，在布罗德（William J. Broad）为费氏作的专访中有所介绍，并被寄予了令人吃惊的同情。布罗德是《纽约时报》记者，文章题为"保罗•费

耶阿本德：科学与无政府主义者"（"Paul Feyerabend：Science and Anarchist"），
刊登在《科学》（*Science*）杂志1979年11月2日，第534–537页。

24.费耶阿本德，《告别理性》，第309页。

25.出处同上，第313页。

26.《艾西斯》（*Isis*），1992年第2卷，第368页。

27.费耶阿本德于1994年2月11日在日内瓦故世。《纽约时报》的讣闻刊登于3月
8日。

28.《不务正业的一生》，保罗·费耶阿本德著（*Killing Time*，Paul
Feyerabend，University of Chicago Press，Chicago，1995）。

29.《末路上的哲学——终结还是革新？》（*After Philosophy*：*End or
Transformation*? edited by Kenneth Baynes，James Bohman and Thomas McCarthy，
MIT Press，Cambridge，1987）。

30.《哲学中的问题》，麦金著（*Problems in Philosophy*，Colin McGinn，
Blackwell Publishers，Cambridge，Mass.，1993）。

31."萨伊尔"（The Zahir），收入博尔赫斯的《自选集》（*A Personal
Anthology*，Jorge Luis Borges，Grove Press，New York，1967）。这一选集中还收
入另外两篇关于绝对真理的悚人故事："葬礼和纪念碑"（Funes，the Memorious）
和"阿列夫"（The Aleph）。

32.出处同上，第137页。

第三章 物理学的终结

1.爱因斯坦的议论，可在约翰·巴洛的《万物至理》一书（*Theories of
Everything*，John Barrow，Clarendon Press，Oxford，U.K.，1991）第88页上查到。

2. 格拉肖的所有评论都在《科学的终结？——攻击与辩护》一书中找到。
（参见第二章注1）

3.可参阅格拉肖和金斯伯格合写的文章《绝望地寻找超弦》（"Desperately Seeking
Superstrings," Sheldon Glashow and Paul Ginsparg），载于《今日物理》（*Physics
Today*），1986年第5期，第7页。

4.见科尔《万物至理》（"A Theory of Everything," K.C.Cole）一文，载于
《纽约时报杂志》（*New York Times Magazine*）1987年10月18日，第20页。该文为
我提供了本章中有关威滕的所有背景材料。我采访威滕的时间是1991年8月。

5.见《科学观察》（*Science Watch*，Published by the Institute for Scientific Information，Philadelphia，Pa.），1991年第9期，第4页。

6.请参阅巴洛《万物至理》一书。

7.请参阅林德利《物理学的终结》（*The End of Physics*，David Lindley，Basic Books，New York，1993）一书。

8.可参阅马多克斯的文章《现在能将所有原理公之于世吗？》（"Is The Principia Publishable Now?" by John Maddox），载于《自然》（*Nature*）杂志，1995年8月3日号，第385页。

9.出自丹尼斯•奥维拜的《寂寞的宇宙之心》（*Lonely Hearts of the Cosmos*，Dennis Overbye，Harper Collins，New York，1992），第372页。

10.出自温伯格的著作《终极理论之梦》（*Dreams of a Final Theory*，Steven Weinberg，Pantheon，New York，1992），第18页。

11.出自温伯格的《最初三分钟》（*The First Three Minutes*，Steven Weinberg，Basic Books，New York，1977），第154页。

12.温伯格，《终极理论之梦》，第253页。

13.《超空间》，加来道雄著（*Hyperspace*，Michio Kaku，Oxford University Press，New York，1994）。

14.《上帝的心智》，保罗•戴维斯著（*The Mind of God*，Paul C. Davies，Simon and Schuster，New York，1992）。决定授予戴维斯"邓普顿宗教促进奖"（Templeton Prize）的评委中，包括乔治•布什和玛格丽特•撒切尔。

15.关于贝特那决定人类命运的计算，首次披露于《终极灾难？》（"UltimateCtastropher？"）一文，载于《原子科学家通报》（*Buletin of the Atomic Scientists*）1976年第6期，第36-37页。该文重印于贝特的论文集《始于洛斯阿拉莫斯之路》（*The Road from Los Alamos*，American Institute of Physics，New York，1991）。我于1991年8月在康奈尔采访了贝特。

16.见戴维•莫明的文章《那些年代出了什么错？》（"What's Wrong With Those Epochs?" David Mermin），载于《今日物理》1990年11期，第9-11页。

17.惠勒的文章、论文已结集出版，题为《在宇宙这个家》（*At home in the Universe*，American Institute of Physics Press，Woodbury，N.Y.，1994）。我于1991年4月采访了惠勒。

18.参见惠勒的文章《信息论，物理学和量子论：寻求其间的纽带》（"Information，Physics，Quantum: The Search for Links"），第5页；该文后收入

《复杂性，熵和信息物理学》（*Complexity*，*Entropy*，*and the Physics of Information*，edited by Wojciech H.Zurek，Addison-Wesley，Reading，Mass.，1990）一书。

19.出处同上，见第18页。

20.这一引文，以及后续文章中有关惠勒在AAAS年会上与灵学家的故事，可参阅《物理学家惠勒：落伍的学者》（"Physicist John Wheeler：Retarded Learner，"by Jeremy Bernstein，*Princeton Alumni Weekly*，October 9，1985：28–41）一文。

21.玻姆的生平，可参阅《玻姆的量子力学替代理论》（"Bohm's Alternativeto Quantum Mechanics，"by David Albert），载于《科学美国人》杂志1994年第5期，第58–67页。此节的部分片段出自我的文章《一个量子异教徒的遗言》（Last Words of a Quantum Heretic），载于《新科学家》（*New Scientist*）1993年2月27日号，第28–42页。玻姆在其著作《整体与隐序》（*Wholeness and the Implicate Order*，Routledge，New York，1983，初版于1980年）中，较详细地阐述了自己的哲学观。

22.有关"E—P—R实验"的论文，玻姆的量子力学新诠释的原始论文，以及关于量子力学的许多有创意的文章，皆可在《量子理论和测量》（*Quantum Theory and Measurement*，edited by John Wheeler and Wojciech H. Zurek，PrincetonUniversity Press，Princeton，N.J.，1983）一书中找到。

23.《科学、秩序与创造力》，玻姆与皮特著（*Science*，*Order and Creativity*，David Bohm and F.David Peat，Bantam Books，New York，1987）。

24.我于1992年8月采访了玻姆，他逝世于10月27日。生前他曾与人合写了一本阐述自己观点的书，此书于他逝世两年后出版，名为《不可分的宇宙》（*The Undivided Universe*，David Bohm and Basil J. Hiley，Routledge，London，1994）。

25.《物理定律的特性》，费曼著（*The Character of Physical Law*，Richard Feynman，MIT Press，Cambridge，1967），第172页（该书最初由BBC公司于1965年发行）。

26.出处同上，第173页。

27."量子力学的诠释：我们站在哪一边？"研讨会，于1992年4月14日在哥伦比亚大学召开。

28.对于玻尔这句话的引用，我曾见过许多不同的版本；这里所引用的说法，是我在拜访惠勒时由他亲口转述的。惠勒曾就读于玻尔门下。

29. 对物理学现状的精辟分析，可参见《物理学、交流与物理学理论的危机》（"Physics，Community，and the Crisis in Physical Theory"）一文，作者 Silvan S.Schweber，载于《今日物理》（*Physics Today*）1993年11期，第34–40页。Schweber 是布兰迪斯大学的杰出物理学史家，认为物理学应向着实际的目标前进，而不应单纯地为知识而知识。我在《粒子形而上学》（"Particle Metaphysics," *Scientific American*，February 1994，第96–105页）一文中，详述了追求统一理论的物理学家所面临的重重困境。在《科学美国人》更早些的一篇题为"量子哲学"的文章（载于1992年第7期，第94–103页）里，我曾对有关量子力学诠释的当前成果进行了综述。

第四章　宇宙学的终结

1. 这次诺贝尔专题讨论会（Nobel symposium）于1990年6月11—16日在瑞典 Graftvallen 举行。霍金在会上的讲演以及会议的其他论文，收集在《我们宇宙的诞生及早期演化》（*The Birth and Early Evolution of Our Universe*，edited by J. S.Jilsson，B. Gustafsson，and B. S. Skagerstam，World Scientific，London，1991）一书中。关于这次会议，我也写了篇报道文字，题为"宇宙的真相"（"Universal Truths"），载于《科学美国人》杂志1990年10月号第108–117页。我到会的第一天，在众人都聚集在一个小树林里准备一个鸡尾酒会时，无意中遇见了斯蒂芬·霍金。当时，我正站在一个看得见摆放着食品饮料的桌子的地方，碰巧霍金的一个护士推着坐在轮椅上的他被堵在了路上，于是她过来问我可否帮她将霍金带到酒会上。当我将霍金从轮椅里抱出来时，我感觉就像是抱着一捆枯柴，直挺挺、硬邦邦的，却非常轻。我用眼睛的余光瞥了他一眼，发现他正满腹狐疑地望着我。突然间，他的脸极度痛苦地扭曲起来，身体剧烈颤抖，同时喉咙里发出"咕咕噜噜"的怪响。对此我的第一个反应就是：一个人正在我怀中死去，多可怕啊！接下来的念头是：斯蒂芬·霍金正在我怀中死去，多么不可思议！这时霍金的护士注意到了他的痛苦状，而我也显得有些沮丧，就挤了过来。其实我是在为自己的这种想法而惭愧。"不必担心，"说着她轻轻把霍金从我怀里抱了去，"他总是这样，不过很快就没事了。"

2. 霍金1980年4月29日所做演讲的摘要，发表在英国杂志《物理学简报》（*Physics Bulletin*）（现在已更名为 *Physics World*）上，1981年1月号第15–17页。

3. 见《时间简史》（*A Brief History of Time*，Stephen Hawking，Bantam

Books，New York，1988），第175页。

4.出处同上，第141页。

5.《斯蒂芬•霍金的科学生涯》（*Stephen Hawking：A Life in Science*，Michael White and John Gribbon，Dutton，New York，1992），此书也记录了霍金由一个物理学家到世界知名人士的历程。

6.可参阅1994年9月号的《科学观察》（*Science Watch*）杂志对霍金的采访。关于霍金对物理学终结的看法，第三章所引用的部分书目中也有所讨论，包括保罗•戴维斯的《上帝的心智》、约翰•巴洛的《万物至理》、史蒂文•温伯格的《终极理论之梦》、丹尼斯•奥维拜的《寂寞的宇宙之心》，以及戴维•林德利的《物理学的终结》。还可参阅乔治•詹森的《心智的火花》（*Fire in the Mind*，George Johnson，Alfred A. Knopf，New York，1995），该书对科学是否能获得绝对真理这一问题作了特别精妙的讨论。

7.施拉姆在宇宙学方面的主要观点，可参阅《创世的痕迹》（*The Shadows of Creation*，Schramm and Michael Riordan，W.H.Freeman，New York，1991）。施拉姆的合著者莱尔丹（Riordan）是斯坦福直线加速器实验室的一名物理学家，他认为，到20世纪末，发现了宇宙膨胀的艾伦•古思由于他的这一工作而获诺贝尔奖。1994年他就这个问题同我打赌，赌注是一箱加利福尼亚葡萄酒。我之所以在此提及此事，是因为我相信自己会赢。

8.林德在《自复制的暴涨宇宙》（"The Self-Reproducing Inflationary Universe"）一文中提出了他的理论，载于《科学美国人》杂志1994年10月号第48–55页。有兴趣的读者还可参阅他的以下两本著作：《粒子物理学与暴涨宇宙学》（*Particle Physics and Inflationary Cosmology*，Harwood Academic Publishers，New York，1990），《暴涨与量子宇宙学》（Inflation and Quantum Cosmology，Academic Press，San Diego，1990）。有关林德的这部分内容，最初发表在我的文章《万能的巫师）（"The Universal Wizard"）上，载于《发现》（*Discover*）杂志，1992年3月号80–85页。1991年4月，我在斯坦福大学采访了林德。

9.1993年2月，我电话采访了施拉姆。

10.1993年11月，我在哈佛采访了乔治。

11.在《祸起萧墙》（*Home is Where the Wind Blows*，University Science Books，Mill Valley，Calif.，1994）一书中，霍伊尔对其动荡的一生作了回顾。1992年8月，我在他家里采访了他。

12.例如，可参见1993年5月13日《自然》杂志第124页的文章，对《我们在宇宙中所处的位置》（*Our Place in the Cosmos*，J. M. Dent，London，1993）作了评论。书中，霍伊尔与其合作者钱德拉•威克拉马辛夫（Chandra Wickramasinghe）认为，宇宙中充满了生命；对此，《自然》杂志的评论员罗伯特•夏皮罗（Robert Shapiro），也是纽约大学的一名化学家，声称此书和霍伊尔最近的几本书，"淋漓尽致地展现了一个富于创造力的人转而去追求奇思异想的变化过程"。由于霍伊尔的离经叛道，长期以来传媒对他很是冷落；而当他的自传出版时，媒体却突然对他抱以极大的热情。例如，可参看马库斯•乔恩（Marcus Chown）的《太空分子人》（"The Space Molecule Man"）一文，载于《新科学家》（*New Scientist*）1994年9月10日，第24–27页。

13.参见"获胜者是……"（"And the Winner is…"）一文，载于《天空与望远镜》（*Sky and Telescope*）杂志，1994年5月号，第22页。

14.可参阅霍伊尔与威克拉马辛夫合著的《我们在宇宙中所处的位置》。

15.可参阅奥维拜的《寂寞的宇宙之心》。

16.出自弗里曼•戴森的文章《叛逆的科学家》（"The Scientist as Rebel," Freeman Dyson），载于《纽约时报书评》（*New York Review of Books*），1995年5月25日，第32页。

17.《宇宙学的发现》（*Cosmic Discovery*，Martin Harwit，MIT Press，Cambridge，1981），第42–43页。由于他督办的一个展览会所招惹的争论，1995年哈威特辞去了华盛顿特区史密斯索尼安学院航空航天博物馆馆长的职务。这个展览会名为"最后行动：原子弹与二战的终结"（"The Last Act：The Atomic Bombandthe End of World War II"），结果被二战老兵和其他一些人指责说，这个展览对美国在广岛和长崎投掷原子弹一事批评过多。

18.同上，第44页。

19.这句话引自约翰•多恩的诗，我是在生物学家劳恩•艾斯蕾（Loren Eisley）的一篇文章的末尾看到的。该文题为"宇宙监狱"（"The Cosmic Prison"），载于《地平线》（*Horizon*）杂志，1970年秋季号，第96–101页。

第五章　进化生物学的终结

1.可参阅1964年哈佛大学出版社出版的《物种起源》（On the Origin of Species），该版的前言作者恩斯特•迈尔（Ernst Mayr），正是现代进化生物学的奠基者之一。

2.岗瑟•斯滕特，《黄金时代的来临》，第19页。

3.玻尔的这一评论，我是在一篇书评上见到的，载于《自然》（Nature）杂志1992年8月6日，第464页。原文是："把深奥的真理还原为细节，这正是科学的任务。"

4.这次集会的时间是在1994年11月，地点在约翰•布鲁克曼（John Brockman）的办公室。布鲁克曼是一位成就辉煌的书商，也是科学家作家们（Scientist-authors）的公共关系顾问。

5.《盲人钟表匠》，道金斯著（The Blind Watchmaker，Richard Dawkins，W. W. Norton，New York，1986），第ix页。道金斯所提到的华莱士，是指Alfred Russell Wallace，他独立地提出了"自然选择"的概念，但在洞察的深度和广度上都与达尔文相去甚远。

6.参阅《均变论是必要的吗？》（"Is Uniformitarianism Necessary？"），载于《美国科学学刊》（American Journal of Science），1985年第263期，第223–228页。

7.可参阅《断续平衡理论：种系渐进论的替代理论》（"Punctuated Equilibria：An Alternative to Phyletic Gradualism，"Stephen Jay Gould and Niles Eldredge），被收录在《古生物学的模型》（Models in Paleobiology，edited by T. J.M. Schopf，W. H. Freeman，San Francisco，1972）中。

8.古尔德著作众多，仅就我所喜欢的推荐：《人类的误测》（The Mismeasure of Man，W. W. Norton，New York，1981），这既是一部关于智力测验史的学术著作，也是激烈反对智力测验的论战性书籍；《奇妙的生命》（Wonderful Life，W. W.Norton，New York，1989），这是陈述其生命是意外产生观点的主要著作。也可参阅古尔德及其哈佛同仁理查德•莱温汀（Richard Lewontin，一位遗传学家，他与古尔德一样，常被指责为具有马克思主义倾向）合写的文章，《圣•马可教堂的拱肩和潘格洛斯范式》（"The Spandrels of San Marco and the Panglossian Paradigm"），载于《皇家学会会刊》（Proceedings of the Royal Society，London），1979年第205卷，第581–598页。这篇文章中的大量批判都是针对心理、行为的达尔文解释中的简单化倾向。对古尔德进化观的同样尖刻的批评，还可参阅《奇妙的生命》一书的书评，作者是Robert Wright，载于《新共和》（New Republic），1990年1月29日。

9.《作为时代产物的断续性平衡理论》（"Punctuated Equilibrium Comes of Age"），载于《自然》（Nature）杂志，1993年11月18日，第223–227页。我是在

1994年11月于纽约城采访的古尔德。

10.转引自道金斯的《盲人钟表匠》，第245页。所引用这段文字的章标题是"揭穿断续平衡理论的真相"（Puncturing Punctuationism）。

11.关于林恩·马古利斯的共生问题研究，若想寻找一种没有废话的表述，可参阅其《细胞进化中的共生》（*Symbiosis in Cell Evolution*，W. H. Freeman，NewYork，1981）一书。

12.可参阅马古利斯被辑入《盖亚：主题，机理与寓意》（*Gaia: The Thesis, the Mechanisms, and the Implications*，edited by P. Bunyard and E.Goldsmith，Wadebridge Ecological Center，Cornwall，UK，1988）一书中的文章。

13.《生命是什么？》（*What is Life?*，Lynn Margulis and Dorion Sagan，Peter Nevraumont，New York，1995，由Simon and Schuster发行）。我采访马古利斯的时间在1994年5月。

14.关于拉夫洛克这一信仰危机的详述，可参阅《盖亚，盖亚——请留步》（"Gaia, Gaia: Don't Go Away"）一文，作者Fred Pearce，载于《新科学家》（*New Scientist*），1994年5月28日，第43页。

15.这里提到的以及另外一些关于马古利斯的傲慢评论，可以在Charles Mann的文章《林恩·马古利斯：科学之任性的大地母亲》（"Lynn Margulis: Science's Unruly Earth Mother"）中找到，载于《科学》（*Science*）杂志1991年4月19日，第378页。

16.在这部分内容中所提到的考夫曼的著述，包括《反混沌与适应）（"Antichaos and Adaptation,"Scientific American，August 1991，第78-84页）、《生物序的起源》（*The Origins of Order*，Oxford Unwerfity Press，New York，1993），以及《在宇宙这个家》（*At Home in the Universe*，Oxford University Press，New York，1995）。

17.可参阅我的文章《在起点上）（"In the Beginning,"*Scientific American*，February 1991，第123页）。

18.古德温的观点可参见其著作《豹是如何改变其色斑的》（*How the Leopard Changed Its Spots*，Charles Scribner's Sons，New York，1994）。

19.史密斯（John Maynard Smith）对佩尔·贝克和斯图亚特·考夫曼两人工作的诋毁性评论，见于《自然》杂志1995年2月16日，第555页；也可参阅评论考夫曼《生命序的起源》的一篇极富洞见的书评，载于《自然》杂志1993年10月21日，第704-706页。

20.我于1991年2月发表在《科学美国人》杂志上的文章（见注17.），评述了关于生命起源问题的一些最有影响的理论。我采访斯坦利•米勒的地点是加利福尼亚大学圣迭戈分校，时间是1990年11月；再度电话采访于1995年9月。

21.岗瑟•斯滕特，《黄金时代的来临》第71页。

22.克里克的"奇迹"说，可在其著作《生命本身》（*Life Itself*，Simon and Schuster，New York，1981）第88页上找到。

第六章　会科学的终结

1.我于1994年2月在哈佛采访了威尔逊。本章所提到的威尔逊的著作包括：《社会生物学》（*Sociobiology*，Harvard University Press，Cambridge，1975，我对于此书的引文出自1980年删节版）；《论人的本性》（*On Human Nature*，Harvard University Press，Cambridge，1978）；《基因、意识与文化》，与拉姆斯登合著，（*Genes*，*Mind*，*and Culture*，with Charles Lumsden，Harvard University Press，Cambridge，1981）；《普罗米修斯之火》，与拉姆斯登合著，（*Promethean Fire*，with Lumsden，Harvard University Press，Cambridge，1983）；《亲生命性》（*Biophilia*，Harvard University Press，Cambridge，1984）；《生命的多样性》（*The Diversity of Life*，W. W. Norton，New York，1993）以及《博物学家》（*Naturalist*，Island Press，Washington，D.C.，1994）。

2.若想了解威尔逊经历中的这一危机时刻的详情，可阅读《博物学家》一书第12章"分子之争"（The Molecular Wars）。

3.威尔逊，《社会生物学》，第300页。

4.这些痛苦经历详细反映在《普罗米修斯之火》中，题为"社会生物学的争端"一章内，见该书第23-50页。

5.威尔逊，《普罗米修斯之火》，第48-49页。

6.这些评论，出自加利福尼亚大学圣迭戈分校的生物学家韦尔斯（Christopher Wells）之手，文章载于《科学》（*Science*），1993年，11／12期，第39页。

7.我个人认为，科学家们在运用遗传学和达尔文理论的术语去解释人类行为这一点上，并未达到像威尔逊所坚信的那种成功地步，可参见我在《科学美国人》杂志上所发表的文章《回顾优生学》（"Eugenics Revisited，"June 1993：122-131），以及《新社会达尔文主义者》（"The New Social Darwinists，"October 1995：174-181。）

8.威尔逊，《社会生物学》，第300–301页。

9.参阅迈尔的《一场持久的论战》（*One Long Argument*，Ernst Mayr，Harvard University Press，Cambridge，1991）。在149页中，迈尔写道："综合进化论的缔造者，经常受到无端的指责，说他们竟然宣称已解决了进化论遗留的所有问题，这一指责是极其荒谬的。就我所知，从未有哪位进化论者作过这样的论断，他们至多也不过是认为，自己已进入了达尔文范式的精致化时期，这一有生命力的范式再也不会受到现有难题的威胁。"其中的"难题"一词，正是库恩用以描述非革命时期，或常规科学时期，科学家们所处理的那些问题的。

10.达尔文《人类的由来》（*The Descent of Man*）中的这段文字，我是在赖特的《道德动物》中发现的（*The Moral Animal*，Robert Wright，Pantheon，New York，1994：327）。赖特是《新共和》（*The New Republic*）杂志的记者，他的这本著作，就我的阅读范围而言，是用达尔文术语来解释人类本性的所有著作中最为出色的。

11.出处同上，第328页。

12.我到麻省理工学院去采访乔姆斯基的时间是在1990年2月，这节文字中所引用的乔姆斯基的话就出自这次采访，其余地方所引用的话，源自1993年2月我对他的电话采访。乔姆斯基的政论文章已结集出版，书名为《乔姆斯基选集》（*The Chomsky Reader*，edited by James Peck，Pantheon，New York，1987）。

13.见《新不列颠百科全书》（*The New Encyclopaedia Britannica*，1992年详编版），第23卷，语言学，第45页。

14.《自然》杂志，1994年2月19日，第521页。

15.《句法结构》，乔姆斯基著（*Syntactic Structures*，Noam Chomsky，Mouton，The Hague，Netherlands，1957）。1995年，乔姆斯基出版了另一本语言学著作，题为《极简抽象派艺术纲要》（*The Minimalist Program*，MIT Press，Cambridge），进一步发展了其早期著作中关于先天的、创造性的语法的观点。与乔姆斯基的许多语言学著作一样，这一部也十分艰涩难懂。若想了解乔姆斯基作为语言学家的经历，请参阅《语言学论战》（*The Linguistics Wars*，Randy AllenHarris，Oxford University Press，New york，1993）。

16.然而麻省理工学院的另一位语言学家平克尔（Steven Pinker）却雄辩地论证说，若通过一个达尔文主义的视角来考察乔姆斯基的著作，反而能更好地理解它。参见默罗的《语言本能》（*The Language Instinct*，William Morrow，New York，

1994）。

17.《语言与知识问题》（*Language and the Problems of Knowledge*，Noam Chomsky，MIT Press，Cambridge，1988），第159页。在这本书中，乔姆斯基也探讨了他的认知局限性观点。

18.岗瑟•斯滕特，《黄金时代的来临》，第121页。

19.《重彩描画：向着文化的解释理论前进》（"Thick Description：Toward An Interpretive Theory of Culture"），辑入格尔茨的文集《文化解释论》（*The Interpretation of Cultures*，Basic Books，New York，1973）。

20.出处同上，第29页。

21.《著述与生平：作为作家的人类学家》，格尔茨著（*Works and Lives：The Anthropologist as Author*，Clifford Geertz，Stanford University Press，Stanford，1988），第141页。

22.《深奥的游戏》（"Deep Play"）一文，后被收入《文化解释论》中，这段引文出自412页。

23.出处同上，第443页。

24.我于1989年5月在高等研究院采访了格尔茨本人，1994年8月再度电话采访了他。

25.《追寻事实》，格尔茨著（*After the Fact*，Clifford Geertz，HarvardUniversity Press，Cambridge，1995），第167–168页。

第七章　神经科学的终结

1.克里克曾著书自述其经历：《狂热的追求》（*What Mad Pursuit*，Basic Books，New York，1988）。他阐述其意识观点的著作题为《令人震惊的假说》（The Astonishing Hypothesis，Charles Scribner's Sons，New York，1994）。

2.可参阅克里克和科克的论文《为了一种意识的神经生物学理论》（"Toward a Neurobiological Theory of Consciousness，"Francis Crick and Christof Koch，*Seminars in the Neuroscience*，vol.2，1990：263–275）。

3.我于1991年11月在索尔克研究所采访了克里克。

4.《双螺旋》，詹姆斯•沃森著（*The Double Helix*，James Watson，Atheneum，New York，1968）。

5.克里克，《狂热的追求》，第9页。

6.克里克，《令人震惊的假说》，第3页。

7.埃德尔曼关于心智的著作，由纽约的Basic Books公司出版，包括《神经达尔文主义》（*Neural Darwmism*），1987；《区域生物学》（*Topobiology*），1988；《值得纪念的现状》（*The Remembered Present*），1989；以及《壮观的旋律，灿烂的火花》（*Bright Air*，*Brilliant Fire*），1992。所有这些著作都十分晦涩难懂，即使是最后一本，虽然试图以一种通俗的文体阐述埃德尔曼的观点，也不例外。

8.《埃德尔曼博士的大脑》（"Dr. Edelman's Brain"），史蒂文•莱维（Steven Levy）著，载于《纽约客》，1994年5月2日，第62页。

9.《炮制一种关于大脑的理论》（"Plotting a Theory of the Briain"），大卫•赫勒斯坦（David Hellerstein）著，载于《纽约时报杂志》，1988年5月22日，第16页。

10.参见克里克针对埃德尔曼的《神经达尔文主义》一书所做的极刺激的书评，《神经的埃德尔曼主义》（"Neural Edelmanism"），载于《神经科学动态》（*Trends in Neurosciences*），1989年12卷7期，第240–248页。

11.丹尼尔•丹尼特评论《壮观的旋律，灿烂的火花》一书的书评，载于《新科学家》（*New Scientists*）杂志，1992年6月13日，第48页。

12.科克是在1994年4月12日至17日于亚利桑那州图森市召开的一次会议上，作出这一评论的，会议的主题是"构建意识的科学基础"（Toward a Scientific Basisfor Consciousness）。

13.埃克尔斯在许多出版物中都阐述了自己这一观点，这些出版物包括：与波普尔合著的《自我及其脑》（*The Self and It's Brain*，Springer-Verlag，Berlin，1977）；《自我怎样控制其脑》（*How the Self Controls It's Brain*，Springer-Verlag，Berlin，1994）；还有他与Friedrieh Beck合写的论文《脑活动的量子观与意识的作用》（"Quantum Aspects of Brain Activity and the Role of Consciousness"），载于《美国国家科学院院刊》（*Procoedings of the National Academy of Science*），第89卷，1992年12期，第11357–11361页。我是在1993年2月通过电话采访埃克尔斯的。

14.我到牛津大学去采访彭罗斯是在1992年8月。彭罗斯关于意识问题的两本书是：《皇帝的新脑》（*The Emperor's New Mind*，Oxford University Press，New York，1989）；《心智的阴影》（*Shadows of the Mind*，Oxford University Press，New York，1994）。

15.对于《皇帝的新脑》的批评，可参阅《行为与脑科学》（Behavioral and Brain Science），第13卷，第4期，1990年12月，这一期收入了许多评论彭罗斯这本书的文章。关于《心智的阴影》一书的刺激性评论文章，可参阅《疑云重重》（"Shadowsof Doubt"），作者是著名物理学家菲利普•安德森（Philip Anderson），刊于《自然》杂志1994年11月17日，第288–289页；以及《所有可能的大脑中最好的一个》（"The Best of All Possible Brains"），作者是著名哲学家普特南（Hilary Putnam），载于《纽约时报书评》，1994年11月20日，第7页。

16.《关于心智的科学》，欧文•弗拉纳根著（*The Science of the Mind*，Owen Flanagan，MIT Press，Cambridge，1991）。感谢丹尼尔•丹尼特，是他让我注意到弗拉纳根这一术语的。

17.《作为一只蝙蝠的感受是什么？》（"What Is It Like to be a Bat?"），耐格尔作，收入《人的问题》（*Mortal Questions*，Cambridge University Press，New York，1979）中。这是一本耐格尔的文集，这段引文出自第166页。1992年6月，我打电话给耐格尔，问他是否认为科学将会有终结的一天。"绝对不会。"他答道。"发现越多，产生的问题也就越多，"他紧接着又补充道，"对莎士比亚作品的批评永远也不会完结，为什么物理学就会完结呢？"

18.我于1994年8月在纽约城拜晤了麦金。关于他的神秘论者的观点之详情，请参阅麦金的著作《意识问题》（*The Problem of Consciousness*，Blackwell Publishers，Cambridge，Mass.，1991）。

19.《意识释义》，D.丹尼特著（*Consciousness Explained*，Daniel Dennett，Little Brown，Boston，1991）。还可参阅《大脑及其边界》（"The Brain and It's Boundaries"）一文，载于《伦敦泰晤士报文学副刊》（*London Times Literary Supplement*），1991年5月10日，丹尼特在这篇文章中驳斥了麦金的神秘论主张。我于1994年4月通过电话与丹尼特探讨了神秘论者的观点。

20."构建意识的科学基础"（Toward a Scientific Basis for Consciousness）大会于1994年4月12—17日，在亚利桑那州图森市召开，会议组织者是哈默罗夫（Stuart Hameroff），一位亚利桑那大学的麻醉学家，他对于微管的研究，曾影响了罗杰•彭罗斯对量子效应在意识中的作用的看法。因而，操纵这次会议的发言者主要是那帮在神经科学中主张意识的量子解释的人们，其中不仅包括罗杰•彭罗斯，还包括布赖恩•约瑟夫森（Brian Josephson），一位诺贝尔物理学奖得主，认为量子效应能够解释玄奥现象甚至心灵现象；安德鲁•韦尔（Andrew Weil），一位医生兼兴奋剂官员，他宣称：关于意识的完整理论，必须考虑南美印第安人的这样一种能力：他们在吞服了治疗精神病的药物后，能产生一种身临其境的幻觉；以及丹纳•佐尔（Danah Zohar），一位"新时代"作家，他宣布人类思想源自"宇宙中真空能

量的量子波动"，它"是真正的上帝"。我曾在《科学能解释意识吗？》（"Can
Science Explain Consciousness？"）一文中报道了这次会议的情况，文章发表于
《科学美国人》杂志1994年7月号，第88-94页。

21.戴维•查默斯亲自撰文阐述了他的意识理论，文章刊登在《科学美国人》杂志
1995年12月号，第80-86页。同期刊发的还有克里克与科克合写的一篇批驳查默斯观
点的文章。

22.我于1993年5月在麻省理工学院采访了明斯基，他告诉我说，1966年的时
候，他曾让一个叫杰拉尔德•苏斯曼（Gerald Sussman）的研究生设计一台能辨认物
体（或能"看"）的机器人，作为暑期科研项目。不用说，苏斯曼没能成功（尽管
他在这一领域继续研究了下去，并成为麻省理工学院的一名教授）。人工视像一直
是人工智能领域里最难攻克的问题之一。若想了解人工智能的更详细情况，可参阅
《AI：探索人工智能的喧嚣历史》（*AI：The Tumultuous History of the Search for
Artificial Intelligence*，Daniel Crevier，Basic Books，New York，1993）。还可参
阅伯恩斯坦（Jeremy Bernstein）以颇为恭敬的态度为明斯基所做的人物传略，发表
于《纽约客》，1981年12月14日号，第50页。

23.《心的社会》，马文•明斯基著（*The Society of Mind*，Marvin Minsky，
Simon and Schuster，New York，1985）。书中给出的论点，泄露了明斯基对于科
学进步后果的矛盾心情。例如，在书的68页上有一篇文章，题为"自知是危险的"
（Self-Knowledge is Dangerous）。明斯基在文中宣称："如果我们能完全控制引发
快感的神经系统，我们就会以此再生出成功的快感，而不必再去博取任何真正的成
就。这将成为一切人类事业的终结。"岗瑟•斯滕特也曾预言，这类兴奋作用在新波
利尼西亚将会普遍地蔓延开来。

24.岗瑟•斯滕特，《黄金时代的来临》，第73-74页。

25.出处同上，第74页。

26.吉尔伯特•赖尔是在其攻击"二元论"的经典著作《心智概念》（*The Concept
of Mind*，Hutchinson，London，1949）中，炮制出"机器中的幽灵"一词的。

27.亨利•亚当斯（Henry Adams）提出了这一有关弗朗西斯•培根的唯物主义观
点。参见《亨利•亚当斯教育文集》，第484页（参阅第一章注释3）。据亚当斯的原
文，培根"力劝社会暂时把宇宙是由一种思想中演化而来的观点放在一边，尝试一
下思想是从宇宙中演化出来的主张"。

第八章　混杂学的终结

1.伊利诺伊大学1993年11月发行了有关的书籍；我在这里谈到的评述迈耶·克雷斯模拟星球大战的文章，题为"非线性思维"（"Nonlinear Thinking"），载于《科学美国人》杂志1989年6月号，第26–28页。还可参阅《国际军备竞赛中的混沌》（"Chaos in the International Arms Race，"by Mayer-Kress and Siegfried Grossman），刊于《自然》杂志，1989年2月23日，第701–704页。

2.紧随格莱克的《混沌——创建新科学》（Chaos：Making a New Science，Penguin Books，New York，1987）一书的轰动效应而出版，并明显表现出受到此书影响的书籍，包括：《复杂性——有序和混沌边缘的新兴学科》（Complexity：The Emerging Science at the Edge of Order and Chaos，M. Mitchell Waldrop，Simonand Schuster，N.Y.，1992）；《复杂性——混沌边缘的生命》（Complexity：Life at the Edge of Chaos，Roger Lewin，Macmillan，N.Y.，1992）；《人工生命——计算机与生物学前沿交叉的报道》（Artificial Life：A Report from the Frontier Where Computers Meet Biology，Steven Levy，Vintage，New York，1992）；《复杂化——用新奇科学来阐释悖论世界》（Complexification：Explaining a Paradoxical World through the Science of Surprise，John Casti，Harper Collins，New York，1994）；《混沌之解体——从复杂世界发现简单》（The Collapse of Chaos:Discovering Simplicity in a Complex World，Jack Cohen and Ian Stewart，Viking，New York，1994）；以及《复杂性的前沿——在混沌世界中寻求秩序》（Frontiers of Complexity：The Search for Order in a Chaotic World，Peter Coveny and RogerHighfield，Fawcett Columbine，New York，1995）。这最后一本书包括了许多格莱克《混沌》一书的材料，与我的基本观点也一致，即公众对混沌和复杂性的看法，实际上已把它们当作一回事。

3.格莱克引用的是庞加莱的话，见《混沌》，第321页。

4.《理性之梦》，海因茨·佩格著（The Dream of Reason，Heinz Pagels，Simon and Schuster，New York，1988）。这句话是我从1989年Bantam版平装本的书封广告中引用来的。

5.《大自然的分形几何》（The Fractal Geometry of Nature，Benoit Mandelbrot，W. H. Freeman，San Francisco，1977），第423页。本段前面的评论，即芒德勃罗集是"数学中最复杂的对象"云云，出自计算机专家A. K. 杜德尼之口，见《科学美国人》杂志1985年第8期，第16页。

6.1994年5月，在圣菲研究所举行的一次研讨会上，我目睹了爱泼斯坦演示其人工—社会程序（第九章我将描述其详细过程）。在1995年3月11日的一次研讨会上，爱泼斯坦宣称，他的这一类计算机模型将带来社会科学的革命性转变。

7.霍兰的这一观点，是在他送给我的一篇未发表的论文中提出的，题目是"Echo-Class模型的目标、粗略定义和思考"（"Objectives, Rough Definitions, and Speculations for Echo-Class Models"）（"Echo"指的是霍兰的遗传算法中的主要一类）；在其著作《隐序——适应如何实现复杂性》（*Hidden Order*：How Adaptation Builds Complexity, Addison-Wesley, Reading, Mass., 1995）的第4页上，他强调了这一观点。霍兰另外还发表了一篇简述遗传算法的文章，刊于《科学美国人》杂志1992年第7期，第66–72页。

8.在1995年3月的一次电话采访中，约克提出了这一论点。格莱克在《混沌》（Chaos）一书中承认"混沌"一词为约克首创。

9.见注释2。

10.请参阅梅拉尼·米切尔等人的文章《返观混沌边缘》（"Revisiting the Edge of Chaos,"by Melanie Mitchell, James Crutchfield, and Peter Hraber），载于《圣菲工作报告》（*Santa Fe working paper*）93—03—014。《复杂性的前沿》（Frontiers of Complexity）一书的"注释2"中，也提及了对"混沌边缘"概念的评论。

11.在这篇文章中，并没有公布出塞思·劳埃德所统计到的全部定义。当我打电话向他询问这些关于复杂性的定义之后，他通过电子邮件传给我下列信息，我统计了一下，有45种而不只31种定义。用作修正或括号中的词是指定义的创始者。下面给出了劳埃德清单中的主要定义，仅略作改动：信息（Shannon）；熵（Gibbs, Boltzman）；算法复杂性；算法信息含量（Chaitin, Solomonoff, Kolmogorov）；费希尔信息；Renyi熵；自描述代码长度（Huffman, Shannon, Fano）；矫错代码长度（Hamming）；Chernoff信息；最小描述长度（Rissanen）；参数个数或自由度或维数；Lempel--iv复杂性；交互信息，或通道容量；演算共有信息；相关性；储存信息（Shaw）；条件信息；条件演算信息含量；计量熵；分形维；自相似；随机复杂性（Rissanen）；混和（Koppel, Atlan）；拓扑机器容量（Crutchfield）；有效或理想的复杂性（Gell-Mann）；分层复杂性（Simon）；树形多样性（Huberman, Hogg）；同源复杂性（Teich, Mahler）；时间计算复杂性；空间计算复杂性；基于信息的复杂性（Traub）；逻辑深度（Bennett）；热力学深度（Lloyd, Pagels）；规则复杂性（在Chomsky层中位置）；Kullbach-Liebler信息；区别性（Wooters,

Caves，Fisher）；费希尔距离；分辨力（Zee）；信息距离（Shannon）；演算信息距离（Zurek）；代码间距；长幅序；自组织；复杂适应系统；混沌边缘。

12.参阅《夸克和美洲虎》（*The Quark and the Jaguar*，W. H. Freeman，New York，1994）第3章。其中，诺贝尔奖获得者盖尔曼（也是圣菲研究所的创立者之一）描述了解决复杂性的算法信息理论等方法。盖尔曼在本书33页上承认，"任何关于复杂性的定义都是依据文本的，甚至是主观的。"

13.1987年于洛斯阿拉莫斯召开的人工生命会议，在《人工生命》（作者Steven Levy）一书（见注释2）中有生动的描述。

14.编者导言，克里斯托弗•兰顿，载于《人工生命》（*Artificial Life*）杂志，1994年第1卷第1期，第vii页。

15.若欲了解这一故事的细节，可以参阅罗杰•卢因的《复杂性——混沌边缘上的生命》、米切尔•沃尔德罗普的《复杂性——在有序和混沌边缘的新兴科学》和史蒂文•利维的《人工生命》。

16.《地球科学中数值模型的验证、确认和证实》（"Verification，Validation，and Confirmation of Numerical Models in the Earth Sciences，"by Naomi Oreskes，Kenneth Belitz，and Kristin Shrader Frechette），发表于《科学》杂志1994年2月4日，第641–646页。也可参阅后来对此文的评论，刊登在1994年4月15日的《科学》上。

17.恩斯特•迈尔在《面向新的生物哲学》（*Toward a New Philosophy of Biology*，Harvard University Press，Cambridge，1988）一书中，讨论了生物学不精确性的必然性。可参阅其中的"生物的因果分析"一章。

18.我于1994年8月采访了贝克。对贝克工作的简介，可参阅《自组织临界性》（"Self-Organized Criticality，"Bak and Kan Chen）一文，刊于《科学美国人》杂志1991年第1期，第46–53页。

19.《平衡中的地球》（*Earth in the Balance*，Al Gore，Houghton Mifflin，New York，1992），第363页。

20.参阅《沙堆的不稳定性》（"Instabilities in a Sandpile，"by Sidney R.Nagel），载于《现代物理评论》（*Reviews of Mordern Physics*），84卷第1期，1992年1月，第321–325页。

21.莱布尼茨对于"无可辩驳的积分学"的信念，即认为它能解决所有的问题，甚至也包括神学问题，在《空中的π》（*Pi in the Sky*，John Barrow，Oxford Uninersity Press，New York，1992）一书中有精彩的论述，参见127–129页。

22.《控制论》，维纳著（*Cybernetics*，Norbert Wiener，was published in 1948 by John Wiley and Sons，New York）。

23.约翰·皮尔斯对控制论的这番评论，是在其《信息论导论》（An Introduction to Information Theory，Dover，New York，1980；初版于1961年）一书第210页提出的。

24.克劳德·香农的论文《通信的数学理论）（"A Mathematical Theory of Communications"），发表在《贝尔系统技术期刊》（*Bell System Technical Journal*）1948年6月号和10月号上。

25.1989年11月，我在香农位于马萨诸塞州温彻斯特的家中采访了他。我也写了一篇关于他的人物传略，刊登在《科学美国人》杂志1990年第1期第22—22b页上。

26.这篇关于托姆理论的著名评论，刊在《伦敦泰晤士报高等教育增刊》（London Times Higher Education Supplement）上，见1973年11月30日。我是在《寻求必然性》（*Searching for Certainty*，John Casti，William Morrow，New York，1990）一书第63—64页的文献中找到线索的。卡斯蒂已就有关数学的主题写了不少优秀的著作，他与圣菲研究所也有联系。托姆《结构稳定性和形态发生学》（*Structural Stability and Morphogenesis*）一书的英译本，发行于1975年（Addison-Wesley，Reading，Mass），法文初版发行于1972年。

27.这些有关突变论的负面评论，在卡斯蒂的《寻求必然性》一书第417页上可找到。

28.《机遇与混沌》，大卫·吕埃尔著（*Chance and Chaos*，David Ruelle，Princeton University Press，Princeton，N.J.，1991），第72页。这是一本由混沌理论开拓者编写的有关混沌的论著，论述平实而深刻。

29.见菲利普·安德森（Phillip Anderson）的文章，《重要的是差异》（"MoreIs Different"），载于《科学》杂志1972年8月4日，第393页。本文被收入安德森的选集《理论物理学生涯》（A Career in Theoretical Physics，World Scientific，RiverEdge，N.J.，1994）中。1994年5月，我在普林斯顿采访了安德森。

30.引自David Berreby的文章《无所不知者》（"The Man Who Knows Everything"），载于《纽约时报杂志》（*New York Times Magazine*），1994年5月8日，第26页。

31.我第一次与盖尔曼打交道是在纽约，时间在1991年11月，对这次经历的描述发表在《科学美国人》杂志1992年第3期，第30—32页。1995年3月，我在圣菲研究所再度采访了盖尔曼。

32.见注释12。

33.见《欢迎加入控制世界：计算机文化的人类学注释》（"Welcome to Cyberia：Notes on the Anthropology of Cyberculture，" Arturo Eseobar），载于《当代人类学》（*Current Anthropology*）35卷第3期，1994年6月，第222页。

34.《从混沌到有秩》，普里高津等著（*Order Out of Chaos*，Ilya Prigogineand Isabelle Stengers，Bantam，New York，1984；初版为法文，1979年）。

35.出处同上，第299–300页。

36.吕埃尔，《机遇与混沌》（*Chance and Chaos*），第67页。

37.费根鲍姆的两篇文章是：《特征函数、不动点和尺度函数动力学》（"Presentation Functions，Fixed Points，and a Theory of Scaling Function Dynamics"），载于《统计物理学杂志》（*Journal of Statistical Physics*），第52卷3、4合期，1988年8月，第527–569页；以及《环状映射的特征函数及尺度函数理论》（"Presentation Functions and Scaling Function Theory for Circle Maps"），载于《非线性》（*Nonlinearity*），1988年第1卷，第577–602页。

第九章　限度学的终结

1.以"科学知识的限度"为题的会议，召开于1994年5月24—26日，地点在圣菲研究所。

2.参阅蔡汀的论文《算术中的随机性》（"Randomness in Arithmetic"），载于《科学美国人》杂志1988年第6期，第80–85页；以及《纯数学中的随机性与复杂性》（"Randomness and Complexity in Pure Mathematics"），载于《国际分岔与混沌杂志》（*International Journal of Bifurcation and Chaos*），1994年第4卷第1期，第3–15页。蔡汀还在国际互联网上散发了一本名为《数学的限度》（*The Limits of Mathematics*）的书。其他与会者的著述（按作者姓氏的字母顺序排列，不包括已经引用过的书目）如下：W. Brian Arthur：《经济学中的正反馈》（"Positive Feed-backs in the Economy"），载于《科学美国人》杂志，1990年第2期，第92–99页；John Casti：《复杂化》（*Complexification*，Harper Collins，New York，1994）；Ralph Gomory：《已知、未知和不可知》（"The Known，the Unknown，and the Unknowable"），载于《科学美国人》杂志，1995年第6期，第120页；Roll Landauer：《计算——一个基本的物理观点》（"Computation: A Fundamental

Physical View"），载于《物理评论》（*Physica Scripta*）第35卷，第88–95页，以及《信息是物理的》（"Information Is Physical"），载于《今日物理》1991年第5期，第23–29页；Otto Rössler：《内在物理学》（"Endophysics"），收入《真正的大脑，人工的心智》（*Real Brains*, Artificial Minds, edited by John Casti and A.Karlqvist, North Holland, New York, 1987），第25–46页；Roger Shepard：《作为世界之映象的感觉—认知宇宙》（"Perceptual-Cognitive Universals as Reflections of the World"），载于《心理计量学公报与评论》（*Psychonomic Bulletin and Review*），1994年第1卷第1期，第2–28页；Patrick Suppes：《解释不可预测性》（"Explaining and Unpredictable"），载于《知识》（*Erkenntnis*）杂志，1985年第22卷，第187–195页；Joseph Traub：《打破不可能性》（"Breaking Intractability"），载于《科学美国人》杂志，1994年第1期，第102–107页（与Henry Wozniakowski合写）。

我在《证据之死》（"The Death of Proof"）一文中，讨论了圣菲会议上提出的某些与数学相关的主题，文章刊于《科学美国人》杂志1993年第10期第92–103页上。我近期读过的关于"知识的限度"方面的最好的一本书，是《心智的火花》（*Fire in the Mind*, George Johnson, Alfred A. Knopf, New York, 1995）。

3.我于1994年9月在哈德逊河畔采访的蔡汀。

4.《历史的终结与最后一个人》，弗朗西斯•福山著（*The End of History and the Last Man*, Francis Fukuyama, The Free Press, 1992）。

5.我曾为《科学美国人》写过一篇关于布瑞安•约瑟夫森的传略，题为"约瑟夫森的心结"（"Josephson's Inner Junction"），刊于1995年第5期，第40–41页。

第十章　科学神学，或机械科学的终结

1.《世界、肉体和魔鬼》，贝尔纳著（*The World, the Flesh and the Devil*, J.K.Bernal, Indiana University Press, Bloomington, 1929），第47页。非常感谢达特茅斯学院的扎斯乔夫（Robert Jastrow），正是他赠给我一册贝尔纳的文集。

2.《特殊智力儿童》，莫拉维克著（*Mind Children*, Hans Moravec, Harvard University Press, Cambridge, 1988）。我对莫拉维克的采访是在1993年12月。

3.《永无止境》，戴森著（*Infinite in All Directions*, Freeman Dyson, Harperand Row, N.Y., 1988），第298页。

4.戴森对于科学的浪漫观点，把他带到激进的相对主义边缘。可参见他的文章

《作为叛逆的科学家》（"The Scientist as Rebel"），载于《纽约书评》（*New York Review of Books*），1995年5月25日，第31页。

5.参见戴森的文章《无尽的时间：开放宇宙中的物理学和生物学》（"Time Without End: Physics and Biology in an Open Universe"），载于《现代物理学评论》（*Reviews of Modern Physics*）1979年第51卷，第447–460页。

6.戴森，《永无止境》，第196页。

7.出处同上，第115页。

8.出处同上，第118–119页。

9.弗里德金的传奇经历（以及Edward Wilson和后来的经济学家Kenneth Boulding的事迹）在《三个科学家和他们的上帝》（*Three Scientist and Their Gods*，Robert Wright，Pantheon，N.Y.，1988）中有精彩的描述。我1993年5月电话采访了弗里德金。

10.《不朽的物理学》，蒂普勒著（*The Physics of Immortality*，Frank Tipler，Doubleday，New York，1994）。

11.《人择宇宙学原理》，蒂普勒和巴罗合著（*The Anthropic Cosmological Principle*，Frank Tipler and John Barrow，Oxford University Press，New York，1986）。

12.皮埃尔·泰哈迪·德·查丁在"人类起源问题续篇：众多有人栖居的世界"一章中，讨论了地外生命的拯救问题。见《基督教与进化论》（*Christianity and Evolution*，Harcourt Brace Jovanovich，New York，1969）。

13.《科学的限度》，梅达沃著（*The Limits of Science*，Peter Medawar，Oxford University Press，New York，1984），第90页。

尾声　上帝的恐惧

1.参阅《上帝的心智》（*The Mind of God*，Paul C. Davies，Simon and Schuster，New York，1992）第9章："宇宙尽头的奥秘"（The Mystery at the End of the Universe）。

2.参阅博尔赫斯的散文《博尔赫斯和我》（"Borges and I"），收入《自选集》（*A Personal Anthology*，Jorge Luis Borges，Grove Press，New York，1967），第200–201页。

3.参见《宗教体验的多样性》（*The Varieties of Religious Experience*，William

James，Macmillan，New York，1961，詹姆斯的著作初版于1902年）中题为"神秘主义"的一章之注释9。

4.参阅《查尔斯·哈茨霍恩哲学》（*The Philosophy of Charles Hartshorne*，edited by Lewis Edwin Hahn，Library of Living Philosophers，La Salle，Ill.，1991）。我电话采访哈茨霍恩的时间在1993年5月。

5.《逻辑哲学论》（*Tractatus Logico-Philosophicus*，Ludwig Wittgenstein，Routledge，New York，1990），第187页。维特根恩坦这本晦涩的著作，初版于1922年。